ATTAINABLE REGION THEORY

ATTAINABLE REGION THEORY

An Introduction to Choosing an Optimal Reactor

By

DAVID MING
University of the Witwatersrand, Johannesburg

DAVID GLASSER and DIANE HILDEBRANDT
Material and Process Synthesis research unit, University of South Africa

BENJAMIN GLASSER
Rutgers, The State University of New Jersey

MATTHEW METZGER
Merck & Co., Inc

WILEY

Published by John Wiley & Sons, Inc., Hoboken, New Jersey
Published simultaneously in Canada

Library of Congress Cataloging-in-Publication Data:

Names: Ming, David, 1985- author. | Glasser, David, 1936- author. | Hildebrandt, Diane, 1961- author. | Glasser, Benjamin John, 1968- author. | Metzger, Matthew, 1983- author.
Title: Attainable region theory : an introduction to choosing an optimal reactor / by David Ming, David Glasser, Diane Hildebrandt, Benjamin Glasser, Matthew Metzger.
Description: Hoboken, New Jersey : John Wiley & Sons, 2016. | Includes index.
Identifiers: LCCN 2016022700 (print) | LCCN 2016025203 (ebook) | ISBN 9781119157885 (cloth) | ISBN 9781119244714 (pdf) | ISBN 9781119244707 (epub)
Subjects: LCSH: Chemical reactors–Design and construction. | Statistical tolerance regions.
Classification: LCC TP157 .M528 2016 (print) | LCC TP157 (ebook) | DDC 660/.2832–dc23
LC record available at https://lccn.loc.gov/2016022700

Cover Image Credit: Courtesy of the Author

Typeset in 10/12pt TimesLTStd-Roman by SPi Global, Chennai, India

10 9 8 7 6 5 4 3 2 1

CONTENTS

PREFACE

The problem that this book sets out to solve can be formulated very simply. Given a set of chemical reactions with known kinetics, what is the best reactor that can be used to carry out the set of reactions. As easy as the question is to ask, the answer is not obvious, as on the face of it there are an infinite number of possibilities. What this preface will do is to outline the 50-year journey to solve this problem. This history also embodies a cautionary tale for those doing (and wanting to do) research.

I did my PhD looking at a reaction and its kinetics. As a result of this, I became interested in chemical reactor theory. I remember, in particular, the excellent books of Denbigh and Levenspiel. Both of them talked about the aforementioned problem and documented some heuristics to help industrialists to come up with some solutions. However, it was recognized that these heuristics were sometimes contradictory.

At that stage, I became interested in optimization in general. Also, on my first sabbatical leave, I was able to work with some of the greats of chemical engineering (Stanley Katz, Reuel Shinnar, and Fritz Horn) and became involved solving some more limited optimization problems, such as using contact times for catalytic reactions, and minimizing holding times for a series of continuous-flow stirred tank reactors (CSTRs). Also it was at this time that I became interested in using Pontryagin's maximum principle (developed for space exploration problems) on chemical reactor problems. This was the stage that I, at first, thought we could solve the main problem by extending residence time distribution theory to nonlinear kinetics. There was the well-known result for segregated systems that one could work out conversions for all kinetics using the residence time distribution. I thought we should be able to extend this theory for all possible reactors and then use the maximum principle to solve the general problem.

On the next two sabbaticals, I worked with Roy Jackson and Cam Crowe to take this idea further, and in the end, much to our joy, Roy Jackson and I solved the problem (after 15 years!). The joy soon evaporated as we realized that even though we had a complete description for all possible reactors with nonlinear kinetics, in order to solve the main problem we had to find a function that could have a non-countable infinity of changes with possibly some of the values going to infinity. This was clearly an impossible problem to solve.

This meant going back to the drawing board! I started to fiddle with some simple problems from the literature by drawing simple two-dimensional graphs. What I soon realized is that if there were concavities on the graphs, they could be filled in with straight lines to make what are called convex hulls. What was really exciting was that these lines were nothing more than mixing between two points on the graph. Out of this came the idea that a reactor was a system that was made up of two processes: reaction and mixing. Each of these processes can be represented on a graph as vectors. Suddenly, the problem I was looking at changed to a geometric one, as I was now looking at making the region in the space as large as possible using the two vector processes. When visiting Martin Feinberg at Rochester, he drew my attention to a paper by Fritz Horn that talked about what he termed attainable regions (ARs). That is, in this context, the largest region in some component space that one could obtain using any processes. He showed that if one had this AR, the optimization problem was relatively simple. Without at first realizing what I was doing, I had found a geometric way of finding the AR for reacting systems. At this stage, I was only doing it graphically on problems that could be represented in two dimensions.

At this stage, it became clear why one could not solve this problem using standard optimization methods. This was because mixing is not a differential process in the ordinary sense. That is, one could not use methods that only looked in the immediate neighborhood as one could mix from any point that was itself attainable, and this could be far from the neighborhood of the point we were examining. An interesting point is that the AR method can solve these non-continuous problems, and so is essentially a new method of optimization, but to the best of my knowledge this idea has not been taken up in other fields of study.

When I arrived back from sabbatical, I looked for a student to work on the problem to extend the results. At this stage, Diane Hildebrandt, who I had taught as an undergraduate, and had done her MSc with me , was looking for a PhD project. I warned her that the project was a little open-ended. However, she was sufficiently interested in the topic (was this brave or foolhardy?) to want to work on it. In the end, out of her work, came the foundations of our work on AR theory. At last, we had a method that could solve the problem I had started out to look at more than 20 years earlier.

What is really interesting (and important) is that we were able to solve a problem that was generally regarded as impossible by thinking about it in a very different way. Instead of trying to answer the problem of finding the optimal reactor in one step, we had first asked what processes are occurring in a chemical reactor, how can we represent them, and how can we then use these processes to ask what are all possible outcomes. Only at this stage do we look at how to perform the optimization. By breaking the problem up into smaller stages, in the end, we were able to solve it.

Publishing the first AR paper turned out to be a bit of a nightmare! As reviewers knew solving this problem was not possible, faith even in scientists was more important than reason. The paper kept being rejected with reviewers telling us there must be something wrong with what we were doing, even though they could not say what it was. In the end, we had to say to the editor if the reviewers could not find the mistake, he should publish it and wait for someone to come up and show us the fallacy. Luckily, in the end he agreed, and we are still waiting for someone to show us why what we did was wrong!

At this stage, my son Benjamin, also a chemical engineer, worked on AR theory for his MSc, for the first industrially interesting problem. Diane then went on sabbatical where she worked with Martin Feinberg to formalize some of the ideas she had started on her PhD. Out of this arose the idea that there are only three types of reactor needed to find the AR in any dimensions, namely plug flow reactors (PFRs), continuous-flow stirred tank reactors (CSTRs), and differential sidestream reactors (DSRs). Suddenly, from there being an infinite number of possible reactors needed in an optimal system, there are only three types. The significance of this is that while initially we needed to find the AR before doing the optimization, it is now, in principle, possible to generate an algorithm to solve large problems without necessarily finding the full AR, which may potentially be a very large problem in higher dimensions.

The method became a topic of research for us, and some others, and we published many papers in the area. Also the method became accepted and now appears as standard in many textbooks on chemical reactor theory and design. In particular, we began to look at ways of calculating ARs in higher dimensions in order to be able to have a successful algorithm that could be used fairly generally. David Ming has worked on this and also extended the ideas from steady-flow reactors to batch reactors. Matt Metzger who spent time with us from the United States found the topic interesting and helped to gather all the information in a logical and consistent way and helped to see that it could be presented in a readable way.

Thus, even though the story at this stage is complete, I believe there is a lesson we can learn from this. I call this the four P's of research. These are as follows:

1. Passion
2. Patience
3. Perseverance
4. Persuasion

I believe the history of this book typifies all of these aspects.

PASSION

If I (and others) had not believed that this project was worthy of spending time on it, it would not have been done.

PATIENCE

I worked patiently on it for 15 years to get an answer that was of no real use.

PERSEVERANCE

Perseverance is the hard work you do after you get tired of doing the hard work you already did. —Newt Gingrich. I believe this speaks for itself in the context of this work.

PERSUASION

It is not sufficient to do good work on its own; one needs to persuade the rest of the world. We really struggled with this! Thomas Grey said it all.

Full many a gem of purest ray serene,
the dark unfathomed caves of ocean bear,
full many a flower is born to blush unseen
to waste its sweetness on the desert air

David Glasser

ACKNOWLEDGMENTS

The dilemma that we face with attainable region (AR) theory is that some aspects of the field are very easy to grasp, whereas other parts are very difficult. Over the years, we have done our best to accommodate for these different aspects, yet there has not been a single, consistent, treatment of AR theory that could be used to address multiple audiences and interest groups. It is our hope that this textbook will not only provide this consistency, but it will also be used by those who wish to understand the field of AR theory for the first time.

The opportunity to write a textbook comes with a mix of excitement and trepidation. On one hand, it is exciting to be given an opportunity to share our ideas—many of which have been developed over *many* years by *many* people—and to articulate them in our own way. On the other hand, it is daunting to think of how best to do this, so that we live up to the expectations of all those who have contributed to these ideas, and also to those prospective readers who are coming to the field for the first time. Indeed, attempting to write a coherent set of work, suitable for many points of view, has been a challenging task, and this textbook would not exist without the help and support from our family, friends, and colleagues who have helped at various stages along the way. In particular, we are very grateful to Cody Bishop and Kwame Owusu-Ansah for helping to proofread (and correct) early versions of the work. We would also like to thank Michelle Low, not only for supporting us throughout the course of this work but also for contributing her time and creativity to take many of the excellent photographs used in this book. Your diligence and continuous enthusiasm throughout this time has been invaluable to us. We are also grateful to Professor Arun Moharir for helping to supply and organize many of the examples and ideas for the companion website. We thank Linda Anthony and Eric Lam, for allowing much of the early stages of this work to be developed at Rutgers University.

Last, we thank our family and friends who have supported and encouraged us throughout this journey. Thank you for understanding and also for sometimes *not* understanding, but still supporting us! In the book, we try to emphasize a common theme of *you can't fix what you don't know*, as a way to explain the importance of understanding performance bounds. In many ways, this theme is mirrored by your support. You have helped us to understand our own limits, and your encouragement has made this task much more achievable. Thank you.

David Ming
David Glasser
Diane Hildebrandt
Benjamin Glasser
Matthew Metzger

PRIOR KNOWLEDGE

This book assumes a basic understanding of chemical reaction stoichiometry and material balances, equivalent to that administered in an undergraduate course in chemical engineering or chemistry. Knowledge of introductory chemical reactor theory is also beneficial, although it is not a requirement.

We shall be interested in a broad, macroscopic, view of reactors—one where elementary mass balances are applied over entire process units. That is, we are not interested in *modeling* the system in detail, and attainable region (AR) theory does not demand the use of microscopic transport equations for instance.

AR theory has a *geometric* flavor. Hence, readers who possess a basic understanding of calculus (derivatives and integration) and linear algebra (vectors and matrices) may find the introductory concepts of AR theory easier to follow. Familiarity with plotting graphs in two and three dimensions on a computer is also helpful. Nevertheless, we shall always strive to provide the necessary theory and tools, when appropriate. This fulfills the dual purpose of introducing the reader to the particular terminology and nomenclature adopted in this book.

HOW THIS BOOK IS STRUCTURED

The material covered in this book is organized into two sections. It may be helpful to refer to Figure P.1 for an overview of the organization of chapters. Section I (Chapters 1–5) focuses on the basics of attainable region (AR) theory. Importantly, this section introduces a different way of viewing chemical reactors and reactor networks. The examples discussed in Section I are of a simpler nature, with an emphasis on describing all problems in two dimensions only. Section I is best read in a sequential fashion.

Section II (Chapters 6–10) focuses on Extended topics in AR theory, and also discusses specific fields of research that may be of interest to select readers. These chapters are more mathematical in nature, and may appeal to readers with a stronger background in computer programming and mathematical modeling. Whereas Section I is intended for senior undergraduates, Section II is more suited for graduate students and working professionals. We hope that these chapters will be used to gain a deeper appreciation for what is an exciting field of study in chemical reactor network synthesis. Although there is more flexibility in how this section can be tackled, we suggest reading Chapters 6 and 7 together.

Chapter 1: In this chapter, we introduce the idea of the reactor network synthesis problem and performance targeting. Ultimately, we attempt to articulate two important messages: *how do you know you have achieved the best? You cannot fix what you don't know.*

Chapter 2: In this chapter, we introduce the idea of concentration and mixing from a geometric perspective. We look at how data may be interpreted in the phase plane (concentration space) and the link between mixing, convex hulls, and attainability. Readers already familiar with vectors, and the

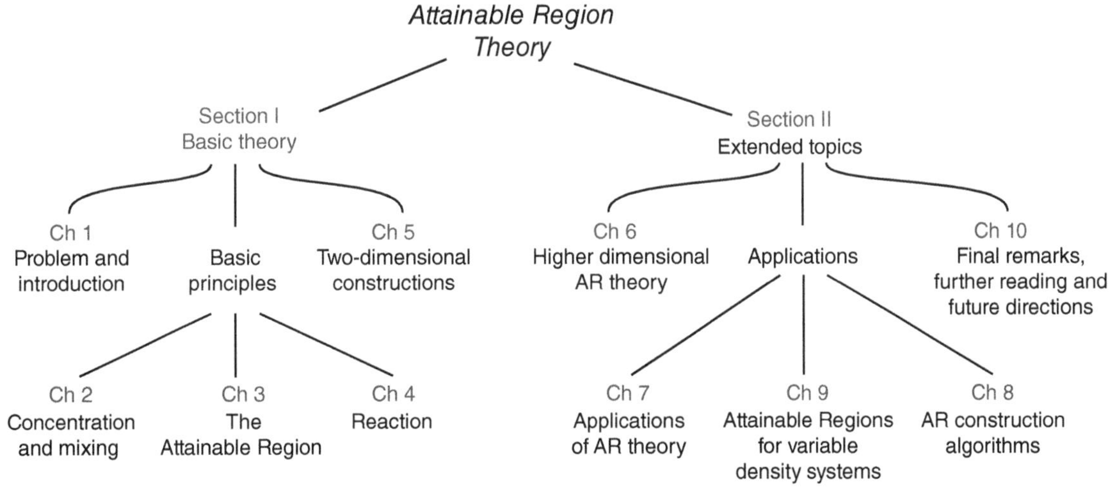

Figure P.1 Organization of chapters.

rudiments of ordinary differential equations, may choose to skip Chapter 2 and move on to Chapter 3, where the AR is described.

Chapter 3: In this chapter, we apply the concepts developed in Chapter 2 to a simple thought experiment and show how system performance may be improved with these ideas. This leads us to introduce, for the first time, the idea of the AR for a reactive system. Basic properties of the AR are also discussed.

Chapter 4: In this chapter, we discuss how reaction may also be viewed from a geometric perspective. We introduce the three fundamental reactor types used in AR theory, and we also discuss additional properties of the AR related to reaction.

Chapter 5: In this chapter, we look to apply our knowledge of concentration, reaction, and mixing to simple systems. We generate the AR for a number of two-dimensional systems and show how the AR approach can lead to insights into the design of chemical reactor networks.

Chapter 6: In this chapter, we describe the nature of the AR boundary in higher dimensional spaces, and we explain a number of important results used in AR theory. We also look at ways of calculating conditions to reside on the AR boundary in any dimension.

Chapter 7: In this chapter, we discuss a number of worked examples for more complex systems, including higher dimensional systems and constructions where there is an equipment constraint. We also discuss how AR theory can be applied to batch reactors.

Chapter 8: In this chapter, we describe a number of AR construction algorithms that may be implemented on a computer.

Chapter 9: In this chapter, we look at how to extend concepts of the AR to variable density systems, such as non-ideal systems in the gas phase.

Chapter 10: We conclude by providing summaries and key take-away points from each chapter. This chapter also provides a small set of literature for further reading as well as potential research interests.

ICONS

Throughout the book, look out for icons to signal an interesting or important aspect of the text. These could be worked examples, important ideas, or simply extra information.

CONCEPT: An idea that may be useful in later discussions.

EXAMPLE: A question, or set of questions, that are intended to test your understanding of an idea, or to extend on a specific topic.

ILLUSTRATION: A worked example or discussion demonstrating an application of AR theory or an extension of an idea.

IMPORTANT: An important concept that should be kept in mind.

OBSERVATION: An idea or a result that may not be obvious that we wish to highlight.

RECALL: A concept or relation developed previously that we wish to highlight again.

SIDE NOTE: Additional information or an interesting fact that is related to the present discussion.

SOFTWARE AND COMPANION WEBSITE

Due to the geometric nature of attainable region (AR) theory, as well as the complexity of the systems considered, we often need software tools to help us interpret and visualize our problems. Rather than attempting to populate a conventional CD-ROM with software and additional examples that cannot be adapted over time, we have decided to release this material on a companion website (http://attainableregions.com), which has been developed for this book. We hope that this approach will allow us to cater to the changing needs of the reader and AR community as a whole, where these software resources can be tailored accordingly over time.

Figures and demonstrations supplied in the book have been produced primarily in MATLAB. However, the use of MATLAB is not a requirement for understanding the examples in this book or using the companion software on the website. Instead, Python (specifically the Anaconda distribution) has been chosen for the website examples and demonstrations. Python may be used without any restriction, even for commercial use. It is freely available on the Internet, and, importantly it is also well supported within the scientific and engineering community. Information on how to install the Anaconda distribution and how to access the website examples can be found both in Appendix C, and on the companion website.

NOMENCLATURE

Commonly used symbols and notation adopted in this book are listed in the following table. Symbols that are only used once for a particular description and that are defined at the point of use may not be listed. Also note that some symbols may be used in different chapters for different situations, and thus more than one definition may apply.

VECTOR, MATRICES, AND SETS

Symbol	Description	Definition
$\mathbf{A}$	Stoichiometric coefficient matrix for a system of reactions	
$\mathbf{C}$	Molar concentration vector	$\mathbf{C} = [c_A, c_B, \ldots, c_n]^T$
$\mathbf{C}^0$	DSR sidestream mixing concentration	
$\mathbf{C}_0$	A point residing on a hyperplane H($\mathbf{n}$, b)	$b = \mathbf{n}^T\mathbf{C}_0$
$\mathbf{C}^*$	Concentration vector from a mixture	$\mathbf{C}^* = \lambda\mathbf{C}_1 + (1-\lambda)\mathbf{C}_2$
$\mathbf{C}^\#$	Equilibrium concentration achieved in a reactor	
$\mathbf{E}$	DSR controllability matrix	$\mathbf{E} = [\mathbf{K}, \mathbf{N}]^T$
$\mathbf{F}$	Set of feed concentrations	$\mathbf{F} = \{\mathbf{C}_{f1}, \mathbf{C}_{f2}, \ldots, \mathbf{C}_{fN}\}$
$\mathbf{G}$	Mass flow rate vector for variable density systems	$\mathbf{G} = [g_1, g_2, \ldots, g_n]^T$
$\mathbf{I}$	The identity matrix	
$\mathbf{J}(\mathbf{C})$	Jacobian matrix of $\mathbf{r}(\mathbf{C})$. Sometimes also given as $d\mathbf{r}(\mathbf{C})$	
$\mathbf{L}$	Set of points belonging to a lineation	
$\mathbf{K}$	Controllability matrix for a critical DSR excluding null space $\mathbf{N}$	
$\mathbf{N}$	Null space of the stoichiometric coefficient matrix $\mathbf{A}$	$\mathbf{N} = \text{null}(\mathbf{A}^T)$
$\mathbf{n}$	Hyperplane normal vector	
$\mathbf{n}_i$	Molar flow rate vector	$\mathbf{n}_i = [c_{Ai}, c_{Bi}, \ldots, c_{ni}]^T$
$\mathbf{z}$	Mass fraction vector	$\mathbf{z} = [z_A, z_B, \ldots, z_n]^T$
$\mathbf{r}(\mathbf{C})$	Rate vector of a function of C	$\mathbf{r}(\mathbf{C}) = [r_A(\mathbf{C}), r_B(\mathbf{C}), \ldots, r_n(\mathbf{C})]^T$
$\mathbf{S}$	Set of points representing the stoichiometric subspace	
$\mathbf{S}\backslash\mathbf{X}$	The complement region of $\mathbf{X}$ to $\mathbf{S}$.	
$\mathbf{t}(\mathbf{C})$	CSTR tangent vector	$\mathbf{t}(\mathbf{C}) = [\mathbf{I} - \tau\mathbf{J}(\mathbf{C})]^{-1}\mathbf{r}(\mathbf{C})$
$\mathbf{v}(\mathbf{C})$	Mixing vector	$\mathbf{v}(\mathbf{C}) = \mathbf{C}_2 - \mathbf{C}_1$
$\mathbf{W}$	Molecular weight matrix	
$\mathbf{X}$	Set of points representing a current set of achievable states	
$\mathbf{z}^{[k]}$	The kth iterated Lie bracket	
ε	Extent of reaction vector	$\boldsymbol{\varepsilon} = [\varepsilon_1, \varepsilon_2, \ldots, \varepsilon_n]^T$
$\mathbf{0}$	The zero vector	$\mathbf{0} = [0, 0, \ldots, 0]^T$

LETTERS

d	Number of independent reactions	
c_i	Molar concentration of component i	
F(t)	Fed-batch feeding rate as a function of reaction time	
G	Total mass flow rate through a reactor or reactor network	$G = \sum g_i$
g_i	Mass flow rate of stream i in a network	
H(**n**, b)	Hyperplane defined by normal vector **n** and scalar b	
k_i	Rate constant i in a rate expression	
M	Total mass	$M = \sum m_i$
m_i	Mass of component i	
n	Number of components participating in a reaction	
P	Reactor pressure	
P_k	Polytope representing the AR at iteration k	
Q	Volumetric flow rate through a reactor or reactor network	
$r_i(\mathbf{C})$	Rate expression for component i as a function of **C**	
T	Reactor temperature	
t	Batch reaction time	
V	Reactor volume or total volume of a reactor network	
V(t)	Fed-batch volume as a function of reaction time	
W_i	Molecular weight of component i	
W_{avg}	Average molecular weight of a mixture	
Y_i	Yield of component i	
Z	The compressibility constant for real gases	$P\hat{V} = ZRT$
z_i	Mass fraction of component i	
$\mathbb{R}$	The set of all real numbers	
$\mathbb{R}^n$	The set of all vectors (n-tuples) with n real numbers (n-dimensional space)	

GREEK SYMBOLS

α	DSR sidestream mixing parameter for constant density systems	
β	DSR sidestream mixing parameter for variable density systems	
ε_i	Extent of reaction for reaction i in a system of reactions.	
θ	Angle between two vectors	$\mathbf{x}^T\mathbf{y} = \|\mathbf{x}\|\|\mathbf{y}\|\cos(\theta)$
$\Lambda(\mathbf{C})$	Critical CSTR determinant surface	
λ	Mixing fraction	
ν_i	Stoichiometric coefficient of a component i in a reaction	
ρ_{avg}	Average density of a mixture (by mass)	
σ	Reactor residence time in mass fraction space	$\sigma = V/G$
Σ_{ij}	Instantaneous selectivity of component i to j	
τ	Reactor residence time (constant density)	$\tau = V/Q$
Υ_i	Instantaneous yield of component i	
$\varphi(\mathbf{C})$	vDelR critical DSR condition for a system in $\mathbb{R}^3$ ($d = 3$)	$\varphi(\mathbf{C}) = [\mathbf{J}(\mathbf{C})\mathbf{v}(\mathbf{C})]^T[\mathbf{v}(\mathbf{C}) \times \mathbf{r}(\mathbf{C})]$

MATHEMATICAL OPERATIONS AND ABBREVIATIONS

∇	Del operator (nabla)	$\nabla = \left[\frac{\partial}{\partial c_A}, \frac{\partial}{\partial c_B}, \ldots, \frac{\partial}{\partial c_n}\right]^T$
$\|\mathbf{x}\|$	The norm of a vector **x**	$\|\mathbf{x}\| = \sqrt[2]{x_1^2 + x_2^2 + \cdots + x_n^2}$
$L_f h(\mathbf{C})$	Lie derivative of a scalar function h(**C**) with respect to a vector function **f**(**C**).	
[**f**, **g**](**C**)	Lie bracket of vector functions **f**(**C**) and **g**(**C**)	$[\mathbf{f}, \mathbf{g}](\mathbf{C}) = d\mathbf{g}(\mathbf{C})\mathbf{f}(\mathbf{C}) - d\mathbf{f}(\mathbf{C})\mathbf{g}(\mathbf{C})$
Det(**A**)	Determinant of a matrix **A**	
conv(**X**)	Convex hull of a set of points **X**	
cl (**X**)	The closure of a set of points **X**	
rank(**A**)	Rank of matrix **A**	
null(**A**)	Null space of matrix **A**	

SUBSCRIPTS

f	Refers to feed streams	$\mathbf{C}_f = [c_{Af}, c_{Bf}, \ldots, c_{nf}]^T$
i, j, k, n	General reference to elements in a list, or index counter	

SUPERSCRIPTS

T	Matrix transpose	
#	Relating to an equilibrium state	$C^{\#}$ is an equilibrium concentration
*	Relating to a mixture	C^{*} is a concentration obtained from a mixture

ABBREVIATIONS

AR	Attainable region
BTX	Benzene–toluene–xylene system
CAS	Computer algebra system
CSTR	Continuous-flow stirred tank reactor
DSR	Differential sidestream reactor
IDEAS	Infinite DimEnsionAl State-space
LP	Linear program or linear programming
MILP	Mixed integer linear program
MINLP	Mixed integer nonlinear program
PFR	Plug flow reactor
RCC	Recursive constant control policy

SECTION I

BASIC THEORY

1

INTRODUCTION

1.1 INTRODUCTION

This book is concerned with a field of study called *attainable regions* (ARs), which is a set of ideas intended to address a generalized problem, often encountered in chemical reactor and process design. Although the problem can become quite detailed if we allow it, the basic idea is simple to understand. This chapter serves to articulate the type of problems AR theory could help address. To gain a sense of the scientific discipline that we are interested in, many (but not all) of the problems we are concerned with can be represented by Figure 1.1.

Assume that we are supplied feed material of a known state (i.e., composition, temperature, and flow rate) and asked to design a process that optimizes the production of a desired product. The system in question is represented by a single block, a "black box." This is a process that accepts a feed and converts it into something of higher economic or social value, chiefly via *reactive* unit operations. The block could be a batch or continuous process. In this book, we will mainly focus on continuous reactors. Products from the block are given by a single effluent stream. In reality, this could come about from a number of intricate recycles and bypass streams, occurring amongst a vast set of unit operations within the block—that is, the physical process represented by the block in Figure 1.1 is likely to be very complex in reality.

For many problems, we might specifically wish to find the best concentration of a desired component exiting the network, or perhaps the total cost of equipment to meet a minimum level of production—at a minimum product quality—within the network. For now, we shall leave these descriptions open so that we can describe the central problem that is related to them all.

A relevant question to ask in this context is the following: How do we best design the internals of the block given in Figure 1.1? Since we are specifically interested with reactive processes, how do we best design a process involving reaction? There is no assumption of a predefined design inside the block. We are concerned with the generalized design of the internals, and not simply the optimization of a set design. (Some readers may recognize this as *synthesis*.) These questions often constitute what is referred to as the *reactor network synthesis problem*. Moreover, once we have synthesized a design, how do we measure a "good" design from other competing designs? Although it may appear as the former problem is our chief concern, it is actually the latter problem that we interested in. Our goal in this book is to understand how we find and measure "good" designs using AR theory.

AR theory is hence a set of ideas in chemical reactor design that aims to understand the reactor network synthesis problem. But to understand this problem will ultimately require us to understand a broader problem of what it means to be "the best," for various designs within the block may achieve the same outcome. This is our primary concern in AR theory. It is related to the reactor network synthesis problem, but it is also distinct.

1.2 MOTIVATION

1.2.1 Toluene Production as a Case Study

To motivate why AR theory is useful, let us begin with a story, involving a team of three young engineers, who are interested in the production of toluene. We would like to gain a sense of some of the typical design considerations encountered by the team, which may also be addressed with AR theory. Sections 1.2.2–1.2.5 describe the team's story, whereas Section 1.2.6 explains the story in relation to AR theory. Sections 1.3 and 1.4 are intended for readers who are already familiar with the rudiments of chemical reactor design.

Sam, Alex, and Donald are promising young engineers working for a large chemical company that produces toluene ($C_6H_5CH_3$) as one of its products. One day, their boss asks

Attainable Region Theory: An Introduction to Choosing an Optimal Reactor, First Edition. David Ming, David Glasser, Diane Hildebrandt, Benjamin Glasser, and Matthew Metzger.

Companion website: http://attainableregions.com

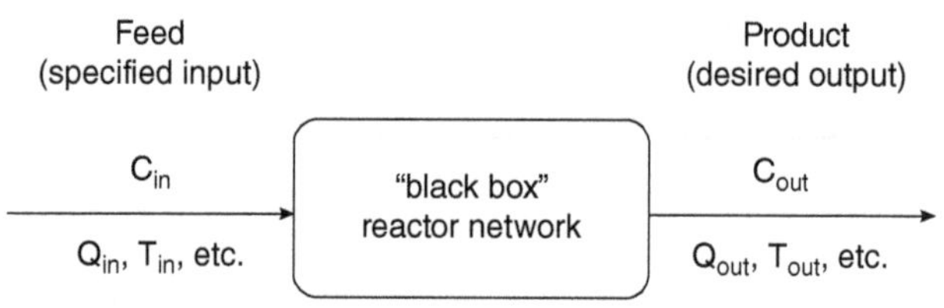

Figure 1.1 Overview of reactor network.

them to investigate how the company could improve the quality of their toluene product. Specifically, he asks:

How can we maximize the amount of toluene produced from a feed of 0.5 moles of ethylene and 1.0 moles of benzene?

Although there are potentially many unit operations that might improve the amount of toluene in the system, we are mainly interested in unit operations that involve reaction. The objective may be refined slightly to say that we wish to find a *reactive* system that maximizes the amount of toluene produced.

The three engineers decide to first conduct a number of experiments in the company's lab, as a way to gain a deeper understanding of the system of reactions. The discussions given in the following text chronicle their experimental process, their discoveries, and ultimately their recommendations in addressing this problem.

1.2.2 Part One: Initial Investigations

Having no prior knowledge or experience with toluene and its production, Sam first consults one of the senior chemists in the company, and gathers information regarding the reaction chemistry. The following set of reactions explain the system, including all significant products and by-products.

$$C_6H_6 + \frac{1}{2}C_2H_4 \leftrightarrow C_6H_5(CH_3) \tag{1.1a}$$

Benzene + $^1/_2$ ethylene ↔ toluene

$$C_6H_5(CH_3) + \frac{1}{2}C_2H_4 \leftrightarrow C_6H_4(CH_3)_2 \tag{1.1b}$$

Toluene + $^1/_2$ ethylene ↔ xylene

$$2C_6H_6 \leftrightarrow C_6H_5C_6H_5 + H_2 \tag{1.1c}$$

2 benzene ↔ diphenyl + hydrogen

Benzene (B) reacts with ethylene (E) to form toluene (T). Toluene and ethylene further react to form xylene (X). At the same time, benzene reacts to form two side-products in a competing side reaction: diphenyl (D) and hydrogen (H). This system may be classified as a series–parallel reaction, with toluene as an intermediate product, which shall hereafter be referred to as the BTX system.

TABLE 1.1 Sam's First Few Experiments

Experiment	Reaction Time (h)	Toluene Concentration (mol/L)
1	5.0	0.0087
2	2.1	0.0373
3	1.1	0.0560
4	0.1	0.0361

For the purposes of the investigation, the BTX reaction is independent of both temperature and pressure. It is assumed that the toluene concentration inside the reactor is simple to measure, so that concentrations are easily recorded, and that it is also possible to initiate and terminate the reaction easily.

Sam then begins her investigation in the simplest manner possible: by experimenting in a batch reactor (a beaker) in the lab, using an initial feed concentration of 0.5 mol/L ethylene and 1.0 mol/L benzene.

She starts by carrying out the experiment for a number of (arbitrary) reaction times using the same initial conditions each time, and then by recording the associated toluene concentration achieved after each run. Sam conducts four experiments in total, and a summary of her results is displayed in Table 1.1.

From the data, there appears to be an ideal reaction time, between 0.1 and 2.1 h, where the toluene concentration is maximized. Using this insight, Sam carries out a series of additional experiments, attempting to isolate the exact reaction time that maximizes toluene concentration. In effect, her approach is to use reaction time as a parameter to find the maximum toluene concentration.

After many experiments, Sam finds that a reaction time of 35.4 min (0.59 h) produces the highest toluene concentration: 0.0619 mol/L. Confident in her results, Sam reports back to her boss with the good news.

OBSERVATION: Initial investigations

Many of us tend to begin our investigations in a similar manner to Sam's approach—with little previous knowledge or known insight, we simply begin with that *is* known and then attempt to improve the current performance of the system using known parameters. This is a reasonable approach, but additional insight into the system might bring about further improvements.

1.2.3 Part Two: Iterative Improvement

1.2.3.1 Accidental Improvement Happy with Sam's initial results, the team decides to run the reaction on a larger scale. The reaction is predictable and simple to operate, and they manage to consistently achieve the toluene concentration expected with little complications. A standard operating regime is eventually developed, involving four steps: filling, reacting, emptying, and cleaning.

One day, whilst preparing the reactor for the following day, Alex forgets to empty the entire contents of the reactor vessel, and a small portion of the reactor product is retained in the reactor overnight. Alex enters the lab the next day, ready to complete a new set of batches, unaware of his mistake the previous day. He fills the reactor with fresh feed material, and runs a new batch.

To his surprise, when he expects to obtain the same toluene concentrations as in the previous experiments, he realizes that the toluene concentration has changed. It has *increased.*

Indeed, his absentmindedness has turned out be something of great interest—not only is it possible to adjust the reaction time of a particular batch, but it is also possible to adjust the starting concentration of the batch, using product from a previous run. With this vision, Alex proposes a new three-step reaction procedure as follows:

1. First, running a fresh batch with feed material for a restricted period;
2. Then, retaining a small fraction of product;
3. Last, refilling the reaction vessel with fresh feed again, thereby altering the starting concentration of the next batch. This batch is then run to the desired exit concentration.

Alex runs the reaction with the new operating procedure (for an arbitrary reaction time and mixture concentration) and finds that he is able to achieve a toluene concentration of 0.0652 mol/L in the following batch, which is an improvement over what Sam could achieve.

1.2.3.2 Experimenting with Different Combinations

Noting Alex's achievement, Donald begins to experiment with different retained fractions, in an attempt to find further improvements. He begins with an arbitrary combination of reaction times and mixing fraction, but he finds that some combinations produce *worse* results, whereas others offer moderate improvements. He writes down his results in a matrix format, which are illustrated in Figure 1.2.

% Reactor Volume retained \ Reaction time	0.1 hours	0.5 hours	1.0 hours	1.5 hours	2.0 hours
20%	0.0633	0.0652	0.0644	0.0630	0.0616
40%	0.0643	0.0677	0.0659	0.0630	0.0603
60%	0.0645	(0.0690)	0.0659	0.0612	0.0571
80%	0.0639	0.0673	0.0624	0.0559	0.0500

Figure 1.2 Donald's investigation, summarized in a matrix for the maximum toluene concentration achieved.

The optimal combination from Donald's finding suggests that 60% of the reactor product volume must be retained and a reaction time of 0.5 h must be used. This gives a toluene concentration of 0.0690 mol/L, or approximately 11% over Sam's investigation. Happy with this result, Sam, Alex, and Donald report back to their boss with the updated operating procedure.

> **OBSERVATION: Refining the approach**
>
> Although Donald, Sam, and Alex have managed to improve the toluene concentration, finding improvements are becoming more complicated in general. This procedure now involves finding an optimal combination of parameters, as opposed to just one parameter with Sam's investigation.

1.2.4 Part Three: Coffee

One morning, Donald walks into the office kitchen to make a cup of coffee. He likes his coffee strong and black. He sits the cup down on the counter and accidentally brushes past a nearby jug of milk. The jug knocks over and some of the milk spills into Donald's cup, mixing with the coffee and changing the color of its contents slightly.

Donald notes the change in the color of the coffee. Rather than discarding it and pouring a new cup, he realizes that his cup is like a beaker, and the addition of milk has changed the beaker's contents. An interesting idea appears in Donald's mind—could material be added into the toluene reactor *during* reaction? He realizes that Sam and Alex have been constraining themselves to the operation of standard batches only. But the reaction could be carried out in a *fed-batch reactor*, such as in Figure 1.3.

Donald sets out to see if he can improve the toluene concentration using fed-batch reaction. He allows material to flow into the reactor and mix with the reactor contents for the entire duration of the run, recording the maximum toluene concentration achieved after each run. After many experiments, Donald believes that he has found the optimum toluene concentration: 0.0761 mol/L; obtained by feeding at a constant rate of 1.0 L/h during reaction. This result is again an improvement over previous methods.

> **SIDE NOTE: New options available**
>
> The introduction of a new mode of operation has opened up a wider choice of potential operating approaches to pursue.

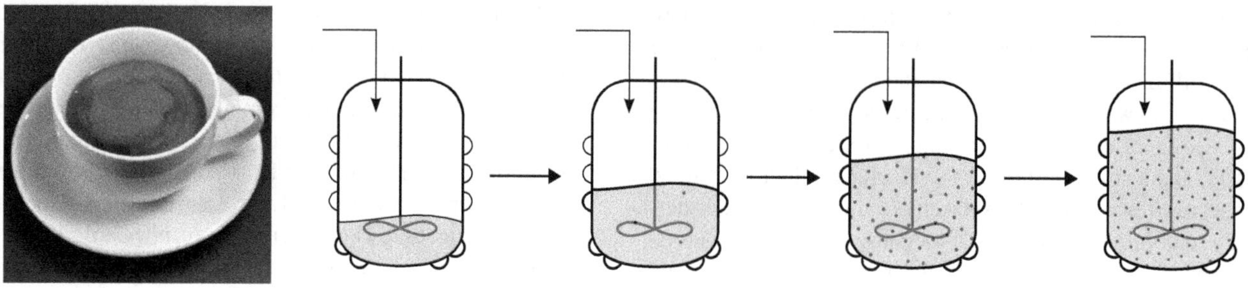

Figure 1.3 Coffee and fed-batch reactors.

1.2.5 Part Four: Additional Improvements

The team believes that they are close to reaching a point where no further improvements can be made. One day, Sam forgets to switch off the reactor just after a batch has reacted. She turns off the feed to the fed-batch reactor but forgets to turn off the reactor itself, and the reaction proceeds with a feeding rate of zero.

To her surprise, an even higher toluene concentration is obtained. Sam has, in effect, created an operating *sequence* involving the following two distinct reaction steps:

1. Reaction period involving fed-batch operation;
2. Reaction period of standard batch operation with no additional feeding.

The team discovers that if the fed-batch portion of the sequence is allowed to run for 1.6 h with a feeding rate of 1.0 L/h, followed by period of no feeding for another 15.6 min (0.26 h) as a standard batch, then a maximum toluene concentration of 0.0793 mol/L may be achieved, which again is an improvement. A summary of all the team's discoveries and recommendations for this investigation is provided in Table 1.2.

The team has managed to improve the toluene concentration by 28%: starting from 0.0619 mol/L, initially obtained by a standard batch reactor, to 0.0793 mol/L, obtained by a combined fed-batch and batch sequence. Satisfied with their progress—as well as a little tired of having to do so many experiments—Sam, Alex, and Donald report their findings as final recommendations.

SIDE NOTE: Additional modifications

Note that concepts from Section 1.2.4 could also be incorporated into this operating sequence. Depending on how long we react, and how much product is mixed, the result might be improved even further. Each time a new discovery is found, many more combinations with existing procedures can be explored.

1.2.5.1 A Change in Objective Sam, Alex, and Donald's boss is so impressed with their findings; he proposes another problem for them to tackle (one that the company has not yet been able to fully understand): instead of seeking to maximize toluene concentration, what can be done to minimize the production of hydrogen as a by-product?

Although the team has developed useful insights from their toluene experiments, a new set of experiments must now be carried out for hydrogen, and they have little understanding that could elicit an immediate recommendation for this new objective. Their boss' request leaves them feeling a bit bewildered, and they walk away hoping that they will be as fortunate with their new investigation as they were with the previous experiments.

TABLE 1.2 All Discoveries and Recommendations Made by Sam, Alex, and Donald

	Toluene Concentration (mol/L)	Associated Method Used
Part One	0.0619	Standard batch reactor. Optimized for reaction time.
Part Two	0.0690	Repeated batch with partial mixing. Optimized for reaction time and starting concentration.
Part Three	0.0761	Fed-batch reactor at 1.0 L/h. Optimized for feeding rate.
Part Four	0.0793	Fed-batch for 1.6 h at 1.0 L/h, followed by a standard batch optimized for 15.6 min (0.26 h).

ILLUSTRATION: Results from a number of random experiments

When the objective of the investigation is modified, a new set of experiments must be carried out. Perhaps the team can use what they have learnt previously for the new investigation, but it is unclear whether the same techniques will also work with hydrogen.

Suppose that a new set of experiments are carried out, in an attempt to understand the hydrogen minimization problem, and that Table 1.3 provides a summary of those experiments. These data originate from the same BTX reaction used in the previous sections.

Which experiments would you recommend for further investigation? How much more improvement do you think is possible?

TABLE 1.3 Experiments for Minimum Hydrogen Production

Attempt	Operating Method	Hydrogen Concentration (mol/L)	Toluene Concentration (mol/L)
1	Batch reactor (0.3 h)	0.4042	0.0563
2	Batch reactor (1.5 h)	0.4396	0.0484
3	Fed-batch reactor with 0.1 L/h, 5 h	0.3682	0.0641
4	Fed-batch reactor with 1.0 L/h, 5 h	0.3677	0.0676
5	Batch reactor (0.55 h), followed by a fed-batch (0.7 h, 0.2 L/h)	0.3621	0.0685
6	Standard batch (0.5 h) mixed with 15% fresh feed at the end	0.3601	0.0522

1.2.6 What this Book is About

1.2.6.1 Attainability The challenges faced by Sam, Alex, and Donald might also be commonly encountered by those who work in a design and experimental environment. Not only is it important for Sam, Alex, and Donald to understand how to improve the reaction, but it is also important for them to understand *what* procedure should be improved and *why* it is needed. Had they been aware of this knowledge from the start, they may have arrived at their final recommendation much quicker and been more confident in their recommendations overall.

What is known in the present moment influences our thinking—in terms of what improvements are currently possible—and how these improvements should be carried out. We may be accustomed to a certain way of thinking, and sometimes our decisions depend on events that were not originally planned. Hindsight is often hard-earned. Furthermore, sometimes we do know what design or procedure should be followed, but they cannot be implemented due to financial or physical constraints. Do we always understand what impact these constraints have in terms of a potential loss, an opportunity cost?

Could we have foreseen these challenges from the beginning? How do we find these limits? And what tasks must be done in order to achieve them? There is a theme of *attainability* that runs through these questions, and AR theory provides a framework for helping to address these questions in chemical reactor networks.

1.2.6.2 Performance Targets AR theory is concerned with problems related to the attainability of certain states. Understanding what states are attainable and what states are not allows for *performance targets* to be established, which may be incorporated into the design process. Examples of common performance targets in reactor design might include the following:

- Minimizing the production of CO_2 or other unwanted by-products from the reactor;
- Finding the smallest reactor volume needed to meet a desired output;
- Achieving the highest production rate of a biological product from a batch reactor;
- Maximizing the profit from the sale of a reactor product;
- Determining the maximum operating temperature within the reactor.

AR theory also helps us to understand what equipment is needed to achieve these targets. In relation to reactive equipment, this means providing insight into what type of reactor should be used, and how different reactor types should be arranged to achieve a certain target, such as concentration, conversion, yield, and profit.

> **SIDE NOTE: Optimal result still unknown**
>
> We have not yet confirmed whether a toluene concentration of 0.0793 mol/L, achieved in Section 1.2.5, is the highest concentration possible. In Chapter 7, we shall solve this problem again and show that a toluene concentration of 0.0807 mol/L is achievable. Moreover, we can also answer the hydrogen minimization problem, described in Section 1.2.5.1, using previous calculations involving the toluene maximization problem. Hence, both questions may be addressed without the need for additional work to be carried out.

1.3 REACTOR NETWORK SYNTHESIS

In Section 1.2, we described how Sam, Alex, and Donald approached the BTX problem from an experimental perspective. How might our approach change if we are given mathematical expressions for the rates of reaction? In the following sections, we wish to describe some common ideas and approaches in theoretically designing a network of reactors (the reactor network synthesis problem), and also describe a central challenge faced in reactor network synthesis, even when mathematical and optimization techniques are available. For example, suppose that kinetics is also available for the BTX reaction and assumed to follow the data in Table 1.4:

TABLE 1.4 Rate Expressions for the BTX System

Component	Rate Expression
Benzene (r_B)	$-k_1c_Bc_E^{0.5} - 2k_3c_B^2$
Ethylene (r_E)	$-0.5k_1c_Bc_E^{0.5} - 0.5k_2c_Tc_E^{0.5}$
Toluene (r_T)	$k_1c_Bc_E^{0.5} - k_2c_Tc_E^{0.5}$
Xylene (r_X)	$k_2c_Tc_E^{0.5}$
Diphenyl (r_D) = Hydrogen (r_H)	$k_3c_B^2$

where $k = 1.0\,L^{0.5}/(mol^{0.5}.h)$, $k_2 = 1.0\,L^{0.5}/(mol^{0.5}.h)$, and $k_3 = 10.0\,L/(mol.h)$; c_i represents the concentration of component i in the system. Since expressions for species reaction rates are known, it is possible to model the BTX reaction, and hence mathematical optimizations may be carried out.

Perhaps, as a first attempt, the BTX reaction could be compared using a number of different, common, reactor models, such a continuous-flow stirred tank reactor (CSTR). One could solve the CSTR equation and then compare the toluene concentration achieved in a CSTR to that achieved in a standard batch reactor (or equivalently in a plug flow reactor or PFR). Figure 1.4 illustrates the results of this approach using the BTX kinetics given in Table 1.4. Different reactor types behave uniquely, and thus each reactor type could be separately optimized for maximum toluene production.

> **SIDE NOTE: Reactor equations**
>
> If the density is assumed to be constant, a steady-state mass balance over a CSTR gives the following expression[1]:
>
> $$Q(c_i^{out} - c_i^{in}) = V_{CSTR}r_i \quad (1.2a)$$
>
> Here, c_i^{in} and c_i^{out} are the inlet and exit concentrations of component i, respectively. Variables Q and V_{CSTR} are the volumetric flow rate and CSTR vessel volume, respectively. Similarly, the equation for a constant volume batch reactor follows the differential equation:
>
> $$\frac{dc_i}{dt} = r_i \quad (1.2b)$$
>
> where t is the reaction time of the batch. Solving Equation 1.2a and 1.2b produces the toluene concentration profiles in Figure 1.4.

In this case, the CSTR achieves a higher toluene concentration, but are there perhaps other reactor types (i.e., a packed bed reactor) that could outperform the CSTR? Additionally, are there *combinations* of reactors—as a *network of reactors*—that could achieve better?

To demonstrate the potential complexity involved when dealing with multiple reactor designs, consider Figure 1.5, which proposes a number of different reactor configurations (reactor structures) that might be used. For simplicity, the configurations are limited to combinations of a maximum of three reactors, using PFRs and CSTRs only.

How do we determine which reactor configuration is best? In this example, we could potentially solve each system and find the best configuration from the size options. However, we can always devise new configurations that may be superior. A number of additional design choices must be made when considering reactor structures, even for simple configurations such as in Figure 1.5. Some of these considerations might include the following:

- How many individual reactors do we consider in each structure?
- What reactor types do we consider?

[1] A summary of fundamental reactor types is provided in Appendix A.

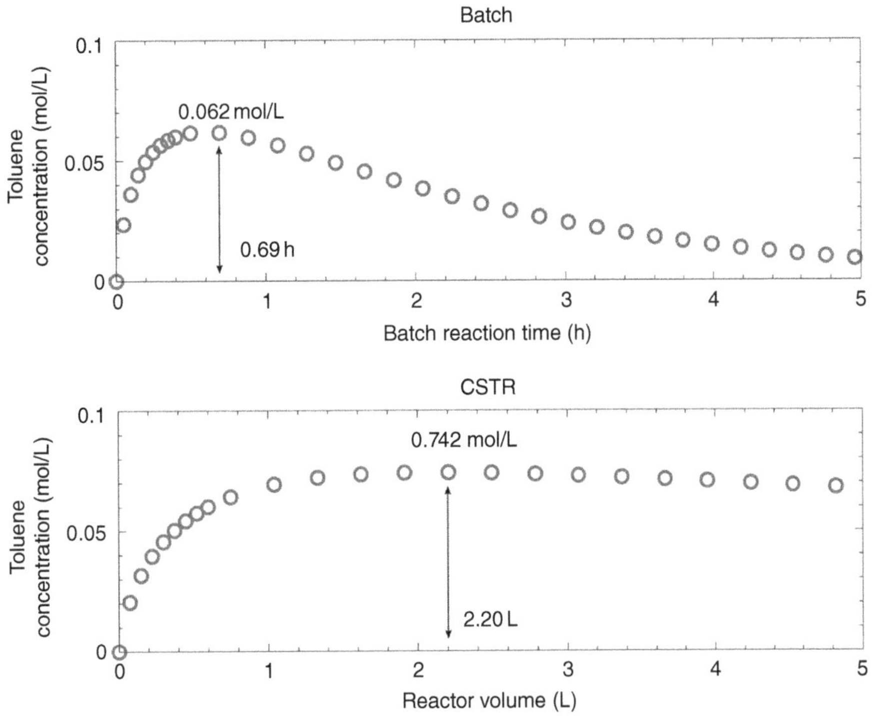

Figure 1.4 Toluene concentration data obtained in a batch experiment and a continuous-flow stirred tank reactor (CSTR).

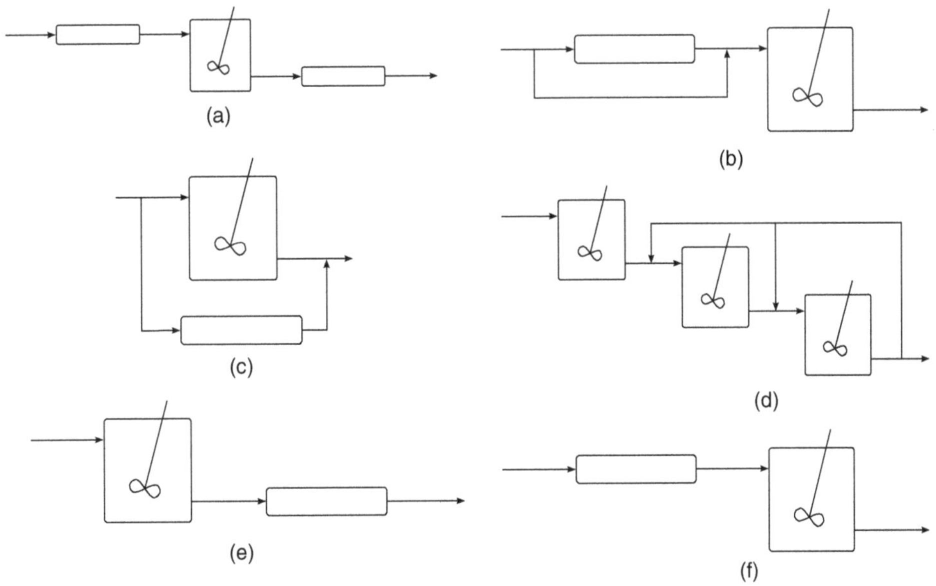

Figure 1.5 Options (a)–(f) represent a number of different reactor configurations. Only PFRs and CSTRs are used with a maximum of three reactors per configuration.

- Do we include recycle and bypass streams? If so, where are they placed within the structure?
- How many parallel reactors are included?
- How many reactor structures should be considered?

How should one select the best reactor structure from a pool of many reactor configurations? Nominating specific combinations in an exhaustive fashion is not feasible because we could always devise a different configuration that may perform better. Yet again, we are presented with a

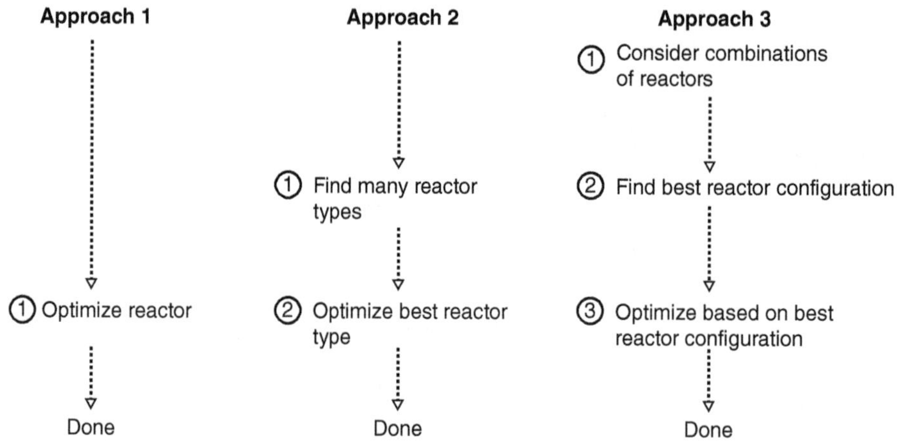

Figure 1.6 Overview of different approaches. Each approach will terminate in an optimal design, although the particular design achieved may differ depending on the order of steps taken.

similar question as in Section 1.2.6—*when do we know we are the best?*

CONCEPT: Reactor structures

By a *reactor structure*, we have in mind a well-defined arrangement of reactors (a reactor network) that produces a particular output. The network might typically be composed of various reactor types. The kinds of problems that we will investigate in this book all fall into this category. We will be interested in looking at how combinations of reactors (as opposed to a single reactor) can often provide meaningful improvements to a problem.

Figure 1.6 proposes a number of general approaches to the reactor synthesis problem, starting from the simplest and most constrained approach, to the most general (and difficult) approach. In AR theory, we are ultimately concerned with problems involving reactor structures that produce the best performance, and thus AR theory falls into Approach 3 of Figure 1.6.

ILLUSTRATION: Are multiple reactors necessary?

Consider the following two reactions:

$$6.21A + 0.145B \rightarrow 1.001C + 10D + 0.1F + 2.15G \tag{1.3a}$$

and

$$\begin{aligned} 2A + C &\rightarrow 1.5B \rightarrow 0.25C \\ 2C + 3.6D &\rightarrow 3D + 6E \\ 2E + F + 0.17B &\rightarrow B + 1.5C + G \end{aligned} \tag{1.3b}$$

How would you build a reactor that maximizes the production of component C according to Equation 1.3a? How would you optimize for the same scenario if Equation 1.3b occurred in the reactor instead?

In Equation 1.3a, the reaction is "simple": although multiple components participate in the reaction, there is only one reaction, and hence there is only one pathway that reactants may proceed to form products (all components are linked by a common extent of reaction). The choice of reactor type *may* have a significant influence on system performance, although it is possible to achieve the same performance with different reactor types.

By comparison, Equation 1.3b contains the same number of individual components, yet these are split over a number of different reaction paths. Equation 1.3b is hence more complex: each path could be individually pursued, resulting in a number of different product mixtures depending on which reactions are dominant, the initial concentration of species, and the intrinsic behavior of different reactor types to these reactions. Consequently, the choice of reactor type may have a significant influence on performance. One reactor type may only achieve a small portion of all possible product states, favoring one reaction over other competing reactions—a CSTR might favor the first two reactions, whereas a PFR might favor the last reaction. Utilizing different reactor types may expose a larger set of possible outputs, but the opportunity for different reactions to occur means that the underlying reactors must be sufficiently generalized to target the operating point of interest. In order to achieve these states, reactor structures are more suitable.

ILLUSTRATION: Biological reactions

Consider the metabolic pathway diagram in Figure 1.7 for the production of nattokinase (and other biological products) by glucose and glycerol. Here, there are *many* reactions occurring. Each arrow displayed in the schematic represents a reaction.

The diagram given in Figure 1.7 is common in biological systems. Complex recycling of intermediate products, and the possibility for compounds to traverse along many divergent paths, means that it is often not straightforward to predict how the flow of material is best arranged along a certain pathway, or what the best pathway is for the production of a desirable intermediate product. Modeling the reactions present in a biological system is complex, and optimizing for a bioreactor that maximizes the production of a specific byproduct is hence a challenging task.

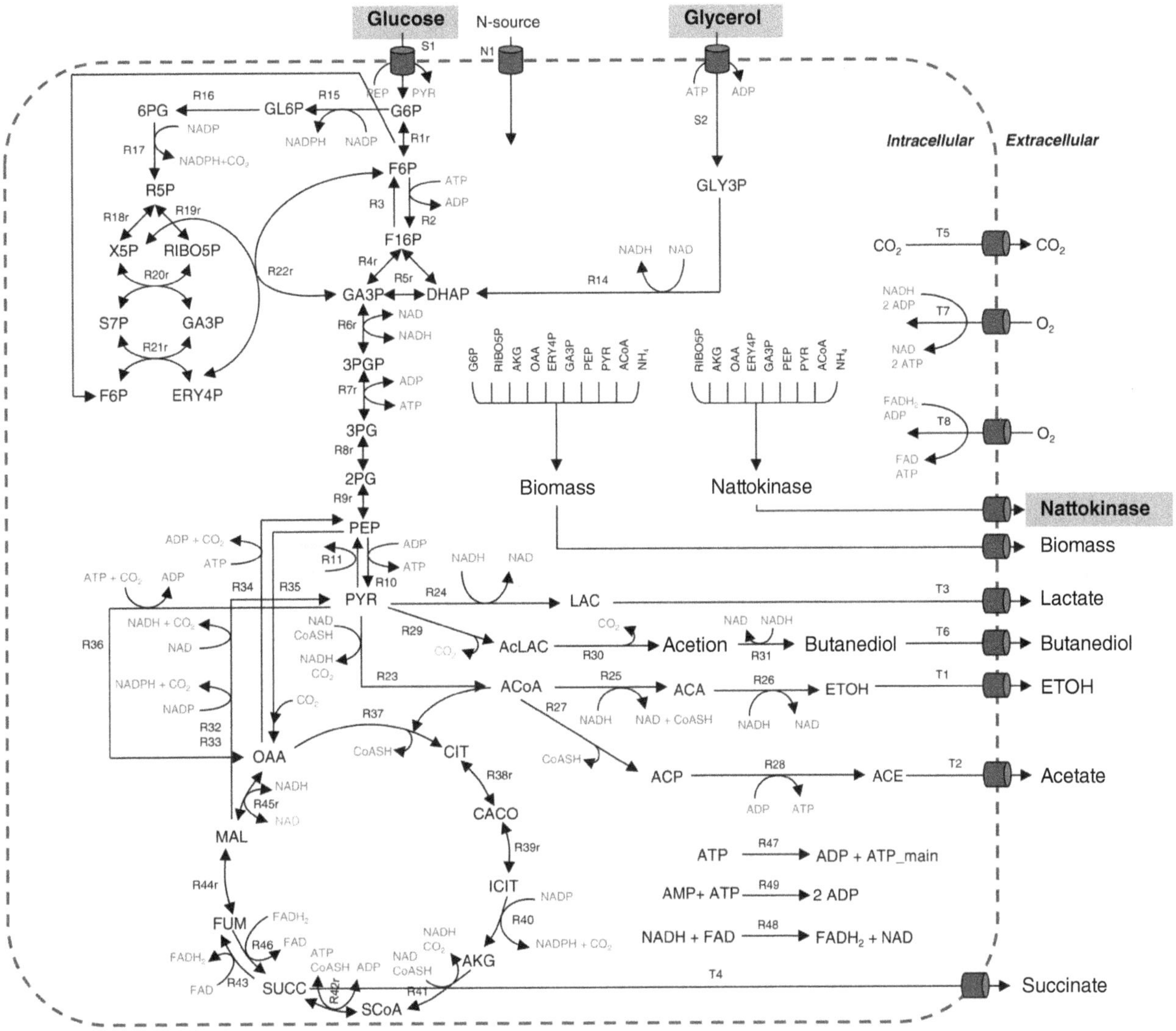

Figure 1.7 Metabolic pathway showing how glucose and glycerol are converted in a series of complex biological reactions to produce a number of different by-products. Unrean and Nguyen (2013). Reproduced with permission of Springer.

There is again an opportunity here to use multiple reactor types to maximize a desired product, yet in practice, simplifications are often applied (i.e., lumped reaction models) to make the problems more tractable. The design of the reactors themselves are also often of a simpler nature, employing a single CSTR (chemostat) or fed-batch reactor with biomass recirculation (Nielsen et al., 2011). Utilizing one reactor in one configuration may limit the possible combination of pathways, which may limit performance as a result.

Although multiple reactors can be used for single reaction systems, they are potentially not required. For systems involving multiple reactions, reactor structures may be *essential*—unlocking certain states that would not be possible with a single reactor type. However, solving reactor network synthesis problems is more difficult than single reactors, because reactor structures introduce more complexity into the design.

SIDE NOTE: Multiple reactions and multiple reactors

We find that methods such as AR theory are required not because of problems involving multiple reactors, but because of problems involving multiple reactions. Many of these methods are unnecessary when the reactions are inherently simple. The reactor network synthesis problem arises often as a result of complexities in the system from multiple reactions.

When the reaction is complex, the best performance is often achieved in a reactor network (a combination of reactors). AR theory deals with reactor problems involving *more than one reactor.*

1.4 SOLVING THE REACTOR NETWORK SYNTHESIS PROBLEM

1.4.1 Reactor Superstructures

1.4.1.1 Description A popular approach to solving the reactor network synthesis problem is by use of *reactor superstructures*. A reactor superstructure is a reactor configuration where many simpler reactor types are arranged in a specific pattern and connected via a network of bypass and recycle streams, so that various outcomes can be achieved from a single structure. Reactor superstructures thus represent a superset of all possible outputs, wherein an optimal answer is a subset of the superstructure.

ILLUSTRATION: Examples of reactor superstructures

Consider how a series of CSTRs might act as potential building blocks that approximate other reactor types.

For PFR approximation, CSTRs can be connected in series, as in Figure 1.8(a). A variety of different outputs may be produced from a single reactor type in the appropriate arrangement. More complex reactor structures have been published in the scientific literature. One example is provided in Figure 1.8(b) (Rooney and Biegler, 2000).

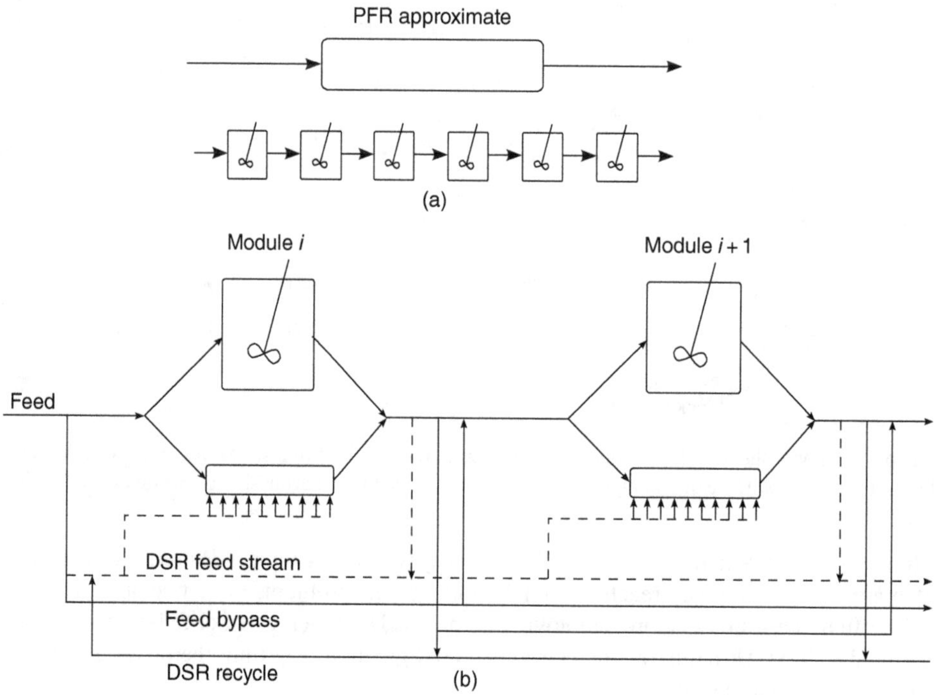

Figure 1.8 (a) A CSTR configuration that approximates a plug flow reactor (PFR). Kauchali et al. (2002). Reproduced with permission of Elsevier. (b) Example of a reactor superstructure. Rooney and Biegler (2000). Reproduced with permission of Elsevier.

Combinations of basic reactor types, such as CSTRs and differential sidestream reactors (DSRs), are arranged in a specific pattern to produce a building block structure for the creation of more sophisticated reactor structures. These superstructures allow for the description of a wider variety of different achievable states.

Using the superstructure approach, nonreactive unit operations (i.e., separation) may be included into the design as well, which allows for a wide range of optimization scenarios to be solved for. The superstructure approach is a powerful method to reactor network design, as a large amount of detail and flexibility can be incorporated into the configuration.

SIDE NOTE: The power of the superstructure approach

Superstructure methods are a powerful and popular approach to the reactor network synthesis problem, as well as to the generalized problem of process synthesis (the design of entire processes). An advantage of superstructures is that they interface easily with existing plants—that is, it is often possible to mathematically model the operation of plants that have already been designed and built, which once represented as a mathematical model, can be optimized with standard optimization methods. This makes the superstructure approach highly useful in industry, as existing plants can be modeled and optimized directly, making results much easier to relate to real-life performance.

1.4.1.2 Solving Optimal Reactor Superstructures With a large number of interlinked bypass and recycle streams, amongst a number of different reactor types, reactor superstructures are expressed in a large mathematical model—a system of differential algebraic equations (DAEs)—which can be solved using mathematical optimization techniques. To solve the model, the parameters of the superstructure model are adjusted and checked against an appropriate objective function to determine optimal performance. Optimization follows the generalized form:

Maximize/minimize:	Objective function (i.e., volume, concentration, time or operating costs)
Subject to:	Parameters and constraints (i.e., time, reactor model equations, flow rates and mixing fractions)

The complexity of modeling and optimizing the superstructure is then related to the complexity of the superstructure itself. Parameters in the model act like switches on a switchboard, and a solution to the optimization problem is a specific combination of switches that achieves the desired result. A balance must be established between practicality and accuracy: the superstructure must be adequately generalized (more complex) to describe many outcomes, yet computationally simple to allow for solution of the associated problem in a reasonable time.

Solutions are based on the objective function specified. A change in the objective function introduces a change in the superstructure configuration, which might require re-solving the problem for the new objective function. (When Sam, Alex, and Donald are asked to minimize the production of hydrogen in the BTX reaction, this signals a change in objective function.)

Two important questions that arise when dealing with the optimization of reactor superstructures thus arise:

1. Are there similar superstructure configurations that achieve the same result? (*Are there multiple solutions?*)
2. Does a better superstructure exist? (*Is the answer globally optimal?*)

More generalized structures, capable of describing a wider variety of solutions, may require additional complexity in the superstructure model. Additional complexity in the design also leads to increased complexity in the solution of the equations describing the superstructure, which may make the problem more difficult to solve. (Figure 1.8(b) might be able to describe more complex problems and solutions based on Figure 1.8(b) may offer better performance than Figure 1.8(a), but it is also more difficult to solve.)

ILLUSTRATION: Word combinations

How many unique English words can you form from the following set of letters? Not all letters in the set need be used:

$$\{C, X, A, H, T\}$$

Suppose now that the letter "E" is included into the set. How many more words can you find? What if letters "D" and "O" are added to the set?

With the help of a computer, we can show that there are approximately 11 unique words that can be formed from the set {C, X, A, H, T}, 32 unique combinations are possible from the set {C, X, A, H, T, E}, and 83 words are possible from {C, X, A, H, T, E, D, O}. Indeed, it is evident that with each addition of a new letter, there are many more words that can be generated. (If we were to add the letter "R" to the set for instance, the number of unique combinations would increase to roughly 211 words!)

Each time additional variations are introduced, many new combinations are possible, and complexity explodes. This phenomenon is also observed with reactor superstructures—with highly generalized superstructures, there is a greater possibility of obtaining many configurations of the base design. Many of these designs may even result in the same output. The mathematical models associated with these superstructures are also more difficult to solve. But there may also be many other solutions for which the superstructure does *not* acknowledge. The set of solutions obtained might therefore only be a *subset* of all possible outcomes in the system.

1.4.2 AR Theory

In AR theory, we have an interest in the reactor network synthesis problem, but it is not our chief interest. It is sometimes not possible to generate an optimal reactor structure using AR theory, but it is still possible to generate important information that may help to understand what the optimal reactor structure might be. Let us refer back to the following two points from Section 1.4.1.2:

1. Are there multiple solutions?
2. Is the solution globally optimal?

In both of these questions, we must begin with knowledge of the final answer *before* the reactor structure is devised—we must know, beforehand, of all possible solutions (for all possible structures) to understand if multiple solutions exist, and the same information must be known to understand if our structure is globally optimal.

This ordering of tasks is depicted graphically in Figure 1.9. In a "classical" (non-AR theory) approach, reactor structures are first formulated, solved, and then optimized to produce a particular output. An output exists for a particular reactor structure, and the exact output obtained is as a result of the specific path taken in the approach. The typical workflow is then from left to right. In contrast, we wish to first find many answers at once with AR theory, and then utilize these answers to optimize the specific problem at hand. This is done without a specific impression of a particular reactor structure in mind. Rather, the reactor structure is an outcome of the answer and not the other way around, and the workflow in this instance then proceeds from right to left. All possible answers to all possible problems are generated first, and only after this task has been accomplished is the reactor structure for a particular duty found.

As a precursor to the following chapters, we wish to leave with a number of scenarios that AR theory could help to address. These scenarios are related to the BTX problem discussed earlier. It is important to keep these questions in mind, for they shape the approach we would like to adopt with AR theory.

1. *A Known Achievable Point*: Consider if it were known (because an oracle tells you) that a toluene concentration of 0.09 mol/L is achievable *without* knowing what particular piece of equipment or method is needed to achieve it. Would it be possible to utilize this piece of information to deduce what equipment or methods would be needed? Simply knowing that 0.09 mol/L is an achievable concentration may incentivize investigation to determine an appropriate reactor that achieves it.
2. *Multiple Solutions*: Suppose that various sets of reactors could all achieve the same toluene concentration of 0.09 mol/L, but that some reactor arrangements are able to achieve this concentration in a smaller total reactor volume. Is your design still appropriate? Understanding the range of different options available to us is useful in selecting the most appropriate reactor configuration for our needs.
3. *Globally Optimal Solution*: Now imagine that the maximum toluene concentration ever achievable is 0.1 mol/L. Would there be any attempt to improve the system further? What if the maximum concentration is 1.5 mol/L instead? This information is useful to the designer, as now appropriate performance targets

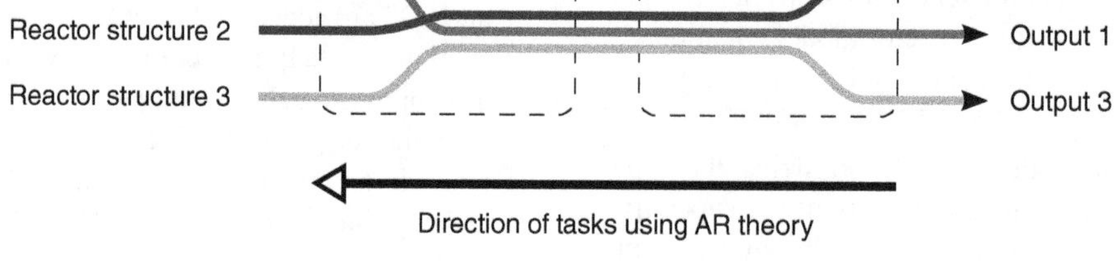

Figure 1.9 Order of tasks using the AR approach.

may be set and used to justify whether additional optimization effort is required.

4. *A Different Global Optimum for a Different Objective*: Imagine further that xylene is now the desired component and not toluene, and that we also wish to minimize the consumption of ethylene. How do we achieve this new objective and is our old design still appropriate? If it is possible to compute a *set* of achievable available concentrations, for many reactor configurations, then a change in objective function is easily incorporated into the new analysis.

1.4.3 Attainability Problems Outside of Reactor Design

We end this chapter with a number of open-ended anecdotes. The topics described are not specific to reactor design, but many of them are simply the same challenge disguised under a different context. All situations seek to understand the same kind of question: *How do we know that we are the best?*

1.4.3.1 Fishing Two friends, Bob and Jim, go fishing on a trout farm and hold a friendly competition to see who can catch the biggest fish by the end of the day. They find a quiet spot by the lake in the morning, settle in, and cast their lines into the lake at the same time. The two spend the rest of the day attempting to outdo each other by catching the largest fish.

The two take their catches back to be weighed and recorded at the end of the day. On this occasion, Bob has managed to catch the largest fish. When they arrive at shore, spectators are astounded by the size of Bob's catch. The lake owner comments that Bob's fish is the largest he had seen in his entire career. He adds that it may even be the largest fish in the entire lake!

But how would we know if Bob's catch is indeed the biggest fish in the lake? Perhaps larger fish exist that have not yet been caught. It is simply not feasible to see all the fish in the lake for one to know with absolute certainty. We would first need to find and measure *all* the fish in the lake. The question is then: how do we find all the fish in the lake?

SIDE NOTE: Colossal squid

Colossal squid—measuring in excess of 10 m in length—are sometimes caught by fishing boats, or are found when they wash up on shore. The only colossal squid that we know of are the few specimens that scientists have observed near the ocean surface, yet they exist in parts of the deep ocean that human society cannot fully explore at present. It is likely that there are even bigger squid that exist deep in the ocean. How would we find these creatures if they reside in a part of the ocean that we cannot reach?

1.4.3.2 Olympic Records On August 16, 2008, during the Summer Olympic Games, Usain Bolt set a new world record for the men's 100 m sprint, achieving a time of 9.69 s. One year later, on August 16, 2009, during the International Association of Athletics Federations or IAAF World Championship in Athletics, Bolt again broke the world record by running a time of 9.58 s (*IAAF World Championships: IAAF Statistics Handbook. Daegu 2011*, 2011).

Indeed, Usain Bolt's successive achievements have made us question something deeper in the sport—what is the shortest time a human being could *ever* run 100 m? Are we close to reaching this point, or will someone run a faster time in the future?

ILLUSTRATION: A maze puzzle

Consider how you might approach solving the maze puzzle in Figure 1.10. The aim is to start at a predefined location and then travel through the maze to reach a desired exit point. A number of potential paths may be pursued, resulting in many final states.

1. Suppose that you wish to terminate your journey at point 1 in Figure 1.10. How would your solution be affected if you started from:
   a. The point labeled "Start" and worked to point 1?
   b. Point 1 and worked backward to the start?
2. Suppose we wish to end at point 7 instead. Would it be easier to start from the beginning or the end of the maze?
3. Imagine that now all the exits are hidden and are only known when they are reached. That is, we do not know that there are a total of eight exit points initially. At the end of each exit, a reward is placed (i.e., a pot of gold). Each exit holds a different reward value, but exit 6 contains the largest reward out of all eight exits (it is the global optimum). If only one exit can be reached on any given journey, how would you go about finding all exit points and simultaneously know that exit 6 is the overall best? If you were to end at exit 3 instead, how would you know if there is a larger reward elsewhere?

This example, although simple, illustrates a fundamental challenge often encountered in reactor network synthesis problems. Each exit in the maze could be considered a final objective. An answer is achieved from an initial state, but various outcomes may be realized from a single starting point. When attempting to solve reactor network synthesis problems, we often do not know how many possible outcomes could be achieved. When an exit is found, how do we know if it is the best? There is certainty only when *all* possible exits have been identified, but this is difficult when we do not start with this information in hand. We seek to understand the *attainability* of all points in a system.

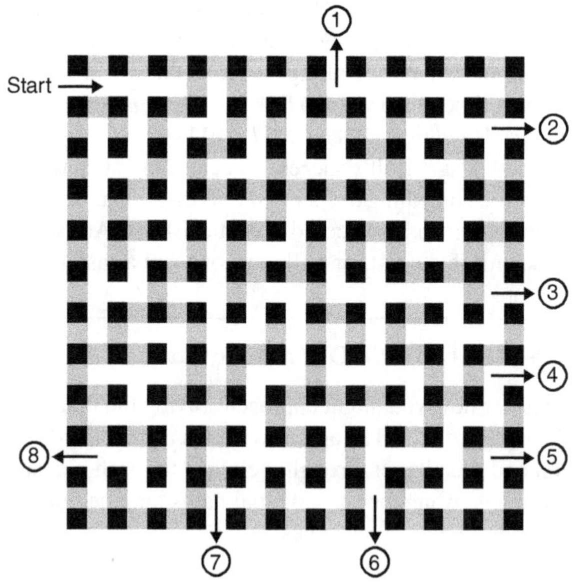

Figure 1.10 A maze with one entry point and multiple exit points.

Note the differences when performing tasks with these two different approaches:

1. In the first (non-AR theory) approach, the answer is obtained by proposing a structure.
2. In the second (AR theory) approach, it is first assumed that a certain answer is achievable, and then we determine what is required to achieve it.

If the best answer is known beforehand, then the same procedure may be carried out to find the corresponding best reactor configuration. It is often difficult to know when a good design has been achieved when we first devise a design and then attempt to find a solution, because it is not clear whether a superior design exists. The order in which tasks are carried out is significant. AR theory attempts to generate answers first. This approach offers us a different way of viewing and solving reactor problems.

IMPORTANT: Attainability and performance targeting

Knowing what is achievable beforehand is useful, for it provides guidance on how one should proceed with a design. With attainability information, it is possible to set appropriate and realistic *performance targets*—you may want not improve your current design if you knew that you were 99% to perfection, but you may wish to do something if you were only 50% of the best instead.

1.5 CHAPTER REVIEW

OBSERVATION: Finding your way in the dark

In the absence of any previous knowledge, we often think that carrying out a design (or conducting an experiment) is similar to how we wonder in the dark: It is straightforward to navigate a room at night when the lights are switched on. It is more challenging when the power goes off, and we are caught in the dark—attempting to avoid the furniture—searching for a flashlight. When the power returns, everything is put into context, and it is again simple to see how everything fits together.

Without any previous knowledge, we often follow a similar procedure in design: only after discovering what works is it clear how different units should fit together, and why certain designs work better than others. Although it is easy to see what to do next once it is known (optimizing for reaction time, feeding rates, etc.), carrying out modifications provides little satisfaction if these improvements are replaced with more effective ones.

We write this book out of a need to understand a very basic question—*when do we know we have achieved the best?* AR theory is an approach that seeks to help answer this question for chemical reactor networks (reactor structures). In this chapter, we described how problems of this nature arise when attempting to optimally produce toluene in the BTX reaction. In general, three approaches to the solution of this problem emerged, which are summarized pictorially in Figure 1.11.

Each of the approaches displayed are useful. One could always begin with a generalized approach, and then focus on specific issues during the investigation. Truly effective designs, we believe, are ones that incorporate all three strategies into the design process, although it is the *ordering* of tasks that is important.

Although we wish to motivate improvements by use of reactor structures (combinations of reactors), there are challenges associated with this approach. It is often not clear how to formulate and optimize systems involving complex combinations, and simply listing many combinations always leads us to ponder if there are other designs that may be better.

With ARs, we want to approach problems from the mindset of looking at the answer first, and then focusing on how

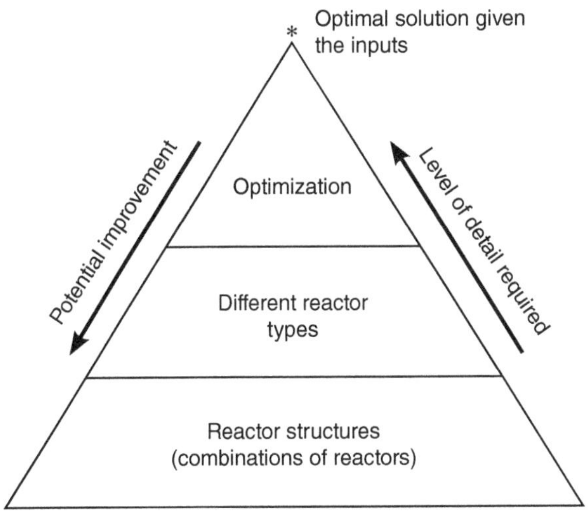

Figure 1.11 Hierarchy of design for reactors.

to achieve this from a known starting point. In this sense, the approach is top-down as opposed to bottom-up, starting from the biggest changes through structure, and refining until the most appropriate design is identified. From then on, optimizations may be carried out with the knowledge that improvements are performed on the best possible reactor structure for the problem.

From this perspective, AR theory should not be viewed as a competing method to optimization or superstructure methods, but rather as complementary tool used to benchmark current designs. Superstructures can be used in conjunction with AR theory to both set reactor performance targets and design the reactors needed to achieve these targets.

We do not wish to stumble in the dark, only identifying our position in the room once we have reached it. Rather, we hope that the ideas of AR theory will act as a flashlight, illuminating aspects of the unknown so that it is easier to advance forward. It is the awareness that these ideas are important that eventually promotes improvement. In short, *we cannot fix what we do not know*. The most useful feature that we would like to generate with AR theory is *hindsight*.

SIDE NOTE: A roadmap for discussing future questions

Throughout this book, we wish to introduce a set of ideas and perspectives that will ultimately allow us to address the challenges faced by Sam, Alex, and Donald. Many of the questions raised by them will be examined in subsequent chapters. We include a number of these questions in the following text, as a roadmap for the reader.

- *How should the problem be interpreted?*

  These fundamental concepts are introduced in Chapter 2. Here, the reader will learn how a change in perspective can allow for more complex concepts, such as the AR, to be described.
- *Is the answer unique? Could I do better? What is the absolute best?*

  Understanding what is the absolute best starts by understanding the AR itself. This concept is described in Chapter 3.
- *What procedure or method is required to achieve a certain performance target?*

  The specific reactor type used, and why it is needed, is described in Chapter 4. A number of worked problems, using AR theory, are given in Chapter 5.
- *How should multiple objectives be handled?*

  A change in perspective does not generally influence how we approach the analysis in AR theory. This concept is described in Chapters 3 and 4. Similar questions arise when there is a change in conditions, such as a change in the reactor feed, or if a different reaction is used. Again, this may give rise to a different recommendation, but the fundamental approach in AR theory remains unchanged. A worked example demonstrating this concept is given in Chapter 5.

REFERENCES

IAAF World Championships: IAAF Statistics Handbook. Daegu 2011, 2011. IAAF Media & Public Relations Department, Monte Carlo.

Kauchali, S., Rooney, W.C., Biegler, L.T., Glasser, D., Hildebrandt, D., 2002. Linear programming formulations for attainable region analysis. Chem. Eng. Sci. 57, 2015–2028.

Nielsen, J., Villadsen, J., Lidén, G., 2011. Design of fermentation processes, in: Bioreaction Engineering Principles. Springer, New York, pp. 383–458. http://link.springer.com/book/10.1007%2F978-1-4419-9688-6 (accessed May 18, 2016).

Rooney, W.C., Biegler, L.T., 2000. Multiperiod reactor network synthesis. Comput. Chem. Eng. 24, 2055–2068.

Unrean, P., Nguyen, N.H., 2013. Metabolic pathway analysis and kinetic studies for production of nattokinase in *Bacillus subtilis*. Bioprocess Biosyst. Eng. 36, 45–56.

2

CONCENTRATION AND MIXING

2.1 INTRODUCTION

2.1.1 Review

In Chapter 1, we looked at a system of reactions for the production of toluene from benzene (the BTX system). A number of different strategies were proposed for approaching the problem of producing the largest amount of toluene possible. Three central strategies were discussed. The final solution obtained varied based on the approach chosen; however, each method is useful in itself.

In this chapter, we wish to focus on laying the foundations for AR theory, with the view of eventually employing reactor structures to improve performance in the BTX system. As a first step, the data given in Chapter 1 will be revisualized and related to a number of simple geometric concepts related to concentration and mixing. These topics are crucial to the proper understanding of AR theory.

2.1.2 Revisualizing Concentration Data

Consider again the original BTX batch data generated from Chapter 1, represented here by Table 2.1.

The data in Table 2.1 are for all components represented by the system of reactions given in Chapter 1. For our current problem, let us look at removing the time component from the data. It is preferable for us to do this as we wish to investigate the performance of the system when time is not a concern, which is useful when we want to understand the best possible system performance that can be achieved. Note that there may be situations where time *is* an important factor, in which case this element should be retained. We shall keep this in mind and come back to problems specifically related to time and reactor volume in later chapters.

Let us isolate the components in the BTX reaction that are of interest to us. Since we are interested in maximizing toluene production, and since benzene is a required feed component, selectivity could be calculated based on the benzene feed and toluene concentrations. Hence, let us plot the benzene and toluene concentrations independently to the other components given before. This visualization is commonly referred to as a *phase plot*, or a *phase diagram*, and is often employed to analyze dynamical systems.

Alternatively, if we already have the concentration profiles generated with time as in Chapter 1, then we could also collect the data using the following procedure:

1. Choose a time given on the abscissa (x-axis) and mark it by drawing a vertical line upward intersecting the concentrations of components. Find the intersections for the components of interest. In this case, it would be benzene and toluene. Refer to Figure 2.4(a) for an illustration demonstrating the procedure.
2. At the intersections with the concentration profiles, read off the ordinate values (benzene and toluene concentration) of interest.
3. Record the values of toluene and benzene in separate columns. Repeat the procedure for all times.
4. Plot the two lists of values against each other.

Figure 2.4(a) shows the concentration profiles for benzene and toluene in concentration–time space. A vertical line is drawn at a reaction time of 0.5 h and the associated benzene–toluene concentration pair is recorded. This procedure is repeated for all concentration pairs in the reaction time range. The resulting plot in benzene–toluene concentration (c_B–c_T) space is displayed in Figure 2.4(b). Toluene has been plotted as the dependent variable on the y-axis versus benzene on the x-axis.[1] Also shown in Figure 2.4(b) is the associated intersection point at a reaction time of 0.5 h from Figure 2.4(a) in c_B–c_T space.

[1] There is no convention for which components should be placed on the graph axes.

Attainable Region Theory: An Introduction to Choosing an Optimal Reactor, First Edition. David Ming, David Glasser, Diane Hildebrandt, Benjamin Glasser, and Matthew Metzger.

Companion website: http://attainableregions.com

TABLE 2.1 Selected Concentration Data for the BTX Data Obtained in a Batch System

Time (h)	Benzene (mol/L)	Ethylene (mol/L)	Toluene (mol/L)	Xylene (mol/L)	Diphenyl (mol/L)	Hydrogen (mol/L)
0.00	1.0000	0.5000	0.0000	0.0000	0.0000	0.0000
0.01	0.8280	0.4968	0.0064	0.0000	0.0828	0.0828
0.05	0.4870	0.4877	0.0236	0.0005	0.2445	0.2445
0.10	0.3182	0.4804	0.0361	0.0015	0.3221	0.3221
0.30	0.1268	0.4636	0.0564	0.0082	0.4043	0.4043
0.45	0.0841	0.4554	0.0609	0.0142	0.4204	0.4204
0.55	0.0677	0.4508	0.0618	0.0183	0.4261	0.4261
0.65	0.0560	0.4466	0.0618	0.0225	0.4299	0.4299
0.70	0.0513	0.4447	0.0615	0.0245	0.4313	0.4313
0.80	0.0437	0.4411	0.0606	0.0286	0.4336	0.4336
0.90	0.0377	0.4378	0.0593	0.0326	0.4352	0.4352
1.00	0.0329	0.4347	0.0578	0.0364	0.4365	0.4365
1.50	0.0185	0.4219	0.0485	0.0538	0.4396	0.4396
2.00	0.0116	0.4125	0.0391	0.0680	0.4407	0.4407
2.50	0.0077	0.4054	0.0310	0.0791	0.4411	0.4411
3.00	0.0052	0.4000	0.0242	0.0878	0.4413	0.4413
3.50	0.0037	0.3960	0.0189	0.0946	0.4414	0.4414
4.00	0.0026	0.3929	0.0146	0.0998	0.4415	0.4415
6.00	0.0007	0.3863	0.0051	0.1111	0.4415	0.4415
10.00	0.0001	0.3834	0.0006	0.1163	0.4415	0.4415

EXAMPLE 1: Visualizing Concentration Data

Question 1:
Table 2.2 represents the data obtained in a batch reaction at different time intervals. Plot the data in the phase plane.

Answer 1:
Representing data in the phase plane involves taking the component values for A and B and plotting them in concentration space, ignoring the time component. This is easily done by reading off the c_A–c_B concentration pairs at each time measurement and plotting these values in c_A–c_B space. A point in space is given by the coordinate (c_A, c_B), which is shown in Figure 2.1.

TABLE 2.2 Concentration Profile Data for a Two-Component System

Time (min)	c_A (mol/L)	c_B (mol/L)
1	0.098	0.113
2	0.169	0.308
3	0.315	0.545
4	0.566	0.338
5	0.907	0.104

Question 2:
Assume now that the reaction involves three components. Table 2.3 gives the relevant data. Plot the data in concentration space.

Answer 2:
Since the data now involve three components, plotting data in the phase plane will result in a three-dimensional plot in c_A–c_B–c_C space. The time column is again ignored to generate this plot, and the results of the plot are shown in Figure 2.2.

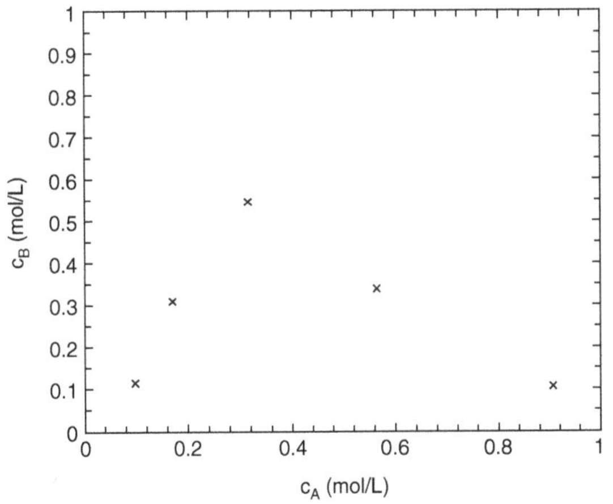

Figure 2.1 Concentration data plotted in the phase plane.

TABLE 2.3 Concentration Profile Data for a Three-Component System

Time (min)	c_A (mol/L)	c_B (mol/L)	c_C (mol/L)
1	0.098	0.113	0.122
2	0.169	0.308	0.337
3	0.315	0.545	0.644
4	0.566	0.338	0.658
5	0.907	0.104	0.926

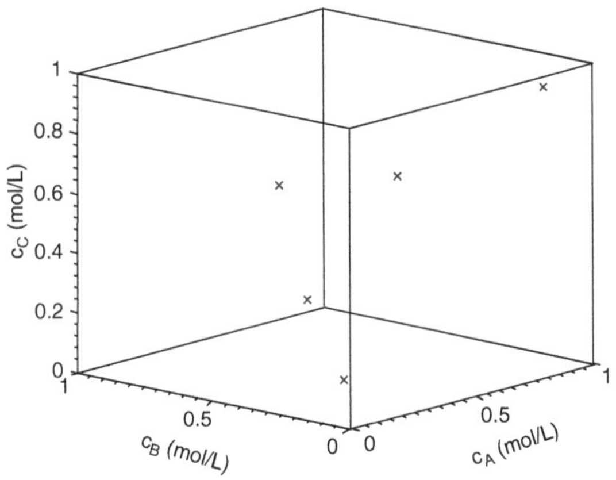

Figure 2.2 Concentration data plotted in three-dimensional space.

Question 3:
Suppose that the reaction now involves five components. The experimental data is given by Table 2.4. What do the data look like in

(a) c_A–c_B space?
(b) c_B–c_D–c_E space?

TABLE 2.4 Concentration Profile Data for a Five-Component System

Time (min)	c_A (mol/L)	c_B (mol/L)	c_C (mol/L)	c_D (mol/L)	c_E (mol/L)
1	1.500	0.164	0.241	1.259	1.500
2	3.000	1.207	1.021	1.979	3.000
3	4.500	2.262	2.251	2.249	4.500
4	6.000	3.210	3.921	2.079	6.000
5	7.500	4.060	6.031	1.469	7.500

Answer 3:

The reaction data now involve five components. It is hence no longer possible to represent the entire data set in the phase plane on a single plot (this would involve a five-dimensional plot in c_A–c_B–c_C–c_D–c_E space). However, it is still feasible to plot subsets of the data as lower dimensional projections into two- and three-dimensional spaces. Hence, for (a), a two-dimensional plot is obtained by taking the concentration data belonging to components A and B and plotting the data whilst ignoring the remaining components. For (b), a three-dimensional plot is generated with columns belonging to B, D, and E. These results are shown in Figure 2.3. The shape of the resulting curves is different depending on what space the data are plotted in.

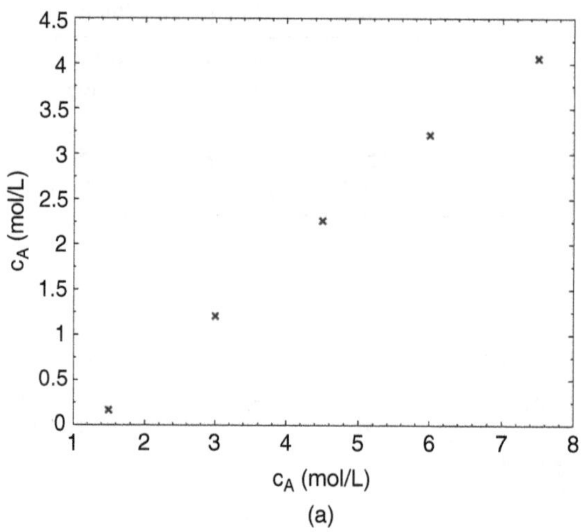

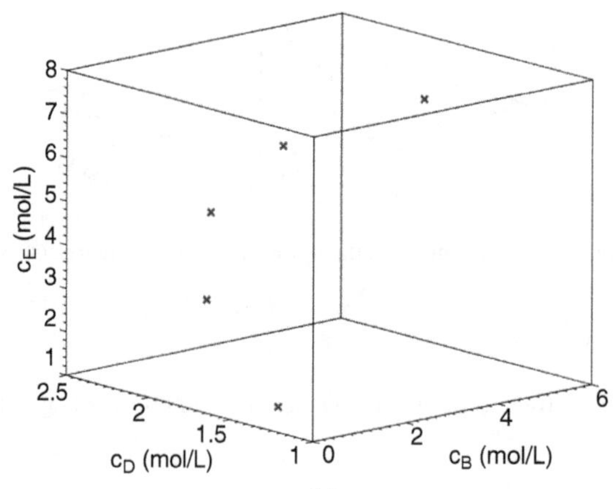

Figure 2.3 (a) Two-dimensional plot of the five component data in c_A–c_B space. (b) Three-dimensional plot of the five component data in c_B–c_D–c_E space.

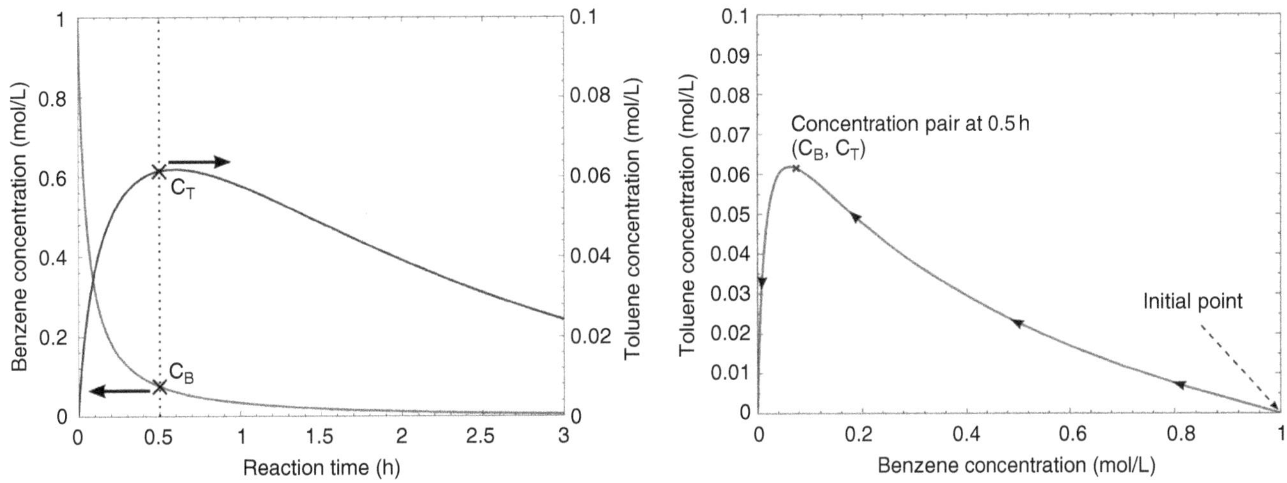

(a) (b)

Figure 2.4 Converting concentration–time data into concentration–concentration data. (a) Concentration profiles for benzene and toluene. Note that ethylene, xylene, diphenyl, and hydrogen are not shown. (b) Benzene and toluene plotted in the phase plane.

This procedure generates a new graph where all aspects associated with reaction time have been eliminated.[2] But it is now clear that the toluene concentration is maximized at low benzene concentrations of approximately 0.04 mol/L. Since benzene is a feed component to the reactor, this plot indicates that the reactor should be run at high benzene conversions. This result is also in support of our initial decision from Chapter 1 to run the reactor at an optimal reaction time.

We have therefore gained some insight into how the BTX reaction should be carried out. And although this result is simple, it highlights an important approach in AR theory: In AR theory, we will seek to interpret all concentration data graphically. Visualizing data in this manner allows us to adopt a different perspective of reactors; but to understand why, we must first introduce a number of fundamental concepts.

> **IMPORTANT: Plotting Concentration Data as a Phase Diagram (Concentration–Concentration Plots)**
>
> Revisualizing data in the phase plane is the first important step to understanding AR theory. Viewing data in this manner often provides a different perspective on the system under investigation.

2.2 CONCENTRATION VECTORS AND DIMENSION

2.2.1 Moving on a Line: Table Salt and Water

2.2.1.1 Initial Concentrations Let us look at a simple example involving table salt (NaCl) in water. Assume that 5 g of NaCl is dissolved into an equal mass of pure water. The total mass of the system is thus 10 g. Mass fractions of NaCl, z_{NaCl}, and water, z_{H_2O}, may be calculated by taking species masses and dividing by the total mass of the system:

$$z_{NaCl} = \frac{m_{NaCl}}{m_{NaCl} + m_{H_2O}} = 0.5$$

$$z_{H_2O} = \frac{m_{H_2O}}{m_{NaCl} + m_{H_2O}} = 0.5$$

To view the data in a graphical manner, we shall borrow an idea from Section 2.1.2 and plot the mass fraction of NaCl and H_2O, each on an axis in two-dimensional mass fraction space. For this particular example, we choose the x-axis to represent the mass fraction of NaCl in solution; the mass fraction of water in solution is then placed on the y-axis.

For *any* amount of NaCl and H_2O added or removed from the system, it is possible to visualize the solution mixture as a point in NaCl–H_2O mass fraction space. That is, any ratio of NaCl and H_2O produces a point in z_{NaCl}–z_{H_2O} space. For the present combination of salt and water supplied before, the system may be represented as the point (0.5, 0.5) in mass fraction space. This mixture is shown as point 1 in Figure 2.5.

2.2.1.2 Additional Water Let us assume that mixing occurs in a beaker. Although an entire space is identifiable in Figure 2.5, we claim that all physically realizable beaker concentrations can only exist on a small portion of space (a subset of points in space). To see this, let us take the original system and add an additional 10 g of water to the solution, resulting in 15 g of H_2O and 5 g of NaCl. The corresponding mass fractions for salt and water are thus 0.25 and 0.75, respectively. It is again possible to plot the new species mass fractions in NaCl–H_2O space. This is represented as the point (0.25, 0.75) and is also shown as point 2 in Figure 2.5. As observed in the plot, we move from point 1 to 2 when additional water is added to the system.

[2]Often, we will annotate phase diagrams with arrows to give an indication of the direction of reaction.

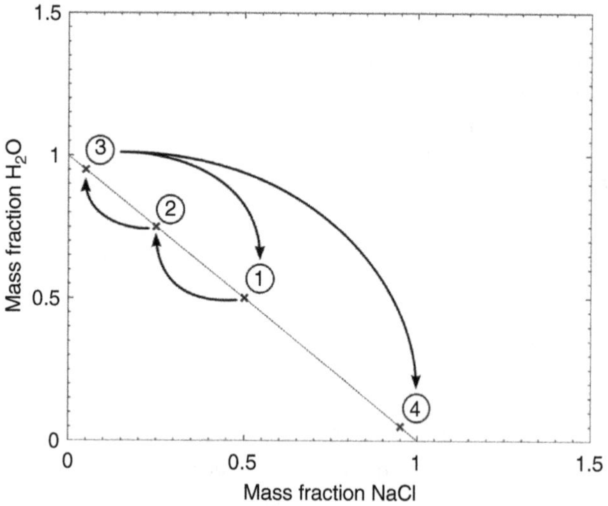

Figure 2.5 Graphical interpretation of the NaCl–H_2O system for different beaker concentrations.

If an additional 80 g is added to the 15 g of water present, then it is easy to compute the mass fractions in the system, giving the point (0.05, 0.95); this mixture is displayed as point 3 in Figure 2.5. Note that by the addition of more water, we have moved from point 2 to 3. It is clear that additional water added to the system reduces the concentration of NaCl in solution, and hence the mixture point will lie closer to the point (1.0, 0.0), which is the coordinate representing pure water.

*2.2.1.3 **Additional Salt*** It is also possible to add more salt to the system. If an additional 90 g of NaCl is mixed into solution, then equal quantities of water and salt are again present in the system. Recall that the point of equal mass fractions in the system is represented by point 1 in the figure. By adding more salt, we therefore move from point 3 back to point 1. Point 1 thus represents two different physical states with the same composition.

EXAMPLE 2: Additional Salt

Suppose instead that from point 3, we wish to generate a solution containing 95% NaCl and only 5% water by mass. What mass of salt must be added to the existing solution and what is the total mass of the resulting solution?

Answer:

A 95% salt, 5% water solution would be represented by the coordinate (0.95, 0.05), and it is point 4 in Figure 2.5. The mass fraction of salt may be expressed as follows:

$$z_{NaCl} = \frac{m_{NaCl}}{m_{NaCl} + m_{H_2O}}$$

Since $z_{NaCl} = 0.95$ and $m_{H_2O} = 95$ g, the mass of salt is the only unknown and can be found by solving the given expression for m_{NaCl}. We find that 1805 g of salt must be added to the 95 g of water. The total mass in the system is then 1900 g.

Note that by plotting the species concentrations in mass fraction space, we can visualize each scenario as a distinct point on the graph. Provided that only combinations of water and salt are added to the system, it is clear that the resulting mixtures are associated with points that lie on a line in mass fraction space. This is easy to interpret once we notice that the species mass fractions must sum to unity:

$$z_{NaCl} + z_{H_2O} = 1$$

OBSERVATION: Moving on a Line

When only two components are present, mass balance requires that species mass fractions lie on a straight line in mass fraction space. This result is also clear by observing that the equation

$$z_{NaCl} + z_{H_2O} = 1$$

is a straight-line equation in z_{NaCl}–z_{H_2O} space.

The system appears to be *one-dimensional* with respect to mixing, even though there are two components present. It is therefore physically impossible to formulate a mixture point, using water and NaCl only, which does not lie on the line passing through points 1–4. The line is thus defined by the two limiting cases of the system: (i) pure water, given by the coordinate (0.0, 1.0) and (ii) pure salt, given by the coordinate (1.0, 0.0), and all mixtures of water and salt must reside on a line joined between these two points.

EXAMPLE 3: Moving on a Plane

Assume that a third component (sugar) is added to the table salt and water system. Can you predict what kind of space this system would represent using mass fractions? Plot a number of points to validate your answer assuming that pure quantities of each component are available.

Answer:

Since there are now three components in the system, the mass fraction of sugar, z_{sugar}, is related by the following expression:

$$z_{sugar} = 1 - z_{NaCl} - z_{H_2O}$$

This provides an additional constraint on the system and the allowable values that z_{sugar} can be assigned to. When there are only two components participating in the system, the mixture concentrations are constrained to a line in two-dimensional mass fraction space. When three components are present, there is an additional degree of freedom, and thus we expect the concentrations to lie on a plane in three-dimensional mass fraction space. The resulting mass fractions have been plotted in z_{NaCl}–z_{H_2O}–z_{sugar} space for sample values of z between 0 and 1 (see Figure 2.6). Observe that the mixture concentrations appear to lie on a plane in the shape of a triangle in mass fraction space.

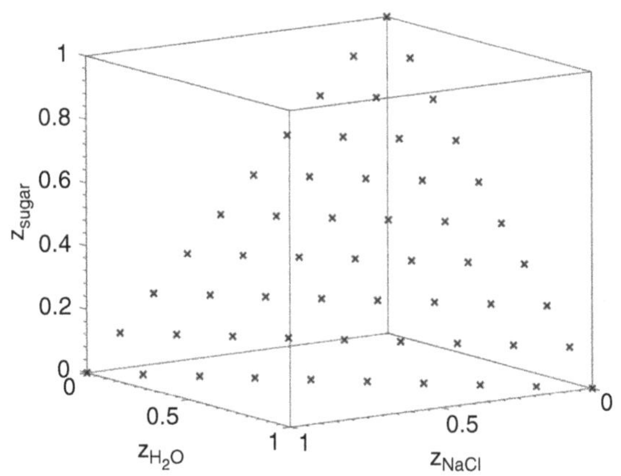

Figure 2.6 NaCl–H_2O–sugar mass fractions.

2.2.2 Moving Freely Through Space

2.2.2.1 Interpretation Suppose now that a three-component mixture is available, containing 5 g of table salt (NaCl) and 10 g of potassium hydroxide (KOH), both dissolved in 1.0 L of pure water.

EXAMPLE 4: Simple Concentration Calculations

If the molecular weights of NaCl and KOH are 58.5 and 56.1 g/mol, respectively, then what are the molar concentrations of these two species in solution?

Answer:

Five gram of NaCl with a molecular weight of 58.5 g/mol gives 0.085 mol of NaCl in solution. Similarly, 10 g of KOH with a molecular weight of 56.1 g/mol indicates that there is 0.178 mol of KOH present in solution. Since the volume of water is 1.0 L, the concentrations of NaCl and KOH are 0.085 and 0.178 mol/L, respectively.

RECALL: Calculating Concentration

The concentration of a component in a mixture may be expressed in different forms. For example, mass and mole fractions express the concentration of a component as a portion of the total stream. Molar concentration is another popular measure of concentration, and it will be used extensively in this book. To calculate the molar concentration of species i in solution, c_i, we divide the number of moles of component i in the mixture, n_i, by the total volume of solution, V:

$$c_i = \frac{n_i}{V}$$

We will again use the idea of plotting the concentration data without a time component. This time, however, *molar concentrations* will be utilized instead of mass fractions. In this instance, we will also replace water with the concentration of KOH on the corresponding axis. Whereas mass fractions require all species to sum to 1, this is not required when interpreting these quantities in terms of concentration. Let us look at how this change affects the plots.

Consider, for example, the original 5 g of NaCl and 10 g of KOH described before. The mixture corresponds to the coordinate (0.085; 0.178) in NaCl–KOH molar concentration space, and is represented by point 1 in Figure 2.7. When an additional 5 g NaCl is added, the mixture moves 0.085 mol/L to the right. The new coordinate is hence given by (0.171; 0.178), which is shown as point 2 in Figure 2.7.

If 10 g of pure KOH are added, the concentration of KOH is doubled. The resulting solution concentration is then given by the coordinate (0.171; 0.357). Point 3 in Figure 2.7 represents this state. Similar to Section 2.2.1, any changes in concentration to the system modify the position of the mixture coordinates in space.

Suppose that an additional 1.0 L of distilled water is added to the solution, we will call this mixture 4. It is not difficult to see that when water is added to the mixture the original concentration associated with point 1 is obtained. Thus,

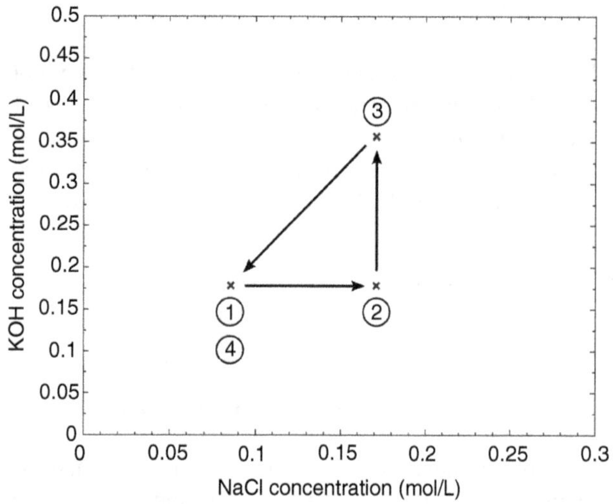

Figure 2.7 Graphical interpretation of molar concentration for the NaCl–KOH–H_2O system.

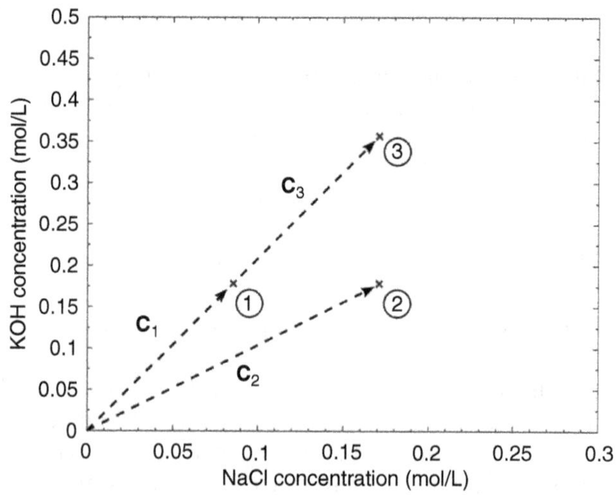

Figure 2.8 Geometric representation of concentration using concentration vectors from the origin.

even though points 1 and 4 both represent physically different states, they are the same when viewed as coordinates in concentration space (because the concentrations in both beakers are identical).

In the three steps taken before, we have traversed along the perimeter of a right-angled triangle and arrived back at the original point. That is, we have traveled from point 1 to 2, then from point 2 to 3, and finally point 3 back to point 1. This process is graphically similar to our original example involving water and salt only. In this instance, instead of moving along a straight line, we are able to move around freely throughout the entire space. The ability to move *freely* through concentration space will be beneficial to us in AR theory.

2.2.2.2 ***Concentration Vectors*** Note that it is convenient to view concentrations in the phase plane because mixtures of many components can be represented compactly by a single variable. To extend this concept further, let us introduce the idea of *vectors*. Multiple mixtures can be represented as vectors from the origin $[0, 0]^T$ in concentration space, where the elements of each vector correspond to species concentrations of the mixture. Figure 2.8 demonstrates this idea for the mixture data in Figure 2.7. Since each mixture is represented as a vector, each is associated with a unique magnitude and direction. The original concentration from Section 2.2.2.1 is currently represented by the coordinate (0.085; 0.178), which is displayed as point 1 in Figure 2.7. However, this mixture may also be viewed as the vector

$$\mathbf{C}_1 = \begin{bmatrix} 0.085 \\ 0.178 \end{bmatrix} \text{ mol/L}$$

IMPORTANT: Concentration as a Vector

Viewing concentrations as coordinates in concentration space is useful in representing the state of a system. Moreover, when this coordinate is interpreted as a vector with a unique magnitude and direction, the geometric properties of vectors may be exploited. This representation will prove to be highly useful in AR theory. Coordinate values will be enclosed in square brackets (in place of parentheses) to indicate vectors.

CONCEPT: Matrix Transpose

Both matrices and vectors are common in AR theory, and hence it is important that we are comfortable with the associated notation. To display vectors compactly, column vectors will often be expressed as the matrix transpose of the equivalent row vector. The superscript T, as used previously, therefore represents the transpose operation from linear algebra.

Observe that the elements of $\mathbf{C}_1$ have units of mol/L. Since $\mathbf{C}_1$ is a vector, we can calculate its magnitude to give

$$\|\mathbf{C}_1\| = \sqrt{0.085^2 + 0.178^2} = 0.197 \text{mol/L}$$

Changes in concentration thus also influence the direction and magnitude of the vector. When 5 g of NaCl is added to the mixture, we arrive at point 2 with the coordinate (0.171,

0.178), which may also be represented by the vector

$$\mathbf{C}_2 = \begin{bmatrix} 0.171 \\ 0.178 \end{bmatrix} \text{ mol/L}$$

Hence, mixture 2 is associated with the vector $\mathbf{C}_2$. In this instance, the magnitude of $\mathbf{C}_2$ is

$$\|\mathbf{C}_2\| = \sqrt{0.171^2 + 0.178^2} = 0.247 \text{ mol/L}$$

which is larger than the magnitude of $\mathbf{C}_1$. When 10 g of KOH are added, the direction of the resulting vector is modified so that it points in the same direction as $\mathbf{C}_1$ (it is collinear with $\mathbf{C}_1$). This vector, although having the same direction as $\mathbf{C}_1$, has a larger magnitude, which is represented by vector $\mathbf{C}_3$ as follows:

$$\mathbf{C}_3 = \begin{bmatrix} 0.171 \\ 0.357 \end{bmatrix} \text{ mol/L}$$

with magnitude given by

$$\|\mathbf{C}_3\| = \sqrt{0.171^2 + 0.357^2} = 0.396 \text{ mol/L}$$

When an additional 1 L of water is added, the concentrations of KOH and NaCl are decreased. We thus have

$$\mathbf{C}_4 = \mathbf{C}_1 = \begin{bmatrix} 0.085 \\ 0.178 \end{bmatrix} \text{ mol/L}$$

and

$$\|\mathbf{C}_4\| = 0.197 \text{ mol/L}$$

The vector magnitude might be interpreted as an indication of the absolute amount of NaCl and KOH in relation to the solvent, whereas the direction indicates the relative proportions of species in the system.

SIDE NOTE: Convention for Representing Vectors and Matrices

Unless stated otherwise, we will assume that vectors are column vectors ($n \times 1$ matrices). We adopt the convention that when symbols appear in boldface, such as **A**, **C**, **v**, and **r**, they represent vectors or matrices. Non-boldface symbols such as t, τ, α, and λ indicate scalar quantities. (Do not be concerned with what these symbols represent yet, for we shall define these quantities and discuss their meaning throughout Chapters 3–5.)

2.2.2.3 Dimension The methodology for describing species concentrations in a mixture is easily generalized for an arbitrary number of n components. That is, for an n component system in concentration space, a column vector of n entries may be formulated accompanying each component in the mixture as follows:

$$\mathbf{C} = \begin{bmatrix} c_1 \\ \vdots \\ c_n \end{bmatrix}$$

Entry c_i is therefore the concentration of component i in solution. We call the vector **C** the *concentration vector*. **C** is significant in AR theory because it may be generalized to represent a number of contexts. For example, **C** could indicate

1. The resulting concentration of a mixture from two separate mixtures.
2. Individual concentrations of two mixtures before they are brought together.
3. Different compositions inside the length of a tubular reactor.
4. Toluene and benzene concentrations within a small-scale beaker reaction experiment.

C might also be used to represent mixtures of gases or solids. Figure 2.9 illustrates some scenarios where **C** might be encountered and what it could represent.

IMPORTANT: The Concentration Vector C

In AR theory, **C** might be used to represent the concentration of various mixtures within a process. **C** might not necessarily even contain concentration data only.

C is hence used as a representation of the *state* of the system, and thus the values contained within **C** must also represent physically realizable states.[3] (For example, since concentrations can never be negative, **C** must always contain positive values.) AR theory will help us to generate a set of values for **C** that represent physically achievable states. The concentration vector concept therefore lies at the heart of AR theory.

EXAMPLE 5: Concentration Vectors A

Four components A, B, C, and D are mixed in 3.0 m^3 of water to form an aqueous solution.

[3]Classic AR texts often refer to **C** as the state vector.

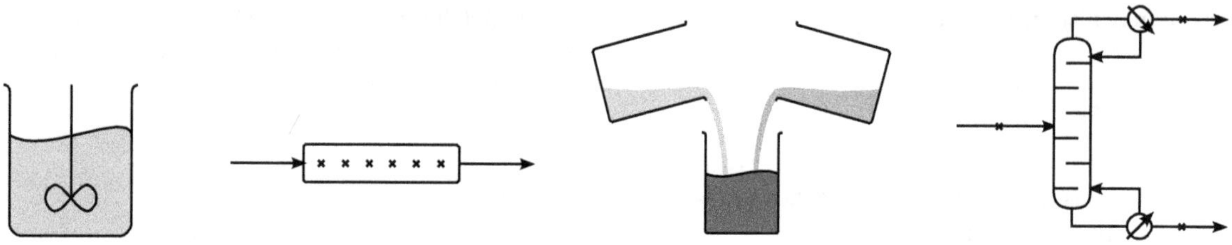

(a) (b) (c) (d)

Figure 2.9 Some examples of what concentration vectors might represent in reality. (a) Reaction in a beaker, (b) points along a reactor, (c) mixtures, and (d) points in a distillation column.

(a) If the molar quantities of each component is given by $n_A = 1$ mol, $n_B = 2$ mol, $n_C = 3$ mol, and $n_D = 4$ mol, what is the corresponding concentration vector for the mixture?
(b) If $n_A = 5$ mol, and the other molar quantities remain the same, what is the new concentration vector?

Answer a:
The concentration vector **C** is formed by calculating the species concentrations for components A–D. The species concentrations are given by $c_A = 0.33\ \text{mol/m}^3$, $c_B = 0.66\ \text{mol/m}^3$, $c_D = 1.0\ \text{mol/m}^3$, and $c_D = 1.33\ \text{mol/m}^3$. The corresponding concentration vector **C** is then given by

$$\mathbf{C} = [c_A\ c_B\ c_C\ c_D]^T = [0.33\ 0.66\ 1.0\ 1.33]^T\ \text{mol/m}^3$$

Answer b:
The molar quantity of A is five times larger than in (a); therefore, the c_A is five times as large if all other components remain the same. Thus $c_A = 1.66\ \text{mol/m}^3$. The resulting concentration vector is then

$$\mathbf{C} = [1.66\ 0.66\ 1.0\ 1.33]^T\ \text{mol/m}^3$$

EXAMPLE 6: Concentration Vectors B

Fifty milligram of CO_2 and 0.031 g of HCl are dissolved in 2 L of water. What is the concentration vector representing the resulting mixture in units of

(a) g/L?
(b) mg/m^3?

Answer a:
The concentration of CO_2 and HCl in units of g/L are

$$c_{CO_2} = \frac{0.05}{2} = 0.025\ \text{g/L}$$

$$c_{HCl} = \frac{0.031}{2} = 0.0155\ \text{g/L}$$

The associated concentration vector is then

$$\mathbf{C} = [0.025\ 0.0155]^T\ \text{g/L}$$

Answer b:
Similarly, species concentrations in units of mg/m^3 are

$$c_{CO_2} = \frac{50}{0.002} = 25,000\ \text{mg/m}^3$$

$$c_{HCl} = \frac{31}{0.002} = 15,500\ \text{mg/m}^3$$

The mixture concentration vector is then

$$\mathbf{C} = [25{,}000\ 15{,}500]^T\ \text{mg/m}^3$$

2.3 MIXING

2.3.1 Introduction

Let us now investigate the geometric properties of mixing. Instead of only holding one beaker as in Section 2.2, assume that an additional beaker is available to experiment with, which also contains NaCl and KOH in water.

It is assumed that no change in density occurs when the two mixtures are brought into contact with each other. This is a fair assumption given that the mixtures are both liquids, and any non-ideal behavior associated with density changes in aqueous solutions is generally small. Furthermore, it is also assumed that any changes in enthalpy are negligible, so that temperature effects may be ignored.

> **SIDE NOTE: Constant density assumption**
>
> It shall always be assumed that density remains constant for mixtures, unless specified otherwise. This assumption is generally acceptable for most liquid mixtures. In Chapter 9, variable density systems will be discussed.

Suppose that the composition of the second beaker is sufficiently different to the first beaker so that any mixtures formed are distinct from either of the originating beaker compositions. Assume that 50 g of KOH and 4 g of NaCl are dissolved into 1.5 L of water in the second beaker. The solution composition contains a substantially larger amount of KOH relative to NaCl. Accordingly, the concentrations of NaCl and KOH must therefore be 0.046 and 0.594 mol/L, respectively. The concentration vector representing beaker 2 is hence given by $[0.046, 0.594]^T$ mol/L. A summary of the two beaker compositions is supplied in Table 2.5.

What occurs when we combine the contents of beakers 1 and 2 together to form a mixture? There are three ways of interpreting the situation, and these are as follows:

1. **Physical**: Two beakers are brought into contact with each other, forming a third mixture. This interpretation provides a qualitative understanding of what is occurring in our system.
2. **Mathematical**: The conservation of mass may be used to calculate the species concentrations in solution. It is then possible to quantify and predict (analytically) the future state of the mixture and the system.
3. **Graphical/geometric**: We now have a third way to interpret system state by plotting data in the phase plane. This is helpful for developing a sense of what mixing means *geometrically*.

It will be our objective to step through each interpretation in the next sections.

2.3.1.1 Physical Meaning We combine the contents of beakers 1 and 2 into a larger vessel and then ensure that the resulting solution is well mixed.

It is clear that by bringing the contents of beakers 1 and 2 together, the concentration of the mixture has changed. Continuous mixing is also possible. Consider a simplified inline mixer obtained by connecting two process streams with a tee piece. One of the lines (line 1) contains a concentration equivalent to that found in beaker 1, whereas the other line is filled with a concentration equal to the concentration in beaker 2. On the other end of the tee, the combined contents from streams 1 and 2 join as stream 3. This arrangement is shown in Figure 2.10.

Although the relative amounts of beakers 1 and 2 will influence the overall mixture concentration, intuition tells us that the mixture must lie somewhere "in between" the concentrations of beakers 1 and 2. To determine the exact concentration of NaCl and KOH in the mixture, a simple mass balance may be carried out.

2.3.1.2 Mathematical Meaning Conservation of mass ensures that the total mass of the mixture is equal to the sum of the individual masses contained in beakers 1 and 2. Since constant density is assumed, the total volume of the mixture is also given as the sum of the individual volumes

TABLE 2.5 Mixture Compositions for Beakers 1 and 2

Beaker No.	NaCl Mass (g)	KOH Mass (g)	Total Mass (g)	Volume (L)	Concentration (mol/L)
1	5.0	10.0	15.0	1.0	$[0.085, 0.178]^T$
2	4.0	50.0	54.0	1.5	$[0.046, 0.594]^T$

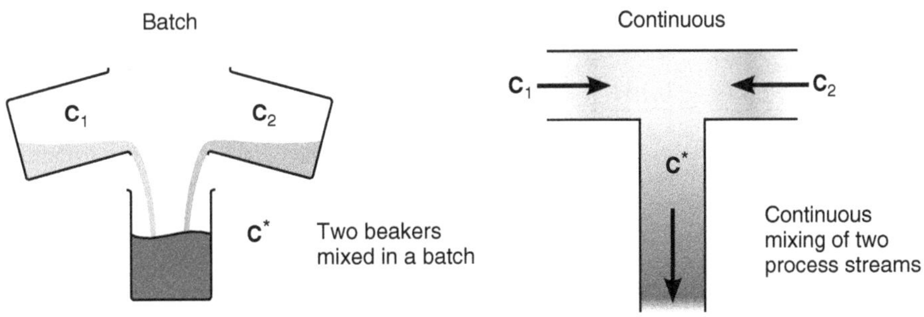

Figure 2.10 Physical interpretation of mixing for batch and continuous processes. (*See color plate section for the color representation of this figure.*)

from beakers 1 and 2. We therefore have

$$m_{tot} = m_1 + m_2$$
$$V_{tot} = V_1 + V_2$$

The subscript "tot" here designates the properties associated with the overall solution. The masses of NaCl and KOH in each beaker are known. Hence, the total moles, n_{tot}, in the mixed solution may be determined using the molecular weights provided earlier:

$$n_{tot,NaCl} = n_{1,NaCl} + n_{2,NaCl} = \frac{5+4}{58.5} = 0.154\,\text{mol}$$

$$n_{tot,KOH} = n_{1,KOH} + n_{2,KOH} = \frac{10+50}{56.1} = 1.07\,\text{mol}$$

The concentrations of the two components are now easily calculated using the total volume of solution, which is 2.5 L in this example.

$$c^*_{NaCl} = \frac{0.154}{2.5} = 0.062\,\text{mol/L}$$
$$c^*_{KOH} = \frac{1.07}{2.5} = 0.428\,\text{mol/L}$$

The asterisk is used to denote the concentrations of the species in the mixed solution. The concentration vector for the mixture, $\mathbf{C}^*$, is then easily determined and given as follows:

$$\mathbf{C}^* = \begin{bmatrix} c^*_{NaCl} \\ c^*_{KOH} \end{bmatrix} = \begin{bmatrix} 0.062 \\ 0.428 \end{bmatrix} \text{mol/L}$$

Alternatively, one could have also determined the concentrations of NaCl and KOH using the previously calculated individual concentrations and volumes from beakers 1 and 2. The expressions for the two are given as follows:

$$c^*_{NaCl} = \frac{c_{1,NaCl}V_1 + c_{2,NaCl}V_2}{V_1 + V_2} = \frac{0.154}{2.5}$$

$$c^*_{KOH} = \frac{c_{1,KOH}V_1 + c_{2,KOH}V_2}{V_1 + V_2} = \frac{1.07}{2.5}$$

Since the expressions are simply two algebraic equations, the system is easily rewritten using vector notation. Concentration pairs may be grouped corresponding to beakers 1 and 2, and expressed as a single vector sum as concentration vectors instead:

$$\begin{bmatrix} c^*_{NaCl} \\ c^*_{KOH} \end{bmatrix} = \frac{V_1}{V_{tot}} \begin{bmatrix} c_{1,NaCl} \\ c_{1,KOH} \end{bmatrix} + \frac{V_2}{V_{tot}} \begin{bmatrix} c_{2,NaCl} \\ c_{2,KOH} \end{bmatrix}$$

$$\mathbf{C}^* = \frac{V_1\mathbf{C}_1 + V_2\mathbf{C}_2}{V_{tot}}$$

Note that V_1 and V_2 are scalar quantities, whereas $\mathbf{C}_1$, $\mathbf{C}_2$, and $\mathbf{C}^*$ are vector quantities, where

$$\mathbf{C}_1 = \begin{bmatrix} 0.085 \\ 0.178 \end{bmatrix} \text{mol/L}$$

$$\mathbf{C}_2 = \begin{bmatrix} 0.046 \\ 0.594 \end{bmatrix} \text{mol/L}$$

2.3.1.3 Graphical Meaning Concentration vectors corresponding to beakers 1 and 2 and the final mixture are plotted in Figure 2.11. The mixture concentration, $\mathbf{C}^*$, lies on the straight line joining $\mathbf{C}_1$ and $\mathbf{C}_2$ (which may not necessarily lie at the midpoint of the line). This result can be confirmed if a straight line is drawn between $\mathbf{C}_1$ and $\mathbf{C}_2$, as displayed in Figure 2.11. It is possible to show this result mathematically by using Equation 2.1 obtained earlier. Recall that

$$\mathbf{C}^* = \frac{V_1\mathbf{C}_1 + V_2\mathbf{C}_2}{V_{tot}}$$

By making the substitution $\lambda = V_2/V_{tot}$, and rearranging the expression, the following result is obtained:

$$\mathbf{C}^* = (1-\lambda)\mathbf{C}_1 + \lambda\mathbf{C}_2$$

This expression can be rearranged to give

$$\mathbf{C}^* = \mathbf{C}_1 + \lambda(\mathbf{C}_2 - \mathbf{C}_1) \tag{2.1}$$

$\mathbf{C}^*$ is therefore a *linear combination* of vectors $\mathbf{C}_1$ and $\mathbf{C}_2$, which are the concentrations of beakers 1 and 2, respectively.

> **(!) IMPORTANT: Mixing on a Straight Line**
>
> When density is assumed constant, then mixing has a special geometric property. Mixtures lie on a straight line joining the two concentrations being mixed in concentration space. Mixing is therefore a *linear* process. This has important consequences in AR theory, as will be seen later.

$\mathbf{C}^*$ is offset in space by the vector $\mathbf{C}_1$, and vector $\mathbf{C}_2 - \mathbf{C}_1$ provides the direction that new mixtures must lie, as shown in Figure 2.12. The magnitude of the difference is scaled

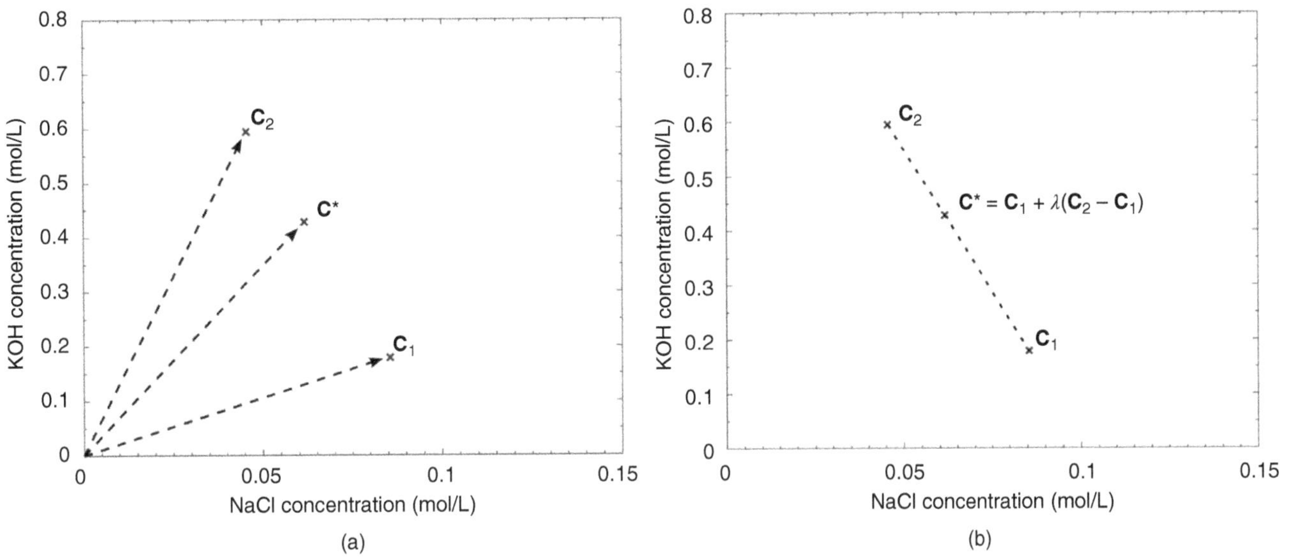

Figure 2.11 (a) Concentrations present in the NaCl–KOH system plotted as concentration vectors from the origin. (b) The graphical meaning of mixing. Mixtures lie on a straight line joining the two concentrations.

using the λ parameter. A linear combination of these two vectors produces the desired mixture concentration. If only the contents of beaker 1 is available, then there is no contribution from beaker 2. Therefore, $V_2 = 0$ and $\lambda = 0$. Conversely, if only beaker 2 is available, then $\lambda = 1$ and $\mathbf{C}^* = \mathbf{C}_2$. For any positive combination of V_1 and V_2, the resulting mixture must generate a concentration coordinate lying on the straight line given by Equation 2.1. It follows that $0 \leq \lambda \leq 1$. This result is also commonly known as the *lever arm rule* (Geankoplis, 1993).

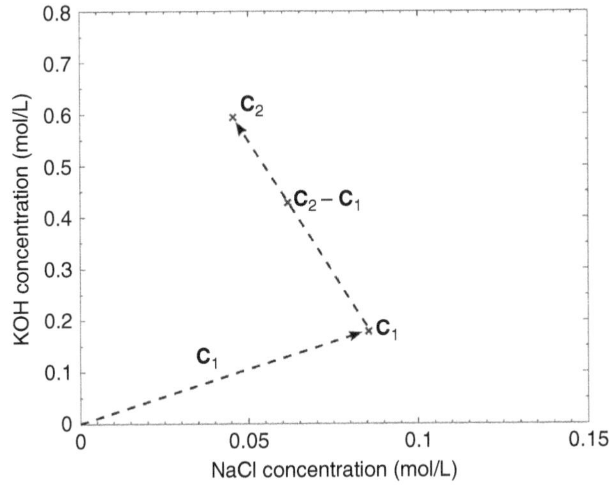

Figure 2.12 Mixture concentrations are formed by a linear combination of concentration vectors.

SIDE NOTE: The Mixing Vector

Mixture concentrations are commonly viewed as a linear combination of vectors. In AR theory, the vector difference $\mathbf{C}_2 - \mathbf{C}_1$ is frequently referred to as the *mixing vector*

$$\mathbf{v} = \mathbf{C}_2 - \mathbf{C}_1$$

Concentrations obtained by combinations of $\mathbf{C}_1$ and $\mathbf{C}_2$ must lie in the direction of $\mathbf{v}$.

EXAMPLE 7: Calculating λ

Suppose that a 1.0 L/s stream (Q_1) containing an equimolar solution of NaCl and KOH at 0.5 mol/L is available. A second stream with a volumetric flow rate of $Q_2 = 300$ mL/s containing a solution of 0.2 mol/L NaCl is mixed with stream 1.

(a) Calculate the value of λ for the mixture. What is the NaCl concentration of the mixture stream?

(b) If it is found that 5.6 mol/L of KOH is present in Q_2, then calculate the concentration vector $\mathbf{C}^*$ of the mixture. Plot the three concentrations in the phase plane.

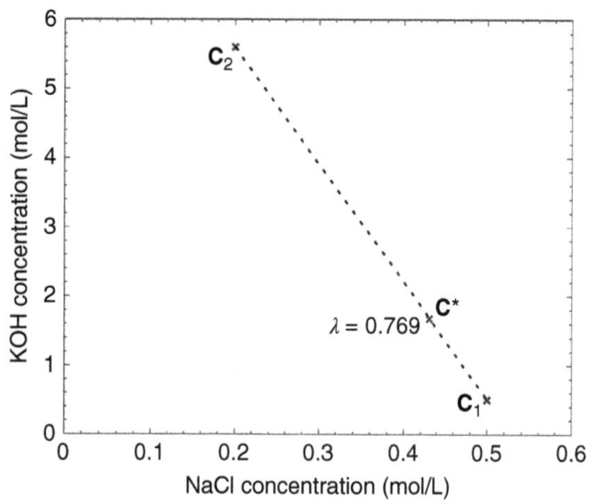

Figure 2.13 Position of the mixture concentration $\mathbf{C}^*$ in relation to $\mathbf{C}_1$ and $\mathbf{C}_2$.

Answer a:

The value of λ is found by dividing Q_1 by the total volumetric flow rate of the mixed stream. If constant density is assumed, $Q_{tot} = Q_1 + Q_2$ giving

$$\lambda = \frac{Q_1}{Q_1 + Q_2} = \frac{1.0}{1.0 + 0.3} = 0.769$$

The NaCl concentration of the mixture is then found either by mass balance or using λ, giving

$$\begin{aligned} c^*_{NaCl} &= \lambda c_{1,NaCl} + (1 - \lambda)c_{2,NaCl} \\ &= (0.769)(0.5) + (1 - 0.769)(0.2) \\ &= 0.431\,\text{mol/L} \end{aligned}$$

Answer b:

To calculate $\mathbf{C}^*$, the mixture of KOH must first be found. This is done in an identical fashion to that given before. Hence,

$$\begin{aligned} c^*_{KOH} &= \lambda c_{1,KOH} + (1 - \lambda)c_{2,KOH} \\ &= (0.769)(0.5) + (1 - 0.769)(5.6) \\ &= 1.68\,\text{mol/L} \end{aligned}$$

Vector $\mathbf{C}^*$ is therefore the mixture concentration of NaCl and KOH:

$$\mathbf{C}^* = [0.431\ \ 1.68]^T\,\text{mol/L}$$

Similarly, concentration vectors corresponding to streams 1 and 2 are given by

$$\mathbf{C}_1 = [0.5\ \ 0.5]^T\,\text{mol/L}$$
$$\mathbf{C}_2 = [0.2\ \ 5.6]^T\,\text{mol/L}$$

A plot of all three concentrations is given in Figure 2.13. Observe how the mixture concentration $\mathbf{C}^*$ lies on the straight line joining $\mathbf{C}_1$ and $\mathbf{C}_2$, as predicted by Equation 2.1.

EXAMPLE 8: Calculating Mixing Fractions

Two process streams are to be mixed continuously. Each stream is an aqueous solution containing three components: A, B, and C. The concentration vectors for streams 1 and 2 are given by

$$\mathbf{C}_1 = [0.1,\ \ 0.5,\ \ 0.7]^T\,\text{mol/L}$$
$$\mathbf{C}_2 = [0.3,\ \ 0.2,\ \ 0.4]^T\,\text{mol/L}$$

(a) If the volumetric flow rates of both streams are equal, and constant density is assumed, what is the value of the mixing fraction λ?

(b) Hence, what is the concentration of the mixture? Express the answer as a concentration vector $\mathbf{C}^*$.

(c) If the volumetric flow rates for streams 1 and 2 were $Q_1 = 1.25$ L/s and $Q_2 = 7.75$ L/s instead, what would be the value of λ and $\mathbf{C}^*$?

Answer a:

Suppose that the volumetric flow rates for stream 1 and 2 are given by Q_1 and Q_2, respectively. If $Q_1 = Q_2$, then the total volumetric flow rate is double that of either stream 1 or 2. If the mixing fraction is given by

$$\lambda = \frac{Q_2}{Q_1 + Q_2}$$

It follows that

$$\lambda = \frac{Q_2}{2Q_2} = 0.5$$

Answer b:

The mixture concentration is calculated using Equation 2.1 as follows:

$$\mathbf{C}^* = \mathbf{C}_1 + \lambda(\mathbf{C}_2 - \mathbf{C}_1)$$

Since, values for $\mathbf{C}_1$, $\mathbf{C}_2$, and λ are known, substitution into Equation 2.1 gives

$$\mathbf{C}^* = \begin{bmatrix}0.1\\0.5\\0.7\end{bmatrix} + 0.5\left(\begin{bmatrix}0.3\\0.2\\0.4\end{bmatrix} - \begin{bmatrix}0.1\\0.5\\0.7\end{bmatrix}\right)\text{mol/L} = \begin{bmatrix}0.2\\0.35\\0.55\end{bmatrix}\text{mol/L}$$

Answer c:

Q_1 and Q_2 no longer have the same value, and thus the mixing fraction λ is not equal to 0.5. Assuming that density is constant, the total volumetric flow of the mixed stream is

$$Q_{tot} = Q_1 + Q_2 = 1.25 + 7.75 = 9.00 \text{ L/s}.$$

From this, λ is calculated as follows:

$$\lambda = \frac{Q_2}{Q_{tot}} = \frac{7.75}{9.00} = 0.861$$

To calculate the following new mixture concentration vector, Equation 2.1 is again used:

$$\mathbf{C}^* = \begin{bmatrix}0.1\\0.5\\0.7\end{bmatrix} + 0.861\left(\begin{bmatrix}0.3\\0.2\\0.4\end{bmatrix} - \begin{bmatrix}0.1\\0.5\\0.7\end{bmatrix}\right)\text{mol/L}$$
$$= \begin{bmatrix}0.272\\0.242\\0.442\end{bmatrix}\text{mol/L}$$

2.3.2 Additional Insights

In the previous sections, we analyzed a simple mixing example and showed how the process of mixing could be viewed from a graphical perspective. Before returning to the BTX problem, it is sensible to discuss why data plotted in concentration space is useful. Consider again the NaCl and KOH beaker experiments from Section 2.3.1. Suppose now a colleague provides you with a third beaker for mixing. What new concentrations might be obtained by the introduction of the third beaker? It is assumed that the beaker contains only NaCl and KOH in water, given by the following concentration vector:

$$\mathbf{C}_3 = \begin{bmatrix}0.050\\0.250\end{bmatrix}\text{mol/L}$$

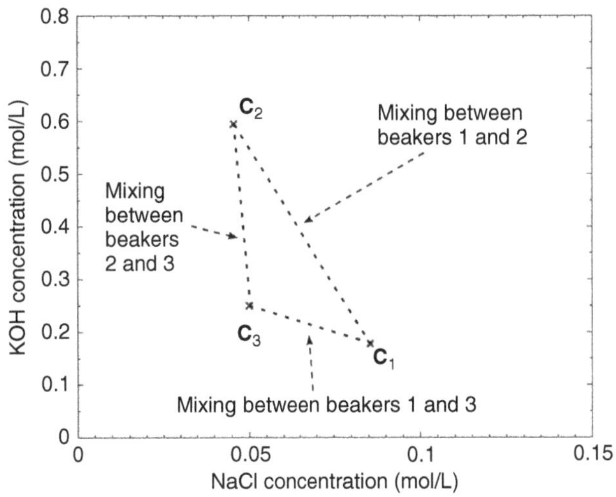

Figure 2.14 The possible set of mixtures achieved when three beaker concentrations are used.

Concentration vectors $\mathbf{C}_1$, $\mathbf{C}_2$, and $\mathbf{C}_3$ may be plotted in space; this is displayed in Figure 2.14.

We are now in a position to mix different amounts from each beaker and interpret the results graphically. Let us conduct a number of simple thought experiments using different combinations of each beaker. To assist in the description of different mixtures arising via combinations of different beaker concentrations, the following notation is adopted: $\mathbf{C}^*_{i,j}$ represents the mixture concentration obtained from combining concentrations *i* and *j*, respectively.

2.3.2.1 *Experiment 1: Mixing Amounts from Beakers 2 and 3*

From Section 2.3.1.3, mixtures obtained from beakers 2 and 3 will result in a mixing vector lying in a direction given by the vector $(\mathbf{C}_2 - \mathbf{C}_3)$. The corresponding vector expression for this mixture is then

$$\mathbf{C}_A = \mathbf{C}^*_{2,3} = \mathbf{C}_3 + \lambda(\mathbf{C}_2 - \mathbf{C}_3)$$

The concentration vector $\mathbf{C}^*_{2,3}$ hence indicates that a mixture concentration has been obtained using a combination of $\mathbf{C}_2$ and $\mathbf{C}_3$, respectively. $\mathbf{C}_A$ is used as an abbreviated notation for $\mathbf{C}^*_{2,3}$; this will become clearer shortly. In Figure 2.14, the possible coordinates of $\mathbf{C}^*_{2,3}$ are plotted by the dashed line joining $\mathbf{C}_2$ and $\mathbf{C}_3$. Mixture concentrations must lie on this line if they formed from beakers 2 and 3.

2.3.2.2 *Experiment 2: Mixing Amounts from Beakers 1 and 2*

This experiment is identical to experiment 1; however, we replace beaker 3 with beaker 1 and repeat the mixing. What concentrations are achievable with this set of beakers? If beakers 1 and 2 are mixed, the resulting concentrations are a combination of $\mathbf{C}_1$ and $\mathbf{C}_2$. The corresponding

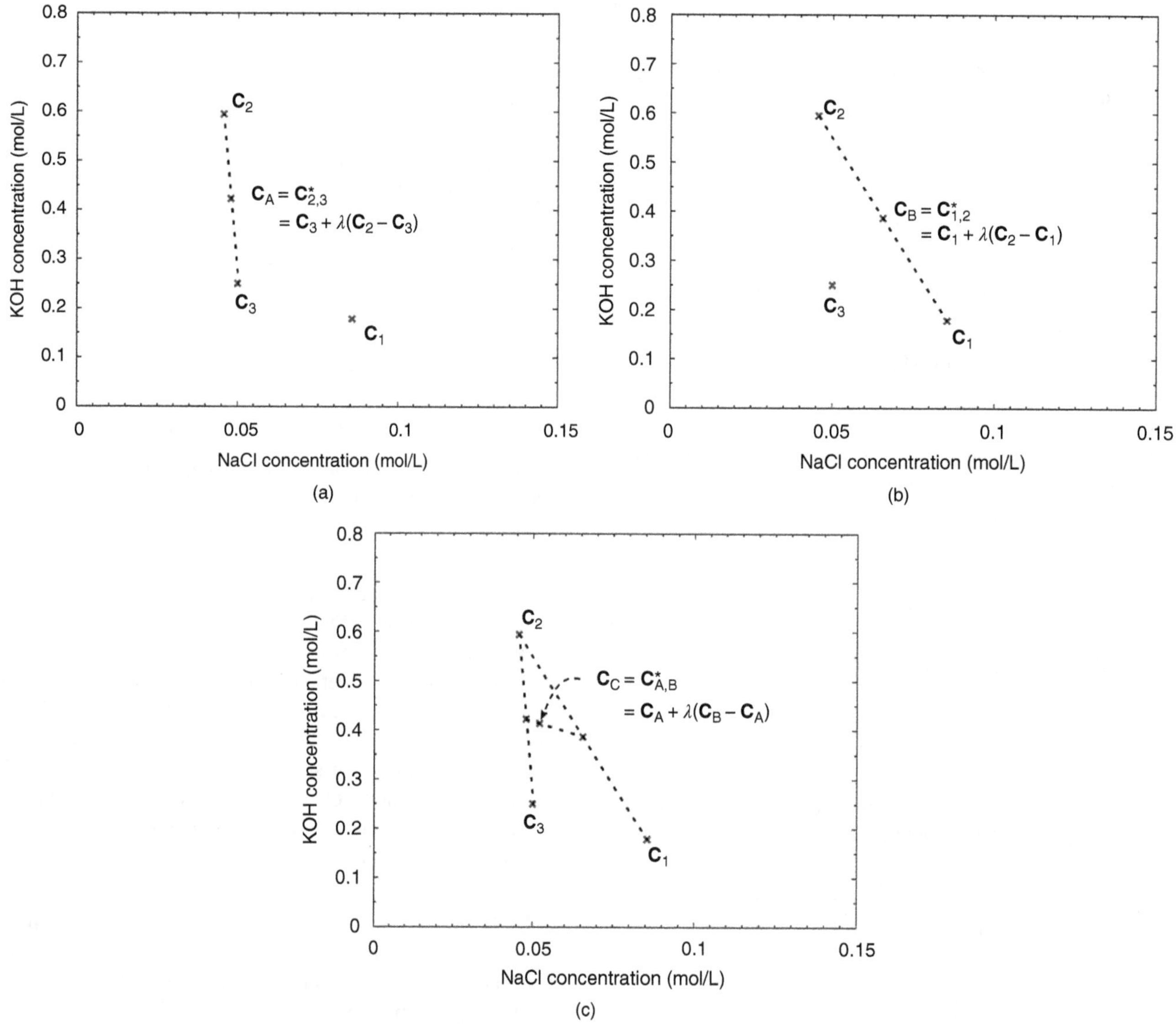

Figure 2.15 Mixtures for experiment 1 (beakers 1 and 3) and experiment 2 (beakers 2 and 3) when equal portions are used. (a) Achieving $\mathbf{C}_A$, (b) achieving $\mathbf{C}_B$, and (c) achieving $\mathbf{C}_C$.

vector expression for the mixture is then

$$\mathbf{C}_B = \mathbf{C}^*_{1,2} = \mathbf{C}_1 + \lambda(\mathbf{C}_2 - \mathbf{C}_1)$$

Clearly, mixtures obtained from this experiment fall on the straight line joining $\mathbf{C}_2$ and $\mathbf{C}_1$.

2.3.2.3 Experiment 3: A Mixture of Mixtures Suppose now that a mixture from experiments 1 and 2 is formed. For illustration, let us mix equal portions of beakers 2 and 3 (from experiment 1). Using the previous expression, the following mixture concentration is obtained:

$$\mathbf{C}_A = 0.5(\mathbf{C}_2 + \mathbf{C}_3)$$

$$= 0.5\begin{bmatrix}0.046\\0.594\end{bmatrix} + 0.5\begin{bmatrix}0.050\\0.250\end{bmatrix}$$

$$= \begin{bmatrix}0.048\\0.422\end{bmatrix}\ \text{mol/L}$$

A plot of $\mathbf{C}_A$ is given in the Figure 2.15(a). It is clear from the plot that the point lies at an equal distance between $\mathbf{C}_2$ and $\mathbf{C}_3$. Next, let us perform the same task for combinations of $\mathbf{C}_1$ and $\mathbf{C}_2$. In the same manner as before, equal parts of $\mathbf{C}_1$ and $\mathbf{C}_2$ are used (50% $\mathbf{C}_1$ and 50% $\mathbf{C}_2$ from Section 2.3.1.2). The resulting mixture is then as follows:

$$\mathbf{C}_B = 0.5(\mathbf{C}_1 + \mathbf{C}_2)$$

$$= 0.5 \begin{bmatrix} 0.085 \\ 0.178 \end{bmatrix} + 0.5 \begin{bmatrix} 0.046 \\ 0.594 \end{bmatrix}$$

$$= \begin{bmatrix} 0.066 \\ 0.386 \end{bmatrix} \text{ mol/L}$$

$\mathbf{C}_B$ is also plotted in Figure 2.15(b). Similar to $\mathbf{C}_A$, the point $\mathbf{C}_B$ lies an equal distance on the straight line joining $\mathbf{C}_1$ and $\mathbf{C}_2$ in NaCl–KOH concentration space.

There are now two beakers, each containing two new concentrations, derived from the original beaker experiments. But $\mathbf{C}_A$ and $\mathbf{C}_B$ may also be mixed together. Since we have now developed a graphical interpretation of this process, we are in a better position to predict the outcome easily via graphical inspection of the plot. In fact, one can plot the mixture concentrations without computing the answer analytically. Since this new mixture must be formed from a combination of $\mathbf{C}_A$ and $\mathbf{C}_B$, the resulting mixture will lie on a straight line joining $\mathbf{C}_A$ and $\mathbf{C}_B$. Algebraically, for the new mixture, we have

$$\mathbf{C}_C = \mathbf{C}^*_{A,B} = \mathbf{C}_A + \lambda(\mathbf{C}_B - \mathbf{C}_A)$$

As a concrete example, assume that we wish to mix 75% of mixture 1 and 25% of mixture 2 (by volume). The concentration vector is given by

$$\mathbf{C}_C = \mathbf{C}^*_A + 0.25(\mathbf{C}^*_B\text{-}\mathbf{C}^*_A) = \begin{bmatrix} 0.0522 \\ 0.4131 \end{bmatrix} \text{ mol/L}$$

To physically synthesize $\mathbf{C}_C$, one might follow the following procedures:

1. Form concentration $\mathbf{C}_A$: Combine beakers 2 and 3 in equal proportions.
2. Form concentration $\mathbf{C}_B$: In a separate container, combine beakers 1 and 2, also in equal proportions.
3. Form concentration $\mathbf{C}_C$: With $\mathbf{C}_A$ and $\mathbf{C}_B$ now obtained, combine the two mixtures in a separate beaker so that the final mixture contains 75% of $\mathbf{C}_A$ and 25% of $\mathbf{C}_B$.

Figure 2.16 shows a schematic of the order of events that accompany these procedures. Concentration, $\mathbf{C}_C$, is thus produced that is different from any other concentration produced before.

2.3.3 Different Ways of Synthesizing $\mathbf{C}_C$

It is possible to obtain $\mathbf{C}_C$ in one less step as that detailed in Section 2.3.2.3. Furthermore, there are actually *infinitely* many combinations between beakers 1, 2, and 3 that might produce a concentration equal to $\mathbf{C}_C$. Although $\mathbf{C}_C$ is unique in terms of the specific concentrations of its constituent species, it is not unique in the manner in which it is synthesized. Let us look at each statement and justify them.

Claim 1: Concentration C_C (the concentration resulting from a mixture of C_A and C_B) may be achieved using only two mixing steps.

There is no formal "proof" as such, and we prefer to simply validate this statement by showing an alternative mixing scenario:

1. First, form a concentration from beakers 2 and 3. The concentrations resulting from this mixture will lie on the line connecting $\mathbf{C}_2$ and $\mathbf{C}_3$. Let us call this concentration $\mathbf{C}^\#_A$, to differentiate it from $\mathbf{C}_A$ in Section 2.3.2.3, (but to still indicate that it is a mixture formed from $\mathbf{C}_2$ and $\mathbf{C}_3$).
2. Using the mixture contents associated with $\mathbf{C}^\#_A$, mix with the contents of beaker 1 (with concentration given by $\mathbf{C}_1$) in the correct proportions to form $\mathbf{C}_C$.

How much of beaker 1, beaker 2, and beaker 3 are required to produce the desired result? We find that in order to obtain the correct value for $\mathbf{C}^\#_A$, the mixture must be composed of approximately 57.18% of $\mathbf{C}_2$ and 42.82% of $\mathbf{C}_3$. We then have

$$\mathbf{C}_1 = \begin{bmatrix} 0.085 \\ 0.178 \end{bmatrix} \text{ mol/L}$$

$$\mathbf{C}_2 = \begin{bmatrix} 0.046 \\ 0.594 \end{bmatrix} \text{ mol/L}$$

$$\mathbf{C}_3 = \begin{bmatrix} 0.050 \\ 0.250 \end{bmatrix} \text{ mol/L}$$

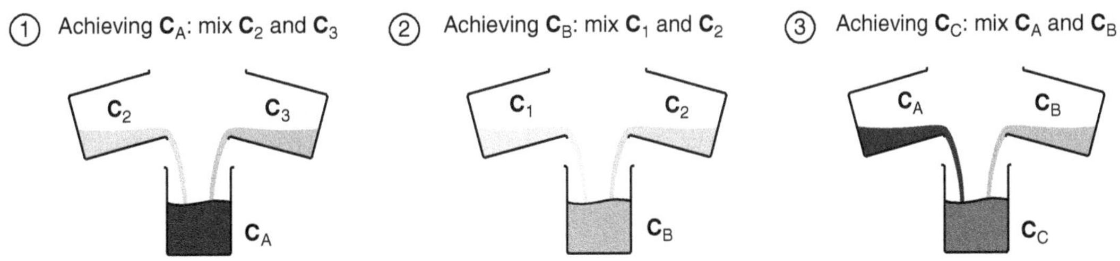

Figure 2.16 Forming a mixture concentration $\mathbf{C}_C$ as a series of three mixing steps. (*See color plate section for the color representation of this figure.*)

Thus,

$$\mathbf{C}_A^{\#} = (0.5718)\mathbf{C}_2 + (0.4282)\mathbf{C}_3$$

$$= (0.5718)\begin{bmatrix}0.046\\0.594\end{bmatrix} + (0.4282)\begin{bmatrix}0.050\\0.250\end{bmatrix}$$

$$= \begin{bmatrix}0.0477\\0.4467\end{bmatrix} \text{ mol/L}$$

Then, in order to produce $\mathbf{C}_C$, 87.5% of the solution should be composed of $\mathbf{C}_A^{\#}$ and 12.5% of $\mathbf{C}_1$. This results in a final mixture concentration of $\mathbf{C}_C$.

$$\mathbf{C}_C = (0.875)\mathbf{C}_A^{\#} + (0.125)\mathbf{C}_1$$

$$= (0.875)\begin{bmatrix}0.0477\\0.4467\end{bmatrix} + (0.125)\begin{bmatrix}0.085\\0.178\end{bmatrix}$$

$$= \begin{bmatrix}0.0524\\0.4131\end{bmatrix} \text{ mol/L}$$

which, after roundoff error, is identical in concentration to that if $\mathbf{C}_C$ were formed in Section 2.3.2.3, by a 75% $\mathbf{C}_A$ and 25% $\mathbf{C}_B$ mixture. Graphically, this procedure is equivalent to constructing a straight line passing through $\mathbf{C}_C$ (which is the desired final concentration) and $\mathbf{C}_1$. The intersection point with the other mixing line, which itself is produced via $\mathbf{C}_2$ and $\mathbf{C}_3$, thus gives the mixture concentration needed in the first beaker mixture experiment, using beakers 2 and 3.

Figure 2.17 demonstrates the mixing process geometrically. The plot has been magnified slightly to clearly show the different mixing lines used to achieve $\mathbf{C}_C$ in two mixing steps.

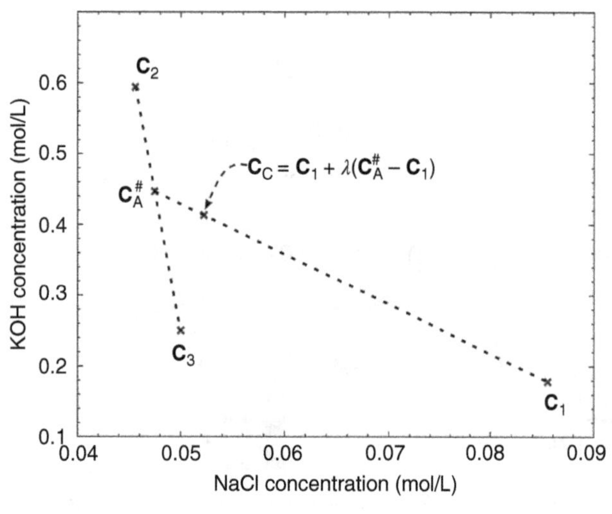

Figure 2.17 Achieving $\mathbf{C}_C$ in only two mixing steps.

ILLUSTRATION: Computing Mixing Fractions for $\mathbf{C}_C$ in Two Mixing Steps

We wish to show how one might go about solving for the mixing fractions (the λ's) used to generate the mixture concentration $\mathbf{C}_C$ given before, in only two mixing steps. In order to terminate at $\mathbf{C}_C$, a mixture of $\mathbf{C}_1$ and $\mathbf{C}_A^{\#}$ must first be formed. The proportion of $\mathbf{C}_A^{\#}$ and $\mathbf{C}_1$ is unknown unfortunately. Moreover, the concentration $\mathbf{C}_A^{\#}$ is also unknown, although to form $\mathbf{C}_A^{\#}$, a mixture of $\mathbf{C}_2$ and $\mathbf{C}_3$ must be used. The proportions of $\mathbf{C}_2$ and $\mathbf{C}_3$ are also unknown.

We start by expressing $\mathbf{C}_A$ as a linear combination of $\mathbf{C}_2$ and $\mathbf{C}_3$ in the usual manner:

$$\mathbf{C}_A^{\#} = \mathbf{C}_3 + \lambda_1(\mathbf{C}_2 - \mathbf{C}_3)$$

λ_1 denotes the mixing fraction between $\mathbf{C}_2$ and $\mathbf{C}_3$ used to form $\mathbf{C}_A^{\#}$. The final mixture concentration $\mathbf{C}_C$ may be expressed in an identical fashion using linear combinations of $\mathbf{C}_1$ and $\mathbf{C}_A^{\#}$ as follows:

$$\mathbf{C}_C = \mathbf{C}_1 + \lambda_2(\mathbf{C}_A^{\#} - \mathbf{C}_1)$$

Here, λ_2 denotes the mixing fraction between $\mathbf{C}_A^{\#}$ and $\mathbf{C}_1$ used to form the desired concentration $\mathbf{C}_C$. Substitution of the expression for $\mathbf{C}_A^{\#}$ into this expression then gives

$$\mathbf{C}_C = \mathbf{C}_1 + \lambda_2[\mathbf{C}_3 + \lambda_1(\mathbf{C}_2 - \mathbf{C}_3) - \mathbf{C}_1]$$

which is a single vector equation in terms of the unknowns: λ_1 and λ_2. The vector equation may be written out as a system of equations in terms of NaCl and KOH as follows:

$$c_{C,\mathrm{NaCl}} = c_{1,\mathrm{NaCl}} + \lambda_2[c_{3,\mathrm{NaCl}} + \lambda_1(c_{2,\mathrm{NaCl}} - c_{3,\mathrm{NaCl}}) - c_{1,\mathrm{NaCl}}]$$

$$c_{C,\mathrm{KOH}} = c_{1,\mathrm{KOH}} + \lambda_2[c_{3,\mathrm{KOH}} + \lambda_1(c_{2,\mathrm{KOH}} - c_{3,\mathrm{KOH}}) - c_{1,\mathrm{KOH}}]$$

This system of equations expresses the mixture concentrations in terms of the individual components participating in the mixture. The values of the concentration variables in this system of equations are known however, and so there are two equations with two unknowns: λ_1 and λ_2.

Values for $\mathbf{C}_C$, $\mathbf{C}_1$, $\mathbf{C}_2$, and $\mathbf{C}_3$ may be substituted into the expression and then λ_1 and λ_2 may be solved simultaneously. For the values given in Section 2.3.2.3, solution of the equations gives $\lambda_1 = 0.5718$ and $\lambda_2 = 0.875$, which are the desired mixing fractions in the two mixtures.

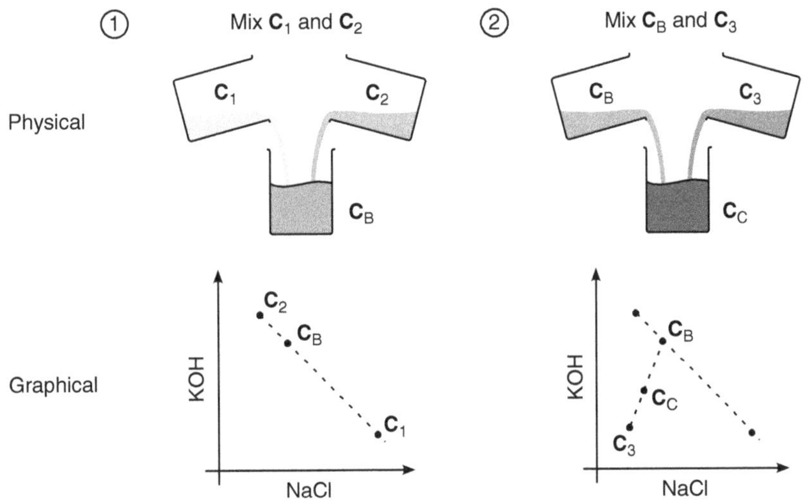

Figure 2.18 Physical steps needed to achieve $\mathbf{C}_C$ as well as the corresponding geometric interpretation in NaCl–KOH space. (*See color plate section for the color representation of this figure.*)

Claim 2: $\mathbf{C}_C$ may be synthesized in arbitrary combinations

Recall that it is possible to generate a line of potential mixture concentrations between two concentration vectors. For the first step, this is achieved with the line connecting $\mathbf{C}_1$ and $\mathbf{C}_2$. For the second step, the goal is to terminate with a final mixture concentration of $\mathbf{C}_C$. This is accomplished by mixing $\mathbf{C}_A$ and $\mathbf{C}_1$ in appropriate amounts. The line generated from the mixture must be linear, and it must therefore pass through concentration $\mathbf{C}_C$ and $\mathbf{C}_1$. The unique straight line passing through $\mathbf{C}_C$ and $\mathbf{C}_1$ may be constructed and used to locate the intersection point with the mixing line obtained from $\mathbf{C}_2$ and $\mathbf{C}_3$ (beaker 2 and beaker 3). The intersection point of the two lines is then the required intermediate concentration $\mathbf{C}_A$ needed to mix with beaker 1. This must be true as $\mathbf{C}_A$ satisfies both mixing relations.

It is not difficult to see that the point $\mathbf{C}_C$ is not unique. For instance, it is also possible to synthesize $\mathbf{C}_C$ by starting with $\mathbf{C}_B$ instead (a mixture of $\mathbf{C}_2$ and $\mathbf{C}_1$). $\mathbf{C}_3$ could then be used to provide the appropriate mixing line to achieve $\mathbf{C}_C$. Figure 2.18 shows this process using both the graphical interpretation and the physical processes required. It is also equally feasible to employ mixtures of mixtures to form $\mathbf{C}_C$. The only requirement being that any two straight-line segments can be constructed that intersect at the coordinate given by $\mathbf{C}_C$. All points lying in the interior of the triangle bounded by points $\mathbf{C}_1$, $\mathbf{C}_2$, and $\mathbf{C}_3$ are therefore achievable via mixing, as shown in Figure 2.19.

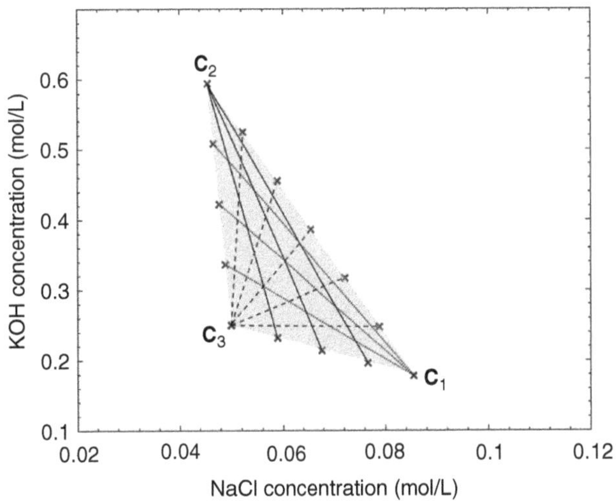

Figure 2.19 There are infinitely many combinations that might be employed to achieve concentrations in the interior of the three mixing lines defined by $\mathbf{C}_1$, $\mathbf{C}_2$, and $\mathbf{C}_3$.

2.3.4 Mixing and Attainability

Suppose that a number of different compositions are given. This might be from, say, a set of beakers available in a laboratory, or a number of process streams available on-site. Suppose further that you are given a list of desired compositions we wish to achieve from the available compositions.

Given what is known about mixing, how do we go about determining the *entire* set of concentrations achievable from the list, with the mixtures available? In other words, what makes one point attainable compared to one that is not, and what are the criteria for determining this? How can we predict which points could be attainable?

EXAMPLE 9: Secret Drinks Formulations

Imagine that you work for a beverage company that produces a number of popular fizzy drinks. The secret to your drinks lie in two proprietary chemicals, which collectively give the drinks their unique characteristics:

1. Chemical X, which gives the drinks their "bang."
2. Chemical Y, which gives the drinks their "zing."

X and Y are both highly secret and expensive to synthesize, but once they have been formed they can be mixed, inexpensively, to produce a number of popular drinks. Your company keeps a list of seven drinks recipes (formulations) that each contain different proportions of X and Y, which are summarized in Table 2.6

TABLE 2.6 List of Possible Concentrations Given by the R&D Department

Drink Recipe No.	Component X "Bang" (mol/L)	Component Y "Zing" (mol/L)
1	3.0	5.0
2	4.0	2.0
3	1.0	6.0
4	9.0	9.0
5	5.0	5.0
6	2.0	7.7
7	2.2	8.0

One day, your boss asks you to try and find ways to create these seven drinks using combinations of only three X - Y formulations. The three formulations are the easiest to produce on a large scale, and your boss is seeking ways to produce the largest range of drinks recipes given a limited set of starting formulations—doing so would be beneficial, as then some products would not need to be synthesized (which is expensive), but rather they could be formed via mixing (which is cheap). The corresponding concentration vectors for the three X - Y formulations are given as follows:

$$\mathbf{b}_1 = [9.0,\ \ 1.0]^{\mathrm{T}}\,\mathrm{mol/L}$$
$$\mathbf{b}_2 = [1.0,\ \ 9.0]^{\mathrm{T}}\,\mathrm{mol/L}$$
$$\mathbf{b}_3 = [2.0,\ \ 3.0]^{\mathrm{T}}\,\mathrm{mol/L}$$

Your aim is to generate the largest range of possible concentrations only using combinations of $\mathbf{b}_1$, $\mathbf{b}_2$, and $\mathbf{b}_3$. You also wish to determine the exact set of beaker experiments that will allow the company to formulate these recipes (if at all). It is your task to carry out the experiments and report back to your boss. How should we go about determining which formulations can be formed?

Answer:

When Table 2.6 is viewed as a list of possible concentrations, it is not immediately obvious which concentrations might be significant. Recipes for drinks 6 and 7, for example, appear to be similar in composition (they have a similar "bang" and "zing" formulation). Drinks 1, 2, and 5 all appear to fall within the same concentration range. Drink 3 appears to fall between $\mathbf{b}_1$ and $\mathbf{b}_3$ in terms of X and Y. Drink 4 appears to have the highest concentration of "bang" compared to all the other potential points, whereas $\mathbf{b}_2$ contains the most "zing." Attempting to physically carry out a large number of experiments may provide better insights into which formulations are achievable; however, we might do better with our knowledge of the geometric properties of mixing and concentration.

Let us therefore plot the concentrations of the available beakers in c_X–c_Y space. This is done in Figure 2.20. Triangle markers indicate the available concentrations formulations provided by your boss, whilst all seven drinks formulations

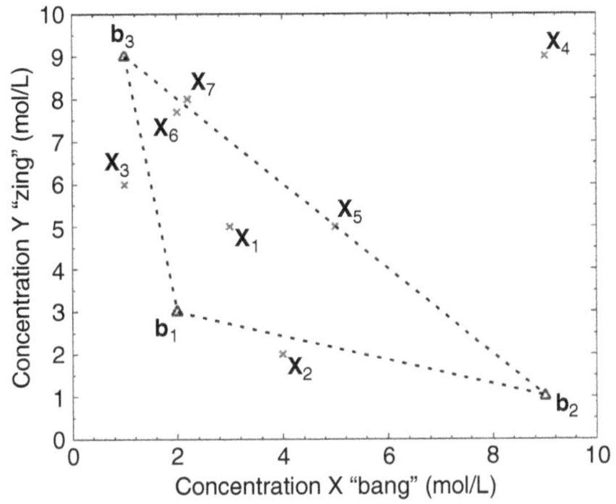

Figure 2.20 Possible drinks recipe formulations (**x**'s) and available operating concentrations (**b**'s) plotted in "bang" and "zing" concentration space.

are displayed as **x**'s in this space. Potential mixture concentrations are also drawn that show the mixture concentrations when $\mathbf{b}_1$ is mixed with $\mathbf{b}_2$ ($\mathbf{b}_{12}$), when $\mathbf{b}_1$ and $\mathbf{b}_3$ are mixed ($\mathbf{b}_{13}$), and when $\mathbf{b}_2$ and $\mathbf{b}_3$ are mixed ($\mathbf{b}_{23}$). These mixtures are given in the figure by dotted lines that trace out the perimeter $\mathbf{b}_1\mathbf{b}_2\mathbf{b}_3$.

Displaying the drinks recipes in this manner makes interpretation of the data significantly easier. It is possible now, simply by graphical inspection, to determine whether any drink of interest is achievable or not. Let us consider each drink in turn and discuss why it may or may not be physically realizable given the mixing constraints:

Recipe 1: Drink 1 appears to lie in the middle of the three available concentrations. It is therefore possible to produce this drink, given at $\mathbf{x}_1$, by only mixing combinations of $\mathbf{b}_1$, $\mathbf{b}_2$, and $\mathbf{b}_3$. As an example, a mixture of $\mathbf{b}_1$ and $\mathbf{b}_2$ could be used to form a mixture lying somewhere on the line $\mathbf{b}_{12}$. It is then possible to mix a portion of $\mathbf{b}_3$ with the newly formed mixture to produce $\mathbf{x}_1$ similar to the procedure given in Section 2.3.3.

Recipe 2: On the surface, it appears as though $\mathbf{x}_2$ may be achievable—both the "bang" and "zing" components of $\mathbf{x}_2$ lie within the X and Y concentration bounds given in $\mathbf{b}_1$, $\mathbf{b}_2$, and $\mathbf{b}_3$. It might appear, then, that given a suitable distribution of proportions between $\mathbf{b}_1$ and $\mathbf{b}_2$, drink 2 might be attainable. In reality, however, this is not true. $\mathbf{x}_2$ does not lie on the mixing line joining $\mathbf{b}_1$ and $\mathbf{b}_2$. Therefore, $\mathbf{x}_2$ cannot be synthesized from a combination of $\mathbf{b}_1$ and $\mathbf{b}_2$. Moreover, unlike $\mathbf{x}_1$, no combinations of $\mathbf{b}_1$, $\mathbf{b}_2$, and $\mathbf{b}_3$ can be formulated to give $\mathbf{x}_2$. Point $\mathbf{x}_2$ simply does not lie on the correct "side" of the line $\mathbf{b}_{12}$ (it contains slightly less "zing" for the same amount of "bang").[4] Drink 2 therefore cannot be formulated from a mixture of the three base formulations.

Recipe 3: It appears as though drink 3 may be achievable. Similar in reasoning to drink 2, the location of $\mathbf{x}_3$ lies on the side of the mixing line between $\mathbf{b}_1$ and $\mathbf{b}_3$ where no other concentrations can be used to form $\mathbf{x}_3$. Drink 3 is therefore unachievable as well (it contains less "bang" for the same amount of "zing").

Recipe 4: Drink 4 contains the highest concentrations in both "bang" and "zing" out of all formulations provided in Table 2.6. The concentration of "bang" in $\mathbf{x}_4$ is equal to that found in recipe 2 whilst the concentration of "zing" is equal to that found in recipe 3. Achieving an equivalent concentration in one component will result in a concentration that is too dilute in the other. Graphically, the three available formulations all lie to the left of $\mathbf{x}_4$ on the graph, and therefore they cannot be used to synthesize a composition in $\mathbf{x}_4$. Drink 4 is therefore unachievable.

Recipe 5: Drink 5 is unique in that it is positioned on the line of achievable mixtures using $\mathbf{b}_2$ and $\mathbf{b}_3$. $\mathbf{x}_5$ can therefore be obtained in a *single* mixing step. This is in contrast to $\mathbf{x}_1$, which requires an additional intermediate mixing step. Thus drink 5 need not be synthesized from scratch, because it can be formed from a mixture of $\mathbf{b}_1$ and $\mathbf{b}_2$. What is more, drink 5 might also be used as an intermediate concentration (a precursor formulation) to make drink 1.

[4]A more precise definition of what is meant by "on the correct side" is discussed at the end of the chapter. For now, it is sufficient to have an idea of this by visually inspecting the plots.

Recipe 6: Drink 6 lies within the mixing lines bounded by the three base formulations. In the same way that drink 1 is achievable, $\mathbf{x}_6$ can be formed by combinations of $\mathbf{b}_1$, $\mathbf{b}_2$, and $\mathbf{b}_3$. For example, one could first mix concentrations $\mathbf{b}_2$ and $\mathbf{b}_3$. The resulting mixture could then be combined with concentration $\mathbf{b}_1$. Drink 6 is therefore achievable from the three base formulations.

Recipe 7: It may be tempting to assume that if drink 6 is feasible, and $\mathbf{x}_7$ is close to $\mathbf{x}_6$ on the graph, then drink 7 may also be feasible. $\mathbf{x}_7$ lies to the right of the line of potential mixtures produced by $\mathbf{b}_2$ and $\mathbf{b}_3$, whereas $\mathbf{x}_6$ lies to the left of the same line—drink 7 contains both slightly more "bang" and "zing" compared to drink 6, but trying to achieve the correct concentration of one of the secret ingredients will result in a smaller concentration of the other. Drink 7 is therefore not achievable.

A summary of our findings is supplied in Table 2.7.

TABLE 2.7 Achievable and Unachievable Points for the R&D Investigation

Attainable	Not Attainable
Drink 1 $\mathbf{x}_1$	Drink 2 $\mathbf{x}_2$
Drink 5 $\mathbf{x}_5$	Drink 3 $\mathbf{x}_3$
Drink 6 $\mathbf{x}_6$	Drink 4 $\mathbf{x}_4$
	Drink 7 $\mathbf{x}_7$

On consideration of the plots, drinks 1, 5, and 6 can be produced via mixing, saving money for the company.

EXAMPLE 10: Slightly More Flexibility

Suppose now that your boss gives you a choice of using an additional concentration from Table 2.6 that could be motivated for consideration as an additional mixing source. He does not know what you have discovered so far, but nevertheless provides three options to you. Only one of the following formulations can be selected:

(a) Concentration 1 ($\mathbf{x}_1$)
(b) Concentration 2 ($\mathbf{x}_2$)
(c) Concentration 4 ($\mathbf{x}_4$)

Based on the findings of the previous example, which concentration would maximize the achievable space of "bang" and "zing"?

Answer a: $\mathbf{x}_1$

$\mathbf{x}_1$ is already known to be achievable through mixing. Hence, there is no need to have an additional (dedicated) source of $\mathbf{x}_1$ as this mixture may be synthesized with combinations of $\mathbf{b}_1$, $\mathbf{b}_2$, and $\mathbf{b}_3$.

Answer b: $\mathbf{x}_2$

In Figure 2.21(a), a shaded region is displayed indicating the extent of new possible concentrations achievable if $\mathbf{x}_2$ is made available for mixing. All of the points within "bang"–"zing" space are achievable by some combination of mixing between $\mathbf{b}_1$, $\mathbf{b}_2$, $\mathbf{b}_3$, and $\mathbf{x}_2$. The set of new possible concentrations, which are made available by the addition of $\mathbf{x}_2$, is given by the triangle $\mathbf{b}_1$–$\mathbf{b}_2$–$\mathbf{x}_2$.

Answer c: $\mathbf{x}_4$

In Figure 2.21(b), the region of potential points achievable using $\mathbf{x}_4$ as the source concentration is displayed. The region filled by the addition of this new point is given by the triangle of points $\mathbf{b}_2$–$\mathbf{b}_3$–$\mathbf{x}_4$. Note that the space is considerably larger than the one generated using $\mathbf{x}_2$ since concentration 4 is a large source of "bang" and "zing." If $\mathbf{x}_4$ is included, then it is feasible to formulate recipe 7 through mixtures as well.

Recommendation

There is no need to include $\mathbf{x}_1$ into the set as this concentration does not expand on the set of available formulations. By comparison, both $\mathbf{x}_2$ and $\mathbf{x}_4$ expand on the set of achievable points. The inclusion of $\mathbf{x}_4$ allows for the synthesis of drink 7. Recall that $\mathbf{x}_7$ is not achievable using only $\mathbf{b}_1$, $\mathbf{b}_2$, and $\mathbf{b}_3$, and thus an additional drink can be gained if $\mathbf{x}_4$ is chosen over $\mathbf{x}_2$. It is therefore best to choose $\mathbf{x}_4$ as this maximizes the set of achievable concentrations.

From the results, we can also conclude that drink 3 will always have to be synthesized for it cannot be made from mixing. If all seven drinks are to be produced, then $\mathbf{b}_1$, $\mathbf{b}_2$, $\mathbf{b}_3$, $\mathbf{x}_2$, $\mathbf{x}_3$, and $\mathbf{x}_4$ must be synthesized. These six formulations all lie on the perimeter traced over the region–that is, these six formulations *form the smallest possible set that encloses the entire set of 10 recipes*.

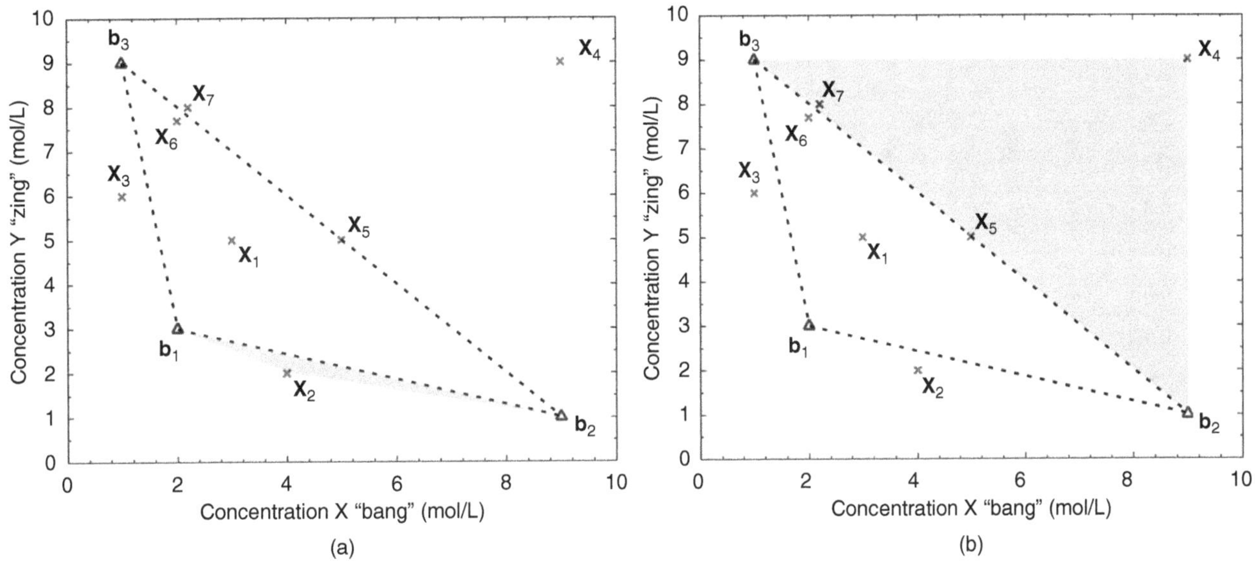

Figure 2.21 (a) The region of new concentrations achieved if $\mathbf{x}_2$ is made achievable and (b) the region of new concentrations achieved if $\mathbf{x}_4$ is made available.

This exercise has highlighted important points about concentration and mixing. Mixing allows us to achieve new points and expand from an initially small fixed set. This idea is central to AR theory—mixing allows one to achieve many solution states from a smaller set of achievable points. Let us extend on the ideas presented so far and formalize them into concrete terminology and definitions.

2.3.5 *n*-Dimensional Mixing and Convex Hulls

2.3.5.1 A Strategy for Many Points Mixing fulfills an important role in filling in regions and distinguishing unique concentrations from redundant ones (concentrations that can be synthesized through mixing). In previous sections, we used a small number of points to simplify discussions, but these concepts can be extended to much larger systems and dimensions.

EXAMPLE 11: Identifying Unique Concentrations

Consider now the set of 50 points provided in Figure 2.22(a). The plot is similar to that provided in the previous sections, although there are now significantly more points in this arrangement. If all points in Figure 2.22(a) are assumed to be achievable, then which points might not be achieved through mixing of other points? How should one go about identifying the unique subset of points that could be used to generate all other points?

Answer:

Identifying the unique subset of points is fulfilled by the following idea: a desired point is achievable through mixing if it lies on a straight line connected by two other points that are individually achievable. Connecting points on the perimeter of the collection with straight lines allow for all points to be enclosed by a small selection of unique points. If we are cautious to only select points on the outermost perimeter/boundary of the set, then the picture in Figure 2.22(b) may be generated. Compared to the set of 50 points initially provided, only 13 lie on the perimeter. The remaining 37 are therefore interior points. The 13 perimeter points are unique, as these allow for the generation of *all* interior points through appropriate mixing of the boundary points. From Section 2.3.3, there are infinitely many ways to obtain interior points. It follows that any point located in the shaded region of Figure 2.22(b) may be generated by using the 13 unique perimeter concentrations only.

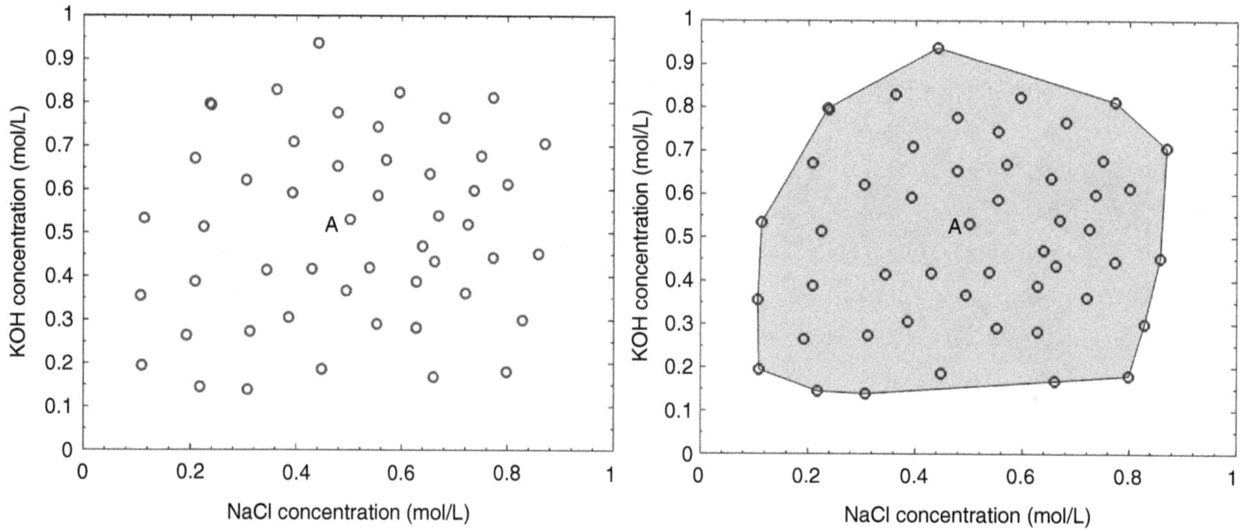

Figure 2.22 (a) A set of many points in a two-dimensional concentration space and (b) the region of achievable points achieved by traversing on the outermost boundary points. The perimeter points generate all points in the interior region.

EXAMPLE 12: Convex Hull Practice on a Computer

Using a scientific package, such as MATLAB or Python, generate a random set of 20 points in $\mathbb{R}^2$ (two-dimensional space). Determine the set of unique points belonging to the convex hull using a convex hull program.[5,6]

OBSERVATION: Interior Points of the Convex Hull

To demonstrate how interior points could be achieved using only boundary points, imagine that point A in Figure 2.23 is a concentration that is to be achieved. Point A might be generated by a number of different straight lines passing through A that is achieved by using the perimeter concentrations $\mathbf{C}_1$ through $\mathbf{C}_{13}$.

Although there are a distinct number of extreme points (there are exactly 13 extreme points), mixing allows for the generation of an infinite number of intermediate points on the perimeter and interior to the region. An interior point of a convex hull can therefore be created in various ways by mixing with the appropriate extreme points.

Given a very large list of points in space, potentially containing thousands of entries, the approach to finding unique points does not change. The idea is to always look for the outermost points that form the the boundary of the region. From these points, all other concentrations can be achieved with mixing in the appropriate proportions.

[5] In MATLAB, the built-in `convhull()` and `convhulln()` functions may be used to compute the convex hull of a set of points. The SciPy Python package is used extensively in the scientific community, containing the `scipy.spatial.ConvexHull()` function that may be used to compute convex hulls.

[6] Example code for this problem is available on the companion website.

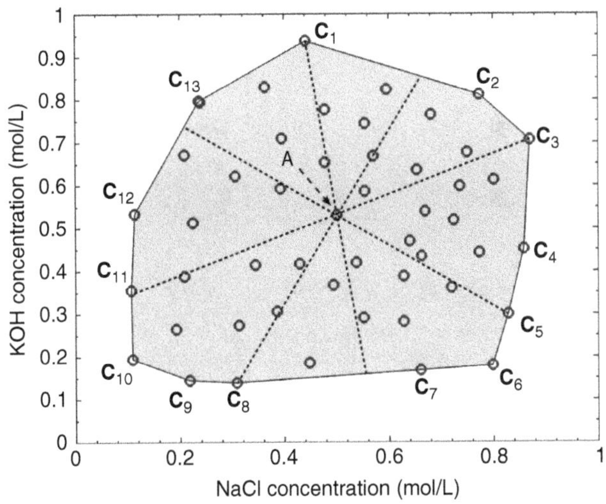

Figure 2.23 Multiple ways of achieving an interior point.

2.3.5.2 ***Higher Dimensions*** Take as another example a set of points given in three-dimensional concentration space. This might occur if a third component, in addition to NaCl and KOH, were introduced into the system. There are now more ways in which to move in space via mixing.

Although it is still possible to visualize these points, as well as any possible mixture concentrations, it is noteably more cumbersome and time consuming to identify the points that lie on the perimeter/boundary. Figure 2.24 illustrates this for 30 points in a three-dimensional space ($\mathbb{R}^3$).

In higher dimensions, the problem is intensified further by the inconvenience of no longer being able to visualize the points graphically in the full space. Mixtures of three or more components are common in many realistic systems however.

For these situations, a reliable method must be available in order to identify such points. This method should be generalized enough so that it can be used for large problems with many points, and in higher dimensions so that we do not need to always rely on having to manually identify the points.

Fortunately, the problem of finding such points is common in a number of mathematical applications (specifically, computational geometry), and thus robust methods exist for finding these points.

2.3.5.3 Convex Hulls

a) Description When one is given a list of concentrations and asked to identify the unique concentrations that may be used to enclose the associated region, one is, in fact, finding the *convex hull* of the set of points. The convex hull of a set of points is, by definition, the smallest subset of points that encloses the original set. Points that form the convex hull are called *extreme points,* whereas points interior to the convex hull are called *interior points*. Identifying unique mixing concentrations in the examples discussed previously is simply an application of convex hulls related to mixing.

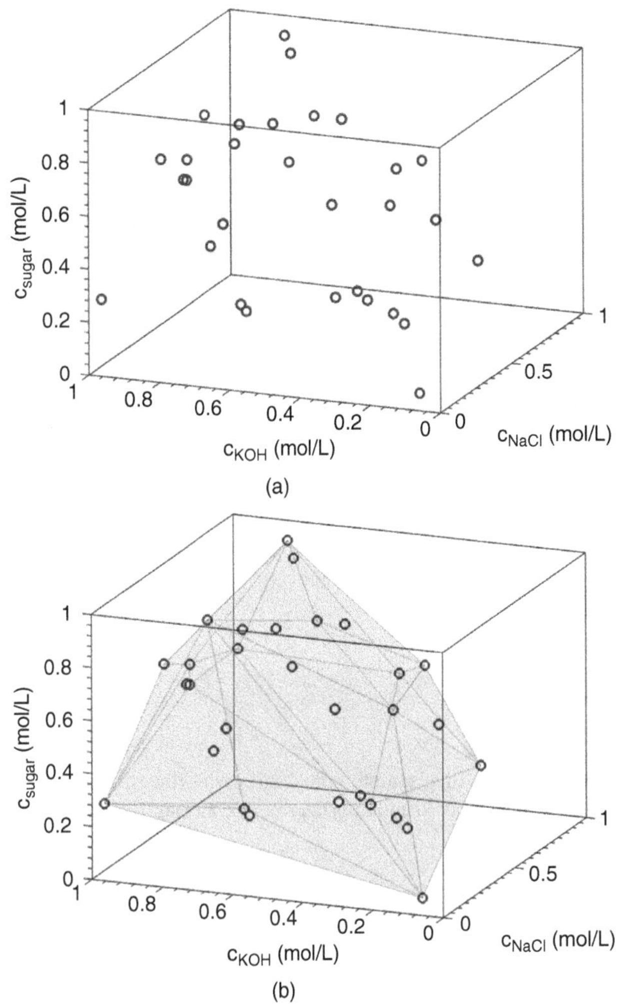

Figure 2.24 (a) A collection of concentrations in NaCl–KOH–Sugar space and (b) the corresponding convex hull for the set of points given in (a).

SIDE NOTE: Computing Convex Hulls

Computing the convex hull is an important task in many fields of science and engineering (including AR theory), and many algorithms are available that numerically compute the convex hull. The most common and widely used is the `quickhull` algorithm (Barber et al., 1996). MATLAB implements a version of the `quickhull` algorithm in its function library in the form of the `convhull()` and `convhulln()` functions.

EXAMPLE 13: Finding the Convex Hull

Consider Figure 2.25. Five points, labeled A–E, are displayed in a hypothetical two-dimensional space. The components that constitute the space obey a linear mixing law. These points might represent concentrations, such as NaCl and KOH, although any other data that obeys a linear mixing law will also be suitable. Points A, B, C, and D lie on the perimeter traced out by the triangle ABC. Point D hence lies on the line segment joining A and B. Point E lies on the interior of the triangle ABC. Of the five points given, can you identify which subset represents the convex hull for the data set?

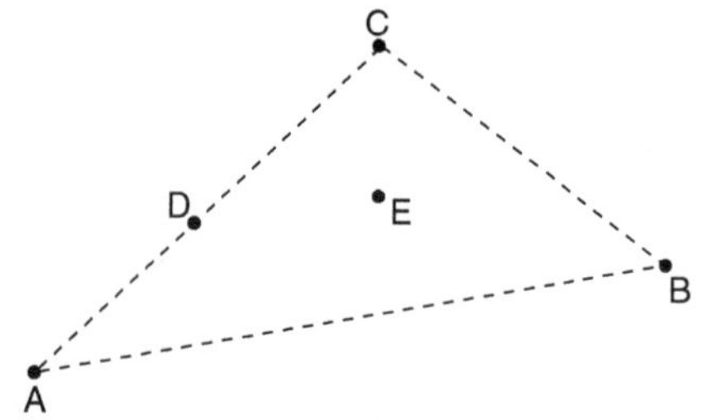

Figure 2.25 Five sample points in a two-dimensional space. The points do not necessarily represent concentrations.

Answer:

Point E clearly lies in the region bounded by ABCD, and thus it does not form part of the convex hull of the set. If points A–E were concentrations, then E would be achievable by mixing with the perimeter points. It may be tempting to think that points A, B, C, and D form the convex hull of the set—these are clearly the points that lie on the perimeter of the triangle. However, since D lies on the line segment joining AB, it is also a linear combination of points A and B. If A, B, and D were concentrations, then D could be formed by mixing points A and B in the appropriate combinations.

It follows that the convex hull of points A–E is actually given by the subset A, B, and C—this is the smallest subset that encloses the original set.

Figure 2.26 provides additional examples of the relation between the convex hull of a set of points (the extreme points of the set) and an initial set of points. Straight lines are used to connect extreme points on the perimeter of the initial set of points. This produces a closed region containing only the extreme points in the set. These points belong to the convex hull of the set of points. It follows that the number of points belonging to the convex hull cannot be larger than the initial set. In the first example in Figure 2.26, all 6 points of the initial set belong to the convex hull, whereas in the second example, only 4 of the initial 12 points forming the letter "E" belong to the convex hull.

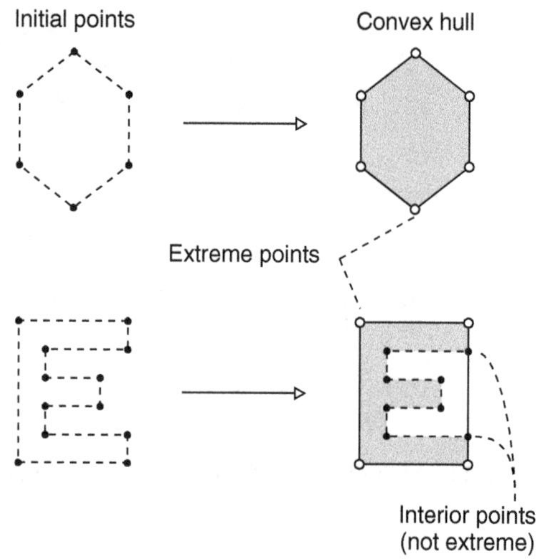

Figure 2.26 Identifying extreme points.

EXAMPLE 14: Convex Hull with Many Components

Table 2.8 shows a set of 15 points in six-dimensional concentration space. Using a convex hull program, determine which points would form part of the convex hull.

Answer:

The six-dimensional data given in Table 2.8 cannot be visualized in its entirety in $\mathbb{R}^3$, and thus a convex hull program must be used to determine the convex hull instead. The `convhulln()` function in MATLAB may be used to compute the convex hull of the points (a similar procedure may performed in Python as well). We find that points 1, 3, 6, 8, 10, 11, and 15 form the set of extreme points in the convex hull. Hence, even though there are 15 mixtures in total, only 7 are distinct. The remaining eight can be formed by combinations of the seven unique mixtures.

Convex hulls are an important property of AR theory, and so they will be used often throughout this book. We will denote the convex hull of a set of points **X** by conv(**X**). It will not be our goal to formalize the convex hull mathematically. It is more important to develop an appreciation for what the convex hull is geometrically, and the meaning with regard to attainability in chemical systems. Let us therefore strengthen the idea of convex hulls with a number of simple (everyday) examples.

TABLE 2.8 Six-Dimensional Concentration Data

No.	Component Concentration (mol/L)					
	c_1	c_2	c_3	c_4	c_5	c_6
1	0.875	0.433	0.486	0.747	0.444	0.781
2	0.889	0.186	0.561	0.738	0.746	0.772
3	0.079	0.649	0.779	0.372	0.789	0.437
4	0.452	0.322	0.737	0.669	0.492	0.331
5	0.836	0.436	0.454	0.705	0.462	0.727
6	0.782	0.032	0.699	0.931	0.229	0.237
7	0.288	0.466	0.756	0.538	0.623	0.378
8	0.494	0.462	0.174	0.331	0.618	0.249
9	0.800	0.275	0.768	0.774	0.207	0.338
10	0.852	0.973	0.966	0.321	0.142	0.628
11	0.895	0.081	0.592	0.735	0.873	0.769
12	0.329	0.575	0.779	0.471	0.729	0.460
13	0.079	0.649	0.779	0.372	0.789	0.437
14	0.875	0.433	0.486	0.747	0.444	0.781
15	0.952	0.391	0.778	0.719	0.578	0.517

ILLUSTRATION: Graphical Description of the Convex Hull

Example A: Garden poles and rope

Suppose you wish to protect a garden by enclosing the area with rope. The rope will be secured tightly around existing light poles as anchor points. A top-down view of the garden layout is given Figure 2.27. Eight garden lights are present.

Crosses indicate the position of the lights in relation to the garden. There are lights both in and around the area. Determining the set of lights to tie rope around is in fact a convex hull problem—we wish to find the minimum set of lights used to enclose the entire region.

Indeed, the process of determining the bounding set of light posts is somewhat like surrounding the area initially in a closed (loose) loop, and then progressively decreasing the circumference of the loop. As the loop gets smaller, the rope slowly begins to take up the outermost lights. Continued tightening of the rope eventually leads to the convex shape given by the dashed lines in Figure 2.27.

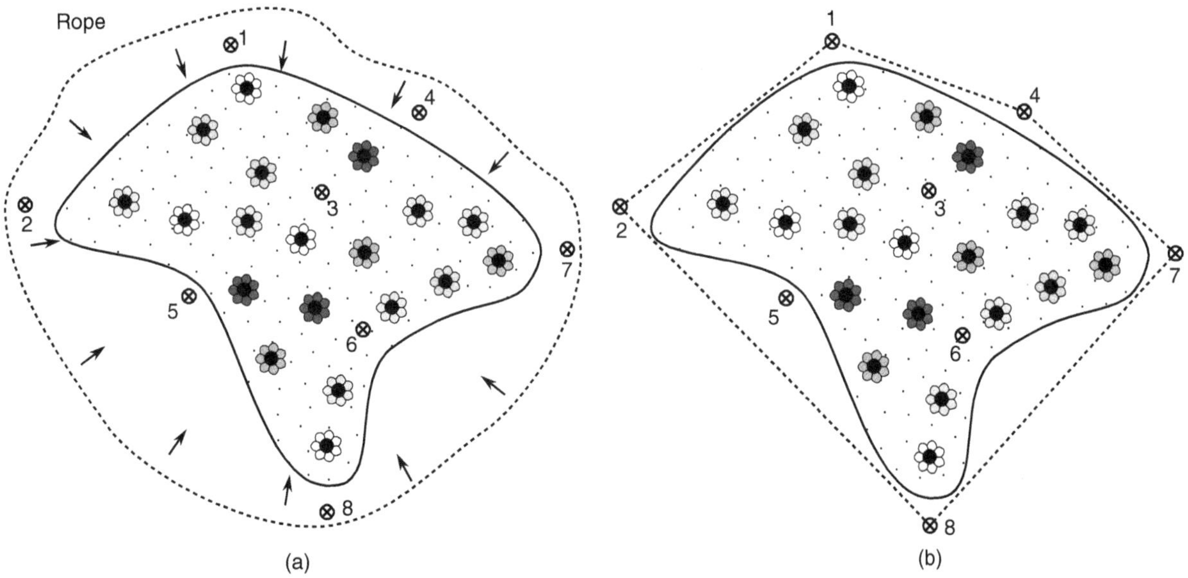

Figure 2.27 Top-down view of a garden. (a) Loose rope enclosing the garden and (b) tightened rope representing a convex hull that encloses all eight lights. (*See color plate section for the color representation of this figure.*)

In this context, the convex hull is the set of poles 1, 2, 4, 7, and 8. Lights 3, 5, and 6 lie inside the perimeter, and are not needed to enclose the region. The rope fulfills the same purpose as mixing lines in Section 2.3, whilst the lights serve the same purpose as the unique beaker concentrations.

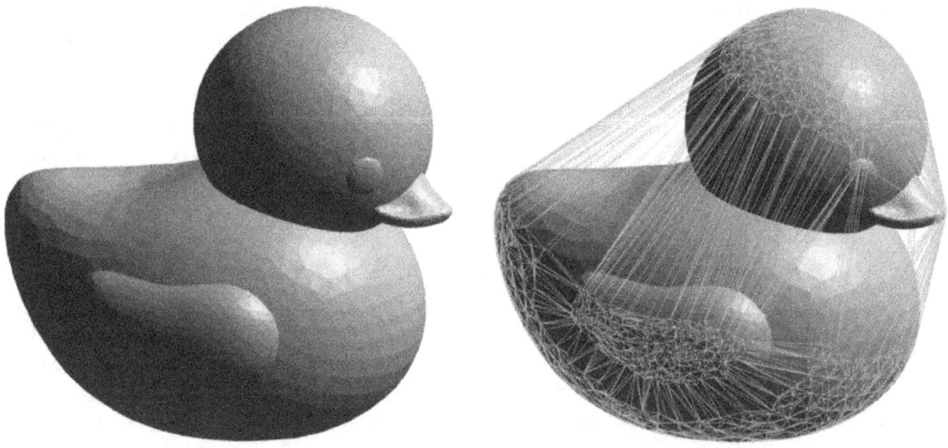

(a) (b)

Figure 2.28 (a) A three-dimensional CAD drawing of a rubber duck and (b) the associated convex hull, given in a wireframe representation. (*See color plate section for the color representation of this figure.*)

Example B: Shrink-wrap

The shape of the convex hull for many three-dimensional objects is often described as something similar to the shape produced if the same object were wrapped in shrink-wrap. Figure 2.28 shows a three-dimensional CAD model of a rubber duck, displayed to demonstrate this effect. The convex hull associated with the model is shown by the gray lines in the illustration.

Note how the lines associated with the convex hull only touch the extremities of the object (the beak, the top of the duck's head, sides of its wings and base). When the convex hull is overlaid on the model, observe how the resulting figure resembles the shape that would be obtained by wrapping it tightly in shrink-wrap.

If it were possible to arrange a set of concentrations in space to resemble the duckling in Figure 2.28(a), then the mixing lines would correspond to the gray lines displayed in Figure 2.28(b).

b) Importance of Convex Hulls It should now be clear that computation of the convex hull, given a set of achievable points, is a useful method when viewing data geometrically. In Figure 2.29, we show an example of the usefulness of the concept when applied to two distinct sets of achievable points. Regions A_1 and A_2 represent two regions of points that are known to be achievable. When the convex hull of the union of the two regions is computed, the resulting figure is that given by the filled region in Figure 2.29(c).

Straight-line sections, which represent mixing lines, connect and join extreme points belonging to concave portions of A_1 and A_2 together, resulting in a final region that is convex. Moreover, the space separating regions A_1 and A_2 are filled by the convex hull computation. The convex hull, and mixing, thus serves two vital roles for attainable sets:

1. *Mixing makes sets convex.* This is achieved by computing the convex hull for the set of points.
2. *Mixing can be used to fill in regions.* Interior points of a set of achievable points may be achieved via mixing of the appropriate points. Hence, the interior of the convex hull of a set of achievable points is also achievable via mixing.

OBSERVATION: Convex Hulls and Mixing

When we compute the convex hull of a set of points, we are finding the unique set of points that may be used for mixing to synthesize all other points. The region generated by the convex hull is therefore a geometric representation of all possible mixing states from the extreme points of the convex hull.

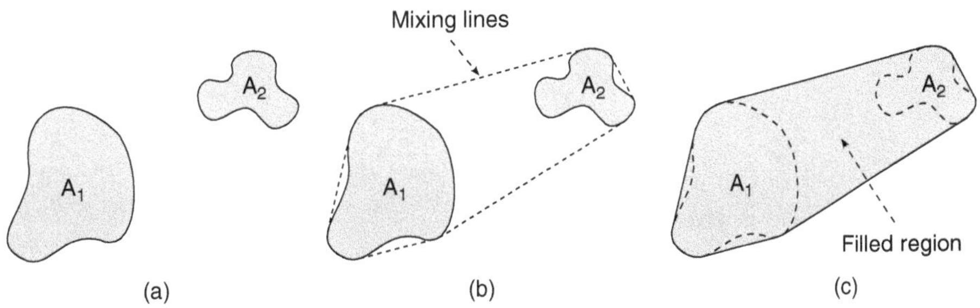

Figure 2.29 Mixing fills in concavities and joins regions together. (a) Two separate (nonconvex) regions A_1 and A_2, (b) mixing between regions A_1 and A_2, and (c) a single (convex) region containing both regions and the space separating A_1 and A_2.

SIDE NOTE: Graphical versus Geometric

We will frequently refer to data as being "from a geometric viewpoint" throughout the book. However, from the examples discussed so far, the same meaning could be applied to "from a graphical viewpoint" instead. We make a distinction between graphical and geometric viewpoints because there is a subtle difference in AR theory:

- Graphical implies that the data and ideas discussed may be represented visually or pictorially. In other words, the term "graphical" implies that it can be *seen*.
- The term "geometric," suggests that the ideas are related to geometry, and although this also implies that these concepts are also visual, they are in fact based on generalized mathematical concepts that might exceed three dimensions.

The term "geometric" hence implies that these ideas are more generalized. Although the examples given are best described graphically, the underlying concepts of concentration, mixing, and reaction are equally valid in higher dimensions.

For this reason, we will refer to AR theory as being a geometric method as opposed to a graphical one, as the ideas are applicable to higher dimensional problems as well.

2.4 CHAPTER REVIEW

This chapter discussed concentration and mixing from a *geometric* viewpoint. These concepts are easy to understand intuitively from a physical perspective. They are also easily expressed mathematically, and hence easily measured and calculated given the necessary data. The geometric perspective, however, is ultimately more interesting and useful in AR theory. When concentration and mixing can be interpreted geometrically, it is possible to develop a deeper appreciation for the attainability of a system.

In this chapter, we also showed that mixing is a *linear process* (when density is assumed constant). As a result, two important tasks can be accomplished with mixing:

1. Mixing fills in regions.
2. Mixing makes a set of concentrations convex.

Mixing is mathematically equivalent to finding the convex hull of a set of points. The convex hull is the smallest subset of a set of points that can be used to form all points through mixing. Achievability is closely linked to mixing and the convex hull of a set of points. This might not be easy to establish without first consideration of the data geometrically (by plotting the data in the phase plane).

These concepts were demonstrated with simple thought experiments in two-dimensional concentration space, although, these ideas are valid for higher dimensions as well. When dealing with higher dimensions, visualizing the full set of data is not always intuitive, and so we must be comfortable with relying on the convex hull to determine unique points in space.

The geometric perspective of a system is an important aspect of AR theory, for it allows us to utilize the fundamental concepts of concentration vectors, mixing, and convex hulls. In Chapter 3, we will return to the BTX beaker experiment and use the graphical concepts described in this chapter to improve the maximum toluene concentration (larger than that obtained in Chapter 1).

REFERENCES

Barber, C.B., Dobkin, D.P., Huhdanpaa, H., 1996. The Quickhull algorithm for convex hulls. Acm Trans. Math. Softw. 22, 469–483.

Geankoplis, C.J., 1993. Transport Processes and Unit Operations, 3rd ed. Prentice Hall PTR, Englewood Cliffs, NJ.

3

THE ATTAINABLE REGION

3.1 INTRODUCTION

In Chapter 2, batch data for the BTX reaction were plotted in a manner that allowed us to infer useful geometric properties related to concentration and mixing. We now understand that mixing allows for two physically separate yet achievable points to be connected by a straight line in concentration space. If more than two unique concentrations are available, it is possible to produce a filled convex region in space. In doing so, concentrations that were previously unachievable can be made achievable via mixing of the appropriate points.

Convex hulls are closely related to mixing. The convex hull assists in differentiating between unique points (those that *cannot* be formed from combinations of other points) and redundant ones (those that *can* be formed from combinations of other points). From this perspective, knowledge of the convex hull for a set of achievable points is all that is required for mixing problems. This result is important for the following two reasons:

1. Interior points of a convex region may be synthesized in infinitely many ways.
2. There is no need to search for all possible achievable points when creating mixtures. Rather, it is sufficient to only identify the unique points that belong to the convex hull—all other achievable points may be obtained from the extreme points of the convex hull.

In this chapter, we seek to expand on these ideas and use them to introduce the concept of the AR. This will be achieved using the BTX system as a motivating case. A number of important AR properties are then described in Section 3.4.

3.2 A MIXING AND REACTION GAME

3.2.1 Introduction and Rules of the Game

In this chapter, we wish to formally introduce the concept of the AR. We shall do this by theoretically building an AR—from scratch—using the BTX kinetics from Chapter 2 in the form of a worked example. The descriptions of the worked example work best when it is framed as a simple game, involving batch experiments, which enforce two rules that are related to the AR. The game is given as follows:

Find the largest possible toluene concentration in the BTX reaction, using only batch reaction experiments and mixing. Different batch profiles may be generated by changing the starting concentration of a batch, using mixtures that have been formed from previously achievable concentrations. The aim will be to indirectly find a starting concentration for a batch reaction that maximizes the concentration of toluene. There are only two rules to the game as follows:

1. *Mixtures can only be formed using previously achievable concentrations.*
2. *Only one batch reaction can be operated at a time.*

To simplify the descriptions slightly, the following conditions are assumed:

- An infinite reserve of feed material is available to carry out the experiments.
- There are no restrictions on the size or the amount of beakers that are available for mixing.
- There is no time limit—we are interested in product quality as opposed to production rate.
- The system operates under constant density.

3.2.2 Filling in the Region

Figure 3.1(a) is a reproduction from Chapter 2 of the batch concentration profiles for benzene and toluene, plotted as a phase diagram. Also marked on the figure are the initial feed point (F) and the final equilibrium point (O). As the reaction proceeds, the concentrations of benzene and toluene evolve so that the concentration vector $\mathbf{C} = [c_B, c_T]^T$ traces out the path given in Figure 3.1(a). The path is initiated at point F and terminates at point O. O is hence the equilibrium point

Attainable Region Theory: An Introduction to Choosing an Optimal Reactor, First Edition. David Ming, David Glasser, Diane Hildebrandt, Benjamin Glasser, and Matthew Metzger.

Companion website: http://attainableregions.com

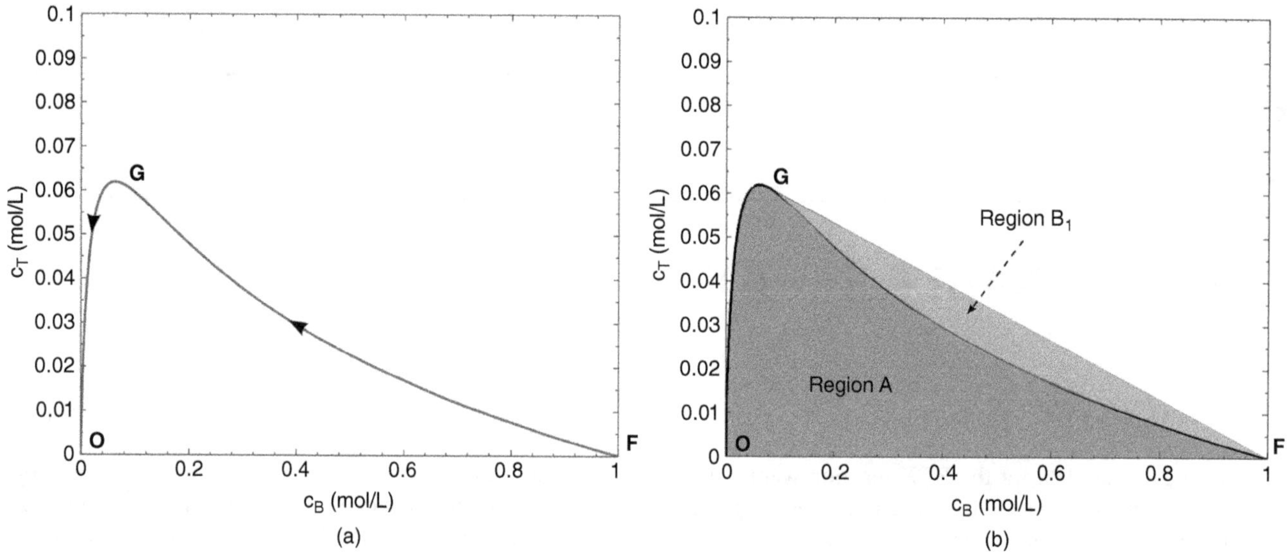

Figure 3.1 (a) Batch concentration profile for the BTX beaker experiment and (b) the convex hull of the set of points for the BTX beaker experiment.

for the batch. The component values at point O are given by the vector $\mathbf{C}^{\#} = [0.0, 0.0]^T$, whereas at the feed they are $\mathbf{C}_f = [1, 0]^T$. Note that point O does not indicate a state where the system is devoid of any material. Rather, this coordinate represents the state where all benzene and toluene have reacted completely into final products, specifically xylene, diphenyl, and hydrogen.

The concepts of concentration and mixing, from Chapter 2, are immediately applicable to the experimental data given here. The set of concentrations given by curve FGO represents more than a one-dimensional line segment—by mixing concentrations on FGO, a *filled* two-dimensional region may be generated as well. Furthermore, since the line segment FG is concave, the set of achievable concentrations can be expanded with mixing by joining point F with point G using a mixing line. All the points within the region traced out by the boundary FGO are then also attainable by appropriate mixing of the boundary points.

In Figure 3.1(b), the resulting convex region associated with the experimental points is shown. Points lying on the boundary of the shaded region represent concentrations wherefrom all interior points may be synthesized. The following two regions of interest have been labeled in Figure 3.1(b):

1. Region B_1 is the collection of all points *belonging to the concave region*, originally defined by curve FG.
2. Region A is the remaining set of achievable points when region B_1 is removed from the entire shaded region FGO. The union of regions A and B_1 constitute the entire set of points.

Let us now seek to improve the maximum achievable toluene concentration, c_T, of the system by searching for new starting mixtures in regions A and B_1 wherefrom successive batches may be run. We shall enforce the restrictions of the game, established in Section 3.2.1.

3.2.3 Scenario 1: Selecting Points from Region A

Let us select a point in the region and generate a batch profile from this point. Point $\mathbf{a}_2$ in Figure 3.2 is representative of such a point in region A. (As will be shown, selecting points for further improvement is somewhat arbitrary.[1]) To achieve point $\mathbf{a}_2$, the following three steps are required:

1. The original batch experiment must be carried out. This is done by starting with fresh feed at a concentration equal to F, and reacting until point $\mathbf{a}_1$ in Figure 3.2. Point $\mathbf{a}_1$ is the particular concentration achieved in the original batch, which is needed for mixing with point F, to produce point $\mathbf{a}_2$.
2. The batch reaction must then be terminated when point $\mathbf{a}_1$ is achieved to prevent any further reaction from proceeding.
3. A portion of the feed, F, must then be mixed with $\mathbf{a}_1$ in the correct ratio so that concentration $\mathbf{a}_2$ is obtained.

We can find the reaction time required to achieve point $\mathbf{a}_1$ by plotting c_T versus reaction time and locating the specific time where the value of c_T is equal to that given by $\mathbf{a}_1$. The line segment $\mathrm{F}\mathrm{G}\mathbf{a}_1$ is therefore representative of the solution trajectory obtained by running the BTX reaction from the feed until the concentration at point $\mathbf{a}_1$ is achieved.

[1] Although some points will give better improvements initially, all the results tend toward the same outcome after a sufficient number of iterations.

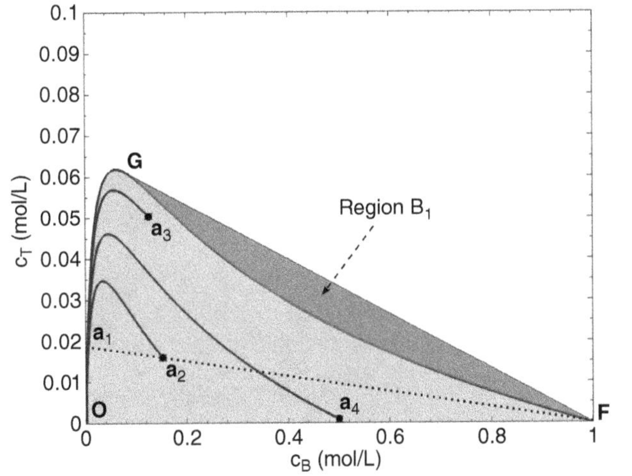

Figure 3.2 Batch profiles for subsequent batches run from points generated in region A from Figure 3.1b. None of the profiles achieve a larger maximum toluene concentration compared to the original experiment.

The straight line $\mathbf{Fa}_1$ in Figure 3.2 shows the set of mixtures obtained by mixing points $\mathbf{a}_1$ and F together. Point $\mathbf{a}_2$ clearly intersects this line, and therefore it is also an achievable mixture concentration.

> **RECALL: Achieving interior points**
>
> Since point $\mathbf{a}_2$ lies in the interior of the convex hull, it is not unique. We have chosen to form concentration $\mathbf{a}_2$ with a straight line passing through the feed and the appropriate concentration in the batch profile FGO, although many other mixing lines could also be used to achieve this point.

Now that $\mathbf{a}_2$ has been obtained, it is possible to run a new batch from $\mathbf{a}_2$ by using $\mathbf{a}_2$ as a starting concentration. Physically, a mixture equal to concentration $\mathbf{a}_2$ must be transferred into a new beaker and allowed to react until equilibrium is approached. Throughout the reaction period, the concentrations of benzene and toluene are again recorded over time, and then plotted in c_B–c_T space.

Since kinetics for the system has already been provided in Chapter 1, the species concentration profiles achieved in the beaker may be computed using the standard batch reactor equation. This is done by numerically integrating the rate expressions with the new initial condition given by concentration $\mathbf{a}_2$. The batch profile resulting from this integration is shown in Figure 3.2, given by the curve $\mathbf{a}_2$O. The profiles have been overlaid with the previous experiment for comparison.

Attempting to improve the toluene concentration by running a subsequent batch experiment from $\mathbf{a}_2$ does not appear to be advantageous, as the corresponding curve is completely contained within region A. As a result, no additional improvement to the toluene concentration is realized.

Is this result simply due to the choice of starting point $\mathbf{a}_1$? Figure 3.2 also displays profiles for an additional two starting concentrations (given in Figure 3.2 by points $\mathbf{a}_3$ and $\mathbf{a}_4$), which also reside in region A. The associated solution trajectories are represented by curves $\mathbf{a}_3$O and $\mathbf{a}_4$O, respectively. Note that these trajectories also do not appear to improve the maximum concentration of toluene.

These profiles suggest that there are no other starting concentrations in region A that could be used to increase the concentration of toluene—in Chapter 4, we shall describe why this behavior is *always* true, which is a useful property used in AR theory. With this, let us rather investigate the concave portion of the region, designated by region B_1.

> **SIDE NOTE: Mixing calculations in this chapter**
>
> Although we shall attempt to keep the following descriptions free of mathematics (for the purposes of clarity), readers interested in performing the mixing calculations may do so by following the examples placed at the end of each discussion. These calculations are not essential at present, but they are supplied to encourage familiarity with the concepts of mixing. These topics shall be revisited and expanded in Chapter 4.

> **OBSERVATION: The effect of subsequent batches from region A**
>
> Running batches with starting concentrations from region A does not appear to achieve a higher toluene concentration. Is this effect also true of region B_1?

3.2.4 Scenario 2: Selecting Points from Region B_1

3.2.4.1 ***Initial Batch*** Points from region B_1 are chosen as starting points for further batches, similar to Section 3.2.3. The specific mixture point used is not important, and so a point located near the center of region B_1 shall be utilized. This is given by point $\mathbf{x}_1$ in Figure 3.3(a).

To produce concentration $\mathbf{x}_1$, the appropriate mixing concentrations must again be found. A unique straight line segment that passes through point F and $\mathbf{x}_1$ may be constructed and extended to locate the point of intersection with the original solution trajectory. There are in fact two such intersection points. Point $\mathbf{y}_1$, shown in Figure 3.3(a), is one possible point; the other point is obtained by the straight line intersecting the curve between points G and O in Figure 3.3(a).

Let us remind ourselves of the physical tasks required. In this instance, the initial batch must first be run for a

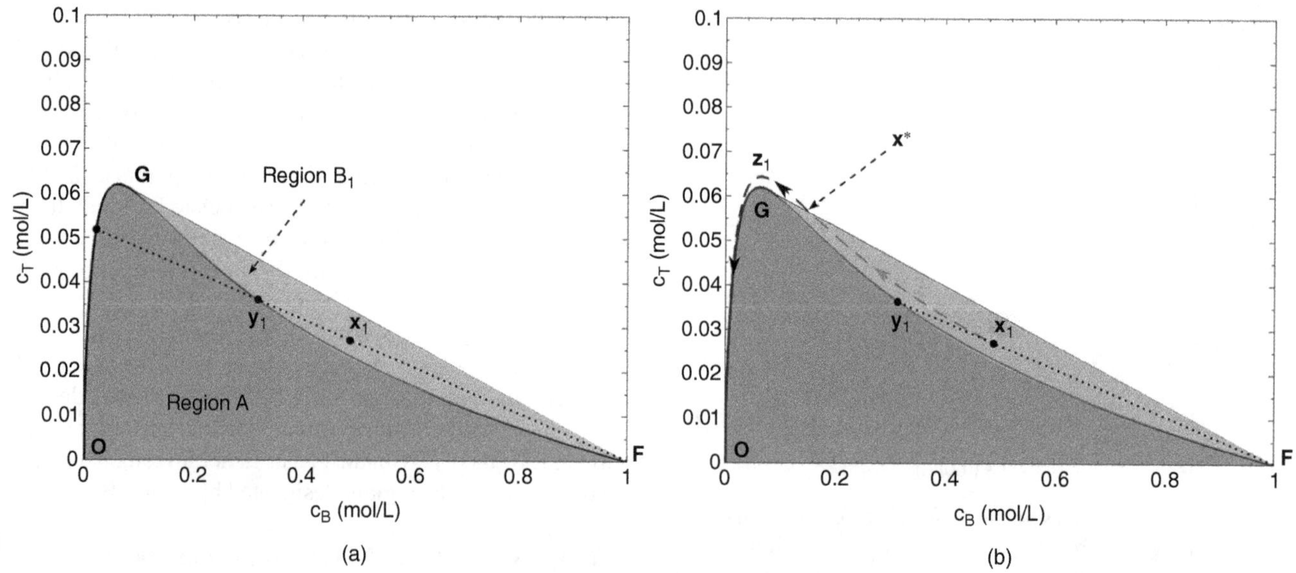

Figure 3.3 (a) Example of a typical mixing line used to achieve point $\mathbf{x}_1$. Point $\mathbf{x}_1$ is located in the concave section of the profile (region B_1). (b) The batch profile obtained when a new experiment is run using a starting concentration equal to point $\mathbf{x}_1$.

time that produces composition $\mathbf{y}_1$. Mixing $\mathbf{y}_1$ with feed then produces point $\mathbf{x}_1$. Recall that concentrations lying in region B_1 are those that are formed from a concavity in the original batch profile, and thus these concentrations are not originally achievable without mixing between points on the curve $F\mathbf{y}_1G$.

EXAMPLE 1: Forming $\mathbf{x}_1$

The concentration given by point $\mathbf{x}_1$ is formed from a mixture of F and $\mathbf{y}_1$. Suppose that F and $\mathbf{y}_1$ are given by the concentration vectors $\mathbf{C}_f = [1,\ 0]^T$ mol/L and $\mathbf{y}_1 = [0.3181,\ 0.0361]^T$ mol/L. Prove that $\mathbf{x}_1 = [0.4886,\ 0.0271]^T$ mol/L if the mixing fraction is $\lambda = 0.25$ (a mixture of 25% $\mathbf{C}_f$ and 75% $\mathbf{y}_1$).

Answer:

Since $\mathbf{x}_1$ is formed by mixing $\mathbf{C}_f$ and $\mathbf{y}_1$ with $\lambda = 0.25$, substituting in values then gives the concentration of $\mathbf{x}_1$:

$$\begin{aligned}\mathbf{x}_1 &= \lambda\mathbf{C}_f + (1-\lambda)\mathbf{y}_1 \\ &= (0.25)\begin{bmatrix}1\\0\end{bmatrix} + (0.75)\begin{bmatrix}0.3181\\0.0361\end{bmatrix} \\ &= \begin{bmatrix}0.4886\\0.0271\end{bmatrix}\ \text{mol/L}\end{aligned}$$

SIDE NOTE: Sam, Alex, and Donald

In Chapter 1, we introduced the story of Sam, Alex, and Donald, and described their quest to find an optimum operating strategy to maximize the production of toluene in the BTX reaction. At one point during their investigation, Alex and Donald discover that a higher toluene concentration can be achieved if the starting concentration of the batch is altered—by retaining a fraction of the product material and mixing with fresh feed.

In effect, the proposed alteration is the same as the batch game described in this chapter. By altering the mixing fraction of the batch, the starting concentration for the next batch is altered, which is equivalent to finding starting concentrations that lie in regions A and B_1 in Figure 3.2. In Chapter 1, Donald realized that for certain pairs of retained fraction and reaction time, the subsequent batch could not achieve a higher toluene concentration—in effect, these pairs produce concentrations that reside in region A of Figure 3.2. Similarly, for those combinations of retained fraction and reactions times where an improvement is observed, this is graphically equivalent to producing a starting concentration residing in the concave region of Figure 3.2 (region B).

3.2.4.2 Second Batch Profile When a mixture concentration equal to point $\mathbf{x}_1$ is used to perform an additional batch experiment, it is again possible to plot the resulting solution trajectory. A similar procedure as in the previous scenario may be carried out. The results obtained are overlaid over the previous plot for comparison, which are displayed in Figure 3.3(b).

The batch profile obtained by the new experiment is designated by curve $\mathbf{x}_1\mathbf{z}_1$O. Arrows are also placed on the curve to help identify the direction that benzene and toluene concentrations advance during the reaction. The initial portion of the reaction is contained within the original concave region B_1 up to point $\mathbf{x}^*$ in Figure 3.3(b). The reaction proceeds in a direction that increases the toluene concentration up to a maximum point. Beyond this point, toluene is consumed to form xylene and the value of c_T then decreases as a result. Toluene continues to be consumed until the equilibrium point, given at point O, is again achieved.

In contrast to Section 3.2.3, starting with point $\mathbf{x}_1$ achieves a higher concentration of toluene than the original experiment. The maximum toluene concentration achieved is now 0.06547 mol/L, which is displayed as point $\mathbf{z}_1$. Compared to the previous maximum of 0.0619 mol/L, this profile generates a 5.8% improvement in the maximum toluene concentration achievable. We have therefore managed to improve the system by using a starting mixture taken from region B_1.

OBSERVATION: The effect of subsequent batches from region B_1

Unlike starting points from region A, using an initial point from region B_1 instead produces a profile with a larger maximum toluene concentration.

SIDE NOTE: Achieving point x*

Not all concentrations residing on curve $\mathbf{x}_1\mathbf{z}_1$O are unique—as the reaction proceeds, an initial portion of curve $\mathbf{x}_1\mathbf{z}_1$O resides *within* region B_1. Ordinarily, points on this section of the curve would have already been achievable due to mixing with the original batch. This subset of concentrations might then be realized in either of the following two ways:

1. Beginning with point $\mathbf{x}_1$ and reacting, as described previously;
2. Mixing directly with points F and G in the correct proportions from the original profile.

It appears that only at point $\mathbf{x}^*$ (the intersection of the second experiment trajectory and the mixing line with the first experiment) is there an observable difference in the beaker concentration obtained by mixing alone. That point $\mathbf{x}^*$ belongs to the solution trajectory of the experiment originating at point $\mathbf{x}_1$, suggests that it is feasible to first synthesize point $\mathbf{x}^*$ by mixing F and G in the correct proportions, and then performing the experiment from $\mathbf{x}^*$. Both methods result in identical outcomes.

3.2.5 Further Improvements

We have managed to produce a batch profile that achieves a larger maximum toluene concentration than a single batch reactor. This is accomplished with no more than a suitable choice of reaction time, and the careful mixing of product and feed concentrations in the correct ratios. Yet, what further improvements are available by continuing with additional reaction steps? Let us define a "reaction step" as one where

- A batch reaction is carried out using an existing (achievable) concentration;
- The convex hull for the associated concentration profile in concentration space is found;
- The concave portion of the convex hull is identified—which is equivalent to finding region B_1;
- Finally, a mixture point residing in the concave region is chosen wherefrom further batch experiments may be run.

Let us now investigate what occurs when additional reaction steps are employed.

Figure 3.4(a) shows the convex hull computed for the new set of points, belonging to the second reaction. This is given by the boundary F$\mathbf{z}_1$O. To reproduce the region shown in Figure 3.4(a), both batch profiles must be used: $\mathbf{x}_1$, from experiment 1, must be synthesized before $\mathbf{z}_1$, from experiment 2, can be achieved.

Furthermore, points on the original profile FGO that were previously unique (those points that constituted the original convex hull) are no longer distinct now. These points are achieved, in infinitely many ways, by mixing with boundary points belonging to the larger convex hull associated with the second batch experiment.

Note the point of mixing along curve $\mathbf{x}_1\mathbf{z}_1$O with the feed F differs slightly to the new maximum toluene concentration given by point $\mathbf{z}_1$. In general, there is no requirement for these two concentrations to correspond. A new concavity,

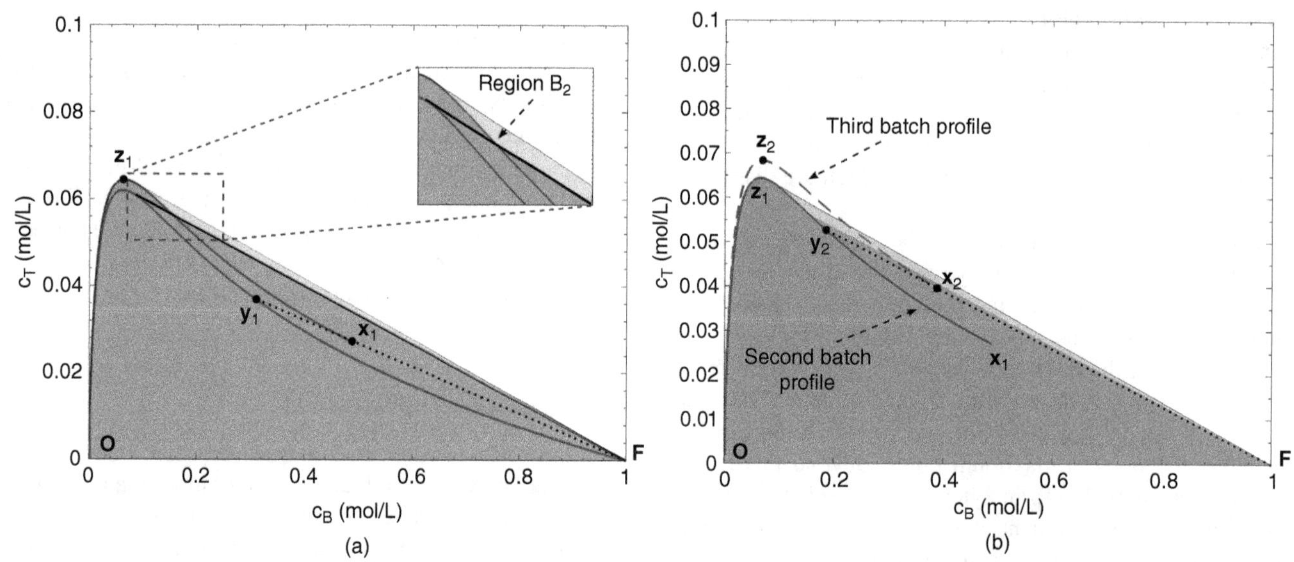

Figure 3.4 (a) The concave region obtained (shown as the filled region B_2) belonging to the updated batch profile initiated at point $\mathbf{x}_1$. (b) Batch profile obtained from a third batch experiment from $\mathbf{x}_2$. Concentration $\mathbf{x}_2$ is formed by mixing concentration $\mathbf{y}_2$ with the feed F.

belonging to the new profile, may also be identified, which is displayed in Figure 3.4(a) by the shaded region B_2. Observe that the area of B_2 is also smaller than the area belonging to region B_1.

3.2.5.1 A Third Batch Figure 3.4(b) shows the same data as before, now with the results of the third experiment also included. The curves from experiment 1 as well as some portions of the curve from experiment 2, have been removed from the figure for clarity. A mixture concentration lying in region B_2 can also be seen, identified by point $\mathbf{x}_2$ in Figure 3.4(b).

ILLUSTRATION: Achieving $\mathbf{x}_2$

In order to achieve point $\mathbf{x}_2$, point $\mathbf{y}_2$ is required. Similarly, to achieve point $\mathbf{y}_2$, a starting mixture of point $\mathbf{x}_1$ (from experiment 2) is required. Concentration $\mathbf{x}_1$, in turn, is formed from a mixture of the feed F and point $\mathbf{y}_1$ (obtained from experiment 1).

In order gain a better sense of the physical interpretation of this process, a schematic showing the steps taken to synthesize a concentration equal to $\mathbf{x}_2$ is provided in Figure 3.5.

Upon consultation of Figure 3.5, we see that the following tasks are required:

Experiment 1

- Reaction step: react fresh feed until the concentration in the beaker is equal to point $\mathbf{y}_1$.
- Mixing step: mix $\mathbf{y}_1$ with feed in the appropriate amounts to produce point $\mathbf{x}_1$.

Experiment 2

- Reaction step: using the starting concentration given by point $\mathbf{x}_1$, react until concentration $\mathbf{y}_2$ is achieved.
- Mixing step: mix $\mathbf{y}_2$ with additional feed to produce point $\mathbf{x}_2$.

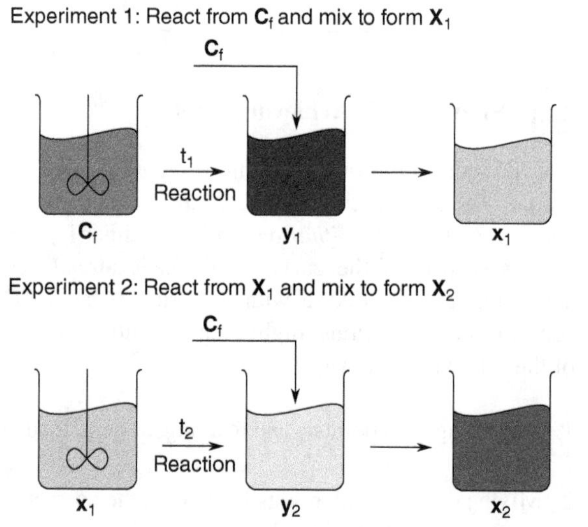

Figure 3.5 The physical steps required to achieve a mixture concentration equal to point x_2.

EXAMPLE 2: Forming $\mathbf{x}_2$

Concentration $\mathbf{x}_1$ is used at the starting concentration for a second batch experiment. A helpful friend performs the experiment for you and informs you that after a 9 min reaction time ($t_2 = 0.15$ h from Figure 3.5), the concentration measured inside the beaker is $\mathbf{y}_2 = [0.1842, 0.0528]^T$ mol/L. What is the concentration of $\mathbf{x}_2$ (the concentration that will be used in a subsequent batch experiment) if $\mathbf{x}_2$ is formed by a combination of 75% $\mathbf{y}_2$ and initial feed concentration $\mathbf{C}_f = [1, 0]^T$ mol/L?

Answer:

The concentration belonging to $\mathbf{x}_2$ is formed from a combination of $\mathbf{y}_2$ and $\mathbf{C}_f$. We know that 75% of $\mathbf{y}_2$ is used to form $\mathbf{x}_2$, the remainder (25%) is from $\mathbf{C}_f$.

$$\begin{aligned}\mathbf{x}_2 &= \lambda \mathbf{y}_2 + (1-\lambda)\mathbf{C}_f \\ &= (0.75)\begin{bmatrix}0.1842\\0.0528\end{bmatrix} + (0.25)\begin{bmatrix}1\\0\end{bmatrix} \\ &= \begin{bmatrix}0.3881\\0.0396\end{bmatrix} \text{ mol/L}\end{aligned}$$

$\mathbf{x}_2$ may now be used in a new batch experiment.

The profiles in Figure 3.4(b) share traits similar to those obtained by experiment 2—there is an initial portion of the curve that lies within region B_2 (indicating that these concentrations are not unique), after which, the trajectory emerges through the mixing line, generating a new set of unique concentrations. A new maximum toluene concentration is obtained, which occurs at point $\mathbf{z}_2$. The toluene concentration associated with $\mathbf{z}_2$ is approximately 6.8% larger than that of $\mathbf{z}_1$ from the previous batch.

Two experiments are now needed to achieve the appropriate starting concentration for the BTX reaction. For each new iteration, an improvement over the current best is obtained, although, additional effort is required in realizing the correct starting concentration (in the form of obtaining the appropriate intermediate mixtures $\mathbf{x}_i$).

3.2.5.2 A Fourth and Fifth Batch We shall repeat the same procedure as before, by selecting a mixture concentration residing in the concave portion of the current region, and track the performance improvement of the system after each additional experiment. Figure 3.6(a) shows the results of experiment 4, with the convex hull calculated using a concentration obtained from experiment 3. A higher maximum toluene concentration is achieved, although even more effort is now required to synthesize point $\mathbf{x}_3$ (which requires three batch reactions). The maximum toluene concentration is achieved at point $\mathbf{z}_3$ in Figure 3.6(a).

After a fifth batch, the set of profiles given in Figure 3.6(b) is obtained. Figure 3.6(b) thus represents the aggregate of all

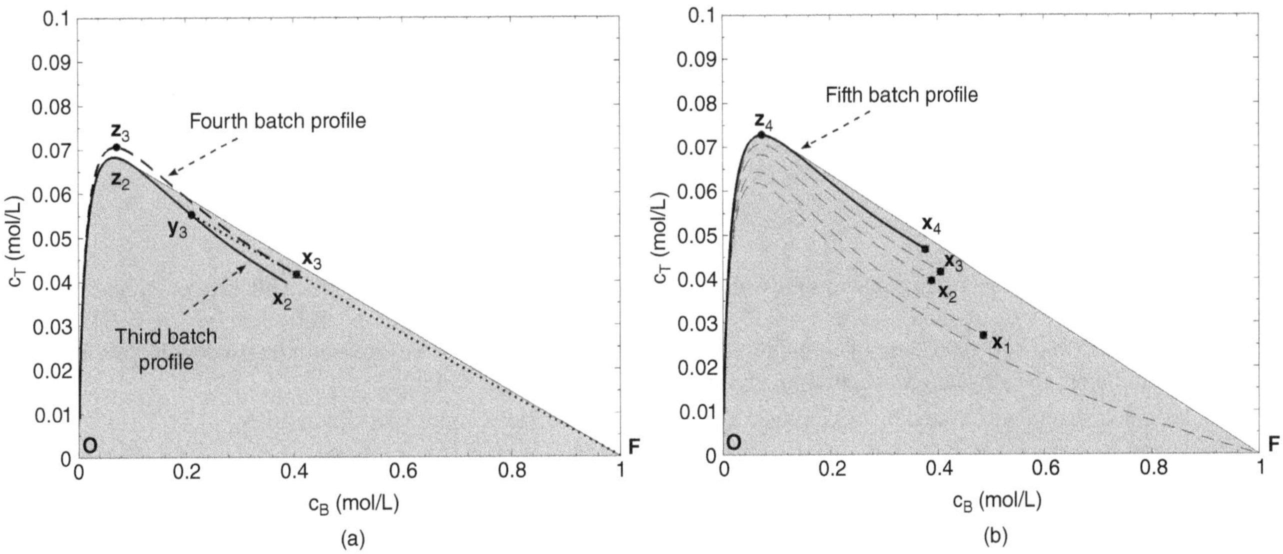

Figure 3.6 (a) Fourth batch experiment and (b) fifth batch experiment. All five previous batch profiles are displayed on a single plot.

the experimental batches, overlaid on a single plot in c_B–c_T space. The mixing points belonging to each starting batch are also displayed. These are numbered according to the batch experiment associated with it ($\mathbf{x_1}$ is the first successive batch, and so on).

EXAMPLE 3: Finding values for $\mathbf{x_4}$ and $\mathbf{x_5}$ given reactor information

A psychic tells you that $\mathbf{y}_3 = [0.2071, 0.0555]^T$ mol/L (the final concentration of the fourth batch) and also predicts that the reaction times for the fifth batch profile is $t_5 = 0.15$ h. What must the concentrations of $\mathbf{x}_4$ and $\mathbf{x}_5$ be if 25% fresh feed is mixed after each new batch?

Answer:

Concentration $\mathbf{x}_4$ is found directly from $\mathbf{y}_3$ and $\mathbf{C}_f$:

$$\mathbf{x}_4 = \lambda \mathbf{y}_3 + (1 - \lambda)\mathbf{C}_f$$

If 25% of fresh feed is used and mixed with 75% $\mathbf{y}_3$, then $\lambda = 0.75$.

$$\mathbf{x}_4 = (0.75)\begin{bmatrix}0.2071\\0.0555\end{bmatrix} + (0.25)\begin{bmatrix}1\\0\end{bmatrix}$$

$$\mathbf{x}_4 = \begin{bmatrix}0.4053\\0.0416\end{bmatrix} \text{ mol/L}$$

To find $\mathbf{x}_5$, which is a combination of the reaction product from reaction 4 and $\mathbf{C}_f$, we must first run a batch using $\mathbf{x}_4$ as the initial concentration. We can integrate the batch equation for the batch reaction time of $t = 0.15$ h,

$$\begin{aligned}\frac{dc_B}{dt} &= r_B\\ \frac{dc_E}{dt} &= r_E\\ \frac{dc_T}{dt} &= r_T\end{aligned} \tag{3.1}$$

where r_B, r_E, and r_T are the rate expressions for benzene ethylene and toluene, respectively:

$$\begin{aligned}r_B &= -k_1 c_B c_E^{0.5} - 2k_3 c_B^2\\ r_E &= -0.5k_1 c_B c_E^{0.5} - 0.5k_2 c_T c_E^{0.5}\\ r_T &= k_1 c_B c_E^{0.5} - k_2 c_T c_E^{0.5}\end{aligned}$$

Note that all three component batch equations must also be integrated since the system depends on all three components. Integrating Equation 3.1 for c_B, c_E, and c_T, after $t = 0.15$ h, we find that the concentration of the fifth batch is $c_B = 0.1697$ mol/L, $c_E = 0.4577$ mol/L, and $c_T = 0.0623$ mol/L, which is the concentration of $\mathbf{y}_4$. Since we are only interested with plots in c_B–c_T space, we can ignore the ethylene concentration, giving $\mathbf{y}_4 = [0.1697, 0.0623]^T$ mol/L.

Concentration $\mathbf{x}_5$ can now be found by mixing feed $\mathbf{C}_f$ with $\mathbf{y}_4$ in a 25–75% proportion, which gives

$$\begin{aligned}\mathbf{x}_5 &= \lambda \mathbf{y}_4 + (1 - \lambda)\mathbf{F}\\ &= (0.75)\begin{bmatrix}0.1697\\0.0623\end{bmatrix} + (0.25)\begin{bmatrix}1\\0\end{bmatrix}\\ &= \begin{bmatrix}0.3773\\0.0467\end{bmatrix} \text{ mol/L}\end{aligned}$$

3.3 THE AR

3.3.1 Ten Experiments

We can continue with this strategy by running additional experiments, computing the convex hull of the experimental points, and finding concave regions in the resulting profiles for additional batches. After approximately nine iterations, it is no longer possible to significantly improve the maximum toluene concentration by mixing with previous experiments.

Table 3.1 summarizes of the percentage improvement and the maximum toluene concentration obtained after each successive experiment. The data given are slightly different to those supplied in Section 3.2.4; however, the overall trend in improvement is indicative of the general performance observed in the system.

After nine iterations, the percentage improvement offered declines to approximately 0.79%, and it appears as though the maximum toluene concentration approaches a limiting value. In Figure 3.7, the resulting region is shown for 10 representative batches. Evidence that the system is approaching equilibrium is observed by the similar shape in batch profiles for experiments 8, 9, and 10.

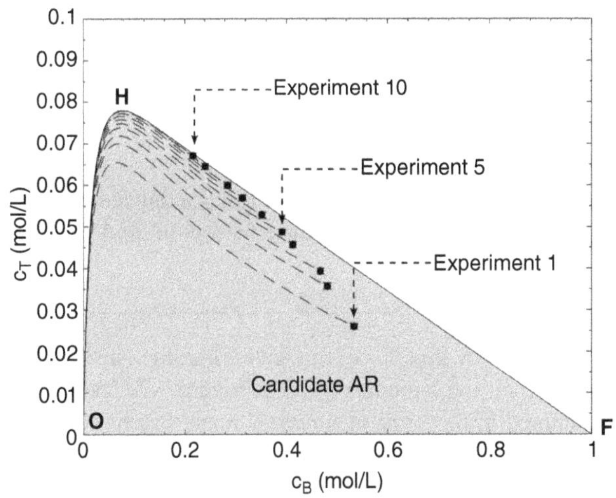

Figure 3.7 The result of many successive batches.

3.3.2 The Limit of Infinitely Many Batches

Let us rationalize, via a thought experiment, what occurs in the limiting case of infinitely many reactions. In what follows, as we continue to run batch experiments, compute the convex hull of the set of points, and then seek concave regions to start new batch experiments, we can envision that a state is attained where no further improvements are achieved. The concave sections of the resulting regions become smaller with each new batch, and therefore it is difficult to locate additional mixture concentrations (the $\mathbf{x}_i$'s) that serve to operate additional batches from. On reflection, the existence of a concave region provides a visual indication of *potential improvement*—the actual improvement is found by taking a point in the region and running a new experiment with that concentration—and measuring the size of the concave region is thus a measure of the remaining set of states that offer potential improvement to the system.

On consideration of this behavior, in the limit of infinitely many iterations, all states for potential improvement are picked, and the following conclusion is obtained:

No combination of reaction and mixing with available concentrations allows for further improvement to the system.

We have arrived at a significant result: the largest shaded region after many such batches is, in fact, the AR for the BTX system.[2] The region represents the *entire* set of concentrations achievable by reaction and mixing alone. No other concentration, besides those contained in or on the boundary of the convex hull, is achievable with the feed concentration specified. This region has been generated via a series of batch

[2]We shall verify that the region in Figure 3.7 is the AR for the BTX system in Chapter 7.

TABLE 3.1 Summary of Batch Experiments and the Associated Maximum Toluene Concentration Achieved

Experiment No.	Mixture Concentration Vector $\mathbf{x}_i$ (mol/L)	Max Toluene Concentration $\mathbf{z}_i$ (mol/L)	Improvement Over Previous Batch (%)
1	$[1.0000, 0.0000]^T$	0.06189	N/A
2	$[0.5346, 0.0261]^T$	0.06547	5.80
3	$[0.4829, 0.0357]^T$	0.06994	6.82
4	$[0.4691, 0.0393]^T$	0.07186	2.74
5	$[0.4124, 0.0456]^T$	0.07372	2.60
6	$[0.3907, 0.0486]^T$	0.07479	1.44
7	$[0.3520, 0.0528]^T$	0.07576	1.31
8	$[0.3131, 0.0569]^T$	0.07659	1.10
9	$[0.2852, 0.0600]^T$	0.07720	0.79
10	$[0.2406, 0.0645]^T$	0.07768	0.63

experiments in an iterative fashion. In later chapters, we will produce the AR using continuous reactive equipment.

> **IMPORTANT: The AR**
>
> The AR is the set of all possible states achieved by mixing and reaction for a specified feed point and reaction kinetics.

For the BTX kinetics discussed earlier, the *candidate AR* is defined by the boundary FHO in Figure 3.7. The true AR is obtained in the limit of *infinitely* many batches. All concentrations within or on this boundary are achievable with the appropriate mixing experiments. Conversely, all points that lie outside the region are not achievable, regardless of what kind of reaction is carried out. This region is specific to the feed point provided, in that the AR given in Figure 3.7 is associated with feed point $\mathbf{C}_f = [1, 0]^T$. If the feed point were to change, a different AR would be produced, representing a different set of (attainable) concentrations.

> **CONCEPT: Candidate ARs and the true AR**
>
> The regions produced in Section 3.2.4 might be considered examples of *an* AR. These are not, however, *the* AR. The true AR is the *entire* set of achievable concentrations. In this example, it is only attained in the limit of performing infinitely many iterations.
>
> The regions prior to this state may also be considered ARs in their own respect, for these regions also represent a subset of the true AR. The regions produced in Section 3.2.4 are better viewed as candidates of the true AR, which are simply referred to as *candidate ARs*.

In addition to finding the AR for the BTX system, a set of steps required to generate the region have also been determined. Currently, the amount of work required to achieve this region is still large and thus prohibitive to implement in reality. In Chapter 4, we shall begin to describe how different continuous reactor types may be employed to obtain the same result that is simpler to implement in practice.

> **ILLUSTRATION: Comparing results to previous recommendations**
>
> In Chapter 1, we described how a CSTR, starting from the feed point, achieves a higher toluene concentration than a similarly sized PFR (or batch reactor) for the BTX system. Let us compare this result to that obtained in this chapter.
>
> In Figure 3.8, the CSTR concentrations, for the same feed point, are overlaid onto the AR for comparison. Observe that the maximum toluene concentration achieved by successive batches (the batch game) is larger than that obtained in the CSTR. In particular, the maximum toluene concentration achieved by the CSTR at point I is approximately 0.0742 mol/L, whereas that achieved by the batch game at point H is approximately 0.078 mol/L.
>
> 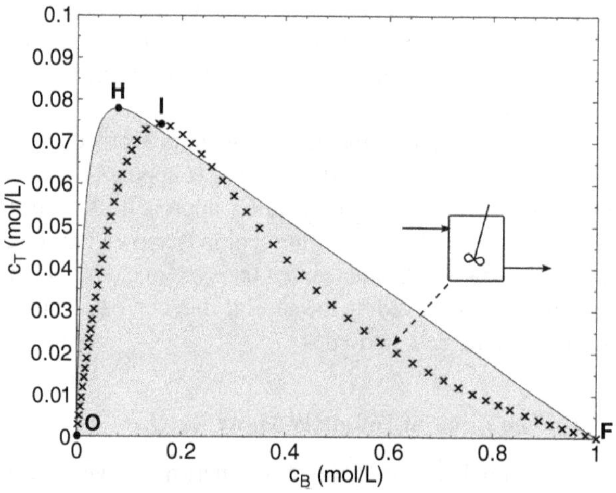
>
>
> **Figure 3.8** A comparison between the maximum toluene concentration achieved via successive batches (point H) to that produced by the optimal CSTR recommendation from Chapter 1 (point I).

3.4 ELEMENTARY PROPERTIES OF THE AR

Now that the concept of the AR has been introduced, a number of basic, yet important, properties of the AR can be described. These properties assist us in identifying when the AR has been found. We shall supplement this list with additional properties when additional AR concepts have been described in later chapters. Many of these properties are adapted from the papers Feinberg and Hildebrandt (1997) and Hildebrandt et al. (1990).

Property 1: The AR is convex From a geometric perspective, mixing allows for regions to be “filled” in state space. If a particular set of attainable points is generated, and the set displays concavities, then it can be made convex by mixing. Identifying the convex hull for the set of points allows us to identify which points may be used for mixing, which converts a distinct set of points into a filled region in concentration space. Figure 3.9 illustrates the familiar picture of a non-convex region being transformed into a convex region

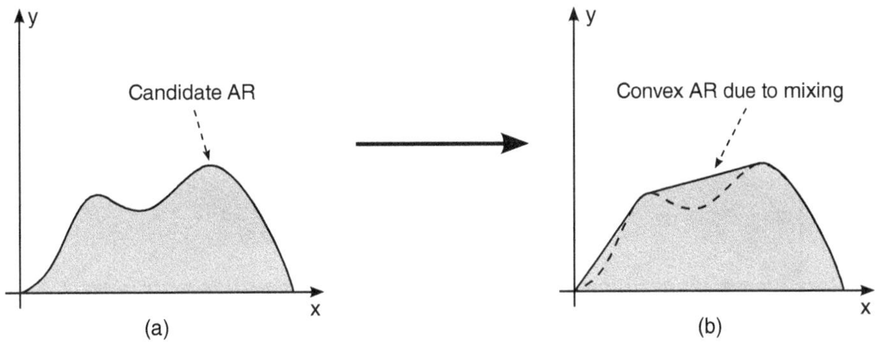

Figure 3.9 Property 1. The AR is convex. (a) A nonconvex region and (b) a convex region generated via mixing operations.

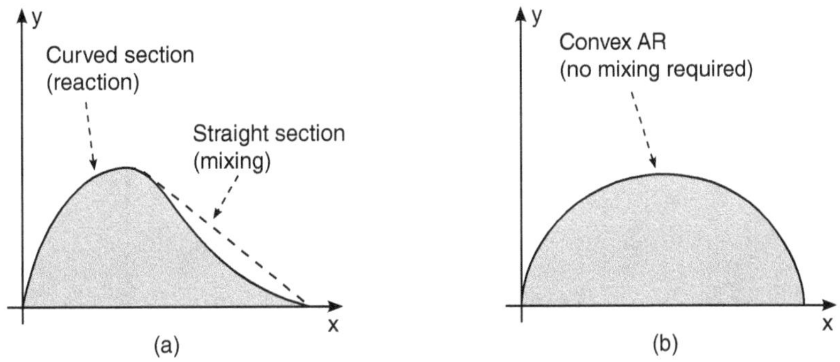

Figure 3.10 Property 2. (a) The AR is generally composed of straight (mixing) and curved (reaction) lines. (b) Sometimes, when the profile is already convex, mixing is not required. Hence, there is no requirement that the ARs contain both reaction and mixing sections.

via mixing. Note that the convex shape of the resulting region arises from the property that mixing is a linear process. For concentration, the linearity of mixing is dependent on the constant density assumption.

Property 2: The AR is composed of mixing and reaction sections We can picture the surface of the AR as being composed of sections that belong to different types of processes. Mixing processes are linear, whereas reaction is generally nonlinear. Flat sections of the AR surface therefore belong to mixing processes, whilst curved sections belong to reactive processes. Concave sections belonging to reaction can always be made convex via mixing, and therefore the boundary of the AR can only be composed of flat or curved sections that are also convex.

Furthermore, there are many systems where reaction generates an entirely convex region from the feed. In this instance, mixing will not produce new concentrations, for the convex hull of the set is identical to the original set of points. An example of this phenomenon is displayed in Figure 3.10(b).

Property 3: The AR must contain the feed By definition, the AR must contain the feed point, for the feed is attainable. If many feed points are present, then the AR must be one that is generated by the convex hull of all available feed points.

Consider then the illustrations in Figure 3.11, wherein two ARs are shown. The AR in Figure 3.11(a) is generated for a single feed point, whereas the AR in Figure 3.11(b) belongs to three available feeds.

When multiple feeds are available, it is possible to generate an AR associated with each point. Since each region is achievable, mixing allows for the union of all regions, into a single convex region. Some portions of the individual ARs may be common and overlap, whereas other portions serve to expand the region in a manner that would not otherwise be possible by a single feed. The combined AR from all feed points cannot be smaller than the individual ARs corresponding to each feed point.

Property 4: The AR is unique, it is simply connected, and it exists as a single region[3] Assume that the AR is not unique. Suppose that for a given feed point, it is possible to generate a number of regions that are believed to be the AR. Each of these regions must be instances of regions

[3]By simply connected, we mean any closed loop in the region can be reduced (squeezed) to a single point. Informally, any region containing "holes"—such as a doughnut—is not simply connected, for it cannot be reduced to a single point. Multiple filled regions are also not considered simply connected as a whole.

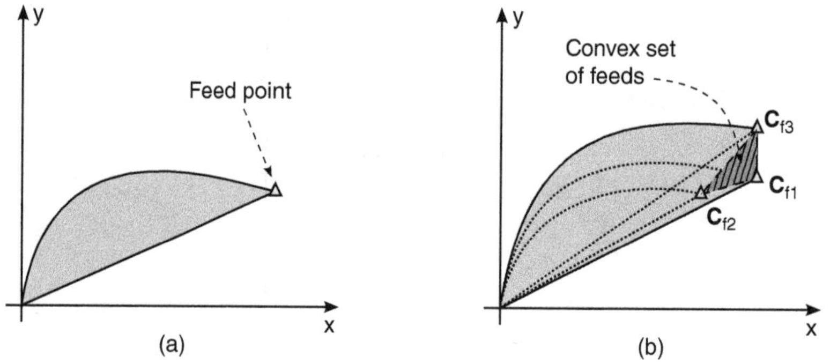

Figure 3.11 Property 3. The AR must contain the feed. (a) An AR containing a single feed point and (b) an AR associated with multiple feeds containing all feed points.

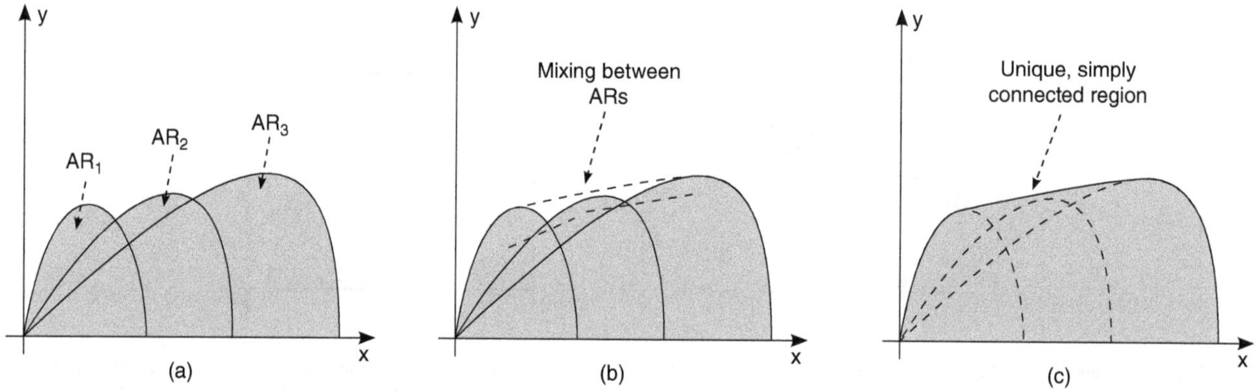

Figure 3.12 Property 4. The AR is unique, simply connected, and a single region. (a) Three separate ARs AR_1, AR_2, and AR_3, (b) mixing between the three regions, and (c) the final (single) region obtained contains all regions and is simply connected.

that are achievable. If each region is achievable, then it is also feasible to join the separate regions by mixing, which produces a larger, single, region in space. This process may be repeated for as many unique regions generated, producing one unique region. It is therefore always possible to produce a single region from many smaller regions using mixing. The AR therefore must be unique, simply connected, and exist as a single region.

In Figure 3.12, this property is illustrated with three candidate regions (AR_1, AR_2, and AR_3). If it is believed that each region is attainable, then they may be connected via mixing. The result is a single region encompassing the entire set of all three candidate regions.

Property 5: The AR exists as a subspace of $\mathbb{R}^n$ The number of components participating in a system determines the dimension of the space that the region is generated in. The AR must therefore also reside in this space.

Consider the reaction $A \rightarrow B$. If 1 mole of pure A is converted into B, then all possible species amounts are bounded between 0 and 1 mole. If the reaction is instead given by $A \rightarrow B + C$, then the total amount of A is distributed amongst B and C, and the possible set of species amounts is constrained further. Therefore, mass balance restrictions, based on the reaction stoichiometry and the amount of feed, provide rules for what the maximum and minimum quantities of each component must be in the reaction.[4]

Whereas concentration space is open—its bounds are $[0; \infty)$—reactions on a given feed are necessarily bounded by mass balance. The AR must therefore be contained within a subset of concentrations in $\mathbb{R}^n$, and although the AR is often thought to exist as an infinite region in state space, it is in fact finite and well defined, as shown in Figure 3.13.

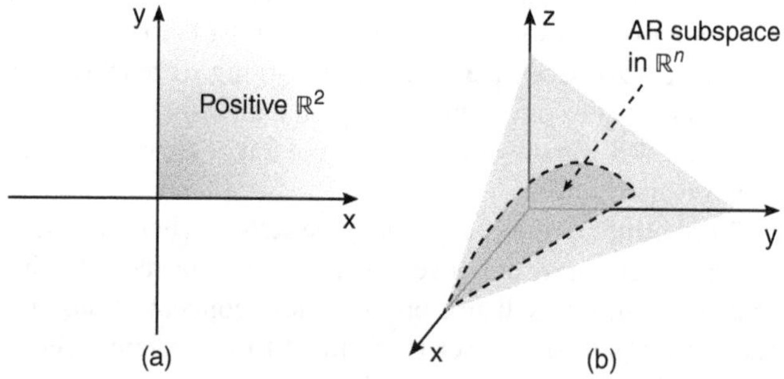

Figure 3.13 Property 5. The AR is a constrained subset of $\mathbb{R}^n$. Mass balance restricts systems from achieving all concentrations in the range $[0, \infty)$, such as that shown (a) for a system in $\mathbb{R}^2$, or similarly for (b) representing a constrained region in $\mathbb{R}^3$.

[4] Aspects of dimension are discussed in Chapter 6.

3.5 CHAPTER REVIEW

In this chapter, we focused on applying the graphical properties of concentration and mixing, initially developed in Chapter 2, to improve the BTX system from Chapter 1. The manner in which improvements are obtained is framed in the form of a game, with two simple restrictions: reactions occur in batch reactors and operated one at a time, and only mixing with previously achievable concentrations is allowed. When these restrictions are enforced, the maximum toluene concentration may be improved in an iterative manner.

Repeated application of this game leads us to a state where no further improvements to the maximum toluene concentration are possible, and in this way, we discovered the AR for the BTX system in two dimensions. The AR is a hence a region in concentration space that represents all possible achievable concentrations for the BTX system, and all processes involving reaction and mixing—even those that we cannot imagine—must lie in or on the AR.

Properties of the AR were also discussed in this chapter. These properties help us to identify when a candidate AR has been found, and many of these properties come about naturally from the geometric nature of mixing.

In Chapter 4, we shall expand our knowledge of the geometric viewpoint to include reaction, and describe the function of three fundamental reactor types that are used in AR theory. With this knowledge, we will be in a position to generate candidate ARs efficiently, and use this information to form optimal reactor networks, which may be applied to optimize complex reactive systems.

REFERENCES

Feinberg, M., Hildebrandt, D., 1997. Optimal reactor design from a geometric viewpoint—I. Universal properties of the attainable region. Chem. Eng. Sci. 52, 1637–1665.

Hildebrandt, D., Glasser, D., Crowe, C.M., 1990. Geometry of the attainable region generated by reaction and mixing: With and without constraints. Ind. Eng. Chem. Res. 29, 49–58.

4

REACTION

4.1 INTRODUCTION

In this chapter, we wish to describe how reaction may be viewed as a geometric process, similar to that of concentration and mixing. This viewpoint will allow us to describe three important reactor models used in AR theory. Importantly, this theory will also assist us in transitioning our early ideas from batch reaction to continuous operation.[1]

4.2 REACTION RATES AND STOICHIOMETRY

Consider again the BTX system:

$$B + \frac{1}{2}E \overset{k_1}{\rightarrow} T \qquad \text{reaction 1}$$

$$T + \frac{1}{2}E \overset{k_2}{\rightarrow} X \qquad \text{reaction 2}$$

$$2B \overset{k_3}{\rightarrow} D + H \qquad \text{reaction 3}$$

It is assumed that the species reaction rates obey mass action kinetics. Let us look at the specific rate expressions obtained for each component in the system in order to better understand the behavior of the kinetics.

4.2.1 Benzene Reaction Rate

Benzene is a reactant in the BTX system. There are two possible avenues for how it may be converted into a product. The first occurs when benzene reacts with ethylene to form toluene by reaction 1. The rate of benzene conversion in this reaction is thus written with respect to both reactants, which is given by

$$r_{B1} = -k_1 c_B c_E^{0.5}$$

[1]In Chapter 7, we describe how AR theory may be applied to batch reactors.

Note that the term is negative indicating that benzene is consumed. k_1 is the rate constant for reaction 1 and c_B and c_E are the concentrations of benzene and ethylene, respectively. The subscript number 1, associated with r_{B1}, signifies that this expression relates to reaction 1. Note there is no special significance to the form of the rate expression given here. The rate of reaction for benzene might assume a number of different forms. By mass balance, one mole of benzene and half a mole of ethylene react to form one mole of toluene.

It is also possible for benzene to simultaneously decompose, in an autocatalytic reaction, given by reaction 3. For every two moles of benzene that react by reaction 3, one mole of diphenyl and hydrogen are produced. The rate of reaction for benzene is again expressed in terms of the reactants that participate in this reaction. In this instance, benzene is the sole reactant, and the following rate expression for this reaction is obtained as a result:

$$r_{B3} = -2k_3 c_B^2$$

Compared to reaction 1, reaction 3 is second order overall.

> **RECALL: Reaction order**
>
> The overall reaction order may be obtained by summing the exponents in the rate expression. Reaction 1 has order of 1.5, whereas reaction 3 is second order overall.

The *overall* reaction rate for benzene in the system is determined by summing the individual reaction rates for benzene:

$$r_B = r_{B1} + r_{B3}$$

Substitution of the expressions for r_{B1} and r_{B3} produces

$$r_B = -k_1 c_B c_E^{0.5} - 2k_3 c_B^2$$

Attainable Region Theory: An Introduction to Choosing an Optimal Reactor, First Edition. David Ming, David Glasser, Diane Hildebrandt, Benjamin Glasser, and Matthew Metzger.

Companion website: http://attainableregions.com

The overall rate of reaction for benzene in the BTX system is hence composed of two terms, one for each reaction in which benzene is a participant.[2]

4.2.2 Toluene Reaction Rate

The rate expression for toluene is determined in a similar manner as that for benzene. The concentration of toluene in the system is a function of two competing reactions: one where toluene is a product, and the other where toluene is a reactant. A one-to-one molar relation exists between the consumption of benzene and the production of toluene in reaction 1, and in general, the reaction rates for all species that participate in reaction 1 may be determined simply by stoichiometry, where the rate of formation for toluene must be equal to the rate of reaction for benzene by mass balance.[3]

$$\begin{aligned} r_{T1} &= -r_{B1} \\ &= k_1 c_B c_E^{0.5} \end{aligned}$$

This expression is the negative complement of the benzene reaction rate for reaction 1.

RECALL: Species reaction rates using stoichiometry

If a rate of reaction is known for a particular component in a reaction, then the rates of reaction for all other species present in the reaction are also known by stoichiometry. If this were not the case, then the rate of a particular species would outbalance the rate of another and violate the conservation of mass in the reactor. Consider the following hypothetical reaction:

$$v_1 A + v_2 B \rightarrow v_3 C + v_4 D$$

The reaction rates for components B, C, and D may be written with respect to component A as follows:

$$r_B = \frac{v_2}{v_1} r_A$$

$$r_C = \frac{v_3}{v_1} r_A$$

$$r_D = \frac{v_4}{v_1} r_A$$

Ethylene reacts in the presence of benzene in a 2 : 1 benzene to ethylene ratio. The rate of ethylene consumption must be half the rate of benzene conversion in reaction 1. The rate of reaction for ethylene in reaction 1 is then expressed in following form:

$$\begin{aligned} r_{E1} &= \frac{1}{2} r_{B1} \\ &= -\frac{1}{2} k_1 c_B c_E^{0.5} \end{aligned}$$

Ethylene may also combine with toluene by reaction 2. An additional half mole of ethylene may be reacted with toluene to form xylene. The form of this expression is specified as follows:

$$r_{T2} = -k_2 c_T c_E^{0.5}$$

Similar to that of benzene, the overall rate of reaction for toluene is found by the summation of the individual reaction rate contributions as follows:

$$\begin{aligned} r_T &= r_{T1} + r_{T2} \\ &= k_1 c_B c_E^{0.5} - k_2 c_T c_E^{0.5} \end{aligned}$$

Note that this expression differs to the rate expression for benzene. The first term is positive, indicating that toluene is formed by reaction 1. The second term is negative, indicating a consumption of toluene to form xylene by reaction 2. Note that the overall rate of reaction for toluene is dependent on both the concentration of benzene, ethylene *and* the concentration of toluene itself.

EXAMPLE 1: Remaining BTX component reaction rates

Using the rate expressions for benzene and toluene, determine the remaining rate expressions for ethylene, xylene, hydrogen, and diphenyl in the BTX system.

Answer:

The remaining species reaction rates are easily found by stoichiometry. Ethylene is consumed in both reactions 1 and 2. Hence,

$$\begin{aligned} r_E &= \frac{1}{2} r_{B1} + \frac{1}{2} r_{T2} \\ &= -\frac{1}{2} k_1 c_B c_E^{0.5} - \frac{1}{2} k_2 c_T c_E^{0.5} \end{aligned}$$

Xylene is a product of reaction 2 only, giving

$$\begin{aligned} r_X &= r_{T2} \\ &= k_2 c_T c_E^{0.5} \end{aligned}$$

Both hydrogen and diphenyl appear as products in reaction 3 only, giving

[2] It is perhaps more correct to include a r_{B2} term in the overall rate of reaction for benzene; however it is clear that there is no benzene contribution in reaction 2.

[3] See chapter 6 in Fogler (2010).

$$r_H = r_D = -\frac{1}{2}r_{B3}$$
$$= k_3 c_B^2$$

EXAMPLE 2: Rate expressions from reaction stoichiometry

Consider the following reaction:

$$0.613A \xrightarrow{k_1} B + 1.44C$$

$$\alpha A \xrightarrow{k_2} \frac{7}{11}D$$

Here, α is a stoichiometric coefficient for A in reaction 2. Reactions involving non-integer stoichiometric coefficients are common in biological reactions. The overall rate expression for A, r_A, is given by the peculiar kinetics (which has been fabricated for demonstration) as follows:

$$r_A = -k_1 \exp(c_A).\sin(c_B^2) - k_2\frac{\log(c_A)}{c_D^2 + 1}$$

Determine the rate expressions for components C and D.

Answer:

Rate expressions for C and D can be found using the overall rate expression for A and the reaction stoichiometry. To simplify the discussion, let us rewrite r_A as follows:

$$r_A = -f_1 - f_2$$

where

$$f_1 = k_1 \exp(c_A).\sin(c_B^2)$$
$$f_2 = k_2\frac{\log(c_A)}{c_D^2 + 1}$$

f_1 and f_2 correspond to the rate expressions for A in reactions 1 and 2, respectively. r_C and r_D are then easily expressed in terms of the appropriate component stoichiometric coefficients, f_1 and f_2. By mass balance, the following relations hold:

$$r_C = \frac{1.44}{0.613}f_1$$
$$r_D = \frac{7}{11\alpha}f_2$$

Substituting the expressions for f_1 and f_2 then gives the desired rate equations for components C and D as follows:

$$r_C = \frac{1.44}{0.613}k_1 \exp(c_A).\sin(c_B^2)$$
$$r_D = \frac{7}{11\alpha}k_2\frac{\log(c_A)}{c_D^2 + 1}$$

SIDE NOTE: Unsuitability of extent of reaction in AR theory

The extent of reaction, ε, is a commonly used quantity in chemical reaction engineering that expresses the consumption or formation of components within a reaction. For those familiar with the theory of the extent of reaction, it may be more natural to express the concentrations of species in terms of extents and an initial feed concentration, rather than to rely on the absolute concentration of a particular species in the system. Although this may simplify matters slightly, extent of reaction is not commonly used in AR theory for two chief reasons:

1. Extents can take on negative values: Extent of reaction is positive for products and negative for reactants. Concentrations, by comparison, may never take on negative values. This property provides a convenient set of bounds for all species present in a mixture. If concentration is used over extent for a given reactive system, the associated AR for the system must then also always lie in the positive orthant of concentration space.[4] Additionally, using reaction stoichiometry and mass balance constraints, it is possible to compute a definite bound (called the stoichiometric subspace) that the AR must reside in. Similar bounds are rather less well-defined when extent is employed, and thus greater care must be taken.
2. Mixing is not well-defined: Mixtures cannot be uniquely expressed in terms of extent. Extent is defined with respect to a reaction, whereas concentration may be formed in a number of different ways. Different combinations of extents may all result in the same final mixture concentration. The use of concentration is thus more convenient for viewing mixing geometrically.

Hence, extent is not a suitable variable for use in AR theory, and thus it will not be discussed in this work.

[4]For two-dimensional systems, this is the positive region in the top right-hand section of a graph: where x and y can take on values in the range [0 ∞).

4.3 REACTION FROM A GEOMETRIC VIEWPOINT

4.3.1 The Rate Vector

4.3.1.1 Description In the same manner that one might define a mixture concentration compactly as a vector of species concentrations, it is also possible to form a vector, of equal dimension to the concentration vector, associated with the rate of reaction. If benzene and toluene are the components of interest to the problem, the concentration vector is $\mathbf{C} = [c_B, c_T]^T$. It is natural to express the corresponding rates of reaction for benzene and toluene as a two-component column vector as well. The *rate vector*, $\mathbf{r}(\mathbf{C})$, in c_B–c_T space is hence defined as follows:

$$\mathbf{r}(\mathbf{C}) = \begin{bmatrix} r_B(\mathbf{C}) \\ r_T(\mathbf{C}) \end{bmatrix}$$

Here, r_B and r_T are the rate expressions belonging to benzene and toluene, respectively. The rate vector $\mathbf{r}(\mathbf{C})$ is distinguished from the overall reaction rate for a particular species i, r_i, by boldface type. Note that $\mathbf{r}(\mathbf{C})$, as opposed to r_i, is a vector containing two elements. The size of $\mathbf{r}(\mathbf{C})$ is determined by the number of species rate expressions supplied. This is analogous to the concentration vector $\mathbf{C}$, where the size of $\mathbf{C}$ is determined by the number of components in concentration space. Whereas r_i is a scalar valued function of the concentration vector $\mathbf{C}$, $\mathbf{r}(\mathbf{C})$ is a vector valued function of the concentration vector $\mathbf{C}$. We have been careful to indicate that the vector $\mathbf{r}(\mathbf{C})$ is a function of $\mathbf{C}$. This is important as the rate expressions for each component in the system is a function of the local composition given at $\mathbf{C}$. To see this, observe that

$$r_B = r_B(c_B, c_E)$$
$$r_T = r_T(c_B, c_E, c_T)$$

For the species identified, the rate vector $\mathbf{r}(\mathbf{C})$ depends on only benzene, ethylene, and toluene. In general, species rate functions (r_i) may be a function of all components in the system, and thus the rate vector is also a function of all components in the system. For an n-component system, the rate vector is

$$\mathbf{r}(\mathbf{C}) = \begin{bmatrix} r_1(\mathbf{C}) \\ r_2(\mathbf{C}) \\ \vdots \\ r_n(\mathbf{C}) \end{bmatrix}$$

4.3.1.2 Evaluating Rate Vectors Let us draw from some familiar properties of vectors discussed in Chapter 2. Values for benzene, ethylene, and toluene may be substituted into the rate expressions for each component, and a unique vector, with a definite magnitude and direction, may be computed. For example, a mixture equivalent to the feed point from Chapter 1, containing 1.0 mol/L benzene, 0.5 mol/L ethylene, and 0.0 mol/L toluene, results in the following rate vector calculation for benzene and toluene:

$$\begin{aligned}\mathbf{r}(\mathbf{C}) &= \begin{bmatrix} r_B(c_B) \\ r_T(c_B, c_T) \end{bmatrix} \\ &= \begin{bmatrix} -k_1 c_B c_E^{0.5} - 2k_3 c_B^2 \\ k_1 c_B c_E^{0.5} - k_2 c_T c_E^{0.5} \end{bmatrix} \\ &= \begin{bmatrix} -k_1(1.0)(0.5^{0.5}) - 2k_3(1.0^2) \\ k_1(1.0)(0.5^{0.5}) - k_2(0.0)(0.5^{0.5}) \end{bmatrix}\end{aligned}$$

If values of the rate constants are known, then the numerical values of the components in $\mathbf{r}(\mathbf{C})$ are easily found. Assuming $k_1 = 1.0\ L^{0.5}/(mol^{0.5}.h)$, $k_2 = 1.0\ L^{0.5}/(mol^{0.5}.h)$, and $k_3 = 10.0\ L/(mol.h)$, this gives

$$\begin{aligned}\mathbf{r}(\mathbf{C}) &= \begin{bmatrix} -(1.0)(1.0)(0.5^{0.5}) - 2(10.0)(1.0^2) \\ (1.0)(1.0)(0.5^{0.5}) - (1.0)(0.0)(0.5^{0.5}) \end{bmatrix} \\ &= \begin{bmatrix} -20.707 \\ 0.707 \end{bmatrix} \text{mol/(L.h)}\end{aligned}$$

As a result, for the concentration vector $\mathbf{C} = [c_B, c_E, c_T]^T = [1.0, 0.5, 0.0]^T$ mol/L, the associated rate vector is $\mathbf{r}(\mathbf{C}) = [-20.707, 0.707]^T$ mol/(L.h).[5]

EXAMPLE 3: Calculating r(C) from rate expressions and C

Suppose that a mixture is formed with the following concentration vector for the BTX system:

$$\begin{aligned}\mathbf{C} &= [c_B, c_E, c_T, c_X, c_H, c_D]^T \\ &= [0.12, 0.35, 0.65, 0.07, 0.95, 1.32]^T\end{aligned}$$

If the rate expressions given in Section 4.2 can be used, compute the rate vector $\mathbf{r}(\mathbf{C})$.

Answer:

Values from the supplied concentration vector may be substituted into the species rate expressions belonging to each component, which are repeated as follows:

[5] Note that an ethylene concentration must still be specified in order to compute the rate vector, even though it is not displayed in c_B–c_T space.

Component	Rate Expression
Benzene (r_B)	$-k_1c_Bc_E^{0.5} - 2k_3c_B^2$
Ethylene (r_E)	$-0.5k_1c_Bc_E^{0.5} - 0.5k_2c_Tc_E^{0.5}$
Toluene (r_T)	$k_1c_Bc_E^{0.5} - k_2c_Tc_E^{0.5}$
Xylene (r_X)	$k_2c_Tc_E^{0.5}$
Diphenyl (r_D) = Hydrogen (r_H)	$k_3c_B^2$

Here, $k_1 = 1.0\,L^{0.5}/(mol^{0.5}.h)$, $k_2 = 1.0\,L^{0.5}/(mol^{0.5}.h)$, and $k_3 = 10.0\,L/(mol.h)$. Substituting in values for $c_B = 0.12\,mol/L$, $c_E = 0.35\,mol/L$, and $c_T = 0.65\,mol/L$ results in the following rate vector:

$$\mathbf{r}(\mathbf{C}) = [-0.359, -0.228, -0.314, 0.385, 0.144, 0.144]^T$$

Observe that $\mathbf{r}(\mathbf{C})$ for the BTX system depends only on the concentrations of benzene, toluene, and ethylene. Had the values of c_X, c_D, or c_H been different to those supplied, the rate vector would have been the same.

SIDE NOTE: One-to-one relation between concentration and reaction rate

Observe that there is a one-to-one relation between the concentration vector **C** and the resulting rate vector **r**(**C**). For every unique **C** supplied as an argument in the rate expression, a single rate vector **r**(**C**) may be generated. However, there are certain instances where it might appear as though multiple rate vectors exist at a given concentration, such as when kinetics involving temperature or pressure is used.

*4.3.1.3 **Visualizing Rate Vectors*** Rate vectors are plotted at the point of evaluation. Since **r**(**C**) depends on **C**, each coordinate in concentration space can be assigned a unique vector given by **r**(**C**). This assignment is useful as rate vectors may then be incorporated with concentration and mixing vectors, where all vectors can be visualized on a single plot in concentration space.

EXAMPLE 4: Plotting rate vectors

Table 4.1 gives a list of arbitrary concentration vectors and associated rate vectors for the BTX system in terms of benzene, ethylene, and toluene ($\mathbf{C} = [c_B, c_E, c_T]^T$ and $\mathbf{r}(\mathbf{C}) = [r_B(\mathbf{C}), r_E(\mathbf{C}), r_T(\mathbf{C})]^T$).

TABLE 4.1 Four Arbitrary (Possibly Feasible) Concentrations and Associated Rate Vectors

No.	Concentration Vector $\mathbf{C} = [c_B, c_E, c_T]^T$	Rate Vector $\mathbf{r}(\mathbf{C}) = [r_B, r_E, r_T]^T$
1	$[1.0, 0.5, 0.00]^T$	$[-20.707, -0.354, 0.707]^T$
2	$[0.2, 0.5, 0.09]^T$	$[-0.941, -0.103, 0.078]^T$
3	$[0.2, 0.5, 0.02]^T$	$[-0.941, -0.078, 0.127]^T$
4	$[0.8, 0.5, 0.09]^T$	$[-13.366, -0.315, 0.502]^T$
5	$[0.5, 0.5, 0.05]^T$	$[-5.354, -0.195, 0.318]^T$

Using the supplied information, answer the following questions:

(a) Plot the rate vectors in c_B–c_T space.
(b) Plot the rate vectors in c_B–c_E space.

To make the visualization easier, display all vectors as the same length.

Answer a:
Observe that each rate vector has a unique direction and magnitude corresponding to the concentration vector of interest. Since we wish to plot the rate vectors in c_B–c_T space, the components belonging to benzene and toluene are used and values for ethylene are ignored.

No.	Concentration Vector $C = [c_B, c_E, c_T]^T$	Rate Vector $r(C) = [r_B, r_T]^T$
1	$[1.0, 0.5, 0.00]^T$	$[-20.707, 0.707]^T$
2	$[0.2, 0.5, 0.09]^T$	$[-0.941, 0.078]^T$
3	$[0.2, 0.5, 0.02]^T$	$[-0.941, 0.127]^T$
4	$[0.8, 0.5, 0.09]^T$	$[-13.366, 0.502]^T$
5	$[0.5, 0.5, 0.05]^T$	$[-5.354, 0.318]^T$

The results are shown in Figure 4.1(a). All rate vectors have been normalized to unit length by dividing each rate vector by its corresponding norm, $||\mathbf{r}_i(\mathbf{C})||$.

Answer b:

Plotting rate vectors in c_B–c_E space is similar to answer (a), but requires the use of benzene and ethylene instead. The components associated with this plot are as follows:

No.	Concentration Vector $C = [c_B, c_E, c_T]^T$	Rate Vector $r(C) = [r_B, r_E]^T$
1	$[1.0, 0.5, 0.00]^T$	$[-20.707, -0.354]^T$
2	$[0.2, 0.5, 0.09]^T$	$[-0.941, -0.103]^T$
3	$[0.2, 0.5, 0.02]^T$	$[-0.941, -0.078]^T$
4	$[0.8, 0.5, 0.09]^T$	$[-13.366, -0.315]^T$
5	$[0.5, 0.5, 0.05]^T$	$[-5.354, -0.195]^T$

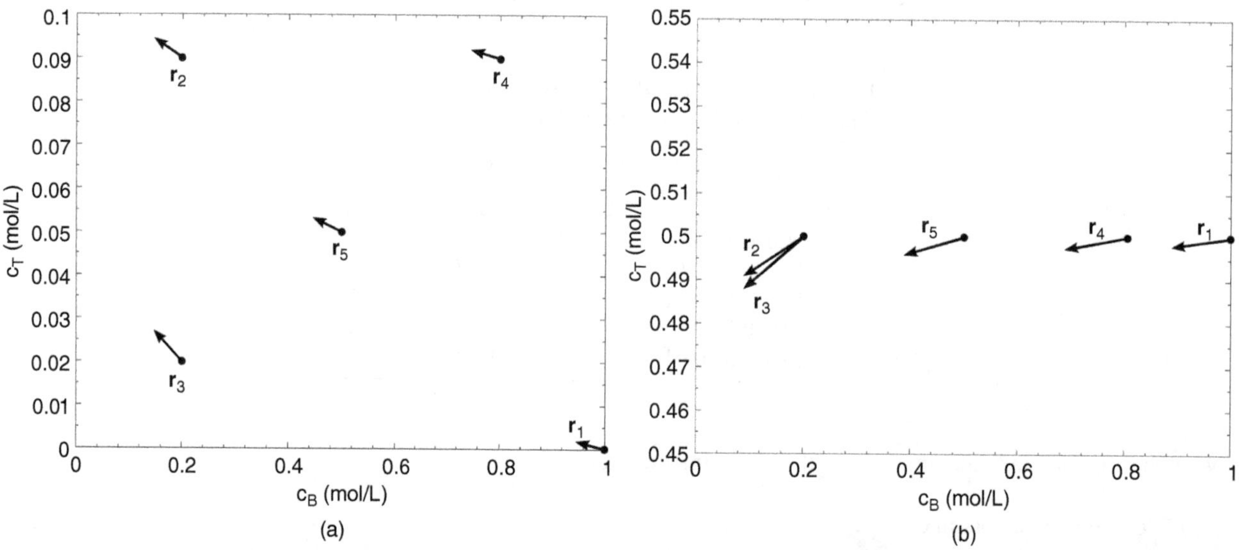

Figure 4.1 (a) Rate vectors in c_B–c_T space and (b) rate vectors in c_B–c_E space.

The corresponding rate vector plot is then shown in Figure 4.1(b).

Note that all of the rate vectors in c_B–c_E space lie on the line $c_E = 0.5$ mol/L.

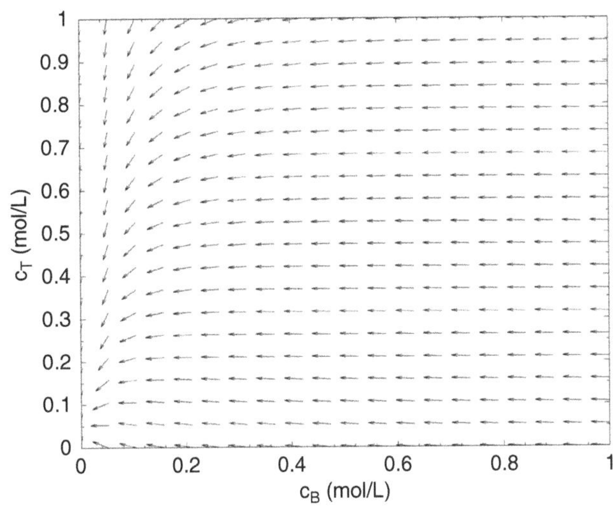

Figure 4.2 Vector field for the BTX system in c_B–c_T space.

4.3.2 Rate Fields

4.3.2.1 Visualizing Vector Fields It is always possible to compute rate vectors in concentration space, irrespective of whether the concentration **C** is achievable or not. (The only situation when this calculation is not feasible is when **r**(**C**) is undefined at **C**.) Since **r**(**C**) is a vector valued function of a vector **C**, it is possible to associate each point $\mathbf{C}_i$ in concentration space with a rate vector $\mathbf{r}_i(\mathbf{C}_i)$. Hence, a system of rate expressions generates a *vector field*. In AR theory, this is termed simply as the *rate field*. The rate field associated with the BTX system is shown in Figure 4.2.

Figure 4.2 shows only the rate vectors associated with positive concentrations for the BTX kinetics as negative concentrations are not physically meaningful. The vectors given in Figure 4.2 are arranged on an evenly spaced 20×20 grid for the range $0.0\ \text{mol/L} \le c_B \le 1.0\ \text{mol/L}$ and $0.0\ \text{mol/L} \le c_T \le 1.0\ \text{mol/L}$.[6] When rate functions are plotted in this manner, the natural shape of the kinetics is easily visualized. Although it is only feasible to visualize rate fields in two and three dimensions, the concept of the rate field is valid for all dimensions.

The 20×20 grid sizing for the rate field generated earlier has been chosen arbitrarily. Although a greater resolution is obtained with more points, there are situations when this hinders visualization of the field. When a large number of grid points are chosen, the magnitude of the vectors is large in relation to the number of points in the grid. In this instance, it is appropriate to scale all vectors on the grid to standard unit length. This is done by dividing all components of the rate vector by its associated magnitude.

$$\hat{\mathbf{r}} = \frac{\mathbf{r}}{\|\mathbf{r}\|}$$

Here, $\hat{\mathbf{r}}$ is the normalized rate vector (normalized to unit magnitude). Note that when scaling is performed, information regarding the magnitude of **r** is lost, which may be important in understanding how large (how fast) reaction rates are relative to each other in space.

[6] In Chapter 3, the y-axis scale ranged between 0.0 and 0.1 mol/L. A larger range is shown here to provide a larger view of the vector field for the BTX system.

EXAMPLE 5: Plotting vector fields

Consider the following two reactions involving components X, Y, and Z:

$$X \rightarrow Y \rightarrow Z$$

For brevity, let $x = c_X$, $y = c_Y$, and $z = c_Z$. The kinetics associated with this system is assumed to be artificial, given by the following rate expressions for components X and Y:

$$r_X(x) = xf(x)$$

$$r_Y(x, y) = f(x)^2 + xy\frac{df(x)}{dx}$$

$$r_Z(x, y) = -xf(x) - f(x)^2 - xy\frac{df(x)}{dx}$$

Here, f(x) and its derivative with respect to x are

$$f(x) = 6x^6 - 6x^5 + 9x^4 - 16x^3 + 9x^2 - 2x$$

$$\frac{df(x)}{dx} = 36x^5 - 30x^4 + 36x^3 - 48x^2 + 18x - 2$$

Answer the following questions:

(a) Plot the rate field for this system in x–y on an evenly spaced 15×15 grid in the range $0\ \text{mol/L} \le c_X \le 1\ \text{mol/L}$ and $0\ \text{mol/L} \le c_Y \le 1\ \text{mol/L}$.

(b) Plot the rate field for the same system in x–z space on an evenly spaced 15×15 grid in the range $0\ \text{mol/L} \le c_X \le 1\ \text{mol/L}$ and $0\ \text{mol/L} \le c_Z \le 1\ \text{mol/L}$. Assume that a feed point of $\mathbf{C}_f = [x_f, y_f, z_f] = [1, 0, 0]^T$ mol/L is available.

Answer a: Vector field in x–y space

We are interested in plotting the rate field in x–y space. The concentration vector is hence defined by $\mathbf{C} = [x, y]^T$. Since the kinetics depends on x and y only, a grid of points in x–y space may be generated over the desired range; the x–y pair at each grid point may then be substituted into the kinetics directly. For example, for each point on the grid, say, $\mathbf{C} = [1, 1]^T$, find f(x) and df(x)/dy:

$$f(1) = 0$$

$$\frac{df(1)}{dx} = 10$$

which may then be used to find $r_X(x, y)$ and $r_Y(x, y)$:

$$r_X(1, 1) = 0$$

$$r_Y(1, 1) = 10$$

This process is repeated for each point on the grid. Once rate vectors are computed at each point and plotted, Figure 4.3(a), for a 15 × 15 grid, is generated.

The vectors are scaled to equal magnitude in order to only indicate the direction of the vectors (the direction of concentration movement).

Answer b: Vector field in x–z space

Plotting the rate field in x–z space is slightly more complicated than that in x–y space. Since the kinetics is not influenced by the concentration of component Z directly, plotting the rate field in x–z space requires that the corresponding concentration is first computed for the x–y pair, and then back calculated by mass balance to find the x–z pair. We can achieve this transformation by writing a steady-state component molar balance in terms of the extent of reaction and the feed point as follows:

$$x = x_f - \varepsilon_1$$

$$y = y_f + \varepsilon_1 - \varepsilon_2$$

$$z = z_f + \varepsilon_2$$

Generating a grid of point in x–z space implies that the values of x and z are known in these equations. Solving for ε_1 and ε_2 in terms of the feed concentration, x and z gives

$$y = y_f + (x_f - x) - (z - z_f)$$

where $x_f = 1$ and $z_f = 0$ from the feed vector. This may be used to find the corresponding y value for the x–z grid point.

Again, suppose that a grid point of x = 1 and z = 1 is used for demonstration. First find f(x) and df(x)/dy.

$$f(1) = 0$$

$$\frac{df(1)}{dx} = 10$$

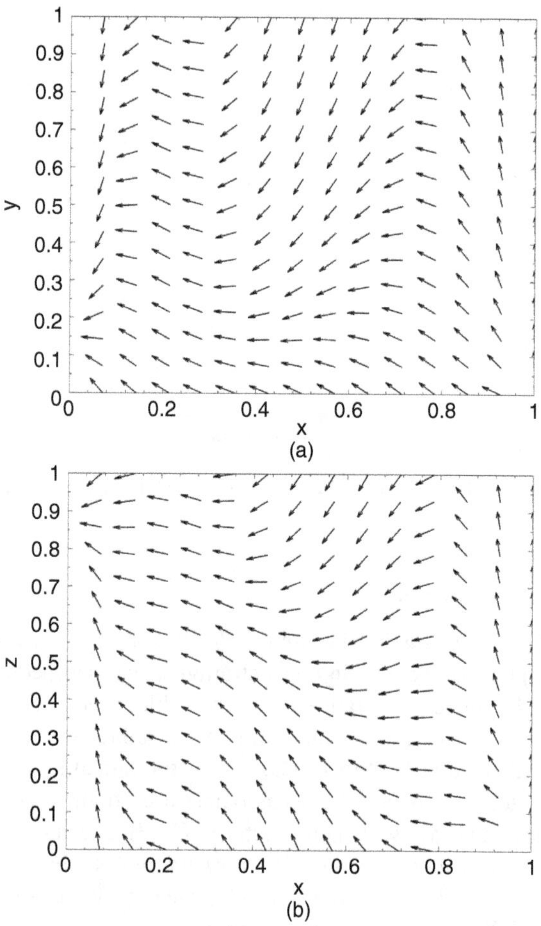

Figure 4.3 (a) Rate field in x–y space and (b) rate field in x–z space.

Then compute that value of y given x = 1 and z = 1:

$$y = y_f + (x_f - x) - (z - z_f) = (0) + (1 - 1) - (1 - 0)$$
$$= -1$$

Using these values, compute $r_X(x, y)$ and $r_Z(x, y)$:

$$r_X(1, \ -1) = 0$$

$$r_Z(1, \ -1) = 10$$

Figure 4.3(b) shows the resulting rate field in x–z space after this procedure has been applied for a 15 × 15 grid in x–z space. Observe that the rate field obtained in this instance is different from that given in part (a) of the example, as expected. The rate fields are however linked by the system of reactions and stoichiometry of the system. Note that there is no assumption that the points or rate vectors displayed in the figures are those that are physically achievable. The rate field is simply a consequence of the system of nonlinear equations describing the reactions under consideration.

4.3.2.2 *Equilibrium Points*

a) Description Equilibrium is attained when the rates of reaction for all components in the system are zero. Geometrically, this is signified by a rate vector with zero entries in all elements. Hence, a concentration $\mathbf{C}^{\#}$ is an equilibrium concentration of the system if

$$\mathbf{r}(\mathbf{C}^{\#}) = \mathbf{0} \tag{4.1}$$

where $\mathbf{0} = [0, 0, \ldots, 0]^T$ is the *zero vector*. Mathematically, these points are simply solutions to the rate expressions of interest. In the BTX system, an equilibrium point is found at $c_B = 0$ mol/L, $c_E = 0$ mol/L, $c_T = 0$ mol/L. (It is simply a coincidence that the zero vector corresponds with an equilibrium point of the BTX kinetics.)

OBSERVATION: Equilibrium points in the rate field

Equilibrium points are points in the rate field where the rate vector is the zero vector.

EXAMPLE 6: Equilibrium points from kinetics

The following species rate functions describe kinetics for a six-component system:

$$r_A = c_A c_B^2 - 5c_A c_E$$
$$r_B = -2c_A c_B^2 - 2c_B c_D$$
$$r_C = c_A c_B^2 - c_C$$
$$r_D = c_C + 10c_A c_E - 2c_B c_D$$
$$r_E = 3c_C - 5c_A c_E$$
$$r_F = 2c_B c_D$$

Given the concentration vectors supplied in Table 4.2, can you identify which of these mixtures corresponds to an equilibrium point? The concentration vector for this system is defined as $\mathbf{C} = [c_A, c_B, c_C, c_D, c_E, c_F]^T$.

TABLE 4.2 A List of Concentration Vectors. Can You Identify Which Vector Is an Equilibrium Point?

No.	Concentration Vector $\mathbf{C}$
$\mathbf{C}_1$	$[1.0, 1.0, 1.0, 1.0, 1.0, 1.0]^T$
$\mathbf{C}_2$	$[0.0071, 0.0088, 0.3974, 2.6842, 2.1056, 1.8427]^T$
$\mathbf{C}_3$	$[0.0, 0.0, 0.0538, 3.0205, 3.1357, 1.8515]^T$
$\mathbf{C}_4$	$[0.0, 0.0, 0.0016, 3.0726, 3.2922, 1.8515]^T$
$\mathbf{C}_5$	$[0.0, 0.0, 0.0, 3.0743, 3.2970, 1.8515]^T$

Answer:

Equilibrium is reached when the rate vector is the zero vector. Since both the kinetics and the concentration vectors have been supplied, computing equilibrium points is a simple task of substituting the concentration vectors into the kinetic expressions to determine whether the associated rate vector produces zero for each component. Table 4.3 summarizes the concentration vector and corresponding rate vectors for this system.

TABLE 4.3 Summary of Corresponding Rate Vectors

No.	Rate Vector $\mathbf{r}(\mathbf{C})$
$\mathbf{C}_1$	$[-6.0, -4.0, 0.0, 9.0, -2.0, 2.0]^T$
$\mathbf{C}_2$	$[-0.0075, -0.0470, -0.3974, 0.3654, 1.1846, 0.0470]^T$
$\mathbf{C}_3$	$[0.0, 0.0, -0.0538, 0.0538, 0.1613, 0.0]^T$
$\mathbf{C}_4$	$[0.0, 0.0, -0.0016, 0.0016, 0.0049, 0.0]^T$
$\mathbf{C}_5$	$[0.0, 0.0, 0.0, 0.0, 0.0, 0.0]^T$

The rate vector $\mathbf{r}(\mathbf{C}_2)$ is the only vector in the list that contains nonzero values for all components. Rate vectors $\mathbf{r}(\mathbf{C}_1)$, $\mathbf{r}(\mathbf{C}_3)$, and $\mathbf{r}(\mathbf{C}_4)$ are all associated with a select number of components that have zero entries. These entries correspond to rates of reaction for the individual components, but not all components in the system. The final rate vector $\mathbf{r}(\mathbf{C}_5)$ is the only vector that is the zero vector, and thus $\mathbf{C}_5$ is an equilibrium point for the system. Note that although the component concentrations of $\mathbf{C}_4$ and $\mathbf{C}_5$ appear to be similar, the corresponding rate vectors are different.

b) Calculating Equilibrium Points For any particular equilibrium concentration $\mathbf{C}^{\#}$ of a reactive system, the overall rate of reaction for *all* species is zero, and thus it is clear that Equation 4.1 must hold. To calculate equilibrium points associated with a given set of kinetics, the roots of a system of nonlinear equations, belonging to the species rate expressions, must be solved. The number of equations in the system is equal to the number of components in space – one rate expression exists for each component of interest.

Computing roots for a system of kinetics is commonly performed numerically. The complexity of rate expressions, even for very small systems, generally prohibits us from computing solutions analytically otherwise.

Numerical solution of nonlinear systems is also rather difficult in practice, and there is, generally, no way to guarantee that all the roots to the system can be obtained (easily) (Acton, 1990; Press, 2007). This knowledge should not deter us however. Provided adequate care is taken to properly understand the system kinetics and its behavior, many systems may be solved for using the appropriate numerical technique.[7] A reliable kinetic expression is generally founded upon equilibrium information, for example. Hence knowledge of equilibrium points for the system are often well understood, more so than the kinetic mechanisms themselves.

EXAMPLE 7: Calculating equilibrium points from rate data (single reaction)

Assuming that a single reaction is given,

$$2A + B \rightarrow C$$

with associated kinetics

$$r_A = -2k_1c_Ac_B$$

where $k_1 = 1$ L/(mol.s). Calculate the equilibrium concentration vector $\mathbf{C}^{\#}$ assuming that a feed vector $\mathbf{C}_f = [c_{Af}, c_{Bf}, c_{Cf}]^T = [1, 2, 3]^T$ mol/L is supplied.

Answer:

This is a single reaction involving three components. The reaction rate of component A is best expressed in terms of the extent of reaction.

$$r_A = -2k_1(c_{Af} - 2\varepsilon)(c_{Bf} - \varepsilon)$$

Equilibrium is obtained when the rate of reaction is zero. Determining the equilibrium extent using the rate of reaction for component A determines the equilibrium concentration for all components in the system. Hence, we have

$$r_A = 0 = -2k_1(c_{Af} - 2\varepsilon)(c_{Bf} - \varepsilon)$$

Two roots solve this expression. Hence, equilibrium is reached when the value of ε is

$$\varepsilon = \frac{c_{Af}}{2} \quad \text{or} \quad \varepsilon = c_{Bf}$$

Although two roots exits, only one is physically possible since both components A and B are reactants and one of the two will be in excess. In this instance, component A is the limiting reactant. It follows that $\varepsilon = 1/2 = 0.5$ mol/L is the equilibrium extent of reaction. The corresponding concentrations at equilibrium are then found by mass balance

$$c_A = c_{Af} - 2\varepsilon$$
$$c_B = c_{Bf} - \varepsilon$$
$$c_C = c_{Cf} + \varepsilon$$

which gives $\mathbf{C}^{\#} = [0.0, 1.5, 3.5]^T$ mol/L.

EXAMPLE 8: Calculating equilibrium points from rate data (multiple reactions)

Assume now that component C from Example 7 is an intermediate product that decomposes to a final stable product D by the following reaction:

$$2A + B \rightarrow C \rightarrow D$$

The rate of reaction for components A and C are given as follows:

$$r_A = -2k_1c_Ac_B$$
$$r_C = k_1c_Ac_B - k_2c_C$$

Calculate the equilibrium points associated with this reaction if the feed vector is the same as that given in question (a).

Answer:

There are now two independent reactions, and thus the equilibrium concentration will be associated with two equilibrium extents. The reaction rates may again be written in terms of extent as follows:

$$r_A = -2k_1(c_{Af} - 2\varepsilon_1)(c_{Bf} - \varepsilon_1)$$
$$r_C = k_1(c_{Af} - 2\varepsilon_1)(c_{Bf} - \varepsilon_1) - k_2(c_{Cf} + \varepsilon_1 - \varepsilon_2)$$

Since the system is a function of components A, B, and C alone, computing the equilibrium extents from r_A and r_C

[7]The `fsolve()` function in MATLAB allows for the solution of a system of nonlinear equations with a user-specified initial guess.

is sufficient to determine the equilibrium concentrations of all components in the system. Equilibrium is again reached when the rate of reaction is zero. In this instance, r_C must also be zero. This gives

$$r_A = 0 = -2k_1(c_{Af} - 2\varepsilon_1)(c_{Bf} - \varepsilon_1)$$
$$r_C = 0 = k_1(c_{Af} - 2\varepsilon_1)(c_{Bf} - \varepsilon_1) - k_2(c_{Cf} + \varepsilon_1 - \varepsilon_2)$$

The two equations are sufficiently simple to solve analytically. Hence, the first term in r_C is the contribution of reaction 1, which at equilibrium is zero. In addition to this, the rate of reaction for component A remains unchanged from Example 7, and thus the same equilibrium value for r_A applies. Solving for ε_1 and ε_2, therefore, gives

$$\varepsilon_1 = \frac{c_{Af}}{2} \text{ and } \varepsilon_2 = c_{Cf} + \varepsilon_1$$

The feed vector associated with this produces an equilibrium concentration vector of $\mathbf{C}^{\#} = [0.0, 1.5, 0.0]^T$. Since the concentration of component D has not been specified, we cannot compute its equilibrium value. This result may be further validated by plotting the concentration profiles obtained in a batch or PFR from the feed. This is given in Figure 4.4.

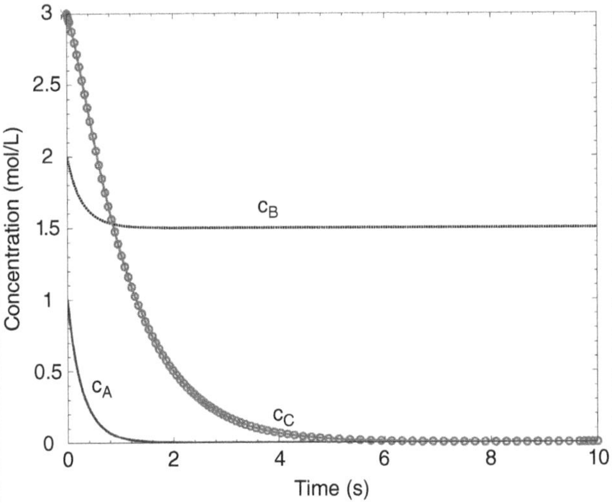

Figure 4.4 Concentration profiles approaching equilibrium.

Note that the final concentration is the equilibrium concentration calculated earlier.

4.4 THREE FUNDAMENTAL CONTINUOUS REACTOR TYPES

4.4.1 Motivation

In much the same way that concentration can be viewed as a vector, it is possible to view a system of reactions as a vector as well. We now wish to explain how well-known reactor models, often studied in undergraduate chemical reaction engineering courses, can also be interpreted geometrically. In this section, we would like to describe the behavior of three fundamental reactor types: PFR, CSTR, and DSR.

Note that the rate field is defined by the kinetics, irrespective of the particular reactor employed. Different reactor types will trace out different paths in concentration space according to the shape of the rate field. As will be shown, each type serves an important purpose in the formation of the AR. Let us begin with the simplest of continuous reactors, the PFR.

SIDE NOTE: Reactor types

A great deal of effort was required in Chapter 3 to generate the AR for the BTX system using batch experiments. Using knowledge and insights gained from this section, we will be in a better position to show how the same result is achievable in a much more efficient manner using continuous equipment.

4.4.2 The Plug Flow Reactor

4.4.2.1 Physical Description

a) Introduction The PFR, or tubular reactor, is a reactor type that is commonly used in industrial chemical processes. PFRs are typically constructed from a tube and in some cases, when the reaction is slow or when the reaction is heterogeneous, it may be necessary to fill the reactor space with catalyst material. PFRs are operated by continuously feeding material at one end and allowing material to gradually react along the tube's length. No material is allowed to enter or leave the reactive space of the tube. As such, material is assumed to travel down the reactor length as a "plug." The plug flow assumption is one in which no axial mixing along the length of the reactor occurs—a differential plug of material is assumed to travel along the length of the PFR, and it is not allowed to interact with the differential plugs adjacent to it. This assumption of *plug flow behavior* is where the PFR derives its name from.

A particular differential segment of material within the PFR, when observed on its own, is assumed to be perfectly uniform in concentration. In this way, a change

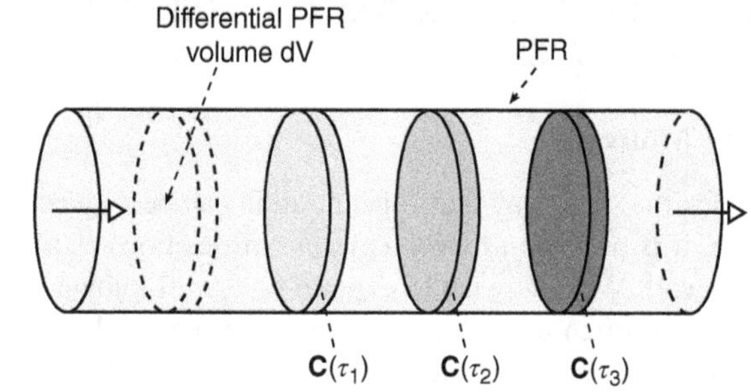

Figure 4.5 Differential segments of material inside a PFR. Each segment is assumed to have constant concentration.

Figure 4.6 (a) Three large-scale PFRs and (b) the symbol used commonly to represent a PFR in process flow diagrams.

in concentration may only be achieved along the PFR's length. However, at any particular point of observation along the length of the PFR, the concentration is constant with respect to time. This behavior is shown in Figure 4.5. In Figure 4.6(a), three large tubular reactors in a vertical orientation are shown that are used to simulate the formation of acid mine drainage, which is contaminated water that is commonly produced in the mining industry. Water is fed at the top of the columns and allowed to slowly pass down the vessels, simulating PFR behavior. The symbol that is commonly used to indicate a PFR is provided in Figure 4.6(b).

b) Residence Time As feed material moves along the length of a PFR, reactants are continually converted into products. It is not difficult to visualize that when the PFR's length is made longer, and the velocity of material is kept constant, reactants spend a longer time in the PFR before exiting. This allows for a higher conversion of reactants to products. Similarly, an identical conversion of material may be achievable if the flow rate of the feed material is decreased to match the time spent of a segment of material in an equivalently longer tube. Both flow rate and reactor volume (which is directly associated with the PFR's length) are two important factors that influence the overall conversion achieved in a PFR. The residence time, τ, of the reactor is thus an important parameter in the design of a PFR. This quantity is defined by the following expression:

$$\tau = \frac{V_{PFR}}{Q} \tag{4.2}$$

Here, V_{PFR} is the volume of the PFR and Q is the volumetric flow rate at the exit of the PFR. τ then has units of (time).

SIDE NOTE: Units of τ

The given description of residence time is the most natural choice for many reactor problems, although, this may not be the sole choice available. Take, for example, if reaction rate is expressed with respect to catalyst mass. An appropriate set of units for the rate might then incorporate the mass of the catalyst contained within the reactor body over reactor volume. In Chapter 9, we will define τ in terms of mass fractions for variable density systems.

c) Analogy of a PFR It is sometimes helpful to envision a PFR as a collection of small batch reactors placed on a conveyor belt.[8] Consider an idealized representation of this scenario, given in Figure 4.7.

The reactor is represented as a collection of beakers traveling on the conveyor at a constant speed S and a fixed length L. At the left side of the belt, a beaker containing the feed concentration is placed on the belt and allowed to traverse along its length. When the beaker reaches the end of the belt, it is removed. It is assumed that each beaker that

[8] Based on the ideas of the late Donald Williams at the University of the Witwatersrand, Johannesburg.

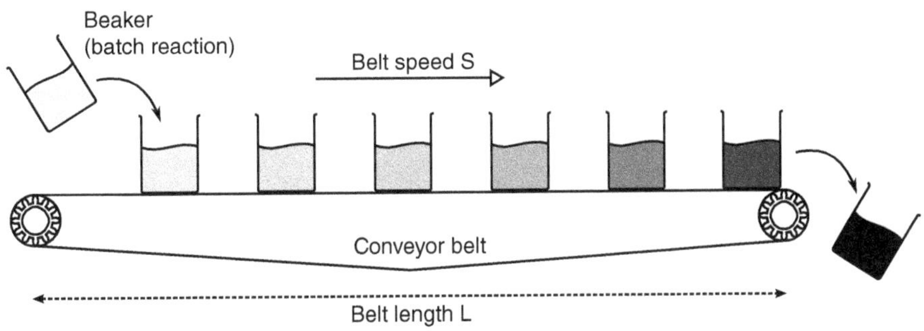

Figure 4.7 Beakers (batch reactors) on a conveyor belt, which approximate PFR behavior. (*See color plate section for the color representation of this figure.*)

is removed is immediately replaced with another beaker, containing the same feed concentration, from the front. In this way, each beaker on the belt serves to imitate a "plug" of material reacting inside the body of a traditional PFR. A single beaker, viewed in isolation to other beakers, is still a batch reactor.

The left side of the belt acts as the feed to the reactor, and the right end may be viewed as the reactor effluent stream. Reactants constantly undergo conversion to form products inside the beaker as they pass down the conveyor belt. Each beaker placed on the conveyor belt spends a finite amount of time before it is removed from the other end. The time spent on the belt is influenced by both the conveyor belt speed S and the belt length L. For a longer belt length and fixed speed S, a beaker spends a longer time reacting before it is removed. The specific time that each beaker spends on the belt is hence equivalent to the residence time of a conventional PFR. From this perspective, the belt speed and beaker volume determine an equivalent flow rate, Q, of material passing through the reactor. These parameters of the conveyor belt determine the final concentration (the effluent concentration) achieved on the other side. Note that there are two different perspectives for viewing concentrations inside this "reactor."

1. **At a fixed position:** At each point along the length of the belt, we may simply stop and observe the concentration *at a fixed location on the belt.* From this perspective, the concentration of the mixture is fixed for a specific position along the conveyor belt. Furthermore, since reaction rate is a function of concentration, reaction rate will also be constant for a fixed location on the conveyor belt.
2. **Traveling on the belt:** Imagine that we step onto the conveyor and travel along at the same speed as the beakers, observing the concentration change *for a particular beaker as it moves down the belt.* From this viewpoint, the concentration of the mixture within a beaker varies as it moves along the belt. A concentration profile with respect to time could be generated in this case. Note that each beaker is simply a batch reactor, and therefore the change in concentration is identical to the batch reactor equation. It follows that reaction rates must vary along the length of the belt in this situation.

> **OBSERVATION: Analogy between a PFR and a batch reactor**
>
> A PFR may be viewed as a series of batch reactors (reacting beakers) traveling along a conveyor belt. The performance of a PFR is identical to a batch reactor as a result. For every point along the length of the reactor, the concentration at the point of observation is constant. As we travel along the length, however, the concentration of species changes.

4.4.2.2 ***Mathematical Description*** Although the mathematical theory of PFRs is well understood, these concepts will be reviewed here from a perspective suitable for use in AR theory. For this to be done in a compact manner, vector notation is preferred. This also allows one to begin thinking how these systems might be incorporated with the graphical concepts learned previously. For demonstration, we shall begin with a single reaction in a batch reactor, and then extend this concept to multiple reactions.

a) PFR Equation for a Single Reaction Consider the following simplified reaction of a single hypothetical component A, reacting to from a product B:

$$\mathrm{A} \rightarrow \mathrm{B} \tag{4.3}$$

The usual assumptions will be in force: it is assumed that the reaction is isothermal and performed under constant density so that no volume change is experienced during reaction. If feed material, containing pure A, is placed into a beaker and allowed to react as a batch, then the well-known batch expression may be employed as follows:

$$\frac{dn_A}{dt} = Vr_A$$

Here n_A is the molar amount of component A (measured in moles), V is the beaker volume, and r_A is the rate of reaction for component A. A similar expression for product B may also be obtained. Since reactor volume is assumed to be constant, V may be divided through the expression to give

$$\frac{dc_A}{dt} = r_A$$

where c_A is the molar concentration of component A, defined by $c_A = n_A/V$. Conversion of this expression from batch operation to an equivalent PFR expression is simple: the concentration profile of the batch is identical to a PFR if reaction time is replaced with residence time in a PFR. It follows that for a PFR, we have

$$\frac{dc_A}{d\tau} = r_A \tag{4.4}$$

Here, τ is the equivalent residence in a PFR from Equation 4.2. This is the PFR expression for a single reaction.

b) PFR Equation for Multiple Components A slightly more complex reaction will now be examined, involving four components in the following two independent reactions:

$$A \rightarrow B$$
$$B + C \rightarrow D$$

In this system, component B is an intermediate product, similar to that of toluene in the BTX system. PFR expressions may be written out for each component, resulting in the following two equations:

$$\frac{dc_A}{d\tau} = r_A$$
$$\frac{dc_B}{d\tau} = r_B$$

Two reactions are present in the system, and hence the concentrations of all species in the system may be expressed in terms of the concentrations of A and B alone by mass balance. The reaction rate for A, r_A, is assumed to be the same as that given in the previous section. The rate expression for component B, r_B, in this instance, should now contain terms relating to the formation of side-product D as well.

Observe that the PFR equations for both component A and B are very similar. It is clear that for each component i in the system, the following generalized equation is in force:

$$\frac{dc_i}{d\tau} = r_i$$

Hence, for n such components present in the system, n PFR equations may be written as follows:

$$\frac{dc_1}{d\tau} = r_1$$
$$\vdots$$
$$\frac{dc_n}{d\tau} = r_n$$

If many reactions with many components are present (i.e., in a biological system), expressing each component explicitly in the form given before may become tedious. It is convenient, when dealing with many components, to employ vector notation instead. This is done by grouping the relevant terms and placing them as components in an associated vector

$$\begin{bmatrix} \frac{dc_1}{d\tau} \\ \vdots \\ \frac{dc_n}{d\tau} \end{bmatrix} = \begin{bmatrix} r_1 \\ \vdots \\ r_n \end{bmatrix}$$

The system is now described by two vectors. One further simplification may be carried out: by factoring out the $d/d\tau$ terms contained in the left vector, the notation is made slightly more compact as follows:

$$\begin{bmatrix} \frac{dc_1}{d\tau} \\ \vdots \\ \frac{dc_n}{d\tau} \end{bmatrix} = \frac{d}{d\tau}\begin{bmatrix} c_1 \\ \vdots \\ c_n \end{bmatrix}$$

OBSERVATION: Notation

Note that when we refer to an expression such as $d/d\tau[c_1, c_2, c_3]^T$, what we are actually indicating is that the $d/d\tau$ operation is applied to *each* element in the vector, so that

$$\frac{d}{d\tau}[c_1,\ c_2,\ c_3]^T = \left[\frac{dc_1}{d\tau},\ \frac{dc_2}{d\tau},\ \frac{dc_3}{d\tau}\right]^T$$

This is done as a convenience to simplify the notation.

When this notation is applied to the generalized PFR expression, the following is obtained:

$$\frac{d}{d\tau}\begin{bmatrix} c_1 \\ \vdots \\ c_n \end{bmatrix} = \begin{bmatrix} r_1 \\ \vdots \\ r_n \end{bmatrix}$$

For the particular example involving components A and B given earlier, an equivalent vector expression for the PFR equation may then be written as follows:

$$\frac{d}{d\tau}\begin{bmatrix} c_A \\ c_B \end{bmatrix} = \begin{bmatrix} r_A \\ r_B \end{bmatrix}$$

The terms on the left-hand side of the equation indicate a standard concentration vector **C** introduced in Chapter 2 $\mathbf{C} = [c_A, c_B]^T$. Similarly, the vector on the right-hand side contains elements representing the rate expressions for A and B in the system $\mathbf{r}(\mathbf{C}) = [r_A, r_B]^T$. Observe that this vector is the rate vector evaluated at **C**.

$$\frac{d\mathbf{C}}{d\tau} = \mathbf{r}(\mathbf{C}) \tag{4.5}$$

The vector version of the equation is identical in form to the single component version for a PFR given in the section "PFR Equation for a Single Reaction." Both describe the behavior of a PFR (as well as a batch reactor). The expression from the section "PFR Equation for a Single Reaction" operates on a scalar, whereas the latter operates on a vector (a system of components)—the vector version of the PFR expression is simply as a generalization of the PFR equation to multiple components.

The vectors in Equation 4.5 bear special significance. They represent the set of concentrations that are obtained, specifically by a PFR. Although the rate vector is generalized in that it is a function of all components, the particular rate vector generated in a PFR is coupled to the particular concentration obtained inside the PFR.

c) *Solving the PFR Equation*

Required Information Note that Equation 4.5 represents a system of ordinary differential equations (ODEs) that relate the rate of change of concentrations inside the reactor with respect to time, to the instantaneous rate of reaction defined by the system kinetics. In order to solve an ODE, an initial condition and integration time must be specified.

Species rate expressions are often complex mathematical expressions coupling many components present in the system together. Rate expressions are also often nonlinear. This makes isolation and analytical solution of a single species component difficult, and it is generally not possible to solve a system of PFR expressions analytically. We are thus often reliant on simultaneous numerical solution of the PFR equation.

Solving Numerical integration of a system of ODEs is common and well documented. The following two pieces of information are required before a numerical solution can be obtained:

1. An initial condition. At a residence time of zero (at the feed of the PFR), an initial feed concentration $\mathbf{C}_f$ must be specified. It follows that the number of initial concentrations needed will be equal to the number of differential equations being integrated. For *n* components, *n* initial concentrations must be supplied to fully specify the problem.
2. A time range is also required over which the PFR equation is solved. Hence, a reactor residence time, or equivalent integration parameter, such as reactor volume or catalyst mass, must be supplied. For larger values in residence time, a higher conversion of reactants is obtained at the reactor exit and a closer numerical approximation to equilibrium is achieved.

Upon specification of this information, it is possible to integrate the PFR expression and solve for a PFR concentration profile.

Output Integration of the system of ODEs results in a list of concentrations achieved by the PFR and corresponding times (similar to the Table 2.1 provided in Chapter 2). This represents the evolution of concentrations along the length of the PFR. These points may be plotted in space (both in time and concentration space), producing a curve representing the solution to the ODE. This is referred to as a PFR *solution trajectory*, or *PFR trajectory* for short.

EXAMPLE 9: Solving a system of PFR equations

Let us compute the PFR trajectory for a previously considered system. It is assumed that the following reaction is available

$$2A + B \rightarrow C \rightarrow D$$

with kinetics for components A to D given by

$$\begin{aligned} r_A &= -2k_1c_Ac_B \\ r_B &= -k_1c_Ac_B \\ r_C &= k_1c_Ac_B - k_2c_C \\ r_D &= k_2c_C \end{aligned}$$

Values of the rate constants are specified as $k_1 = 1$ L/(mol.s), $k_2 = 1\ s^{-1}$. Plot the PFR trajectory in c_A–c_C space for the following feed points and PFR sizes:

(a) $\mathbf{C}_f = [c_{Af}, c_{Bf}, c_{Cf}]^T = [1, 2, 3]^T$ mol/L. The PFR volume is 10 L and the volumetric flow rate to the PFR is 2.5 L/s.

(b) $\mathbf{C}_f = [2, 2, 0]^T$ mol/L. The PFR volume is 20 L and the volumetric flow rate is 25 L/s.

Answer a:

To plot the PFR trajectory, the following system of differential equations must be solved:

$$\frac{dc_A}{d\tau} = r_A = -2k_1c_Ac_B$$

$$\frac{dc_B}{d\tau} = r_B = -k_1c_Ac_B$$

$$\frac{dc_C}{d\tau} = r_C = k_1c_Ac_B - k_2c_C$$

with initial conditions given by: at $\tau = 0$, $c_A = 1.0$ mol/L, $c_B = 2.0$ mol/L, and $c_C = 3.0$ mol/L. This is expressed compactly in vector form as

$$\frac{d\mathbf{C}}{d\tau} = \mathbf{r}(\mathbf{C}) = \begin{bmatrix} -2k_1c_Ac_B \\ -k_1c_Ac_B \\ k_1c_Ac_B - k_2c_C \end{bmatrix} \quad (4.6)$$

where the concentration vector is defined accordingly as $\mathbf{C} = [c_A, c_B, c_C]^T$. A PFR solution trajectory is obtained when an integration range is specified. This is obtained from the reactor sizing. Since it is given that the PFR volume is V = 10 L and that the volumetric flow rate to the PFR is Q = 2.5 L/s, the PFR residence time is calculated as follows:

$$\tau = \frac{V}{Q} = 4.0\,\text{s}$$

From this residence time, the set of equations defined by Equation 4.5 may be integrated numerically using a standard ODE solver.

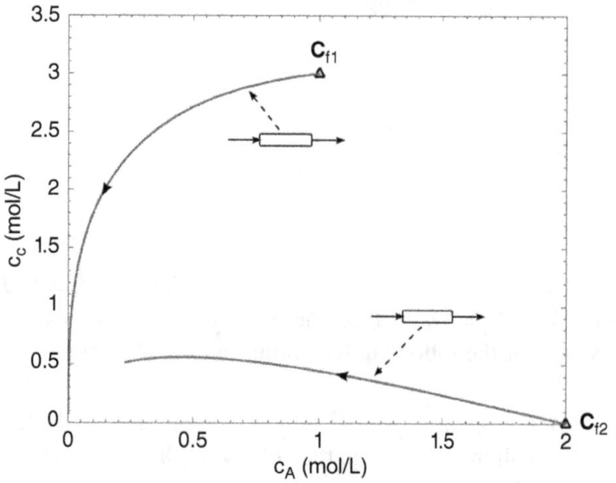

Figure 4.8 PFR trajectories for two different initial conditions (feed points) and integration ranges (residence times).

Answer b:

Computing the solution trajectory for part b of the problem follows an identical procedure. However in this instance, the initial conditions are given as at $\tau = 0$, $\mathbf{C} = [2, 2, 0]^T$ mol/L. Note that we have employed vector notation here to simplify the notation slightly. The PFR residence time is also different in this case, which is given by

$$\tau = \frac{20}{25} = 0.8\,\text{s}$$

Integration of Equation 4.6 with the two initial conditions and integration times produces two distinct PFR trajectories in c_A–c_C space. This is shown in Figure 4.8.

EXAMPLE 10: PFR trajectory for BTX kinetics

Consider again the familiar BTX kinetics. The system of reactions and kinetics are supplied again for convenience as follows:

$$B + 0.5E \rightarrow T$$

$$T + 0.5E \rightarrow X$$

$$2B \rightarrow D + H$$

and

$$r_B = -c_Bc_E^{0.5} - 20.0c_B^2$$

$$r_E = -0.5c_Bc_E^{0.5} - 0.5c_Tc_E^{0.5}$$

$$r_T = c_Bc_E^{0.5} - c_Tc_E^{0.5}$$

$$r_X = c_Tc_E^{0.5}$$

$$r_D = 10.0c_B^2$$

We shall be interested in plotting a PFR trajectory from the feed conditions, specified in Chapter 1, in c_B–c_E–c_T space. This example also wishes to demonstrate how integration limits, for the PFR equation, may be estimated from graphical inspection of the species concentration profiles.

The concentration and feed vector are given by $\mathbf{C} = [c_B, c_E, c_T]^T$ mol/L and $\mathbf{C}_f = [1.0, 0.5, 0.0]^T$ mol/L. Plot a PFR trajectory to the equilibrium point for the BTX kinetics.

Answer:

To compute the PFR solution trajectory to the equilibrium point, the PFR equation must be integrated using $\mathbf{C}_f$ as the initial condition. Equilibrium is only achieved in the system in the limit of an infinitely large reactor (at $\tau = \infty$), and so it is not theoretically possible to achieve true equilibrium. We may however approximate this state arbitrarily closely by choosing a sufficiently long residence time in the numerical integration. Figure 4.9(a) shows a plot of the concentration profiles for benzene, ethylene, and Toluene as a function of PFR residence time.

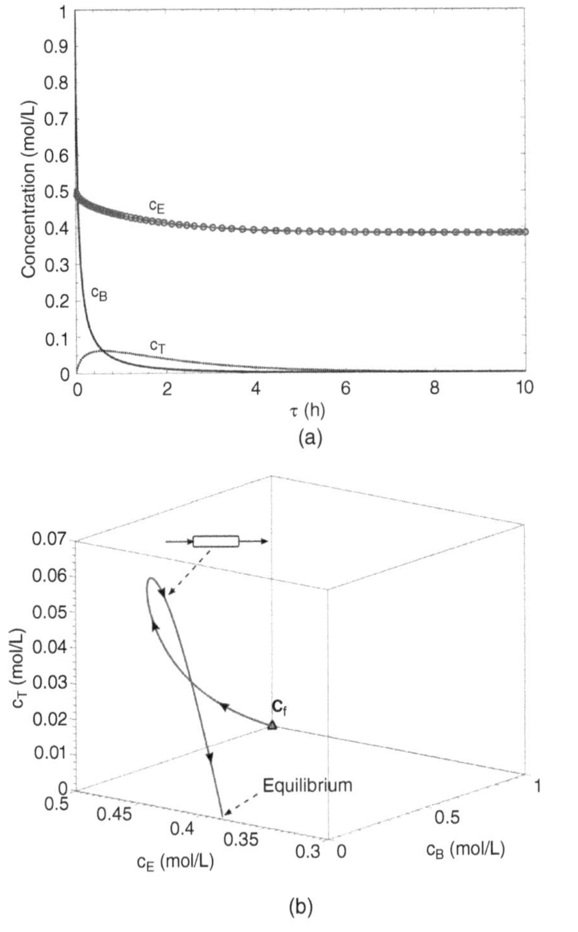

Figure 4.9 (a) Concentration profiles for the BTX system in a PFR and (b) PFR trajectory for the BTX system from the feed point to equilibrium.

Observe that the change in concentrations of each species appear to slow down after approximately 8.0 h. Hence, a PFR residence time of 8.0 h would be sufficient to achieve a close approximation of the equilibrium state for the BTX system.

From the data generated in Figure 4.9(a), we may plot the concentrations in the phase plane in c_B–c_E–c_T space. This is done in Figure 4.9(b).

We should note, with particular reference to Chapter 3, that although we have managed to improve the toluene concentration by generating batch data in a c_B–c_T space (a two-dimensional space), the BTX system is in fact a three-dimensional system. In Chapter 7, we shall return to the BTX system and plot the true AR in all three dimensions.

IMPORTANT: Retaining the full set of concentrations

When solving PFR equations, one is often only interested in the effluent concentration (i.e., the product concentration) achieved by the PFR. In AR theory, we are often interested in understanding how the full set of concentrations, obtained inside PFR, change as material traverses along the length of the reactor. In this way, we are required to retain the entire set of concentrations obtained throughout the entire integration range, and not simply the final effluent concentration. There are only specific points along a PFR solution trajectory that may allow for further expansion of the set of achievable points. Plotting solutions in the phase plane allows us to identify when these opportunities arise easily.

4.4.2.3 Geometric Interpretation An elegant relation exists between the PFR trajectory and the rate vector at a point **C**, on the PFR trajectory. To understand this relation, consider again the vector version of the PFR equation as follows:

$$\frac{d\mathbf{C}}{d\tau} = \mathbf{r}(\mathbf{C})$$

We can integrate the system of ODEs in the usual manner, to produce a set of points that represent a solution function $\mathbf{C}(\tau)$, which may be plotted in concentration space. This is given in Figure 4.10, where a hypothetical two-dimensional solution trajectory for $\mathbf{C}(\tau)$ in c_A–c_B space is shown.

a) Approximating the Gradient of the PFR Trajectory Also shown in Figure 4.10 are two points on the profile, $\mathbf{C}_1 = \mathbf{C}(\tau_1)$ and $\mathbf{C}_2 = \mathbf{C}(\tau_2)$. Both of these points are vectors. $\mathbf{C}_1$ is an arbitrary point, evaluated at a residence time τ_1 along the profile. Similarly, $\mathbf{C}_2$ is a point located near $\mathbf{C}_1$, which is evaluated at a later point in time, equal to $\tau_2 = \tau_1 + \Delta\tau$. The difference between $\mathbf{C}_2$ and $\mathbf{C}_1$ is itself a vector that can be defined as follows:

$$\Delta\mathbf{C} = \mathbf{C}_2 - \mathbf{C}_1$$

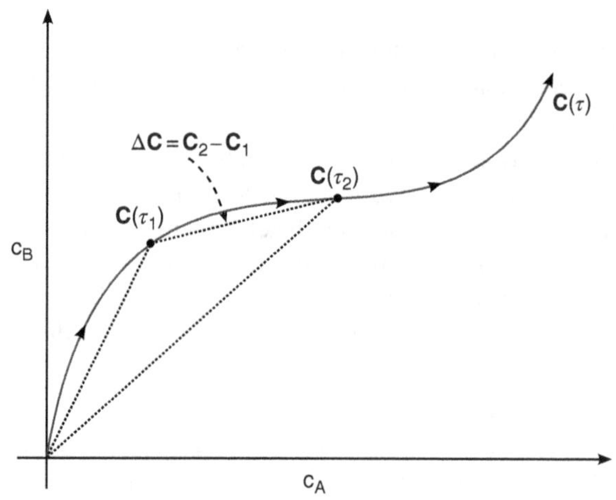

Figure 4.10 Approximating the gradient of a PFR trajectory.

For small time differences, $\mathbf{C}_2$ must lie closer to point $\mathbf{C}_1$, and thus $\Delta\mathbf{C}$ must be small in magnitude relative to the magnitude of $\mathbf{C}_1$ and $\mathbf{C}_2$. We can convince ourselves that the quantity $\Delta\mathbf{C}/\Delta\tau$ is an approximation of the derivative of the curve $\mathbf{C}(\tau)$ by examining the PFR vector equation. $\mathrm{d}\mathbf{C}/\mathrm{d}\tau$ is simply a system of derivatives, each of which can be approximated by an Euler-style finite difference

$$\frac{\mathrm{d}\mathbf{C}}{\mathrm{d}\tau} = \begin{bmatrix} \dfrac{\mathrm{d}c_A}{\mathrm{d}\tau} \\ \\ \dfrac{\mathrm{d}c_B}{\mathrm{d}\tau} \end{bmatrix} \approx \begin{bmatrix} \dfrac{c_A(\tau + \Delta\tau) - c_A(\tau)}{\Delta\tau} \\ \\ \dfrac{c_B(\tau + \Delta\tau) - c_B(\tau)}{\Delta\tau} \end{bmatrix}$$

Approximation of each component in the vector differential equation must lead to the following vector equation:

$$\frac{\mathbf{C}^{t+\Delta t} - \mathbf{C}^{t}}{\Delta t} \approx \mathbf{r}(\mathbf{C})$$

This result shows that $\Delta\mathbf{C}/\Delta\tau$ is an approximation to the gradient of the curve, and that the term $\mathrm{d}\mathbf{C}/\mathrm{d}\tau$ is the true gradient of the line $\mathbf{C}(\tau)$. The gradient of the PFR solution trajectory is thus linked closely to the rate field specified by the system kinetics $\mathbf{r}(\mathbf{C})$, which is the rate vector evaluated at the same point $\mathbf{C}$.

b) What Does This Mean? The rate vector at a given point $\mathbf{C}$ on a PFR trajectory is equal to its gradient at the point of interest. This has an important geometric implication for PFRs in AR theory:

For every point $\mathbf{C}$ *on the PFR trajectory, the gradient of the trajectory is equal to the rate vector* $\mathbf{r}(\mathbf{C})$ *evaluated at* $\mathbf{C}$*. Rate vectors are thus* tangent *at all points along the PFR trajectory.*

The illustration provided in Figure 4.11 demonstrates this property of PFRs graphically. Figure 4.11(a) shows a rate field generated with an arbitrary set of two-dimensional kinetics, and a number of PFR trajectories initiated at different feed points. Figure 4.11(b) illustrates rate vectors at different points along a hypothetical PFR trajectory. The rate vectors evaluated at all points along the path are always tangent to the trajectory.

c) Geometric Significance Since the shape of the PFR trajectory is characterized by the rate field, and since the rate field is unique for a specified set of system kinetics, a single rate vector $\mathbf{r}(\mathbf{C})$ exists at every point $\mathbf{C}$ in space. And given that rate vectors form part of a PFR trajectory, *it is not possible for PFR trajectories to intersect each other*. Consider what might occur if this result were not true. This behavior would imply that multiple rate vectors exist at the intersection point, which cannot occur if $\mathbf{r}(\mathbf{C})$ is uniquely defined at $\mathbf{C}$, which is shown in Figure 4.12. This result shall be used as a basis for establishing further properties of the AR.

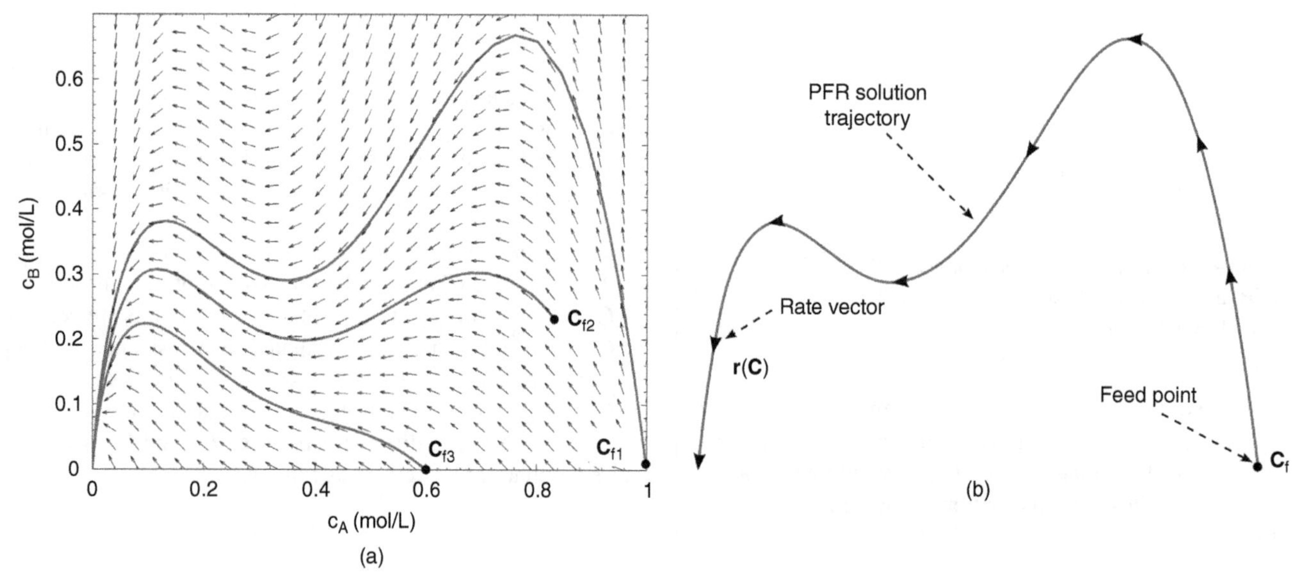

Figure 4.11 (a) Geometric interpretation of the PFR trajectory and (b) rate vectors are tangent to the PFR trajectory. (*See color plate section for the color representation of this figure.*)

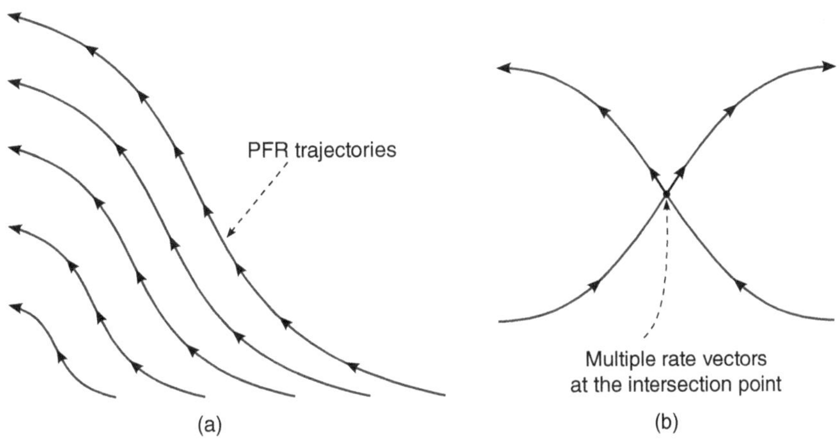

Figure 4.12 (a) PFR trajectories can never intersect each other, and (b) this phenomenon would imply that multiple rate vectors exist at the intersection point.

SIDE NOTE: PFR trajectories

PFRs are the simplest of all three fundamental reactor types in AR theory. This is clear when the close link between the kinetics and PFR solutions trajectories can be shown. In Chapter 6, we will discuss why almost all optimal reactor structures must terminate with a PFR.

SIDE NOTE: Russian nesting dolls (matryoshka dolls)

Growing candidate ARs can be viewed similar to the effect of nesting Russian dolls (matryoshka dolls). A region grown from a previous candidate must be larger than the original; however it must also contain all parts of the previous region, as displayed in Figure 4.13. The new region must fully enclose the old region in other words. This property is due to the fact that PFR trajectories may never cross.

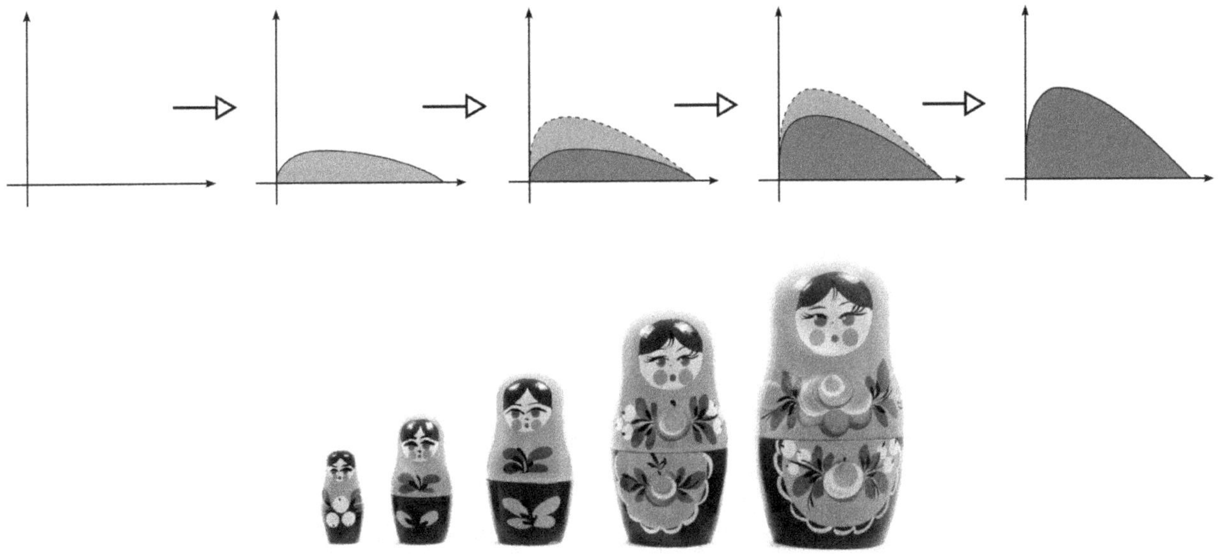

Figure 4.13 *Matryoshka dolls*. https://pixabay.com/en/doll-russian-russia-moscow-game-314345/. CC0 1.0 Universal (CC0 1.0) Public Domain Dedication https://creativecommons.org/publicdomain/zero/1.0/deed.en.

RECALL: Choosing points from a concavity from Chapter 3

Now that a better understanding of PFR trajectories has been established, a more technical justification for improvements in the BTX batch game, from Chapter 3, can now be given. Recall that the toluene concentration was improved by selecting points in the concave section of a PFR trajectory. In Chapter 3, this was denoted by region B of the current convex hull. By comparison, when mixing points from region A were chosen, the resulting batch trajectories (PFR trajectories) did not result in an expansion of the candidate region.

On reflection, we can now see why this choice is appropriate given the nature of PFR solution trajectories. For an expansion to have occurred in the batch mixing game, a starting concentration must have been chosen that resulted in a PFR solution trajectory lying *above* the current boundary. But PFR trajectories are never allowed to cross for the same kinetics, and thus this situation can only occur when a mixture concentration chosen from region B is utilized (region B in Chapter 3 was associated with the concave section of the current region). Figure 4.14 shows a graphical description of this effect.

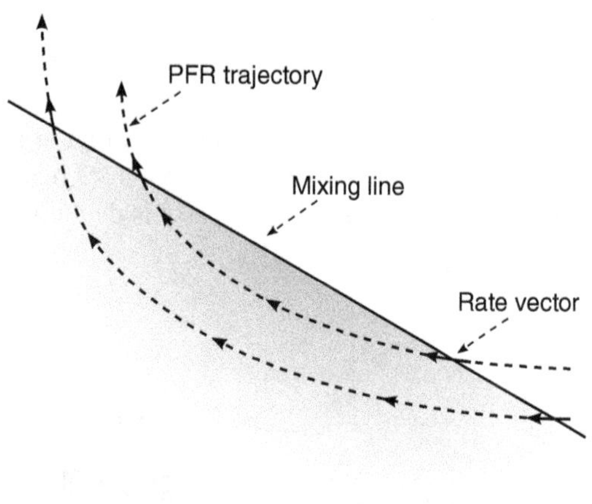

Figure 4.14 PFR trajectories on the boundary of a mixing line. Trajectories are not allowed to cross for the same kinetics.

4.4.3 The Continuous-Flow Stirred Tank Reactor

4.4.3.1 Physical Description

a) Introduction The CSTR[9] is a continuous reactor that operates essentially as the mirror opposite to the PFR with respect to mixing in that *complete and perfect mixing* is assumed. Hence, reaction is assumed to occur at a single, uniform concentration in a CSTR, and there is no distinguishable difference in concentration throughout the entire vessel volume.

CSTRs are typically fabricated out of a large tank in practice, and fitted with a mechanical agitator to facilitate sufficient mixing. Sometimes, alternate forms of agitation are also employed, such as the bubbling of gas through a liquid mixture using gas spargers (Lehrer, 1968). Perfect mixing is fairly difficult to achieve in practice, and often baffles must be fixed to the tank to promote mixing, as well as to ensure that there is no variation in concentration within the tank during operation.

A consequence of the perfect mixing assumption is that the concentration of the effluent stream of a CSTR must be identical to the concentration found in the bulk fluid inside the tank. Knowledge of the effluent concentration provides insight into the conditions inside of the reactor, and vice versa. Only one product concentration is produced in a CSTR as a result. This is significantly different from the PFR, where concentration is assumed to change continuously down the length of the reactor.

Figure 4.15(a) shows an image of a laboratory scale CSTR, which is used for liquid phase reactions. The vessel wall is clear so that the contents of the reactor can be seen. A heating coil is also installed in the CSTR for temperature control. A large agitator is installed to ensure that the vessel contents is well mixed. Figure 4.15(b) shows the symbol that is commonly used to represent a CSTR in reactor engineering texts.

SIDE NOTE: Perfect mixing in a CSTR

For every element of feed that enters a CSTR, the perfect mixing assumption implies that the entering element is *instantaneously* converted to the concentration of the bulk volume. Conversion of reactants might then be viewed to occur as a result of mixing and dilution rather than from reaction alone.

RESIDENCE TIME A measure of the *average* reaction time of a fluid element inside a CSTR is often interpreted in an analogous way to that of the residence time of a fluid element inside a PFR. We hence define the CSTR residence time τ as follows:

$$\tau = \frac{V_{CSTR}}{Q}$$

Here, Q is the volumetric flow rate of the CSTR effluent stream, whereas V_{CSTR} is the volume of the CSTR vessel.

[9]The CSTR is also often abbreviated as CFSTR.

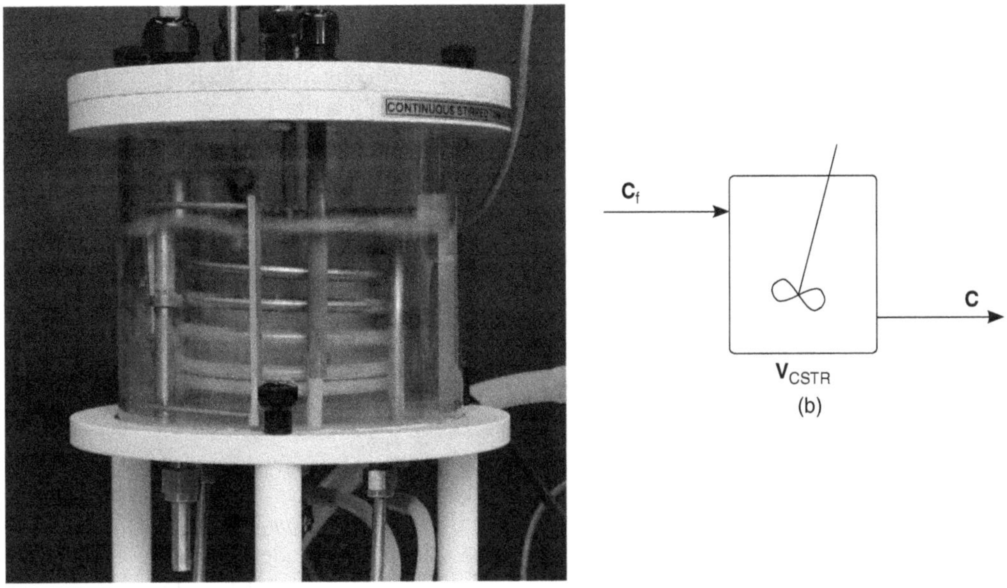

(a)

Figure 4.15 (a) A laboratory-scale CSTR and (b) symbol used to represent a CSTR.

EXAMPLE 11: Calculating CSTR residence times

A CSTR is constructed from a cylindrical tank with a base area of 1.5 m^2, and is used to convert a liquid stream containing component A into a desired product B.

$$A(l) \rightarrow 2B(l)$$

The height of the CSTR vessel is 2.0 m high. The operating level in the tank is controlled at 1.8 m, measured from the base of the tank. If the volumetric flow rate Q into the tank is 2.25 L/s, answer the following questions. You may assume constant density.

(a) What is the residence time of the CSTR?

(b) What is the residence time in a PFR for the same volumetric flow rate Q if the PFR has a radius of r = 0.5 m and length of L = 2.5 m?

(c) How long would a batch reactor need to be operated for in order to produce the same concentration as the PFR in question (b) given earlier?

Answer a:

The CSTR residence time is calculated by the ratio of reactor volume to volumetric flow rate $\tau_{CSTR} = V/Q$. Since constant density may be assumed for the system, the volumetric flow rate of the effluent stream is equal to the inlet volumetric flow rate. The volume used is the volume occupied by the reactor and not the total vessel volume. Hence, for a base area of $A = 1.5\ m^2$ and an operating level of $h = 1.8$ m, the reactor volume is $V_{CSTR} = Ah = 2.7\ m^3 = 2700$ L. A volumetric flow rate of $Q = 2.25$ L/s, therefore, gives a CSTR residence time of $\tau = 2700/2.25 = 1200\ s = 20$ min.

Answer b:

The PFR residence time is calculated in a similar manner to a CSTR. Hence, $\tau = V_{PFR}/Q$. The volumetric flow rate is specified at $Q = 2.25$ L/s. The reactor volume V_{PFR} may be calculated by computing the volume of a cylinder corresponding to the dimensions of the PFR. Therefore, for a cylinder radius of $r = 0.5$ m and length $L = 2.5$ m, $V_{PFR} = L \times (\pi r^2) = 1.9635\ m^2 = 19.635$ L.

From this, PFR residence time is calculated as $\tau = V_{PFR}/Q = 872.7\ s \sim 14.5$ min.

Answer c:

The mathematical behavior of a batch is analogous to that of the PFR in a continuous setting. Hence, residence time in a PFR is equivalent to batch reaction time in a batch for the same initial conditions. It follows that a reaction time of $t \sim 14.5$ min is required to achieve the same final concentration as the PFR described in question (b).

EXAMPLE 12: Residence time with gases

Consider now what might happen if the reaction is assumed to occur in the gas phase.

$$A(g) \rightarrow 2B(g)$$

In this case, the constant density assumption is inappropriate, and thus species mass fractions, z_i, must be used instead. Assuming that a total mass flow rate into the CSTR is 5 kg/s, and that there is 0.1 kg of catalyst, what is the residence time of the CSTR?

Answer:

A steady-state mass balance using mass fraction is given as follows:

$$G\mathbf{Z} = G\mathbf{Z}_f + W\mathbf{r}(\mathbf{Z})$$

Here, $\mathbf{Z}$ is the *mass fraction vector*, $\mathbf{Z} = [z_A, z_B]^T$, and is analogous to a concentration vector for each component in the system. G is the total mass flow rate flowing through the CSTR (G is therefore always constant). It is also assumed that the rate vector is now expressed in terms of mass fractions and catalyst mass W, so that the units of $\mathbf{r}(\mathbf{Z})$ are (mass of component reacted/(mass of catalyst × time)).

Dividing by G gives

$$\mathbf{Z} = \mathbf{Z}_f + \frac{W}{G}\mathbf{r}(\mathbf{Z})$$

From this, it is clear that an equivalent residence time for the CSTR (using mass fractions) must incorporate the ratio of catalyst mass to CSTR feed rate. Hence, in this instance

$$\tau = \frac{W}{G}$$

The units of τ are also (time). More precisely, the units of τ are ((catalyst mass × time)/(total mass flow rate)).

4.4.3.2 Mathematical Description

a) Single Component For a single component i, the outlet concentration of the CSTR may be calculated by performing a steady-state component mass balance over the CSTR as follows:

$$Q_{out}c_i = Q_{in}c_{if} + V_{CSTR}r_i(\mathbf{C})$$

c_{if} and c_i are the feed and exit concentrations of component i in the CSTR, whereas $\mathbf{C}$ is the concentration vector corresponding to the CSTR effluent concentration. Note that the rate of reaction for component i, $r_i(\mathbf{C})$, is hence evaluated *at the conditions in the tank.*

When constant density is assumed, the inlet volumetric flow rate Q_{in} equals the outlet volumetric flow rate Q_{out}. The CSTR equation may be simplified by dividing through by Q and combining with V_{CSTR} in terms of residence time τ. This results in the following expression:

$$c_i = c_{if} + \tau r_i(\mathbf{C}) \tag{4.7}$$

Here, $\tau = V_{CSTR}/Q$. This is the well-known CSTR equation for a single component i.

b) Multiple Components We can extend the CSTR equation to include multiple components by writing out each component balance as follows:

$$c_1 = c_{1f} + \tau r_1(\mathbf{C})$$
$$\vdots$$
$$c_n = c_{nf} + \tau r_n(\mathbf{C})$$

It is convenient to express the system of equations in terms of a vector equation—similar to the PFR, giving

$$\mathbf{C} = \mathbf{C}_f + \tau\mathbf{r}(\mathbf{C}) \tag{4.8}$$

Observe that all components in the system are related by a single, common residence time τ.

OBSERVATION: Approximating PFRs

For small residence times, the performance of a PFR is similar to a CSTR for an equivalent feed point and residence time. In deriving the differential equation for a PFR, one must perform a mass balance over a differential plug of material that is assumed to contain an even composition (in effect, a small CSTR). In Chapter 8, a number of different AR construction methods are described that approximate PFR solution trajectories by many small CSTRs in series.

c) Solving the CSTR Equation

Required Information Solution of the CSTR equation requires largely the same prerequisite information needed for the solution of the PFR expression. To obtain a numerical value for the effluent concentration, the following information is needed:

1. A measure of reaction time. A CSTR volume and flow rate, usually combined as a residence time τ.
2. An initial feed condition. In this instance, the composition of the feed stream, $\mathbf{C}_f$, entering the CSTR.

The CSTR expression may then be used to solve for the remaining unknown species effluent concentrations when these two parameters are specified.

SOLUTION The methods employed to solve the CSTR equation (a system of nonlinear algebraic equations) differ to the solution of the PFR equation (a system of ordinary differential equations). Formally, the CSTR expression is solved by computing the roots that satisfy the following equation, for a specified residence time τ and feed concentration $\mathbf{C}_f$:

$$\mathbf{C}_f + \tau\mathbf{r}(\mathbf{C}) - \mathbf{C} = \mathbf{0}$$

For n components in the system, n CSTR equations may be written corresponding to each component. Difficulties in solution arise when nonlinear rate expressions are used, since then multiple solutions may exist for a fixed τ and $\mathbf{C}_f$. This is due to the fact that the rate vector is evaluated at the exit concentration $\mathbf{C}$ and not at the feed concentration $\mathbf{C}_f$.

EXAMPLE 13: Solving the CSTR equation

Given the following single reaction:

$$A \rightarrow B$$

with kinetics for component A given by

$$r_A = -\frac{k_1 c_A^2}{1 + c_A}$$

where $k_1 = 2.5$ L/(mol.s), compute the following:

(a) Assuming a CSTR with residence time $\tau = 8$ s is available, and the concentration in the product stream is given by the concentration vector $\mathbf{C} = [0.104, 0.396]^T$, what is the feed concentration $\mathbf{C}_f$ to the CSTR?
(b) Suppose now that the feed vector is specified as $\mathbf{C}_f = [1.0, 2.0]^T$. Assuming that residence time of the CSTR is unchanged from part (a), calculate the concentration vector $\mathbf{C}$ at the exit of the CSTR.

Answer a:

The CSTR equation may be written for in vector form, giving

$$\mathbf{C} = \mathbf{C}_f + \tau\mathbf{r}(\mathbf{C})$$

Since the concentration vector at the product stream is specified, the rate vector $\mathbf{r}(\mathbf{C})$ is also known and thus the feed vector $\mathbf{C}_f$ may be calculated directly from the CSTR expression as follows:

$$\mathbf{C}_f = \mathbf{C} - \tau\mathbf{r}(\mathbf{C})$$

From stoichiometry, the rate of reaction for component B is

$$r_B = -r_A = \frac{k_1 c_A^2}{1 + c_A}$$

The rate vector may then be calculated using the values given in $\mathbf{C}$, which gives $\mathbf{r}(\mathbf{C}) = [-0.0245, 0.0245]^T$ mol/(L.s). Solving for the feed vector therefore gives

$$\begin{aligned}\mathbf{C}_f &= \mathbf{C} - \tau\mathbf{r}(\mathbf{C})\\ &= \begin{bmatrix}0.104\\0.396\end{bmatrix} - (8.0)\begin{bmatrix}-0.0245\\0.0245\end{bmatrix}\\ &= \begin{bmatrix}0.3\\0.2\end{bmatrix}\ \text{mol/L}\end{aligned}$$

Note that this is a straightforward calculation since $\mathbf{C}_f$ is easily isolated from the remaining terms in the CSTR expression.

Answer b:

This is a somewhat more challenging calculation. To compute the exit concentration of the CSTR, we must again solve the CSTR equation. However this time, the rate vector is no longer known as it relies on the concentration vector $\mathbf{C}$. Writing out the rate CSTR equation for component A gives

$$c_A = c_{Af} - \tau\frac{k_1 c_A^2}{1 + c_A}$$

c_{Af}, τ, and k_1 are known in this case, and thus we may solve for c_A. This results in a quadratic expression in c_A as follows:

$$c_A^2(1+\tau k_1)+c_A(1-c_{Af})-c_{Af}=0$$

which may be solved analytically to give

$$c_A=\frac{-(1-c_{Af})\pm\sqrt{(1-c_{Af})^2+4c_{Af}(1+\tau k_1)}}{2(1+\tau k_1)}$$

Substituting the given values into this expression thus gives two solutions. The feasible solution is $c_A = 0.2182$ mol/L. Substituting this value of c_A into the CSTR expression for c_B gives $c_B = 2.7818$ mol/L. The CSTR exit concentration is hence given as $\mathbf{C} = [0.2182, 2.7818]^T$ mol/L.

Observe that this computation was significantly more challenging to perform, as it involved the solution of a system of nonlinear equations.

Given a residence time τ and feed concentration $\mathbf{C}_f$, *analytic* solution of the CSTR equations may not be possible for nonlinear kinetics. This is true even in the case of a single reaction. Consequently, we are again compelled to employ *numerical* methods for the solution of the CSTR equation.

In contrast to the solution of PFRs, solving the CSTR equation is considerably more challenging.[10] There is generally no guaranteed method for computing CSTR solutions for all kinetics.[11] We thus often only obtain satisfactory solution after investing time in understanding the behavior of the CSTR equations for the kinetics of interest.

(!) IMPORTANT: Multiple CSTR steady states

Since the CSTR equation is simply a system of nonlinear algebraic equations, it is possible to obtain multiple CSTR steady states *for a fixed feed concentration and residence time*. For example, if there are polynomial terms in $\mathbf{r}(\mathbf{C})$, then more than one concentration will exist as roots to the equation. Even if these roots do not bear any physical significance to the system, they must still be known in order for the appropriate construction of the AR to be carried out. More complex expressions are also valid (and common) in modern-day rate expressions. Be wary of this when attempting to solve for CSTR solutions. The presence of multiple steady states presents a considerable challenge when constructing the AR. This will be shown in later chapters.

Output and Visualization Upon specification of a feed point $\mathbf{C}_f$ and CSTR residence time τ, a CSTR effluent concentration $\mathbf{C}$ (or many concentrations in the case of multiple steady states) can be obtained. Unlike the PFR, which operates over a *range* of achievable concentrations, the CSTR operates at distinct concentrations for a fixed $\mathbf{C}_f$ and τ. Specification of a different $\mathbf{C}_f$ or τ results in a different value for $\mathbf{C}$ as a result.

CSTR Locus Comparison of CSTR performance to PFR performance can be achieved by specifying a range of CSTR residence times for a common $\mathbf{C}_f$, and then solving for the corresponding CSTR effluent concentrations $\mathbf{C}$. In this way, a number of CSTR solutions are obtained, similar to that of a PFR trajectory. When a collection of CSTR solutions is obtained for a fixed feed and plotted in space, each point on the profile indicates a physically different CSTR scenario corresponding to a unique residence time. The collection of CSTR points for a range of residence times is referred to as a *CSTR locus*.

In Figure 4.16, a CSTR locus is generated and used to compare with a PFR trajectory for the same kinetics and feed point. The end result appears similar, but the information represented by the two profiles is different. Note that although a PFR produces a continuum of solutions for a fixed feed and reactor volume, the points on a CSTR locus corresponds to a physically distinct scenario in the CSTR.

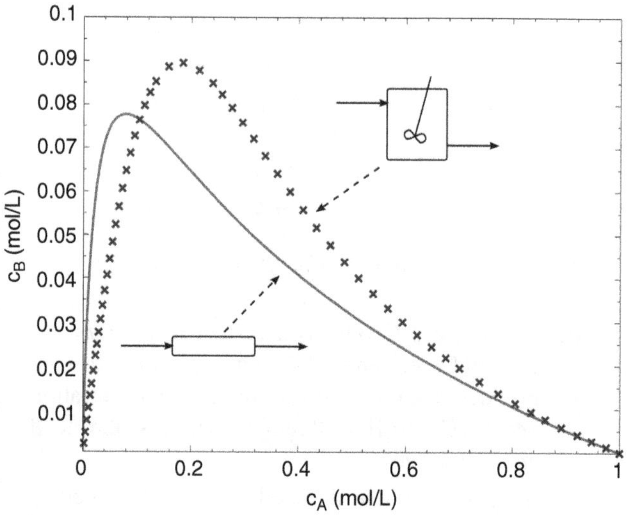

Figure 4.16 A CSTR locus and PFR trajectory from the feed point.

[10] This often arises, even when identical kinetics is used for both solution of the PFR and CSTR.

[11] Acton (1990) and Press (2007) provide two excellent descriptions why.

EXAMPLE 14: Comparing CSTR and PFR performance

Let us review some of the basic CSTR concepts with an example involving a single reaction. We would like to compare the difference in outlet concentration between a PFR and a similarly sized CSTR. That is, we will assume that we possess a CSTR of equal residence time to a PFR.

Question 1

Single reaction with linear kinetics

This reaction is a simple degradation of component A to component B as follows:

$$A \rightarrow B$$

The kinetics is assumed to be first order in A

$$r_A = -k_1 c_A$$

with $k_1 = 1.0\,s^{-1}$. What conversion of A is achieved if a feed point $\mathbf{C}_f = [5.0, 0.0]^T$ is available and the CSTR and PFR residence time is $\tau = 6.5$ s?

Answer:

Finding the PFR exit concentration involves the integration of the PFR differential equation, whereas the CSTR effluent concentration is obtained by a solution of a system of equations. In this example, the kinetics is linear in c_A, and thus is simple to solve for the exit concentration in the CSTR. The CSTR equation for component A is

$$\begin{aligned} c_A &= c_{Af} + \tau r_A \\ &= c_{Af} - \tau k_1 c_A \end{aligned}$$

Solving for c_A and c_B gives

$$\begin{aligned} c_A &= \frac{c_{Af}}{1 + \tau k_1} \\ c_B &= c_{Bf} + \tau k_1 \frac{c_{Af}}{1 + \tau k_1} \end{aligned}$$

which gives the concentration vector $\mathbf{C} = [0.667, 4.333]^T$. By comparison, integration of the PFR equation

$$\frac{d\mathbf{C}}{d\tau} = \mathbf{r}(\mathbf{C})$$

With initial conditions given by $\mathbf{C}_f$ results in a product concentration of $\mathbf{C}(\tau = 6.5) = [0.0075, 4.9925]^T$.

Question 2

Single reaction, nonlinear kinetics

In this example, we will again keep with a single reaction for simplicity, but now the rate is described by a different kinetic expression. Assume that the kinetics are now second order with respect to c_A so that

$$r_A = -k_1 c_A^2$$

where $k_1 = 1.0\,s^{-1}$. Whereas in the previous question the kinetics was described by first-order terms, we now wish to investigate the effect that nonlinear functions have on the solution. Hence, compute the effluent compositions, stated as concentration vectors, for a PFR and CSTR with the same residence time and feed point as in Question 1.

Answer:

The rate expression now involves a nonlinear term. The CSTR expression for component A is hence given as follows:

$$c_A = c_{Af} - \tau k_1 c_A^2$$

Solving for c_A in this instance involves the solution of a quadratic equation, similar to that given in the previous example.

$$c_A = \frac{-1 \pm \sqrt{1 + 4\tau k_1 c_{Af}}}{2\tau k_1}$$

The feasible solution is given as $c_A = 0.8035$ mol/L. Now that the value of c_A is known, c_B may be solved for; this gives $c_B = 4.1965$ mol/L. The concentration vector associated with the CSTR is hence $\mathbf{C} = [0.8035, 4.1965]^T$ mol/L.

Integration of the PFR expression for the same feed point and residence time results in an effluent concentration $\mathbf{C} = [0.1493, 4.8507]^T$ mol/L. Observe that in this instance it is still possible to solve for the CSTR expression, although its solution is more complex. Solving for the PFR exit concentration is the same as that performed in Question 1.

Question 3

Multiple reactions with nonlinear kinetics

As a final example, let us inspect a nonlinear system involving two independent reactions. The system of reactions is now given by an autocatalytic reaction (Brooks, 1988) as follows:

$$A + 2B \rightarrow 3B$$
$$B \rightarrow C$$

The rate expressions for components A, B, and C are

$$r_A = -k_1 c_A c_B^2$$
$$r_B = k_1 c_A c_B^2 - k_2 c_B$$
$$r_C = k_2 c_B$$

where $k_1 = 1.0\,L^2/(mol^2.s)$ and $k_2 = 1.0\,s^{-1}$. Determine an expression for the concentration of component B as a function of the feed vector $\mathbf{C}_f = [c_{Af}, c_{Bf}]^T = [5.0, 0.25]^T$ mol/L and residence time of $\tau = 3.0$ s in a CSTR.

Answer:

Observe that the system is a function of components A and B only. It is therefore possible to solve two simultaneous (nonlinear) equations involving components A and B. The concentration of component C is then determined by mass balance. Expressing the CSTR equations for components A and B gives

$$c_A = c_{Af} - \tau k_1 c_A c_B^2$$
$$c_B = c_{Bf} + \tau(k_1 c_A c_B^2 - k_2 c_B)$$

Calculation involves the solution of two nonlinear equations. We may solve for component A in terms of B and then substitute the expression into the CSTR equation for B. This results in

$$c_A = \frac{c_{Af}}{1 + \tau k_1 c_B^2}$$

$$c_B = c_{Bf} + \tau k_1 \left(\frac{c_{Af}}{1 + \tau k_1 c_B^2} \right) c_B^2 - \tau k_2 c_B$$

Collecting terms involving powers of c_B thus gives

$$f(c_B) = c_B^3 \tau k_1 (1 + \tau k_2) - c_B^2 \tau k_1 (c_{Af} + c_{Bf}) + c_B(1 + \tau k_2) - c_{Bf} = 0$$

which is a cubic equation in c_B. The solution of this expression is best handled by numerical nonlinear solver. A plot of this expression as a function of c_B is given in Figure 4.17. A horizontal line at $f(c_B) = 0$ is drawn to help identify the values of c_B that are roots to $f(c_B)$.

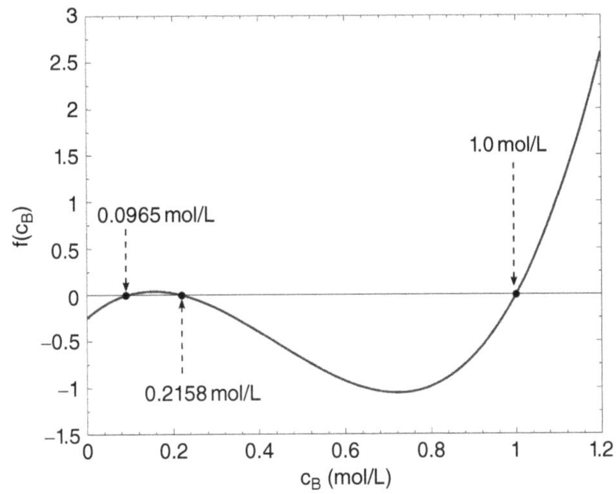

Figure 4.17 A plot of the cubic expression for $f(c_B)$ versus c_B for a residence time of $\tau = 3.0$ s and feed point $\mathbf{C}_f = [5.0, 0.25]^T$.

For the feed point, residence time and kinetics specified, there are now *three* solutions in the range $0.0 \le c_B \le 1.2$ mol/L: 0.0965, 0.2158, and 1.0 mol/L. Using the expression for c_A, the corresponding c_A values are 4.864, 4.387, and 1.25 mol/L. Concentration vectors associated with the CSTR effluent are hence as follows:

$$\mathbf{C} = [4.864, 0.0965]^T \text{ mol/L}$$
$$\mathbf{C} = [4.387, 0.2158]^T \text{ mol/L}$$
$$\mathbf{C} = [1.25, 1.0]^T \text{ mol/L}$$

All three effluent concentrations satisfy the CSTR equation, and thus all three are valid solutions. The actual concentration attained if a CSTR were to be operated with the same conditions would depend on other factors, such as the concentration in the CSTR at start-up, or whether one of the calculated solutions are stable operating conditions.[12]

By comparison, integration of the PFR expression for the same kinetics, feed point, and residence time results in a single effluent concentration of $\mathbf{C} = [0.0019, 1.4897]^T$ mol/L.

Note that the kinetics supplied in this example resulted in having to employ a numerical nonlinear solver in order to solve the CSTR equation. To produce a locus of CSTR solutions, the aforementioned procedure would need to be repeated for a range of different τ values. This may result in a large amount of computation. The PFR by comparison is easy to solve for, even when nonlinear kinetics is used. Unfortunately, many real-world systems involve multiple reactions with nonlinear kinetics. Furthermore, due to the nature of the CSTR equation, multiple solutions may exist for the same feed point and residence time.

[12]Determination of stable and unstable operating points can be found by computing the *eigenvalues* of the system Strogatz (2001).

SIDE NOTE: The difficulties of (reliably) finding CSTR loci

Although the formulation of the CSTR equation is simple, do not let this mislead you into thinking that they are easy to solve. A large portion of this discussion has been dedicated to communicate the idea that CSTR solutions are in fact quite difficult to solve in practice. The solution of the CSTR equation, and the difficulties involved with this, is an important cause for why ARs are difficult to construct.

4.4.3.3 *Geometric Interpretation*

a) Description In Section 4.4.2.3, we described how all rate vectors evaluated along a PFR trajectory must be tangent to the trajectory. Rate vectors evaluated on the CSTR locus share a unique geometric property of their own. Recall the CSTR vector equation

$$\mathbf{C} = \mathbf{C}_\text{f} + \tau\mathbf{r}(\mathbf{C}) \tag{4.8}$$

The expression may be rearranged as follows:

$$(\mathbf{C} - \mathbf{C}_\text{f}) = \tau\mathbf{r}(\mathbf{C})$$

Note that parentheses are placed around the left-hand side terms to emphasize the fact that the quantity $(\mathbf{C} - \mathbf{C}_\text{f})$ is itself a vector. Sometimes, it is convenient to refer to this as the mixing vector $\mathbf{v}$.

$$\begin{aligned}\mathbf{v} &= \mathbf{C} - \mathbf{C}_\text{f}\\ \mathbf{v} &= \tau\mathbf{r}(\mathbf{C})\end{aligned} \tag{4.9}$$

If the value of the effluent concentration $\mathbf{C}$ is known, the associated rate vector $\mathbf{r}(\mathbf{C})$ may be evaluated. From Equation 4.9, it is clear that $\mathbf{r}(\mathbf{C})$ and $\mathbf{v}$ are scalar multiples of each other by τ. Moreover, since τ can only assume positive values, $\mathbf{r}(\mathbf{C})$ and $\mathbf{v}$ must both point in the same direction.[13] Hence for any point $\mathbf{C}$ satisfying Equation 4.9, this point must exist as a CSTR solution for the feed point $\mathbf{C}_\text{f}$. We therefore arrive at the following result for CSTRs:

For a specified feed point $\mathbf{C}_f$ and rate expression $\mathbf{r}(\mathbf{C})$, the point $\mathbf{C}$ is a solution to the CSTR equation if the vector $\mathbf{C} - \mathbf{C}_f$ is collinear with the rate vector $\mathbf{r}(\mathbf{C})$, evaluated at point $\mathbf{C}$.

Figure 4.18 demonstrates the geometric behavior of the CSTR equation. For each point $\mathbf{C}$ on the CSTR locus, the rate rector $\mathbf{r}(\mathbf{C})$ is collinear with the vector $\mathbf{v} = \mathbf{C} - \mathbf{C}_\text{f}$. This is in contrast to the rate vector for a PFR, which is tangent to every point on the PFR trajectory.

SIDE NOTE: A geometric method of finding CSTR solutions

The geometric interpretation of the CSTR allows for a convenient method for solving the CSTR equation: instead of solving a nonlinear system of equations by standard numerical methods (i.e., Newton's method), we can find CSTR solutions by forming the vector $\mathbf{v} = \mathbf{C} - \mathbf{C}_\text{f}$ and then testing the rate vector at $\mathbf{C}$ for colinearity between $\mathbf{r}(\mathbf{C})$ and $\mathbf{v}$.

$$\mathbf{r}(\mathbf{C})^\text{T}\mathbf{v} = \|\mathbf{r}(\mathbf{C})\|\,\|\mathbf{v}\|$$

This result it true for any dimension. This approach to finding CSTR solutions is commonly employed in AR theory.

[13] This has a consequence on the properties of the AR as we shall see later.

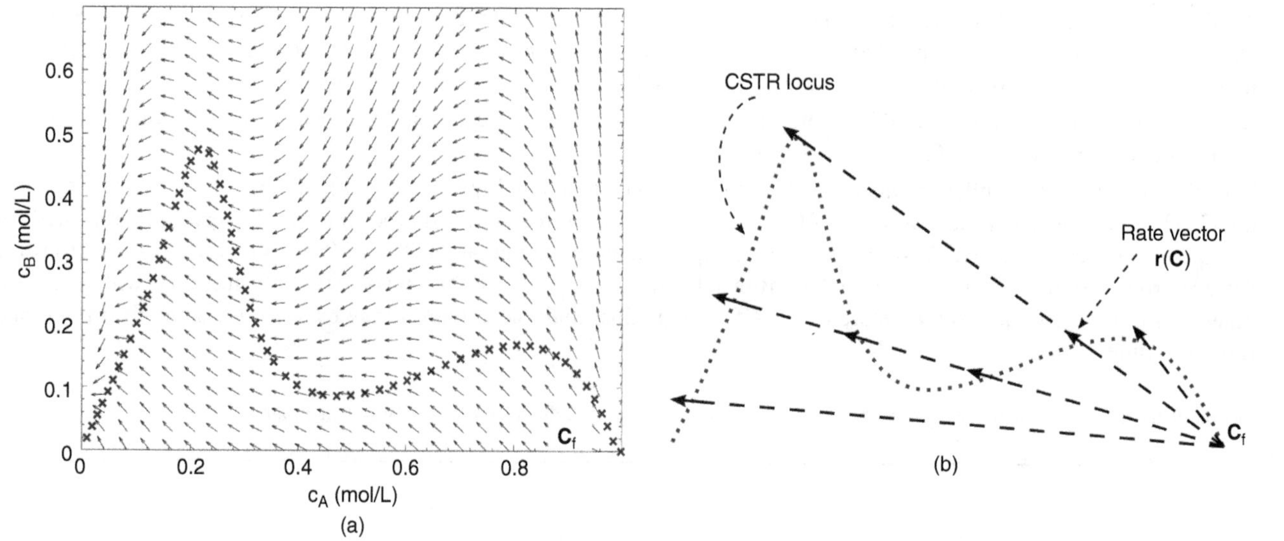

Figure 4.18 (a) Geometric interpretation of the CSTR. (b) CSTR solutions are collinear with the rate vector evaluated at that point and the feed. (*See color plate section for the color representation of this figure.*)

ILLUSTRATION: Determining CSTR points geometrically

Let us now look at how to compute CSTR solutions using the geometric properties of the CSTR vector equation. Consider again the autocatalytic reaction given by Brooks(1988) as follows:

$$\mathrm{A} + 2\mathrm{B} \rightarrow 3\mathrm{B}$$
$$\mathrm{B} \rightarrow \mathrm{C}$$

with kinetics for components A and B

$$r_A = -k_1 c_A c_B^2$$
$$r_B = k_1 c_A c_B^2 - k_2 c_B$$

We have shown previously how it may be difficult to solve for CSTR solutions analytically (and even numerically). This is due to the fact that a system of nonlinear equation must be solved for.

Solution via the geometric interpretation follows a somewhat different methodology. We shall use the fact the mixing vector between the CSTR effluent concentration and the feed vector $\mathbf{C_f}$ is collinear with the rate vector evaluated at the exit concentration $\mathbf{r}(\mathbf{C})$. That is we have

$$(\mathbf{C} - \mathbf{C}_\mathrm{f}) = \tau \mathbf{r}(\mathbf{C})$$

or

$$\mathbf{v} = \tau \mathbf{r}(\mathbf{C}) \tag{4.9}$$

where $\mathbf{v} = (\mathbf{C} - \mathbf{C}_\mathrm{f})$. A colinearity condition may be formed by taking the inner product between $\mathbf{v}$ and $\mathbf{r}(\mathbf{C})$. Hence, a concentration vector $\mathbf{C}$ is a CSTR solution if the following colinearity condition is met

$$\mathbf{r}(\mathbf{C})^\mathrm{T}\mathbf{v} = \|\mathbf{r}(\mathbf{C})\|\,\|\mathbf{v}\| = 0$$

The function $f(\mathbf{C})$ describes a surface in $\mathbb{R}^n$ which the locus of CSTR concentrations must intersect. From this, solving for CSTR solutions involves searching the n-dimensional space of concentrations that satisfy $f(\mathbf{C}) = 0$. This may be automated by computer. For plots in $\mathbb{R}^2$ this procedure may be demonstrated graphically. In Figure 4.19(a), we show the rate field and feed point for the autocatalytic system in c_A–c_B space.

Also displayed in the figure are a number of straight lines drawn over the rate field. For each straight line, we may search along the line and evaluate the rate vectors along the line. Rate vectors that are collinear with the line are hence those that also satisfy the CSTR condition. In Figure 4.19(a), we show these points for the lines given as blue crosses. This procedure may be performed throughout the entire space—essentially a straight line may be swept through concentration space to identify points of colinearity. Figure 4.19(b) shows the result of an automated search for the system.

Note that the CSTR locus forms almost a closed loop. Each point on the locus represents a CSTR solution for the feed point specified. Observe that the scalar quantity used to scale $\mathbf{v}$ and $\mathbf{r}(\mathbf{C})$ in the CSTR expression is the CSTR residence time τ. Hence, when solving for CSTR solutions in this manner, the corresponding residence time is lost in the computation. It is nevertheless easy to recover the list of residence times associated with the locus generated by substituting the CSTR effluent and feed concentrations back into the CSTR expression and solving for τ. The results of this computation are displayed in Figure 4.19(c).

Figure 4.19(c) shows the concentrations of components A and B as a function of computed residence time. Note that for certain residence times multiple CSTR solutions exist. Solving for these concentrations in a traditional manner using a nonlinear numerical solver may prove challenging, particularly if the initial guess is not chosen correctly—we may only compute one of many CSTR solutions for a given feed point and residence time. Computing the CSTR locus geometrically serves as an alternate means of addressing these systems.

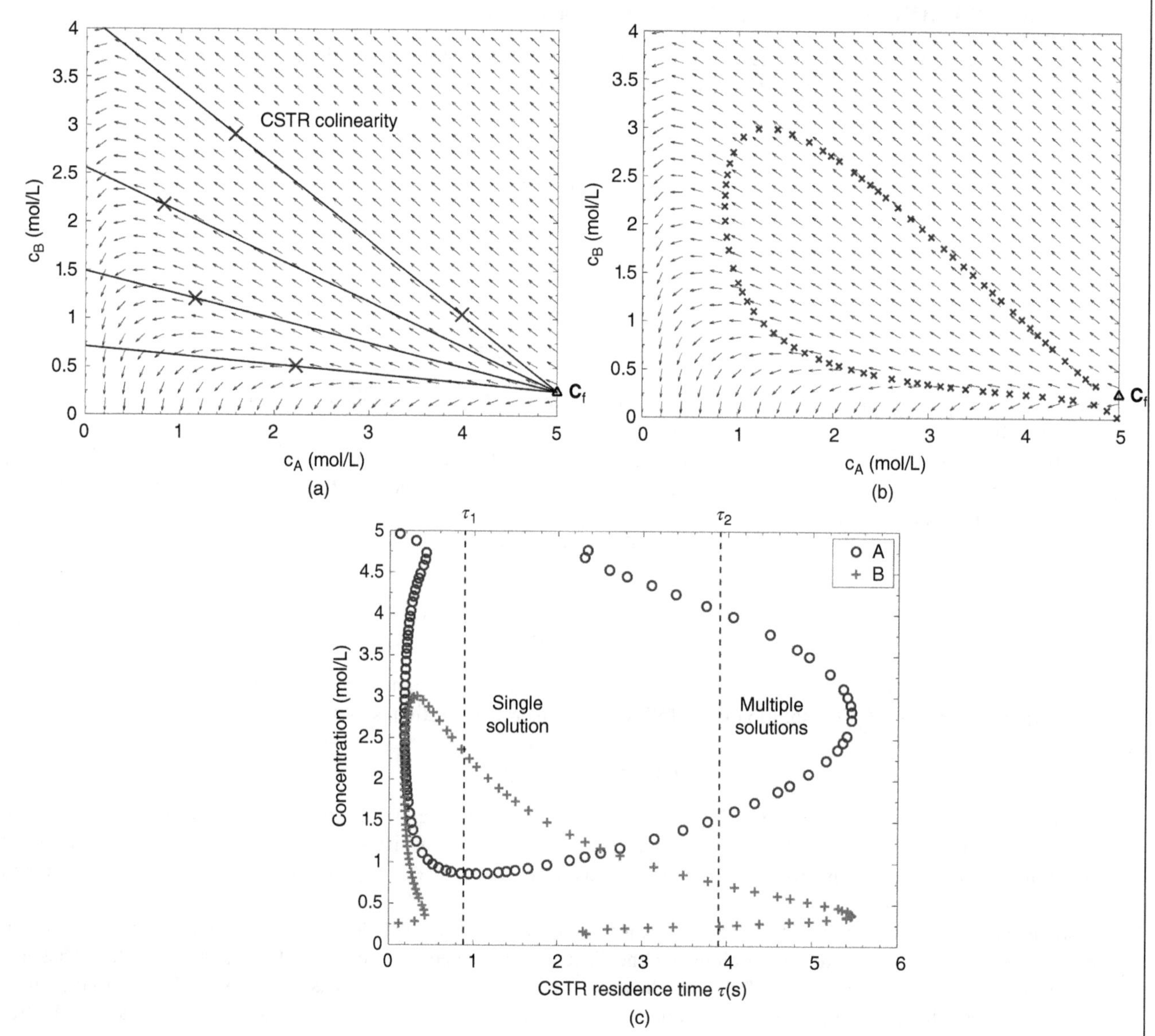

Figure 4.19 (a) Computing CSTR solutions in the autocatalytic system by searching for colinear points, (b) full CSTR locus for the autocatalytic system, and (c) concentration profiles for components A and B achieved in the CSTR for different residence times. (*See color plate section for the color representation of this figure.*)

b) What Does This Mean? Since CSTR solutions occur at distinct points in space, CSTRs allow for the connection of two disjoint points without the need for a definite (continuous) path between the points. In this sense, CSTRs allow for "jumps" through concentration space. Two, seemingly unrelated, concentrations in space may be joined together by a CSTR given an appropriate residence time.

Although the unique behavior of CSTRs is useful in construction of the AR, this also presents challenges when the kinetics is complex. For a sufficiently complex set of nonlinear kinetics, it is reasonable to expect that multiple steady states may manifest in the system through the CSTR equation. If multiple steady states are present, then construction of the AR may result in a candidate region that is smaller than the true AR, but one which still appears to be a feasible AR.

4.4.3.4 CSTRs on the Boundary of a Candidate Region

The geometric nature of CSTR of points provides an interesting discussion regarding a locus of CSTR points on the boundary of *any* candidate AR. Imagine what might occur if the convex hull for a set of points is computed when a large

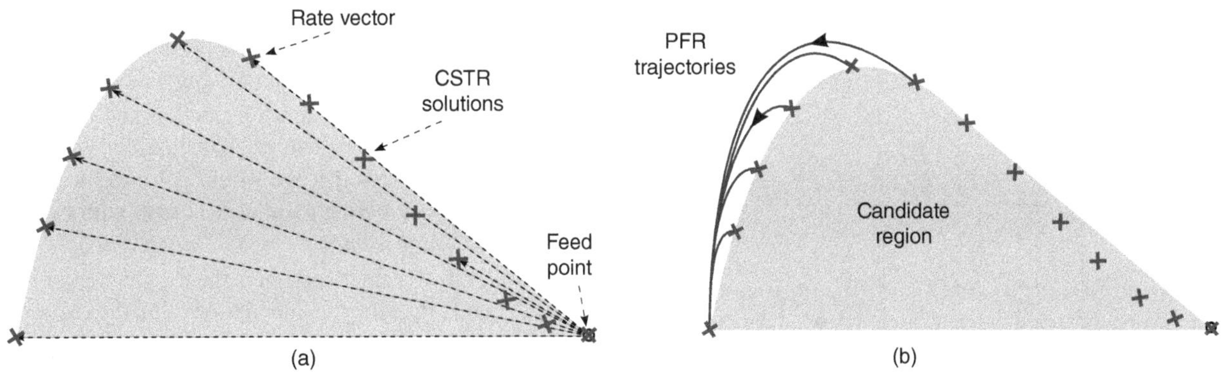

Figure 4.20 (a) A CSTR locus forming the majority of a candidate AR boundary and (b) PFR trajectories from CSTR points on the boundary. The region is often expanded by PFRs when sections of the CSTR locus are found on the region boundary.

section of the convex hull boundary is composed of a CSTR locus, such as that given in Figure 4.20(a).

CSTR solutions are shown as ×'s from a hypothetical feed point. Although the region in Figure 4.20(a) is convex, it still cannot be the true AR. This is because CSTR points must be collinear with the rate vector and the feed point, and thus rate vectors evaluated at the CSTR points may point out of the region.

If rate vectors on the boundary point out of the region, then it is also possible to extend the region using PFR trajectories. This situation is illustrated in Figure 4.20(b). Hence, an important conclusion arises as follow:

*CSTR solutions can only reside on the true AR boundary if rate vectors, evaluated at the CSTR points on the AR boundary, are also **tangent** to the boundary.*

On reflection, it is very difficult for entire sections of a CSTR locus to lie on an AR boundary. In practice, only distinct points of the CSTR locus will touch (be tangent to) the AR boundary, otherwise the remainder of the locus might be extended with PFRs.

OBSERVATION: A section of the CSTR locus on the boundary

If a situation is encountered where a CSTR locus forms part of the candidate AR boundary, then the corresponding region is usually not optimal. All CSTR points on the AR boundary must be associated with rate vectors that are tangent to the AR boundary only. Otherwise, the region can be expanded by running a PFR directly after the CSTR.

SIDE NOTE: The basics of AR theory completed

Although it may not appear as such, we now know enough fundamental reactor theory to begin addressing fairly complex reactor problems using AR theory. The role of the DSR (discussed in Section 4.4.4) is only required for systems involving three or more independent reactions. For readers interested in getting started with AR problems, the remaining sections can be skipped.

EXAMPLE 15: Interpreting boundary structures

Suppose the CSTR and PFR equations are solved for a two-dimensional system involving components c_A and c_B and the plot in Figure 4.21 is generated. What reactor structures would be required to achieve

(a) Point A?
(b) Point C?
(c) Point F?

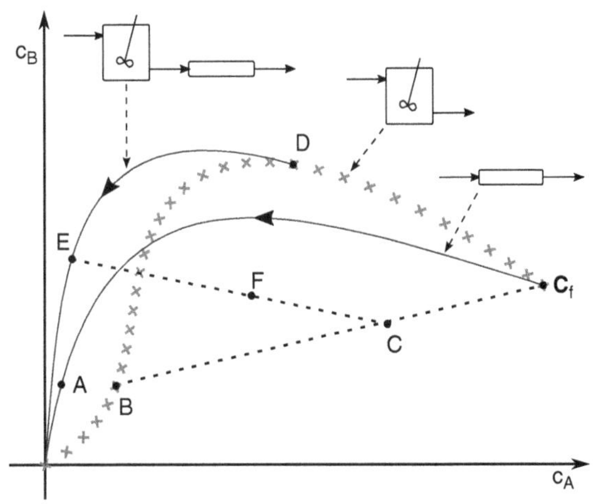

Figure 4.21 Reactor structures for a hypothetical system. (*See color plate section for the color representation of this figure.*)

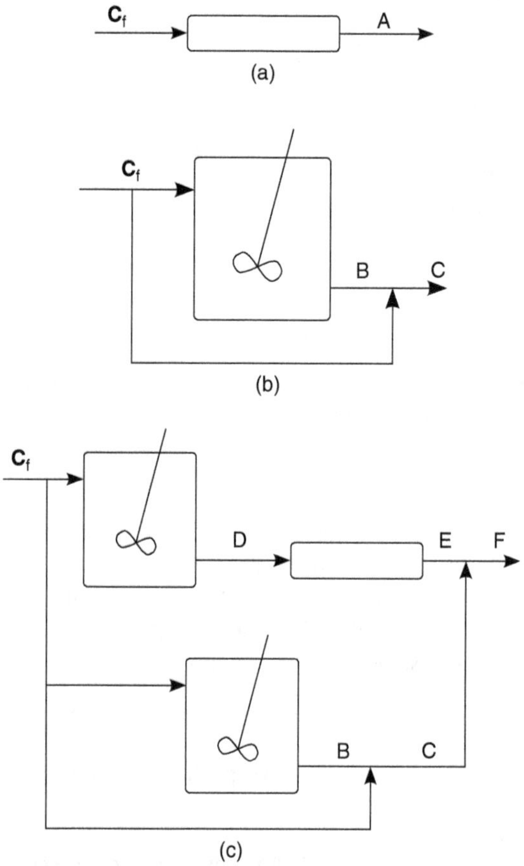

Figure 4.22 (a) Reactor structure required to achieve point A, (b) reactor structure required to achieve point C and (c) reactor structure required to achieve point F.

Answer a:

Point A lies on the PFR trajectory that is initiated at the feed point $\mathbf{C}_f$. Therefore, point A must be achieved via a PFR from the feed, as shown in Figure 4.22(a).

Answer b:

Point C lies on a mixing line that is connected between $\mathbf{C}_f$ and point B. Point B lies on the CSTR locus from the feed. Hence, to produce point C, we must first achieve point B and then mix the CSTR effluent with fresh feed via a bypass, as shown in Figure 4.22(b).

Answer c:

Point F is located on a mixing line joining points C and E. We already know that point C is achieved by a CSTR with bypass from answer (b). Point E lies on a PFR trajectory, although this trajectory is not the same as in answer (a). A CSTR operating up to point D followed by a PFR in series is needed to achieve point E. Point F is therefore a mixture between points E and C, and thus F is obtained by mixing concentrations from a CSTR with bypass of feed and a CSTR–PFR, as shown in Figure 4.22(c).

Using the geometric nature of reaction and mixing, we can interpret reactor structures geometrically, which often results in structures that we would not have been imagined initially, involving parallel structures with bypass as in Figure 4.22(b) and (c).

Note that if we were to compute the convex hull of points A, C, and F, then they would all lie within the region, and thus these points could be synthesized in infinitely many ways. The particular reactor structures described before are simply one such combination that achieved the desired points.

ILLUSTRATION: Recycle reactors

Although we have not explicitly described how recycle reactors work, we can already begin to understand how a typical recycle reactor *should* behave from a geometric viewpoint. Consider first Figure 4.23(a), which shows a PFR with bypass of feed to the effluent stream. A representative PFR trajectory is also drawn in two-dimensional concentration space. Geometrically, bypassing the feed $\mathbf{C}_f$ and mixing with the PFR effluent concentration $\mathbf{C}$ will produce a mixture concentration $\mathbf{C}^*$ on a straight lining joining $\mathbf{C}_f$ and $\mathbf{C}$.

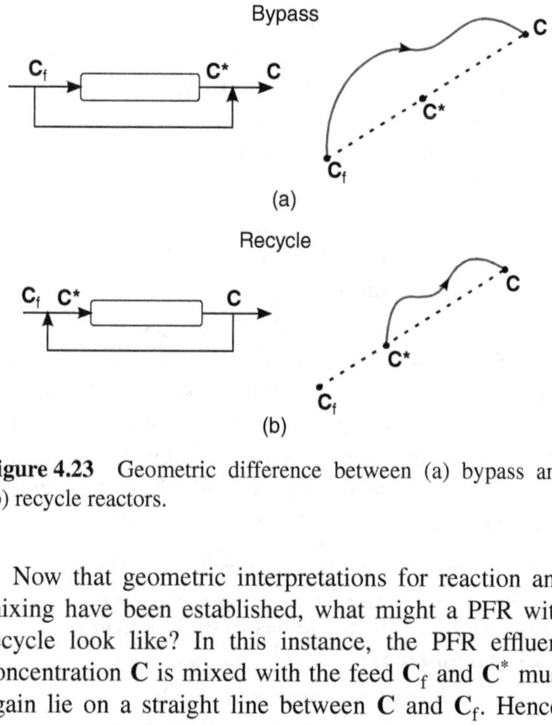

Figure 4.23 Geometric difference between (a) bypass and (b) recycle reactors.

Now that geometric interpretations for reaction and mixing have been established, what might a PFR with recycle look like? In this instance, the PFR effluent concentration $\mathbf{C}$ is mixed with the feed $\mathbf{C}_f$ and $\mathbf{C}^*$ must again lie on a straight line between $\mathbf{C}$ and $\mathbf{C}_f$. Hence, the geometric behavior of all recycle reactors must be similar to that shown in Figure 4.23(b). Whereas the PFR trajectory in Figure 4.23(a) contains all three concentrations $\mathbf{C}_f$, $\mathbf{C}^*$, and $\mathbf{C}$, the PFR trajectory in a recycle reactor only contains $\mathbf{C}^*$ and $\mathbf{C}$, and $\mathbf{C}_f$ must lie beyond the PFR feed concentration $\mathbf{C}^*$ in order for it to be mixed with $\mathbf{C}$. The convex hull of achievable points for a recycle reactor is then different to the convex hull for a bypass arrangement.

Recycle reactors are peculiar because their geometric behavior is harder to predict than reactors with bypass—we must find a steady-state reactor effluent concentration $\mathbf{C}$ that when mixed with $\mathbf{C}_f$ the mixture concentration $\mathbf{C}^*$ is also the feed to the reactor that produces $\mathbf{C}$ again. (For PFR trajectories, this result is equivalent to drawing a straight line passing through $\mathbf{C}_f$ and $\mathbf{C}$ that cuts the PFR trajectory at two points.) This trait of recycle reactors is true for any dimension.

In Chapter 3, we used partial emptying and refilling to improve the concentration of toluene in the BTX reaction. This arrangement of retaining a fraction of product and mixing with feed in the batch is equivalent to running a PFR with recycle.

4.4.4 The Differential Sidestream Reactor

ARs involving two independent reactions may be found using combinations of PFRs and CSTRs alone. However, many problems of interest tend to be higher dimensional in nature. In order to adequately address these problems, the DSR is required.[14] In Chapter 6, we will discuss how DSRs contribute a unique role in the formation of ARs in higher dimensions. The DSR is also useful because it has the ability to mimic the behavior of both CSTRs and PFRs under appropriate operating conditions. Understanding the behavior of DSRs is therefore useful in understanding aspects of AR construction algorithms.

4.4.4.1 Physical Description

a) Introduction From a mixing perspective, the DSR may be viewed as a mixture between a PFR (no mixing) and a CSTR (complete mixing). In practice, this effect can be achieved by allowing for a sidestream of feed material to enter along the length of a PFR. In this way, the DSR may be regarded as a hybrid PFR with sidestream addition used to control mixing, where adjustment of the sidestream flow rate and composition influences the degree of mixing achieved along the length of the reactor.

[14] DSRs are also commonly known as distributed sidestream reactors.

DSRs find particular benefit in applications where fine control of species concentrations is required. Concentrations may be "tuned" with the sidestream addition of material.

It is convenient from a theoretical standpoint to view the DSR as being completely accessible to mixing along the entire length of the reactor. That is, we assume that a continuum of side entry ports is available for the addition of material down the length of the tube. In reality, this behavior must be approximated at discrete positions along the reactor's length. Approximation of DSRs may also be achieved by operating many PFRs in series. Sidestream addition is then realized by mixing points at the PFR effluent streams and flow control valves to adjust the distribution of material down the length of the reactor. Figure 4.24(a)

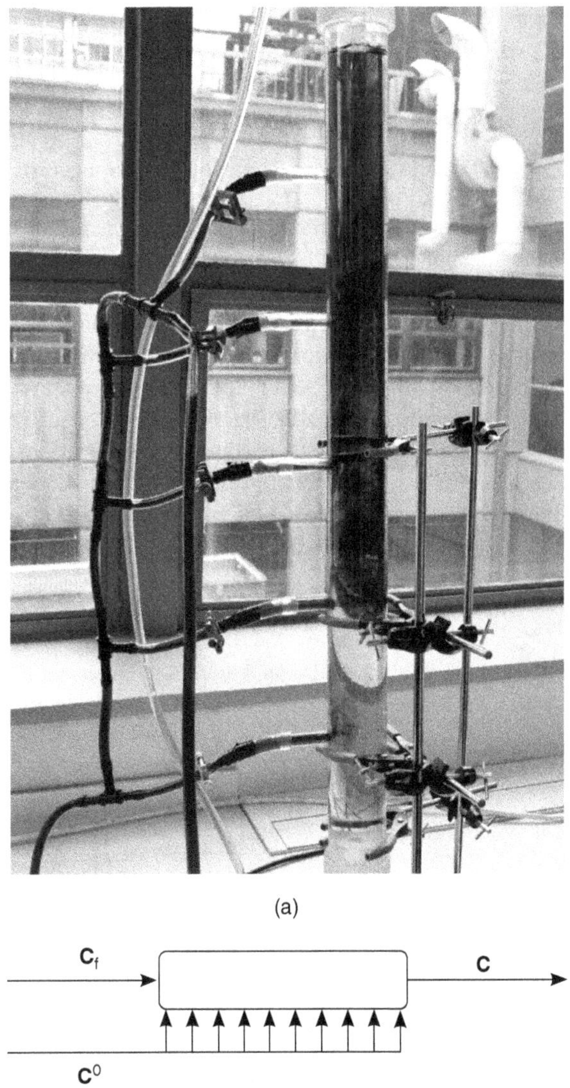

Figure 4.24 (a) A laboratory approximation of a DSR and (b) symbols used to represent a DSR.

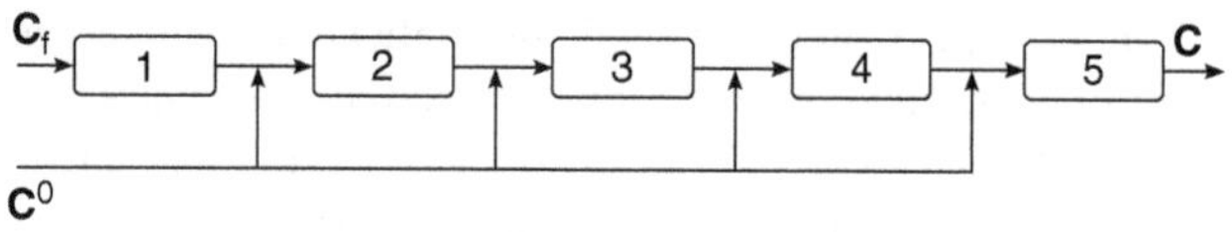

Figure 4.25 Five PFRs in series with sidestream addition that approximate a DSR.

shows a custom-made DSR for a laboratory experiment. Sidestream addition is approximated by five entry ports along the length of the DSR. The flow rate of each sidestream is controlled by a separate valve connected to the port inlet that allows for different sidestream addition policies to be realized. Figure 4.24(b) shows the symbol commonly used to represent a DSR. In Figure 4.25, an approximated DSR is shown using five smaller PFRs. It is important to note that the total mass of material increases down the length of a DSR, which has an effect on the interpretation of the DSR residence time.

SIDE NOTE: Approximating DSR operation

A well-known approximation to DSR operation in reactor design is the cold-shot cooling reactor network. Cold-shot cooling reactors are often employed in exothermic equilibrium limited reactions. A cold-shot of material (that is at a lower temperature than the stream it is mixed with) is fed between a series of exothermic reactors in order to cool the intermediate product and achieve a higher overall conversion. The reactor arrangement is similar to that shown in Figure 4.25. In Chapter 7, the AR for an adiabatic exothermic reaction involving cold-shot cooling will be discussed.

b) Conveyor Belt Analogy Revisited In Section 4.4.2.1, we described how a PFR can be viewed as a series of beakers on a conveyor belt. Let us revisit this interpretation to help describe the physical nature of a DSR. Consider then the modified conveyor belt arrangement shown in Figure 4.26.

The physical arrangement is similar in appearance to the PFR description, although for each beaker, a retort stand and burette containing material held over the beaker are also provided. The purpose of the retort stand and burette is to allow for the feeding of material into the beaker as it travels along the belt. We again assume that the material is perfectly mixed as it enters the beaker. From this standpoint, whereas the PFR can be viewed as a series of batch reactors, the individual beakers here resemble small fed-batch reactors.

The burette need not contain a mixture at the feed composition $\mathbf{C}_f$. In fact, very little will be assumed about the composition in the burette—the mixture concentration might be a constant, or it might change with time as if the material also undergoes a reaction itself. Furthermore, the flow rate of material from the burette might also change with time. The flow rate and composition of material added to each beaker is then assumed to change throughout the duration of the beaker's journey.

There are again two views from which one could interpret the concentration of a beaker on the conveyor belt.

1. One could stand next to the conveyor belt, at a fixed position along its length, and simply observe the concentration in the beaker as it passes the point of observation. There is no change in concentration when viewing beakers as they pass from this perspective.
2. One could also travel on the belt and look at the beaker concentration as it moves down the belt's length. From this perspective, the concentration in the beaker changes with time. The change in reaction may be due to standard reaction, although it might also change as a result of the additional material added by the burette. Furthermore, since material is continually added to the batch, the volume of the beaker increases with time. A larger volume of product is thus obtained

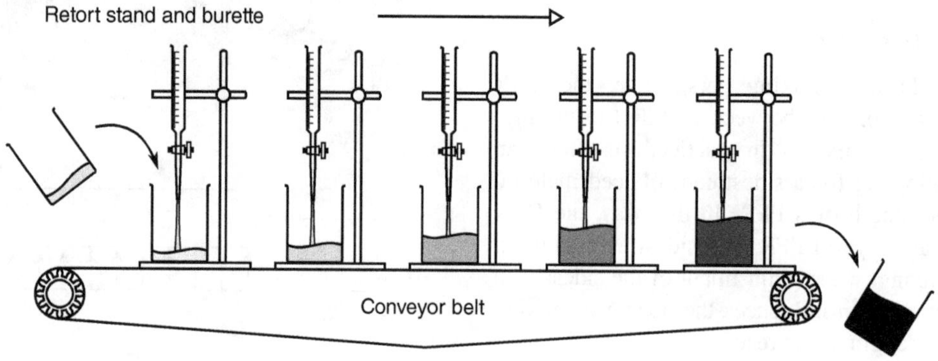

Figure 4.26 A DSR may be viewed as a beaker and retort stand combination on a conveyor belt. (*See color plate section for the color representation of this figure.*)

at the end of the belt compared to what is present in the beaker at the start of the belt.

Assume that the belt is held at a fixed speed. Then a larger production rate is obtained as a result of the burette volume as opposed to a change in belt speed itself. The DSR is in essence a mix between two competing reactor models as follows:

1. If, on the one hand, no additional material flows from the burette (corresponding to adding no material down the length of the DSR tube), then the arrangement is identical to that given for the PFR. In this instance, there is zero influence of mixing on the reaction.
2. If, on the other hand, a very large amount of material is added instantaneously to the beaker, then the reaction is small in comparison to the amount of material added, and the resulting mixture assumes a concentration close to the side stream concentration. In this instance, mixing dominates the reaction.

4.4.4.2 Mathematical Description It is useful to show the derivation of the DSR expression. Both the PFR and CSTR are common reactor models used in chemical reaction engineering, and thus there is little need to formulate these equations from first principles. The DSR, by comparison, may not be familiar to some readers.

ILLUSTRATION: Deriving the DSR equation

It will be useful to keep the physical view of the DSR from Section 4.4.4.1 in mind when performing the derivation. Consider Figure 4.27, showing a differential slice of a DSR.

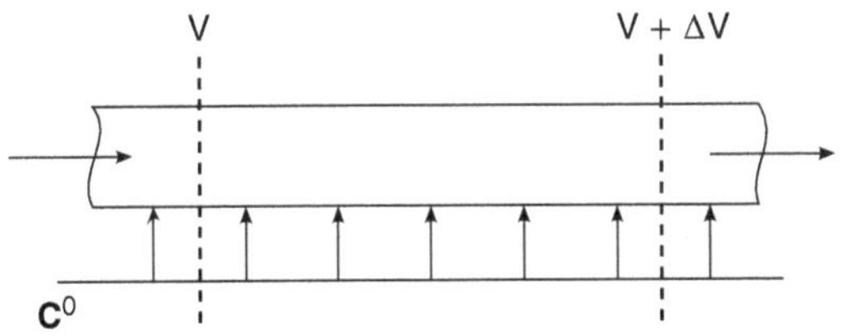

Figure 4.27 Mass balance over a differential segment of a DSR.

A molar balance for component i over the differential boundary segments V and V + ΔV is performed giving

$$n_i^{V+\Delta V} = n_i^V + r_i(\mathbf{C})\Delta V + \alpha c_i^0 \Delta V$$

Here $r_i(\mathbf{C})$ if the rate expression associated with component i, c_i^0 is the sidestream composition to the DSR and α is the rate of sidestream addition per unit volume of reactor (the units of α are [volume of sidestream/(volume of reactor × time)]). It is assumed that concentration remains constant throughout the length of the DSR; however, the flow rate of sidestream material may change through manipulation of α. Superscripts are used to denote the molar flow rate of i at boundaries V and V + ΔV respectively. Dividing through by ΔV and taking the limit as ΔV → 0, we obtain

$$\frac{dn_i}{dV} = r_i(C) + \alpha c_i^0 \tag{4.10a}$$

We may express n_i as the product of the concentration of i and the volumetric flow rate Q at point V along the reactor (recalling that Q is not constant in the DSR).

$$n_i = c_i Q(V)$$

Differentiating with respect to V and invoking the chain rule gives

$$\frac{dn_i}{dV} = c_i \frac{dQ}{dV} + Q\frac{dc_i}{dV} \tag{4.10b}$$

Furthermore, a relation for the volumetric flow rate Q as a function of the DSR reactor volume can be derived as follows:

$$\frac{dQ}{dV} = \alpha \tag{4.10c}$$

Substitution of Equation 4.10b and 4.10c into Equation 4.10a gives

$$Q\frac{dc_i}{dV} = r_i(\mathbf{C}) + \alpha(c_i^0 - c_i)$$

which describes the concentration of component i in a DSR as a function of the sidestream, the rate vector and the DSR reactor volume. For applications in AR, we prefer to parameterize the concentration profiles in terms of a residence time τ. Since the volumetric flow rate varies along the length of the DSR, a conventional definition for residence time (i.e., one that is defined as $\tau = V/Q$) is no longer appropriate.

Instead, we propose the following coordinate transformation that for every reactor volume V, we may associate an equivalent residence time τ. If we let $dV/d\tau = Q$, then substitution of this relation into the DSR expression gives the desired DSR expression in terms of residence time as follows:

$$\frac{dc_i}{d\tau} = r_i(\mathbf{C}) + \alpha(c_i^0 - c_i)$$

Observe that this is for a single generalized component i. An expression for each component in the system may be derived in an identical manner. The resulting system of differential equations may then be summarized in vector notation to give

$$\frac{d\mathbf{C}}{d\tau} = \mathbf{r}(\mathbf{C}) + \alpha(\mathbf{C}^0 - \mathbf{C})$$

Recall from the conveyor belt schematic for a DSR that the volume inside each beaker increases as it moves down the length of the belt. For a constant cross-sectional area, the fluid velocity within the DSR tube increases as a result.

Hence, the DSR expression is given as follows:

$$\frac{d\mathbf{C}}{d\tau} = \mathbf{r}(\mathbf{C}) + \alpha(\mathbf{C}^0 - \mathbf{C}) \tag{4.11}$$

It is clear that the form of the DSR is similar to that of a PFR: both are differential equations, and hence the solution of either will result in a solution trajectory in concentration space. The resulting solution trajectory is accordingly termed a *DSR trajectory*. However, Equation 4.11 contains an extra term relating the instantaneous concentration of the DSR, $\mathbf{C}$, to the sidestream composition, $\mathbf{C}^0$. The α parameter is an important variable used to control the relative proportion of reaction and mixing inside the DSR. When $\alpha = 0$, this indicates that no feed material is fed along the reactor length, and the DSR expression simplifies to the PFR equation given in Section 4.4.2. α is directly proportional to the DSR sidestream flow rate; and as a result, this allows control over the amount of sidestream material added along the length of the DSR. Specifically, we will want to use α to manipulate the resulting DSR trajectory in a manner that achieves concentrations different to a CSTR or PFR alone.

a) *Solving the DSR Equation*

Required Information Since Equation 4.11 represents a system of ODEs, the same information is required to solve the DSR expression as with the PFR. Similar difficulties arise in attempting to analytically integrate the DSR equation also, and so there is again a strong reliance on numerical integration methods for the solution of many realistic DSR problems in practice.

To solve for a DSR trajectory given by Equation 4.11, an initial condition and integration range is needed. The integration parameter again is the DSR residence time τ, whereas the initial condition is the feed composition $\mathbf{C}_f$ at the inlet of the reactor. The sidestream compositions $\mathbf{C}^0$ and flow rate must also be specified along the DSR's length. The sidestream flow rate is typically represented by an α policy along the reactor length. In general, it will be assumed that $\mathbf{C}^0$ is constant down the length of the DSR, although this assumption does not affect our results.

CONCEPT: α policy

In AR theory, the *α policy* is used to refer to how the value of α changes with respect to residence time in a DSR. In Chapter 6, we will describe how it is possible to calculate an α policy corresponding to a DSR that lies entirely on the AR boundary.

Output Successful integration of the DSR expression results in the generation of a list of concentrations and residence times describing a solution trajectory in concentration space. Different combinations of feed conditions, sidestream concentrations, and α policies produce different DSR trajectories for the same kinetics in space.

EXAMPLE 16: Plotting DSR trajectories

The following reaction scheme

$$A + B \rightarrow 2C$$
$$C + B \rightarrow D$$
$$C + A \rightarrow E$$

has kinetics given by

$$r_A = -k_1c_Ac_B - k_3c_Ac_C$$
$$r_B = -k_1c_Ac_B - k_2c_Bc_C$$
$$r_C = 2k_1c_Ac_B - k_2c_Bc_C - k_3c_Ac_C$$

where $k_1 = 1.0$ L/(mol.s), $k_2 = 1.0$ L/(mol.s), and $k_3 = 1.0$ L/(mol.s). The system involves three independent reactions. Let us specify the concentration vector as $\mathbf{C} = [c_A, c_B, c_c]^T$. Plot the following DSR trajectories in c_A–c_B–c_C space using parameters in Table 4.4. For consistency, use a DSR residence time of $\tau = 50.0$ s in your integration.

TABLE 4.4 Parameters for Different DSRs

	$\mathbf{C}_f$	α Policy	$\mathbf{C}^0$
a	$\mathbf{C}_f = [1, 1, 1]^T$	$\alpha = 0$	$[1, 1, 1]^T$
b	$\mathbf{C}_f = [1, 0, 0.1]^T$	$\alpha = \exp(-\tau)$	$[1, 1, 1]^T$
c	$\mathbf{C}_f = [1, 0.5, 0.2]^T$	$\alpha = c_Bc_C$	$[1, 0, 0.1]^T$

Answer:

To plot a DSR trajectory in c_A–c_B–c_C space, the DSR expression must be integrated. The particular solution trajectory obtained is dependent on the feed concentration to the DSR ($\mathbf{C}_f$), sidestream concentration $\mathbf{C}^0$ and α policy.

In Figure 4.28, we show the resulting DSR trajectories for the three specific situations specified in Table 4.4. In the first DSR, the parameters specify a constant α DSR given by $\alpha = 0$. This is in fact a PFR with initial conditions given by $\mathbf{C}_f = [1, 1, 1]^T$. Since there is no sidestream in a, the particular concentration vector used for $\mathbf{C}^0$ is irrelevant. The α policy in the second DSR is specified in terms of the DSR residence time τ and is given by $\alpha = \exp(-\tau)$. This suggests a decrease in the sidestream flow rate along the length of the DSR at the sidestream concentration given by $\mathbf{C}^0 = [1, 1, 1]^T$. In the third DSR, the associated α policy is specified in terms of species concentrations and not in terms of τ. This indicates an α policy whereby the sidestream flow rate is adjusted according to the instantaneous state of the DSR; the expression for α can take on various forms.

Note also that because the feed vectors for each DSR are different, the starting points in space for each trajectory are different as well. A large number of different DSR trajectories can be obtained by appropriate specification of the DSR parameters for the same kinetics.

b) Approximating CSTR and PFR Behavior If the value of $\alpha = 0$ over the entire residence time range, then the PFR expression is obtained. It is thus possible to mimic a PFR in practice by simply not allowing sidestream addition to occur in a DSR. It is also possible, in theory, to operate the DSR as a CSTR. To see this, consider again the form of the DSR equation. If the DSR is integrated for a sufficiently long residence time, then it may be shown that the DSR concentration approaches an equilibrium value. There is no change in concentration in the DSR over time (the derivative with respect to τ approaches zero), and the following expression is obtained:

$$\frac{d\mathbf{C}}{d\tau} = 0 = \mathbf{r}(\mathbf{C}) + \alpha(\mathbf{C}^0 - \mathbf{C})$$

(for long residence times)

The form of this expression is similar to the CSTR equation (Equation 4.8). Certainly, if the substitution $\tau = 1/\alpha$ is made, then the resulting expression is identical to the CSTR equation with feed point given by $\mathbf{C}^0 = \mathbf{C}_f$. It is therefore possible for both CSTR and PFR behavior to be achieved in a DSR under the correct conditions. The generalized DSR expression is thus useful in approximating CSTR and PFR concentrations. Although these approximations might appear impractical, particularly for CSTRs, they find use as a theoretical tool. This dual-natured behavior of DSRs is also useful in many AR construction algorithms.

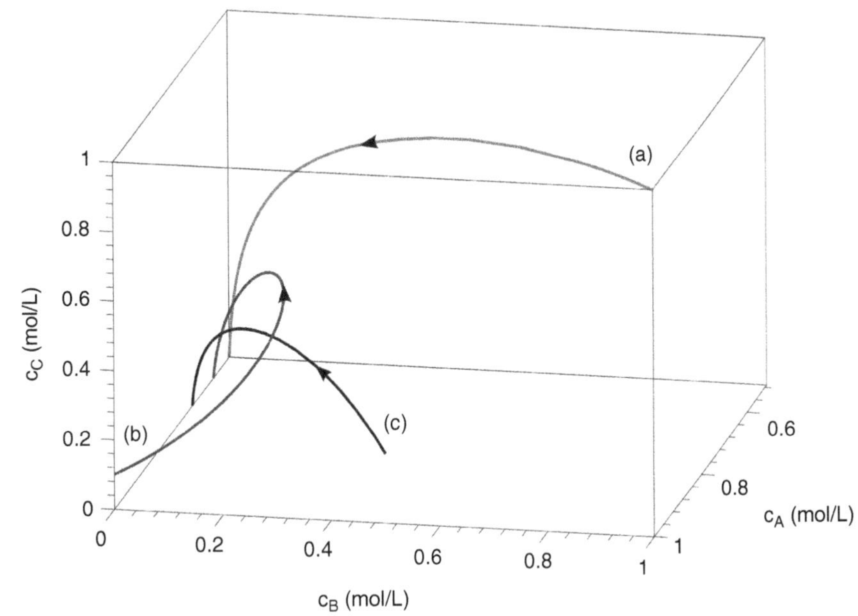

Figure 4.28 DSR trajectories obtained for different parameters for the same kinetics.

EXAMPLE 17: Plotting DSR equilibrium points and CSTRs

Consider the following reaction:

$$A \rightarrow B$$
$$2B \rightarrow C$$

with kinetics given by

$$r_A = k_1 c_A$$
$$r_B = k_1 c_A - 2k_2 c_B^2$$
$$r_c = k_2 c_B^2$$

where $k_1 = 1.0\,s^{-1}$ and $k_2 = 1$ L/(mol.s). Assuming the feed point is given by $\mathbf{C_f} = [c_A,\ c_B]^T = [1,\ 0]^T$ mol/L, answer the following questions:

- Plot DSR trajectories in c_A–c_B space for the following constant α values: $\alpha = 0$, 0.2, 0.5, and 1.1. Assume the sidestream concentration is equal to the feed point. Ensure that the integration time is sufficiently large to approximate the DSR equilibrium concentration.
- Plot the PFR trajectory and CSTR locus from the feed.

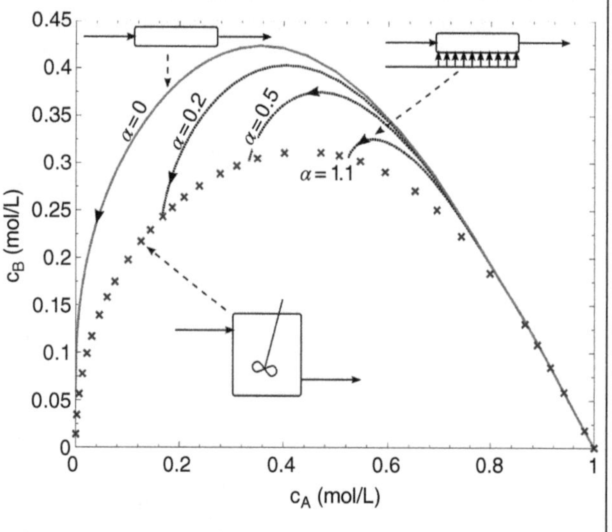

Figure 4.29 DSR trajectories for constant values of α over the integration range. (*See color plate section for the color representation of this figure.*)

Answer:

To plot DSR trajectories, an initial feed $\mathbf{C_f}$, sidestream concentration $\mathbf{C^0}$, and α policy are required. Since $\mathbf{C^0} = \mathbf{C_f}$ and the four constant α policies have been provided, we can plot four DSR trajectories in c_A–c_B space each corresponding to a value of α from the feed point. Integration of the DSR equation can be done with any typical numerical integration technique. After solving the DSR expression for the α values specified, the following curves are produced.

Figure 4.29 also shows the PFR trajectory and CSTR locus from the feed. Observe that the DSR trajectory corresponding to $\alpha = 0$ coincides with the PFR trajectory. DSR residence times have been chosen large enough so as to approximate equilibrium. For values of $\alpha > 0$, the DSR trajectory appears to terminate at points on the CSTR locus. The particular termination point is specific to the value of α in the DSR. This is expected since for a constant α policy, the DSR equation at equilibrium is identical in form to the CSTR equation.

4.4.4.3 *Geometric Interpretation*

a) Description Geometric interpretation of the DSR solution trajectory may be handled in a similar manner to the PFR. The $d\mathbf{C}/d\tau$ term in Equation 4.11 can be viewed as the tangent vector to the DSR trajectory evaluated at point $\mathbf{C}$.

$$\frac{d\mathbf{C}}{d\tau} = \mathbf{r}(\mathbf{C}) + \alpha(\mathbf{C^0} - \mathbf{C})$$

In contrast to the PFR expression, the DSR equation contains two terms on the right-hand side as follows:

1. The rate vector evaluated at the point $\mathbf{C}$, $\mathbf{r}(\mathbf{C})$
2. The difference vector between the sidestream concentration $\mathbf{C^0}$ and concentration $\mathbf{C}$, $(\mathbf{C^0} - \mathbf{C})$

The final direction that the DSR trajectory travels in space is hence controlled by a linear combination of two vectors. If α is zero at any particular point along the reactor, then the final direction is governed completely by the direction and magnitude of $\mathbf{r}(\mathbf{C})$. If, instead, the value of α is made very large so that the term $(\mathbf{C^0} - \mathbf{C})$ dominates the expression, the resulting direction will be toward the sidestream mixing concentration $\mathbf{C^0}$. For any intermediate value of α, where $\mathbf{r}(\mathbf{C})$ and $(\mathbf{C^0} - \mathbf{C})$ are of the same order of magnitude, the final direction will be a linear combination of both $\mathbf{r}(\mathbf{C})$ and $(\mathbf{C^0} - \mathbf{C})$.

Different proportions of $\mathbf{r}(\mathbf{C})$ and $(\mathbf{C^0} - \mathbf{C})$ result in a different final $d\mathbf{C}/d\tau$ term. Many combinations of vectors may be chosen resulting in a unique solution trajectory; however, all of these directions must be confined to within the bounds dictated by $\mathbf{r}(\mathbf{C})$ and $(\mathbf{C^0} - \mathbf{C})$.

Linear combination of two independent vectors allows for a spanning of a subspace. In the case of the DSR, this subspace is spanned by vectors $\mathbf{r}(\mathbf{C})$ and $(\mathbf{C^0} - \mathbf{C})$. Since there

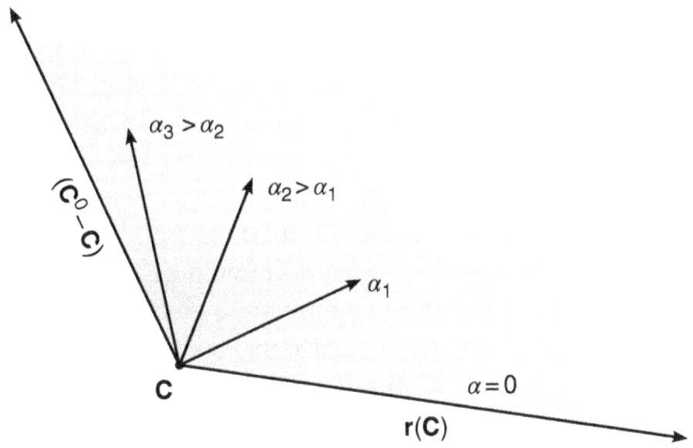

Figure 4.30 A graphical representation of the possible directions a DSR could pursue. The DSR direction is defined by the span of the vectors $\mathbf{C}^0 - \mathbf{C}$ and $\mathbf{r}(\mathbf{C})$.

are only two vectors that are involved in the DSR expression, the dimension of this subspace cannot exceed two.[15] Furthermore, since α is directly proportional to the sidestream flow rate and reactor volume, both of which are positive quantities, α must similarly also always be positive. It is therefore only possible to take positive combinations of $\mathbf{r}(\mathbf{C})$ and $(\mathbf{C}^0 - \mathbf{C})$. Figure 4.30 provides a simple illustration of this property in terms of $\mathbf{r}(\mathbf{C})$ and $(\mathbf{C}^0 - \mathbf{C})$.

b) What Does This Mean? The value of α used in the integration of the DSR is significant in determining the final DSR trajectory produced in space. Although it is possible to set the value of α to a constant over the entire length of the reactor, it is often beneficial in AR applications to have an understanding of how to appropriately manipulate α continuously over the integration period in an optimal manner. Optimal, in the sense of AR theory, is one in which the resulting profile allows expansion of the AR boundary in the largest manner possible.

ILLUSTRATION: DSR trajectories

Consider the plots given in Figure 4.31. The DSR trajectory may be affected by different specifications of the sidestream composition and α policy. In Figure 4.31(a), we show a number of DSR trajectories integrated from a common initial point for different *constant* values of α, labeled α_1 through α_3, respectively. The system kinetics is those belonging to the van de Vusse reaction scheme.[16] The system is three-dimensional, although a projection onto c_A–c_B space is given to show the effect of different DSR solution trajectories associated with different constant α values. The special trajectory associated with $\alpha = 0$ is also given. This corresponds to no sidestream addition, and thus it is in fact a PFR trajectory from the feed point $\mathbf{C}_f$. The sidestream composition used in the generation of these plots is also the feed point, that is $\mathbf{C}^0 = \mathbf{C}_f$. Note that the resulting trajectories obtained for each value of α causes a unique solution in space. One might now envision how the unique trajectories produced by the DSR may be useful in expanding the set of achievable concentrations.

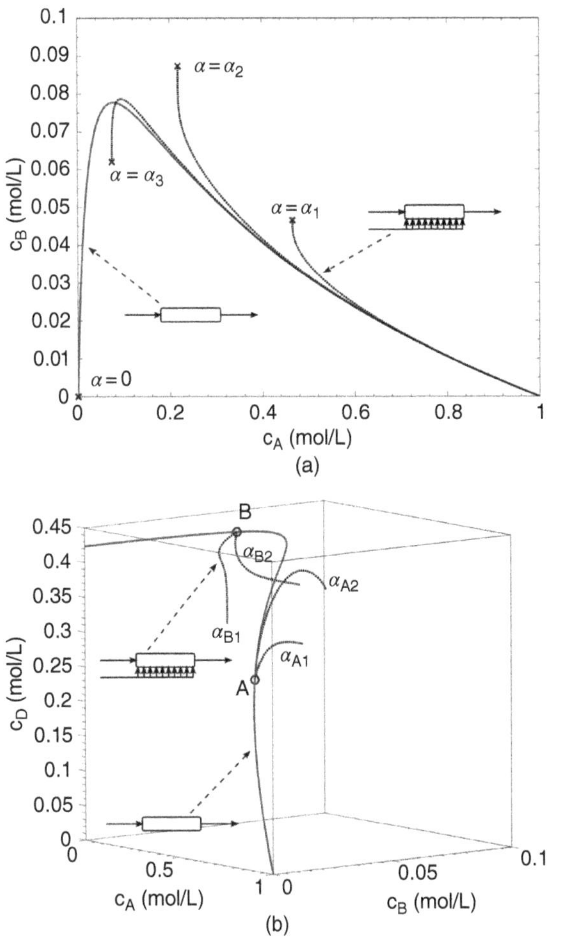

Figure 4.31 (a) DSR trajectories obtained with different α policies and (b) a simple illustration showing how different DSR α policies result in different solution trajectories. Some of the policies serve to grow the region, whereas others do not.

In Figure 4.31(b), the same DSR trajectories as in Figure 4.31(a) are given, however, now in three-dimensional space. In this illustration, the range of possible directions are shown when the value of α is changed at particular points along a base trajectory. The base trajectory in this instance is when $\alpha = 0$ (a PFR trajectory from the feed point). Two points are shown,

[15] See Lay (2012) and Strang (2003).

[16] The van de Vusse system will be studied in detail in Chapter 5 (two-dimensional constructions) and also in Chapter 7 (three-dimensional constructions).

given by points A and B in Figure 4.31(b). Each point shows the resulting trajectories obtained when at two different values of α, namely α_{A1} and α_{A2} for point A and α_{B1} and α_{B2} for point B. The particular values of α in this instance are not important, but it is useful to observe how each trajectory serves to expand the set of achievable concentrations different to that of the base trajectory. There are potentially various curves that may have resulted throughout the entire integration interval for a range of different α values. Some trajectories may expand the base profile, whereas others may be encompassed by the convex hull for the original curve.

In Chapter 8, we shall discuss how constant α DSR trajectories may be employed as a basis for numerically constructing the AR.

4.5 SUMMARY

Now that the physical, mathematical, and geometric nature of all three fundamental reactor types used in AR theory has been detailed, it is useful to provide a summary of the results. Table 4.5 may be used as a point of reference for the problems that are described in later chapters. The ideas described in previous sections form the foundations of AR theory, and hence it is important that these concepts are understood well.

4.6 MIXING TEMPERATURES

4.6.1 Motivation

In the following chapters, it will be useful to have an understanding of how temperature may be applied to systems when constructing candidate regions. Concentration has already been shown to obey a liner mixing law, and in Chapter 5, we will describe how residence time obeys a similar principle. Mixture of these quantities is therefore beneficial when computing convex sets, and an important property for computing the AR. Temperature does not generally obey a linear mixing law unless certain operating assumptions are met, and thus it generally cannot be used in the same manner as concentration and residence time. In this section, we wish to briefly demonstrate why. Consider the following scenario given in Figure 4.32.

Two beakers are available that each contains a unique concentration of species (c_A, c_B, and so on) of a specific mass and temperature. Beaker 1 contains a total mass m_1 that is held at a temperature T_1, and similarly beaker 2 is of mass m_2 held at temperature T_2. When beaker 1 and beaker 2 are mixed together, the resulting mixture has total mass $m^* = m_1 + m_2$ and temperature given by T^*.

4.6.2 Adiabatic Energy Balance

The final mixture temperature T^* may be expressed as a function of the properties of the individual beakers by performing an adiabatic energy balance for the process. For simplicity, assume that the specific heat capacity (Cp) of the components in beakers 1 and 2 and the mixture may be described by a single bulk Cp term. That is, the Cp of the bulk mixture for beakers 1 and 2 and the total are given by Cp_1, Cp_2, and Cp^*, respectively. Assume also that the Cps are not functions of temperature. A simplified adiabatic energy balance may be written over the system, giving the following result:

$$\Delta H = 0 \tag{4.12}$$

$$\begin{aligned} H_3 - H_1 - H_2 &= 0 \\ m_1Cp_1(T_1 - T_0) + m_2Cp_2(T_2 - T_0) &= m^*\Delta H_{mix} \\ &\quad + m^*Cp^*(T^* - T_0) \end{aligned} \tag{4.13}$$

The term ΔH_{mix} represents the enthalpy of mixing for the mixture at the mixture temperature.[17] T_0 is a basis temperature, from which the enthalpy of each beaker is measured (typically 25 °C). A similar substitution may be performed, as described in Chapter 2, to express the mixing process in terms of the ratio of masses in beakers 1, 2, and 3. If the substitution $\lambda = m_1/m^*$ is made, then the resulting expression is obtained:

$$\lambda Cp_1(T_1-T_0) + (1-\lambda)Cp_2(T_2-T_0) = \Delta H_{mix} + Cp^*(T^*-T_0) \tag{4.14}$$

The goal here is to express T^* as a linear combination of T_1 and T_2 in the following form:

$$T^* = \lambda T_1 + (1-\lambda)T_2 \tag{4.15}$$

This may be achieved if it is assumed

1. That the bulk heat capacities of the individual beakers are identical to the mixture, that is, $Cp_1 = Cp_2 = Cp^*$;
2. The enthalpy of mixing of the mixture may be ignored ($\Delta H_{mix} = 0$).

From these assumptions, it is evident that expressing T^* as a linear combination of T_1 and T_2 only occurs under extraordinary constraints. It is therefore unlikely that temperature may be mixed in the same manner as concentration or residence time for the purpose of AR constructions. The construction of candidate regions using temperature are generally not considered with temperature on a dedicated axis, such as with concentration and residence time. The role

[17] Note that each beaker may be a mixture with its own enthalpy of mixing term. The ΔH_{mix} term in Equation 4.13 then represents a lumped mixing term taking into consideration the enthalpy of mixing for each beaker.

TABLE 4.5 Summary of the Three Fundamental Reactor Types Used in AR Theory, Based on Physical, Mathematical, and Geometric Properties

Fundamental Reactor Type	PFR	CSTR	DSR
Symbol	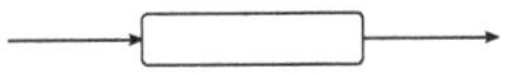	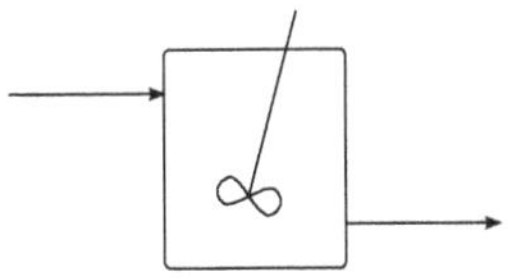	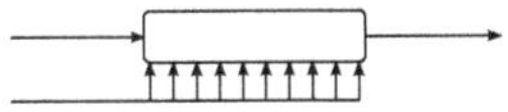
Physical characteristic	No axial mixing (plug flow). Beakers on a conveyor belt.	Complete mixing.	Intermediate mixing (a mix of a PFR and a CSTR). Fed-batch reactors on a conveyor belt.
Mathematical equation	System of ODEs. $\frac{d\mathbf{C}}{d\tau} = \mathbf{r}(\mathbf{C})$ Requires initial condition ($\mathbf{C} = \mathbf{C}_f$ at $\tau = 0$) and integration range $0 \le \tau \le \tau_{max}$.	System of nonlinear equations. $\mathbf{C} = \mathbf{C}_f + \tau\mathbf{r}(\mathbf{C})$ Requires feed point $\mathbf{C}_f$ and CSTR residence time τ.	System of ODEs. $\frac{d\mathbf{C}}{d\tau} = \mathbf{r}(\mathbf{C}) + \alpha(\mathbf{C}^0 - \mathbf{C})$ Requires initial condition ($\mathbf{C} = \mathbf{C}_f$ at $\tau = 0$) and integration range $0 \le \tau \le \tau_{max}$, as well as α policy and sidestream concentration $\mathbf{C}^0$.
Solution type	Integration of ODEs (Runge–Kutta, etc.). Results in a solution trajectory—a curve in space.	Roots of a system of nonlinear equations (Newton's method, etc.). Results in a distinct set of points that satisfy the CSTR equation for a residence time. Many residence times give a CSTR locus.	Integration of ODEs (Runge–Kutta, etc.). Results in a DSR solution trajectory.
Solution results	Many concentrations obtained for a residence time and feed point.	(Potentially many) distinct solutions for a residence time and feed point.	Many concentrations obtained for a residence time. Different α policies give different trajectories for the same feed point.
Solution difficulty	Straightforward to solve.	Generally difficult to solve because of multiple steady states.	Straightforward to solve.
Geometric interpretation	Each point on the trajectory is tangent to the rate vector at that point.	Each point on the CSTR locus is collinear (and points in the same direction) with the mixing vector $(\mathbf{C} - \mathbf{C}_f)$.	The tangent to each point on the DSR trajectory lies in a direction somewhere in between the rate vector and the mixing vector $(\mathbf{C}^0 - \mathbf{C})$. A PFR if $\alpha = 0$ and a CSTR for long integration times ($d\mathbf{C}/d\tau = 0$).

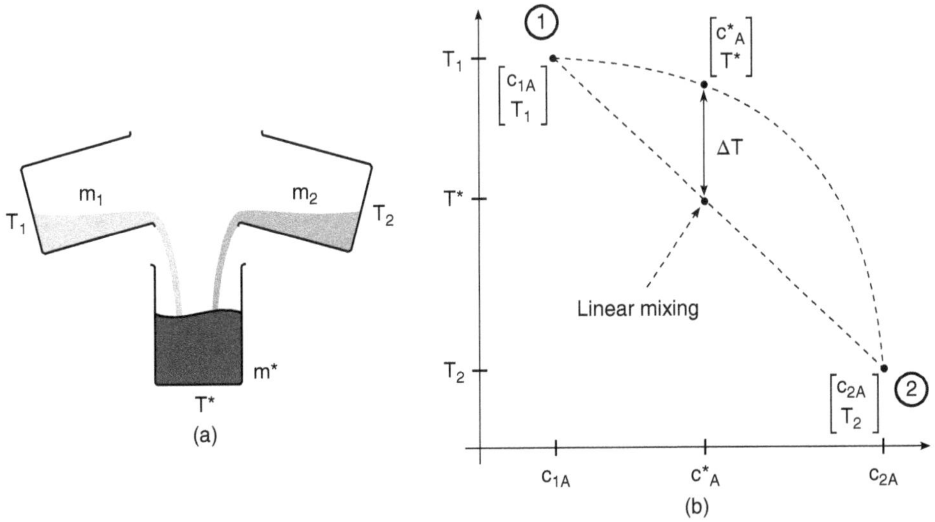

Figure 4.32 (a) Physical interpretation of mixing two beakers of different temperature and (b) geometric interpretation of mixing temperatures.

of temperature must be handled indirectly, such as through the energy balance and rate expressions.

SIDE NOTE: Working with temperature in AR theory

The approach to constructing *adiabatic* ARs with temperature is slightly more complicated than isothermal constructions. The energy balance generally does not allow for temperature to obey a linear mixing law, and as a result temperature cannot generally be treated as a pseudo component in the state vector **C** (which is possible with residence time τ). If temperature is to be incorporated, T must usually be introduced into the analysis via an energy balance and treated as an extra parameter in a rate expression, in the form **r**(**C**, T).

r(**C**, T) is thus a vector in concentration space with its magnitude and direction influenced by its position in concentration space as well as the system temperature, which is a function of the energy balance. Geometrically, this form of the rate vector indicates that multiple rates may be present at a single point in concentration space depending on the temperature specified.

4.6.3 Non-adiabatic Energy Balance

It is possible for temperature to obey a linear mixing law if the ability for a heating/cooling term, q, is introduced to adjust the temperature of the final mixture. To see this, consider again the scenario given in Figure 4.32(a). Since the system is no longer adiabatic, a new energy balance can be written including a heat transfer term q as follows:

$$\begin{aligned} q &= \Delta H \\ q &= H_3 - H_1 - H_2 \\ q &= \Delta H_{mix} + Cp^*(T^* - T_0) - \lambda Cp_1(T_1 - T_0) \\ &\quad - (1 - \lambda)Cp_2(T_2 - T_0) \end{aligned}$$

In order for T^* to operate in a linear fashion, Equation 4.15 must be satisfied. Substituting Equation 4.15 into the expression provides a relation for the heating or cooling duty for linear mixing to be obeyed.

$$q = \Delta H_{mix} + Cp^*[\lambda T_1 + (1 - \lambda)T_2 - T_0] \quad (4.16)$$
$$- \lambda Cp_1(T_1 - T_0) - (1 - \lambda)Cp_2(T_2 - T_0) \quad (4.17)$$

EXAMPLE 18: Mixing temperature

Two beakers of mass $m_1 = 1$ kg and $m_2 = 1.5$ kg are mixed together. The temperature of beaker 1 is $T_1 = 25\,°C$, and the temperature of beaker 2 is $T_2 = 45\,°C$. What is the heating duty required in order for the final mixture temperature, T^*, to obey a linear mixing law?

Assume that a basis temperature of $T_0 = 25\,°C$ is chosen, the specific enthalpy of mixing (per mass of final mixture) is $\Delta H_{mix} = -135.5$ kJ/kg and specific heats of the two beakers and the mixture are $Cp_1 = 3.2$ kJ/(kg.K), $Cp_2 = 4.1$ kJ/(kg.K), and $Cp^* = 3.7$ kJ/(kg.K), respectively.

Answer:

Calculating the energy requirements is straightforward since all relevant information is given. The mixing fraction is determined from m_1 and m_2 as follows:

$$\begin{aligned} \lambda &= \frac{m_1}{m_1 + m_2} = \frac{1}{1 + 1.5} \\ &= 0.4 \end{aligned}$$

Converting temperatures to units of Kelvin gives

$$T_1 = 25 + 273.15 = 298.15\,K$$
$$T_2 = 45 + 273.15 = 318.15\,K$$
$$T_0 = 25 + 273.15 = 298.15\,K$$

Substituting these values into Equation 4.17 and solving for q then gives

$$\begin{aligned} q &= \Delta H_{mix} + Cp^*[\lambda T_1 + (1 - \lambda)T_2 - T_0] \\ &\quad - \lambda Cp_1(T_1 - T_0) - (1 - \lambda)Cp_2(T_2 - T_0) \\ &= (-135.5) + (3.7)[(0.4)(298.15) + (0.6)(318.15) \\ &\quad - 298.15] - (0.4)(3.2)(298.15 - 298.15) \\ &\quad - (0.6)(4.1)(318.15 - 298.15) \\ &= -140.3\,kJ/kg \end{aligned}$$

A negative value indicates a net release of energy. Therefore, 140.3 kJ of cooling is required per kg of mixture in order for temperature to obey a linear mixing law.

Figure 4.32(b) is representative of temperature profile obtained in a nonlinear system. Due to the form of the energy balance, there is often a deviation away from the straight line. The temperature difference between the linear and nonlinear case is therefore proportional to the amount of heating or cooling required.

It is therefore unlikely that the temperature will obey the lever arm rule under adiabatic mixing conditions. If we are to use this method, then we must employ heating or cooling as well. Nicol et al. (1997, 2001) provide descriptions for how this could be incorporated.

> **IMPORTANT: Mixing temperature**
>
> It is generally not possible to mix temperature in the same fashion as with concentration or residence time. This is because temperature generally does not obey a linear mixing law (unless the energy balance is simple).
>
> It is therefore not often that an AR may be plotted with temperature on an independent axis (as is the case with concentration, mass fraction, and residence time), and temperature must be accommodated in a different manner.

4.7 ADDITIONAL PROPERTIES OF THE AR

Chapter 3 outlined a number of fundamental properties of the AR, which are founded primarily on the principles of concentration and mixing. Now that a better understanding of reaction has been established, a number of additional AR properties can be described.

Property 6: All rate vectors evaluated on the boundary of the AR must either point into the current candidate region, or be tangent to the boundary.

If a rate vector exists on the AR boundary that points out of the region, then it would be possible to expand the region by initiating a PFR from that point. Figure 4.33 shows three points on a hypothetical AR boundary in two-dimensional concentration space.

1. Point 1 has a rate vector that points into the region. If a PFR trajectory were to be initiated from this point, the resulting solution trajectory would need to travel, at least initially, into the region. This point may or may not serve to extend the region further. Nothing further may be concluded.
2. Similar to that of point 1, point 2 may also lie on a mixing line. If a PFR were to be initiated with a feed concentration from point 2, the resulting trajectory would travel, at least initially, *out* of the region. This trajectory would hence serve to further expand the set of achievable points.
3. The rate vector associated with point 3 is exactly tangent to the boundary. Point 3 must therefore be a concentration belonging to a PFR trajectory. Portions of the resulting curve therefore either lie on the AR boundary, or inside the region.

Property 7: All points in the complement of the region must contain rate vectors such that, when extended backward, these points do not intersect the AR boundary.

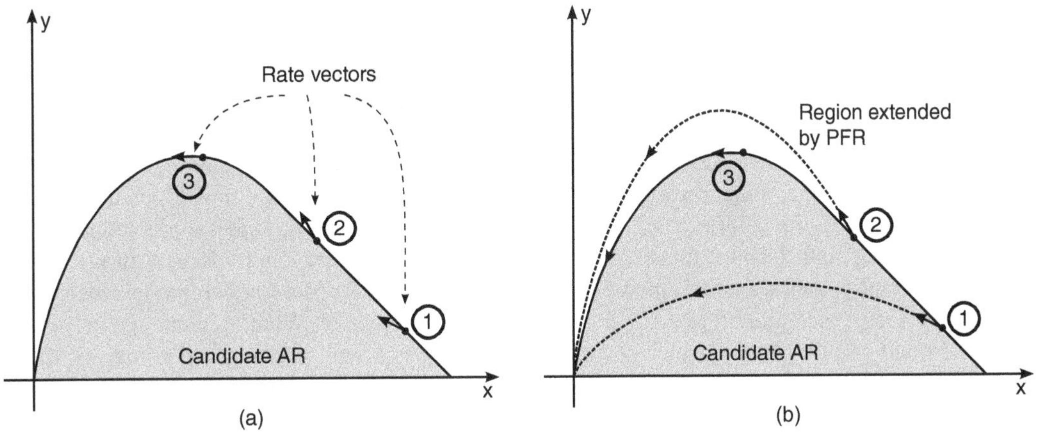

Figure 4.33 (a) Hypothetical AR boundary showing three rate vectors evaluated at specific points on the boundary and (b) the resulting PFR trajectories that are obtained from the three points of interest.

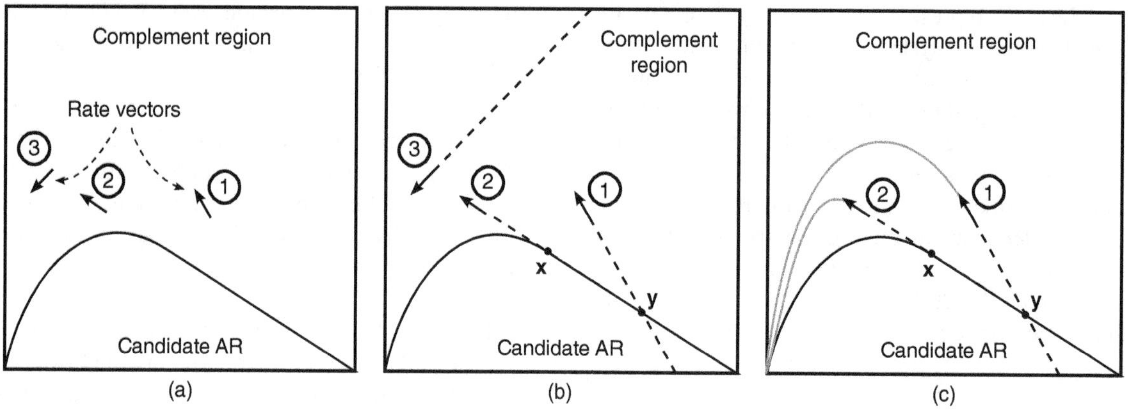

Figure 4.34 (a) Three rate vectors residing in the complement region; (b) rays extended backward from the three points of interest; and (c) only rate vectors extended backward, which intersect the candidate AR boundary, serve to expand the region (via CSTR).

> **CONCEPT: Complement region**
>
> The *complement region* to the AR is the region in space that does not include the AR (it is the remaining space excluding the AR).

All points on a straight line drawn backward from a rate vector are the set of potential CSTR feed concentrations that have effluent concentrations compatible with the point of interest. This is clear when one considers the CSTR equation

$$\mathbf{C} = \mathbf{C}_f + \tau\mathbf{r}(\mathbf{C}) \tag{4.8}$$

or

$$\mathbf{C}_f = \mathbf{C} - \tau\mathbf{r}(\mathbf{C})$$

If the straight line intersects a candidate AR, then the point must be achievable by the CSTR equation.

Consider the hypothetical candidate AR given in Figure 4.34(a). Three points numbered 1–3 are shown, lying in the complement region to the candidate AR. Straight lines may be extended *backward* from the points shown, creating three *rays*, which are shown in Figure 4.34(b).

The rays associated with points 1 and 2 touch the candidate region at points **x** and **y**, and are thus achievable by a CSTR. (Point **x** is the feed concentration to a CSTR that achieves point 2, and point **y** is the feed concentration to a CSTR that achieves point 1.) Note that the ray associated with point 3 does not intersect the candidate region, and as a result it is not directly achievable by a CSTR.

> **CONCEPT: Rays**
>
> A *ray* is defined by the equation
>
> $$\mathbf{C} = \mathbf{C} + t\mathbf{v}$$
>
> where t is a scalar quantity. Thus, rays are vectors containing a definite initial point but no end point. Rays may intersect one or more points in space. For two dimensions, the ray equation may be rearranged to give the equation of a straight line, which is similar in form to the CSTR equation. (The vector form of the CSTR equation can be viewed as a ray equation.)

4.8 CHAPTER REVIEW

Viewing reaction geometrically is useful because we can use the properties of vectors and matrices and integrate this thinking with concentration vectors, mixing, and convex hulls. A geometric understanding of reaction, in conjunction with concentration and mixing, forms the foundations of AR theory, and we are now a position to apply these ideas to reactor network synthesis problems. We will continue to provide new ideas and concepts in the following chapters, although these ideas are simply extensions to the concepts described here. What remains is to now gain practical experience with creating and using AR theory. In Chapter 5, we will look at finding the AR for a number of simple two-dimensional systems. Chapter 5 also marks the end of Section I of the book.

REFERENCES

Acton, F.S., 1990. Numerical Methods That Usually Work. MAA, Washington, DC.

Brooks, B.W., 1988. Semi-batch reactors and their use for autocatalytic reactions. Chem. Eng. Sci. 43, 1287–1290.

Fogler, H.S., 2010. Essentials of Chemical Reaction Engineering, 1st ed. Prentice Hall, Upper Saddle River, NJ.

Lay, D.C., 2012. Linear Algebra and Its Applications. Addison-Wesley, Boston.

Lehrer, L., 1968. Gas agitation of liquids. Ind. Eng. Chem. Process Des. Dev. 7, 226–239.

Nicol, W., Hildebrandt, D., Glasser, D., 1997. Process synthesis for reaction systems with cooling via finding the attainable region. Comput. Chem. Eng. 21, S35–S40.

Nicol, W., Hernier, M., Hildebrant, D., Glasser, D., 2001. The attainable region and process synthesis: Reaction systems with external cooling and heating. The effect of relative cost of reactor volume to heat exchange area on the optimal process layout. Chem. Eng. Sci. 56, 173–191.

Press, W.H., 2007. Numerical Recipes: The Art of Scientific Computing, 3rd ed. Cambridge University Press, Cambridge.

Strang, G., 2003. Introduction to Linear Algebra. SIAM, Philadelphia.

Strogatz, S.H., 2001. Nonlinear Dynamics and Chaos: With Applications to Physics, Biology, Chemistry, and Engineering. Westview Press, Cambridge, MA.

5

TWO-DIMENSIONAL CONSTRUCTIONS

5.1 INTRODUCTION

In previous chapters, we developed a theoretical framework for interpreting concentration, mixing, and reaction from a geometric perspective. Although these concepts are simple in nature, they form the foundations of AR theory. We are now ready to begin applying these concepts to address reactor network synthesis problems.

This chapter aims to provide several worked examples of how to generate candidate ARs. All of these examples are two-dimensional in nature, which is useful for demonstration purposes because the results are easily visualized. Do not let this constraint mislead you into thinking that the examples are insignificant. By the end of this chapter, readers should be able to solve reactor network problems with multiple reactions involving selectivity, yield, conversion, and minimum reactor volume.

The specific method used to generate a candidate region will depend on the complexity of the system studied, but there are tasks that commonly arise for all systems, irrespective of complexity. Before the examples are discussed, a framework is provided that may be useful for solving AR problems for the first time. For readers interested in tackling problems straightaway, Section 5.2 may be skipped.

5.2 A FRAMEWORK FOR TACKLING AR PROBLEMS

The use of AR theory is somewhat different from many other optimization techniques, and hence it is useful to provide a guideline for how to approach reactor network synthesis problems from this viewpoint. This framework outlines five key steps, with the level of difficulty involved in each step placed in parentheses.

1. **Preparation**. Define the reaction dimension and feed set (easy): Given a system of reactions and associated kinetics, the first task is to define the system in terms of a unique set of initial achievable points. Before it is possible to construct the AR, one must first determine the space wherein the AR must reside (by choosing unique species components in the reactions that will represent the AR). The set of feed points are, by definition, attainable. Hence, the convex hull of the feed points might be considered the smallest candidate AR when reaction is not available. For the problems discussed here, a single feed point is assumed.

   It is also beneficial to have an understanding of the position of equilibrium points in the system. If, during construction, points with zero rate vectors are found, this may interfere with the overall construction process. Knowledge of equilibrium points is not a requirement however.[1]
2. **AR construction**. Generate the AR using combinations of the three fundamental reactor types and mixing (often difficult): Utilizing concepts developed in previous chapters, it is possible to begin constructing the AR, specific to the designated reaction kinetics and feed conditions. AR construction typically begins by determining the PFR trajectory and CSTR locus from the feed. Depending on the complexity of the kinetics, it might only be possible to construct a candidate AR that is a subset of the true AR. Even if only a subset of the true AR is utilized, this still allows a better understanding of the operating limits of the system. AR construction is the most difficult and time-consuming step of the entire problem process, yet it also provides the most information about the system.
3. **Boundary interpretation**. Interpret the AR boundary in terms of reactor equipment (sometimes easy,

[1] Recall from Chapter 4 that finding equilibrium points involves solving for the roots to the rate function expressions, which is often a system of nonlinear equations. This may be a difficult task in itself.

Attainable Region Theory: An Introduction to Choosing an Optimal Reactor, First Edition. David Ming, David Glasser, Diane Hildebrandt, Benjamin Glasser, and Matthew Metzger.

Companion website: http://attainableregions.com

sometimes not possible): The AR boundary is defined in terms of reactor structures, composed of combinations of the three fundamental reactor types and mixing. Each reactor type exhibits a unique geometric interpretation. Determining the reactor configurations that form the AR relates to interpreting the surfaces that form the AR boundary. This procedure is straightforward in practice if reactor structures have been used to build the AR. Conversely, it is generally not possible to find AR boundary structures if the AR has been computed numerically by a computer algorithm, which does not employ reactor configurations in the construction scheme. Interpretation of the AR boundary is therefore dependent on the complexity of the system as well as the manner by which the AR is computed.

4. **Overlay objective function**. Define the objective function and overlay this onto the AR to see where it intersects the AR boundary (easy): Once the AR has been determined, the limits of achievability by the system—for the kinetics and the feed point—are known. This information may be used to answer one or more design or optimization questions related to the system. Sections of the objective function that intersect the AR are optimal points, relative to the objective function specified. This task is often straightforward to carry out in practice.
5. **Optimize**. Knowledge of the optimal reactor structures allows for the determination of the specific reactor configuration required to achieve the intersection point (difficult if the optimal structure is not known): If the desired objective function points lie in the interior of the AR, then from Chapter 2, there are infinitely many ways to achieve the objective function using mixing. If optimal points lie on the boundary, then there is generally a unique path to the point using the appropriate optimal reactor structure. Depending on whether the optimal reactor structures are known for the AR (as in the case of point 3), it may be difficult to know how to achieve the intersection points. Often simply understanding the location of the intersection points in relation to the AR boundary is sufficient to formulate appropriate performance targets, which could improve current designs via a traditional optimization methodology.

Since the AR is fixed for a given feed point, multiple objective functions may be formed relative to a single AR, and thus multiple optimizations may be performed. This makes analysis of several optimization scenarios easy, as there is no need to perform further optimizations when the objective function is changed. In the following sections, we will demonstrate how this framework is used via several well-known reactive systems, which are often studied in AR theory.

5.3 TWO-DIMENSIONAL VAN DE VUSSE KINETICS

5.3.1 Introduction

5.3.1.1 Motivation In Chapter 3, we introduced the AR for the BTX problem from an *experimental* perspective. We now wish to find the AR for a similar system using concepts developed in Chapter 4. Only isothermal systems with no volume change on reaction and mixing are examined.[2]

The formal set of reaction kinetics we shall consider is that usually called the Van de Vusse example (Van de Vusse, 1964). The system has been studied extensively in AR theory literature and is often used for demonstration purposes. In this way, the system makes for a fitting "template" problem.

5.3.1.2 Reactions The Van de Vusse reactions involve a parallel decomposition of a hypothetical species A to side-products B and D. Component B is an intermediate product (similar to toluene in the BTX system), which further decomposes to a final product C.

$$\mathrm{A} \rightarrow \mathrm{B} \rightarrow \mathrm{C} \tag{5.1a}$$

$$2\mathrm{A} \rightarrow \mathrm{D} \tag{5.1b}$$

Observe that this set of reactions is similar to the BTX system presented in Chapter 1, if ethylene and hydrogen are omitted. The kinetics of both reactions in Equation 5.1a is first-order irreversible, whereas Equation 5.1b is second-order irreversible. Thus, for component A,

$$r_A = -k_1c_A - k_3c_A^2 \tag{5.2a}$$

The rate of formation of B is

$$r_B = k_1c_A - k_2c_B \tag{5.2b}$$

Similarly, we can write the rate of production for species C and D as follows:

$$r_C = k_2c_B \tag{5.2c}$$

$$r_D = k_3c_A^2 \tag{5.2d}$$

The system is flexible in many respects—particular specification of the rate constants (k_1, k_2, and k_3) results in different ARs, and thus different optimal reactor structures may be generated from a single set of reactions. In the following sections, three distinct scenarios shall be discussed that will facilitate how the AR may be constructed from the theory developed in the previous chapters. From the information generated by the AR, informed decisions can be made regarding the most appropriate reactor structure for a desired duty.

[2]These assumptions will be relaxed in Chapters 8 and 9.

5.3.1.3 *Problem Specification and Simplifying Notation*

It will be useful to consider dimensionless rate constants in the following sections, which are defined as follows:

$$a_1 = \frac{k_3 c_{Af}}{k_1}$$

$$a_2 = \frac{k_2}{k_1} \tag{5.3}$$

Assume that the feed is pure in A with a feed concentration of $c_{Af} = 1.0$ mol/L. Since the kinetics is a function of only components A and B, it is possible to construct the AR in two-dimensional concentration space ($c_A - c_B$). Hence, we aim to construct the AR in $c_A - c_B$ space using a feed vector of $\mathbf{C_f} = [c_{Af}, c_{Bf}]^T = [1.0, 0.0]^T$ mol/L for three distinct scenarios using dimensionless rate constants a_1 and a_2:

1. $a_1 = a_2$
2. $a_1 < a_2$
3. $a_1 > a_2$

AR constructions for these three scenarios will be discussed next, which are facilitated by the AR construction framework discussed in Section 5.2.

5.3.2 Scenario 1: $a_1 = a_2$

5.3.2.1 Step 1: Preparation The construction process is initiated by defining the system kinetics and feed point. This information has already been discussed as part of Sections 5.3.1.2 and 5.3.1.3: we wish to determine the AR for the two-dimensional Van de Vusse system, in $c_A - c_B$ space, using a feed concentration of $\mathbf{C_f} = [c_{Af}, c_{Bf}]^T = [1.0, 0.0]^T$ mol/L.

In this section, we wish to investigate the scenario when $a_1 = a_2$. Rate constants supplied in Table 5.1 satisfy this criterion.

5.3.2.2 *Step 2: AR Construction*

a) PFR Trajectory from the Feed Equation 5.2a through 5.2d may be used along with rate constants given in Table 5.1 to plot concentration profiles of each species as a function of PFR residence time τ. The results of the integration are shown in Figure 5.1.

TABLE 5.1 Rate Constants for Van de Vusse Kinetics in the Equal Rate Case

Rate Constant	Value
k_1	1.0 s^{-1}
k_2	1.0 s^{-1}
k_3	1.0 L/(mol.s)

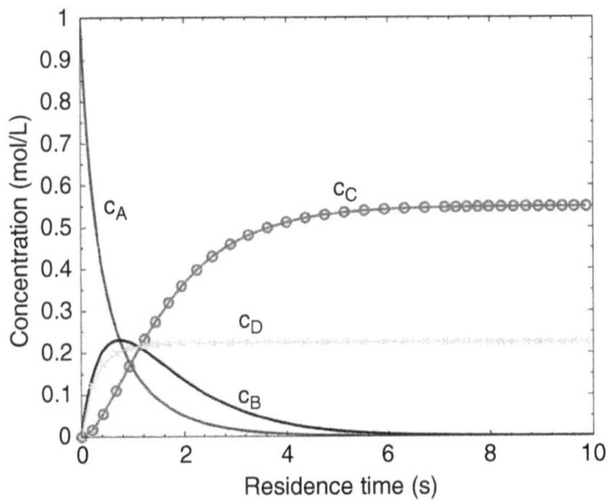

Figure 5.1 Species concentrations as a function of PFR residence time for Van de Vusse kinetics, with $a_1 = 1$ and $a_2 = 1$.

> **RECALL: Obtaining PFR data from the PFR equation**
>
> Recall that the concentration profiles displayed in Figure 5.1 are obtained by integrating the PFR equation
>
> $$\frac{d\mathbf{C}}{d\tau} = \mathbf{r(C)}$$
>
> using the feed point $\mathbf{C_f} = [c_{Af}, c_{Bf}]^T = [1.0, 0.0]^T$ mol/L as the initial condition to the PFR (at $\tau = 0$ s, $\mathbf{C} = \mathbf{C_f}$), and integrating using the PFR residence time τ as the integration parameter. From Figure 5.1, the integration range used is $0.0 \le \tau \le 10.0$ s.

Note that the concentration of component A continually decreases, whereas that of species C and D increases. Component B passes through a maximum value near $\tau = 0.76$ s.

These data can be plotted in the phase plane using the two variables of interest, c_A and c_B, by removing the residence time element from Figure 5.1. The PFR trajectory in this instance is a *convex* curve as shown in Figure 5.2(a). The PFR trajectory is initiated at the feed point $\mathbf{C_f}$ (point F), advances through a maximum point, and then decreases to the equilibrium point $\mathbf{C^\#} = [0.0, 0.0]^T$ mol/L (point O). Note that operations involving mixing with the PFR have been accounted for by displaying the convex hull of the PFR trajectory, which is represented as a shaded region in Figure 5.2(a).

b) CSTR Locus and Optimal Reactor Structure The CSTR locus from feed point F is also displayed in Figure 5.2(a). This locus is found by solving the CSTR equation for a range of CSTR residence times and then plotting the locus of effluent concentrations in $c_A - c_B$ space.

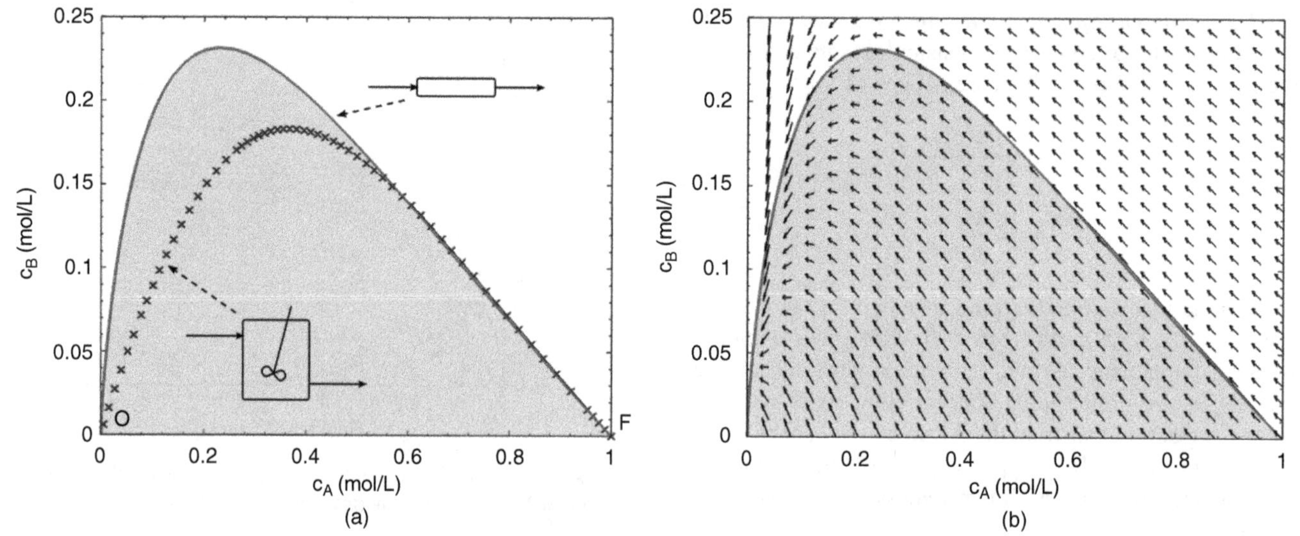

Figure 5.2 AR for the two-dimensional Van de Vusse system when $a_1 = a_2$. (a) The entire AR boundary is generated from a PFR from the feed point F. (b) When the rate field for the system is plotted over the region, we can visually inspect the AR boundary to show that no rate vectors on the AR boundary point out of the region, indicating that the AR has been found.

Note that the data are presented as a collection of points and not as a single solid line because each residence time corresponds to a different operating scenario corresponding to a unique reactor volume and flow rate combination.

> **RECALL: Obtaining CSTR locus from the CSTR equation**
>
> Recall that the CSTR locus displayed in Figure 5.2 is obtained by solving the nonlinear system of equations belonging to the CSTR equation
>
> $$\mathbf{C} = \mathbf{C}_f + \tau\mathbf{r}(\mathbf{C})$$
>
> where τ in this case is the CSTR residence time. For each value of τ, the CSTR equation may be solved (see Chapter 4), generating a distinct CSTR effluent point (or many CSTR effluent points in the case of multiple solutions).

Observe that the CSTR locus is contained entirely within the region generated by the PFR, and hence a CSTR from the feed does not serve to expand the region of achievable points further. From Chapter 4, it is known that PFR trajectories cannot cross.[3] Hence, running PFRs from the CSTR locus will not expand the region—any PFR trajectory extended from points on the CSTR locus in Figure 5.3 are contained within the PFR trajectory from the feed since the entire CSTR locus itself is contained in the region.

[3]See Section 4.4.3.2 as well as the side note on Russian nesting dolls *(Matryoshka dolls)*.

*5.3.2.3 **Step 3: Interpreting the Boundary*** The region displayed in Figure 5.2(a) represents the AR for the Van de Vusse system, when $a_1 = a_2$. The boundary of the region is composed entirely of a PFR from the feed; a CSTR locus from point F does not serve to expand the region. Hence, we may conclude that the optimal reactor structure for this situation is simply a PFR from the feed.

To validate that the region in Figure 5.2(a) is indeed the true AR, the region may be checked against the properties of the AR for consistency. Although we have discussed seven AR properties in Chapters 3 and 4, the following three properties in particular are of most interest here:

1. Feed point: The region contains the feed point F.
2. Convexity: The region in Figure 5.2(a) is convex by definition of the convex hull.
3. Rate vectors on the boundary: In Figure 5.2(b), the rate field is shown and plotted over the region in Figure 5.2(a). Observe that all rate vectors on the boundary either point into or are tangent to the region, and therefore there are no combinations of reactors that extend the region further.

*5.3.2.4 **Step 4: Objective Function*** Suppose that we are interested in maximizing the concentration of component B. The objective function of interest may be viewed geometrically as a horizontal line crossing the c_B axis in $c_A - c_B$ space. The optimal operating point is the intersection point of the objective function with the AR.

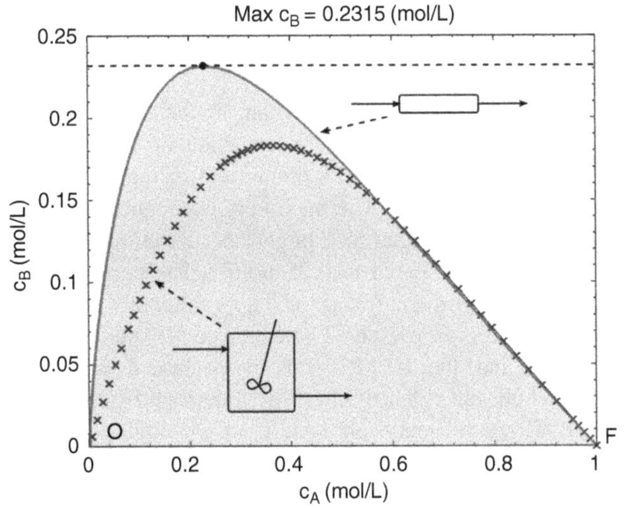

Figure 5.3 Maximizing for c_B using the AR when $a_1 = 1$ and $a_2 = 1$.

5.3.2.5 Step 5: Optimization Figure 5.3 shows the result of applying AR theory to the Van de Vusse system for the case when $a_1 = a_2$, and therefore the region represented in Figure 5.3 is the AR for the specific feed point F and kinetics. Using this region, a maximum possible concentration of $c_B = 0.2315$ mol/L is achievable by locating the point on the AR boundary that achieves the largest value of c_B.

Geometrically, this is achieved by moving the horizontal line (the objective function) until it intersects a point on the boundary of the AR. Furthermore, to achieve this concentration, a PFR with a feed concentration specified by point F must be employed. This reactor structure is obtained by consulting the optimal reactor structures that form the boundary of the AR in Figure 5.3. Using the maximum c_B value, Figure 5.1 may be used to determine the residence time of the PFR required. In this scenario, a residence time of 0.76 s is needed to obtain this concentration.

SIDE NOTE: Analytic solution

Had an analytic solution of the form $c_B(c_A)$ been available (concentration of B in the PFR in terms of concentration of A in the PFR), the optimal point could have been obtained by setting $dc_B/dc_A = 0$ and solving for the root of the resulting expression. This might be viewed as a more traditional approach to maximizing the concentration of component B. The benefit of using AR theory is that we do not need to use calculus to perform the optimization, and we can visually confirm that a CSTR does not provide better performance over a PFR.

EXAMPLE 1: Optimizing for c_B using an alternative objective function

Suppose that the objective function were given by the form

$$c_B = 0.25c_A + 0.15\,\text{mol/L}$$

What is the optimal operating point relative to the objective function and the associated reactor structure? Assume that the reaction kinetics and feed point are the same as that given in scenario 1.

Answer:

Since the system remains unchanged, the AR generated in Figure 5.3 can be used. The only aspect of the problem that differs now is the form of the objective function. Since the objective function is written in terms of c_A and c_B, the function may be plotted and overlaid onto the AR. The points of intersection between the new objective function and the AR boundary correspond to optimal operating points relative to the new objective function of interest.

Figure 5.4 displays the results of the new objective function. The new objective function may be written as a straight-line equation as a function of c_A. The line has a gradient of 0.25 and an intercept of 0.15 mol/L. Thus, the line intersects the boundary at two points, labeled $\mathbf{x}_1$ and x2, respectively, in Figure 5.4. Since these points coincide with the AR boundary and the objective function, they are the optimal operating points. Hence there are multiple optima for the given objective—there are in fact an infinite number of optimal operating points for this system, for all concentrations on the mixing line joining $\mathbf{x}_1$ and $\mathbf{x}_2$ also satisfy the objective function. Point $\mathbf{x}_3$ is a representative point that satisfies this relation. The following three optimal reactor structures may be formulated that satisfy the objective function:

1. A PFR from the feed operated to $\mathbf{x}_1$
2. A PFR from the feed operated to $\mathbf{x}_2$
3. Two PFRs in parallel: one operated at $\mathbf{x}_1$ and the other operated at $\mathbf{x}_2$, with a combined mixture of both exit streams forming $\mathbf{x}_3$.

The particular reactor structure chosen will depend on other factors appropriate to the designer. Note that since $\mathbf{x}_2$ is reached before $\mathbf{x}_1$, the PFR associated with $\mathbf{x}_2$ will have a smaller residence time compared to the PFR belonging to $\mathbf{x}_1$. Choice (ii) may be the simplest structure to employ excluding other factors such as selectivity or yield.

Note that the change in objective function alone does not require a new AR construction. This is because the feed point and kinetics have not changed. There is no need to generate a new AR and "re-solve" the problem for the new objective function. This allows for multiple optimizations with different objective functions to be investigated from a single construction—the AR represents the solution to *all* feasible optimization problems for a given feed.

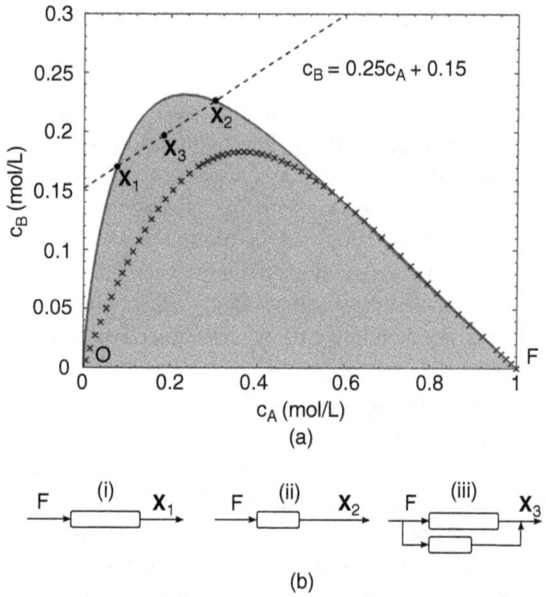

Figure 5.4 Optimizing for a new objective function. (a) Points intersecting the AR that satisfy the objective function. (b) The required optimal reactor structures that produce concentrations $\mathbf{x}_1$ (i), $\mathbf{x}_2$ (ii), and $\mathbf{x}_3$ (iii).

5.3.3 Scenario 2: $a_1 < a_2$

Now let us consider the case where $a_1 < a_2$, specifically when $a_1 = 1$ and $a_2 = 10$. The feed is again $\mathbf{C}_f = [1.0, 0.0]^T$ mol/L, and specific values of the rate constants are given in Table 5.2.

We again start by generating the PFR trajectory from the feed point. We can plot the concentrations of each of the species as a function of time and then convert that data into the same format as Figure 5.3, which is shown in Figure 5.5(a).

The system behavior is similar to the scenario when $a_1 = a_2$. However, a lower concentration of component B is now obtained. The CSTR locus is also included in the figure for comparison. The maximum concentration of B is obtained in a PFR at a much larger concentration of species A (0.7 mol/L compared to 0.25 mol/L). The corresponding PFR residence time is found to be approximately $\tau = 0.18$ s. Again, the region presented in Figure 5.5(b) is essentially convex,[4] and the CSTR locus is contained within the convex hull generated by the PFR trajectory from the feed point. There is hence no benefit in employing PFRs in series with the CSTR as the resulting PFR trajectories will not extend the region. Thus, the maximum amount of species B produced, for the case when $a_1 = 1$ and $a_2 = 10$, is approximately 0.062 mol/L, and achieved in a PFR with a residence time of approximately 0.18 s.

TABLE 5.2 Rate Constants for the $a_1 < a_2$ Case for Van de Vusse Example

Rate Constant	Value
k_1	$1.0\ s^{-1}$
k_2	$10.0\ s^{-1}$
k_3	1.0 L/(mol.s)

OBSERVATION: No change to reactor structure

Observe that the optimal reactor structure has not changed in this instance even though the kinetics and associated AR have changed. However, this result is unique to the kinetics. Generally, a change in the kinetics may affect the AR and hence the optimal reactor structure associated with it. This behavior is shown next.

5.3.4 Scenario 3: $a_1 > a_2$

Let us now investigate the scenario when $a_1 > a_2$, in particular, when $a_1 = 20$ and $a_2 = 2$. As in the previous sections, construction of the AR is achieved by first generating PFR trajectories from the feed point. Figure 5.6(a) provides species concentration profiles for the specific values of the rate constants supplied in Table 5.3; the corresponding PFR trajectory in concentration space is shown in Figure 5.6(b).

Component B progresses through a maximum near a PFR residence time of $\tau = 0.25$ s and then approaches equilibrium (point O). When $a_1 > a_2$, the area in Figure 5.6(b) enclosed between the PFR trajectory (curve FTO) and the *x*-axis is no longer convex. The straight line FZ can be formed to generate the convex hull of the region. Note point Z does *not* correspond to the maximum concentration of component B (the maximum value of c_B is actually achieved at point T).

[4]For the rate constants provided, there is a very small section on the AR boundary, between $0.0 \le c_A \le 0.2$ mol/L where the PFR is slightly concave, but this concavity has little effect on the recommended reactor structure, and thus it is not included here for that reason.

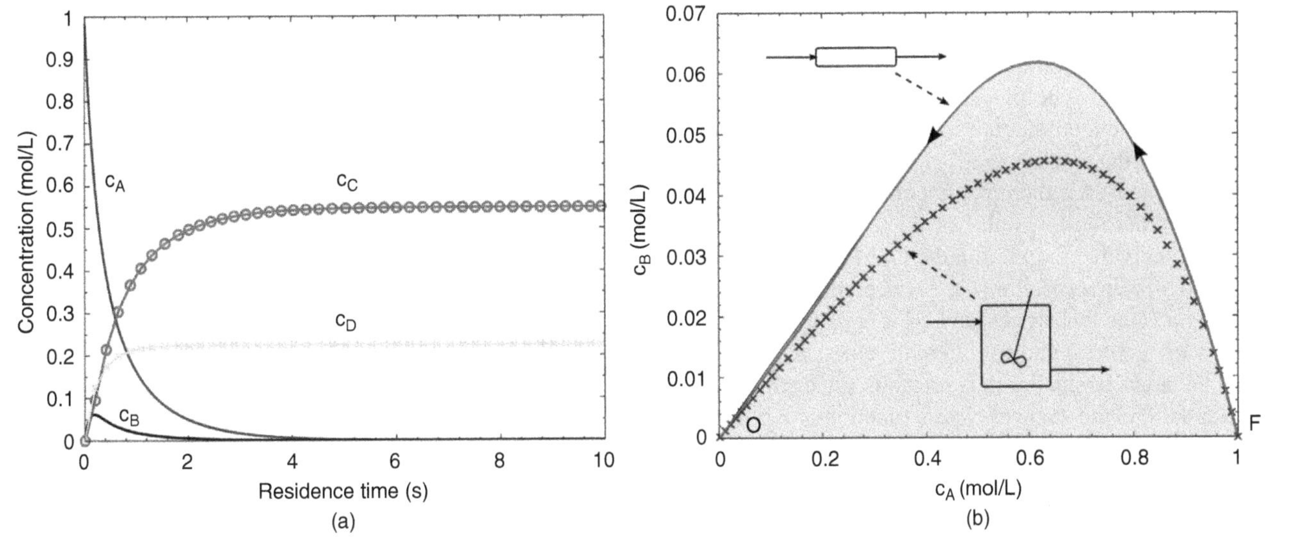

Figure 5.5 (a) Species concentrations as a function of PFR residence time and (b) PFR trajectory in $c_A - c_B$ space for with $a_1 = 1$ and $a_2 = 10$.

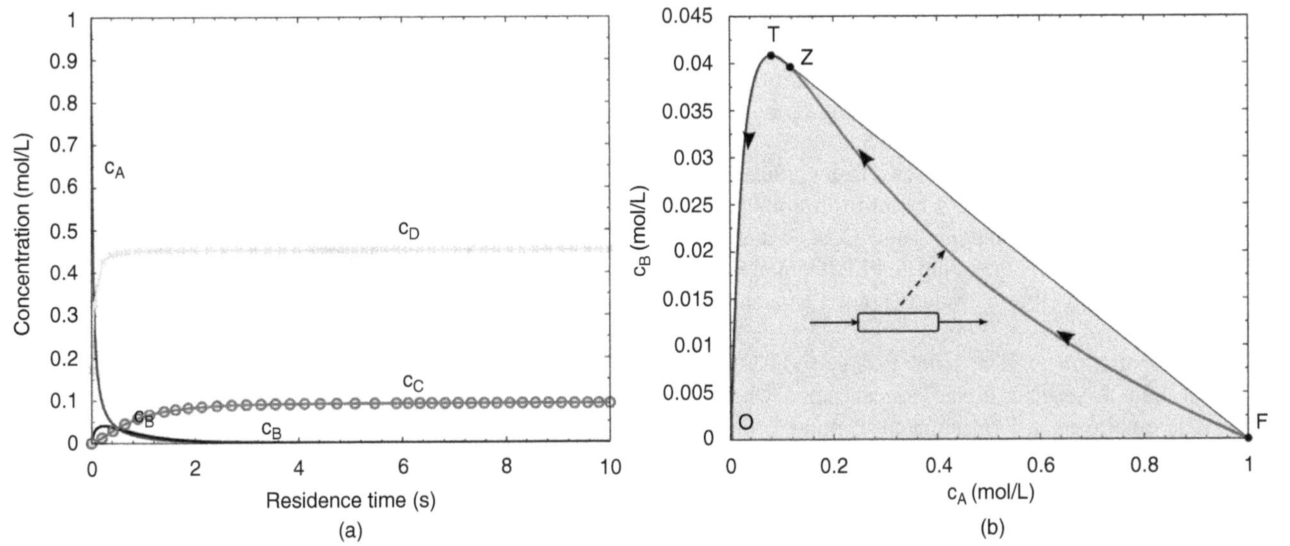

Figure 5.6 (a) Concentration profiles of all species as a function of time and (b) PFR trajectory in $c_A - c_B$ space when $a_1 = 20$ and $a_2 = 2$.

TABLE 5.3 Rate Constants for $a_1 > a_2$ Case for Van de Vusse Kinetics

Rate Constant	Value
k_1	1.0 s^{-1}
k_2	2.0 s^{-1}
k_3	20.0 L/(mol.s)

To determine what occurs if a CSTR is introduced into the analysis, the CSTR locus from the feed point is generated next. Species concentration profiles in the CSTR are displayed in Figure 5.7.

SIDE NOTE: No consumption of component D in a CSTR

Observe that the concentration of species D in Figure 5.7 passes through a maximum and then decreases, behaving in a similar manner to component B. However, the set of reaction kinetics in Equation 5.1a and 5.1b shows that there is no reaction to consume D, and thus there should be no decrease in the concentration of D. But this is not the correct way to interpret the reaction.

Recall that the operation of a CSTR is different to that of a PFR. The CSTR operates at a *single* point for a specific feed point and residence time—specifying both parameters results in a distinct solution (a distinct state) to the system of nonlinear equations given by the CSTR. By varying the value of τ in the CSTR, many distinct states are generated that are each physically different to the neighboring solutions.

Component D is therefore not consumed by the reactor, but rather the low value of c_D for longer residence times is the value required in the CSTR for the particular residence time specified—the concentration of D must take on a value that solves the CSTR equation. Different values of τ correspond to different values of c_D, which may be larger or smaller than previous values of c_D. The rise and fall in the value of c_D as a function of τ is dependent on the CSTR solution required by mass balance.

Note that the CSTR locus is not convex. Mixing line FG connects the feed point to the point of tangency on the CSTR locus, filling in the concavity observed in the CSTR locus. Physically, any concentration along this line may be obtained by mixing an amount of fresh feed with product from a CSTR at point G in a ratio prescribed by the lever arm rule. The required equipment to achieve these concentrations is thus a CSTR with bypass, as shown in Figure 5.8(b).

Seeing as the CSTR locus achieves a larger concentration of component B, there may be a benefit to running a PFR from the CSTR locus. Obtaining point G in Figure 5.8(a) with a CSTR, and then operating a PFR from this point to the maximum possible c_B value (point P), is shown in Figure 5.9(a).

The concentration of B at point P is $c_B = 0.052$ mol/L. Although point L yields a higher concentration when *only* a CSTR is employed, a PFR initiated at point L does not achieve the highest concentration of B.

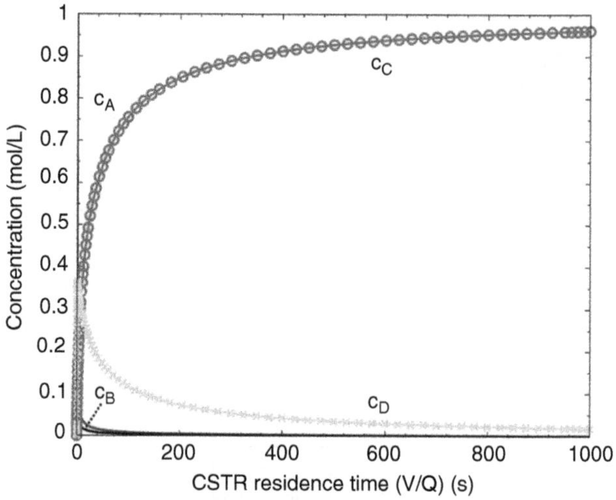

Figure 5.7 Species concentrations profiles achieved in a CSTR for Van de Vusse kinetics, with $a_1 = 20$ and $a_2 = 2$.

The resulting boundary given by points FGPO is convex. Any point within the region is hence attainable via mixing. Rate vectors evaluated at points on the boundary of FGPO would either point into the region (along the mixing line FG), or be tangent to the boundary along the PFR trajectory GPO. This region represents the full AR for the Van de Vusse system when $a_1 > a_2$.[5] The optimal reactor structure, when $a_1 > a_2$, is thus a CSTR followed by a PFR, as shown in Figure 5.9(b). No other reactor structure can do better, and all possible achievable points, for the feed specified, may be generated by this reactor configuration.

[5] As mentioned at the beginning of this chapter, the Van de Vusse system has been studied extensively in the scientific literature. The configuration of a CSTR followed by a PFR was also found to be optimal (Chitra and Govind, 1985) for the case when $a_1 > a_2$.

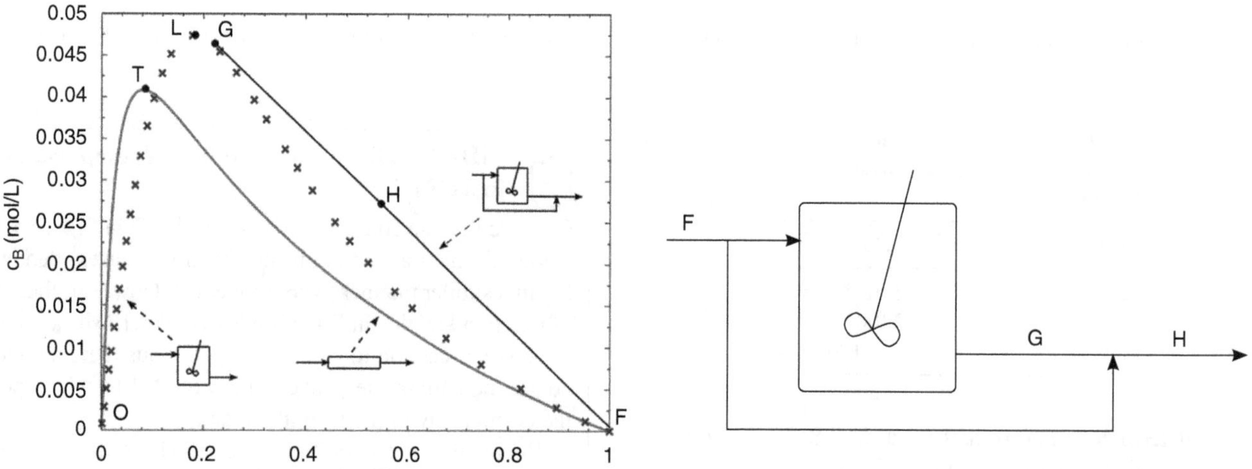

Figure 5.8 (a) PFR trajectory and CSTR locus from the feed point with $a_1 = 20$ and $a_2 = 2$. (b) CSTR with bypass of the feed.

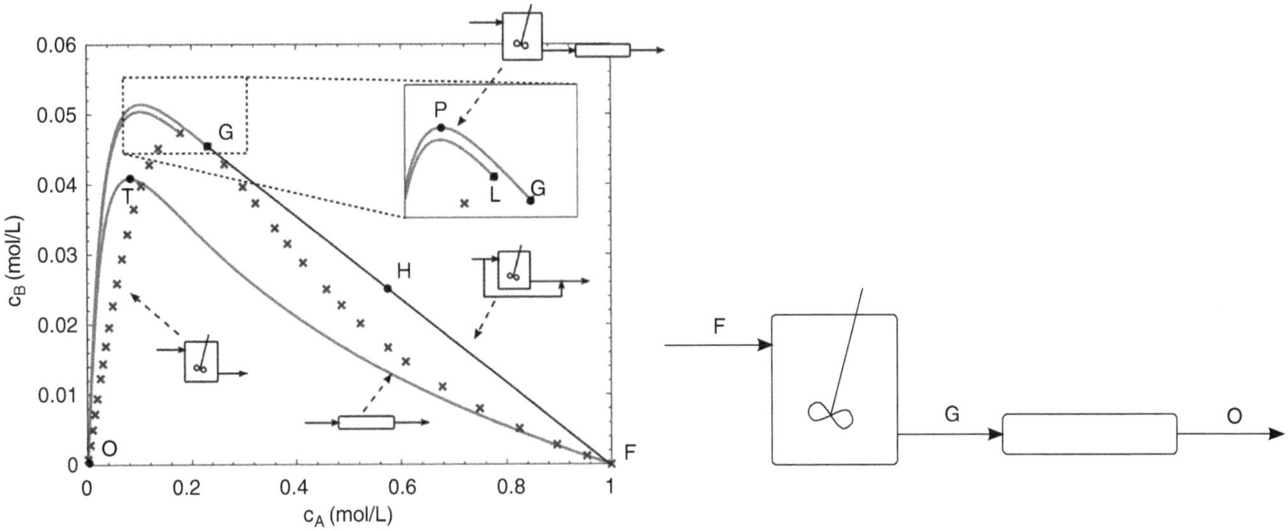

Figure 5.9 (a) Optimal reactor AR boundary structures (*See color plate section for the color representation of this figure.*) and (b) optimal reactor network required to generate the AR.

EXAMPLE 2: Overall yield of B

Now that the AR has been determined, optimization of the system may be carried out if a suitable objective function is supplied. The Van de Vusse reaction involves the production of a number of products (B, C, and D). Since component B is assumed to be the desired product, it may be useful to understand the yield of component B achievable in the system, therefore

(a) Write an expression for the overall yield of B, Y_B.
(b) Derive a graphical interpretation for the yield of B in $c_A - c_B$ space.
(c) Determine the concentration that maximizes Y_B in the system.

Answer a:

We define the overall yield as the number of moles of desired product, divided by the number of moles of reactant consumed. In the case of component B, this is

$$Y_B = \frac{n_{B,\,produced}}{n_{A,\,react}}$$

Since the optimal reactor is in fact a network of reactors, it may be difficult to understand how the yield of B may be computed from a complex reactor network. Hence, consider Figure 5.10.

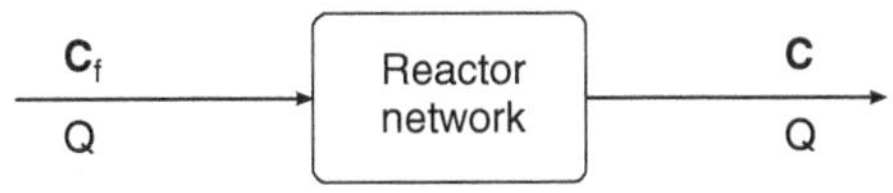

Figure 5.10 Overall reactor network.

Figure 5.10 shows a generalized representation of the reactor network synthesis problem.[6] It is assumed that a single feed is available. The volumetric flow rate of the feed is given by Q. On the other end of the network, a combined product stream of concentration **C** exits the network, which may be composed of potentially many mixtures of product streams within the network. Since the constant density assumption is enforced, the volumetric flow rate of the product stream is also Q. The goal is to determine the specific reactor configuration within the box that optimizes **C**.

[6]All of the reactor structures discussed in this book reside within the block in Figure 5.10.

The overall yield is determined by the product concentration vector **C** relative to the feed vector $\mathbf{C}_f$. An expression for Y_B specifically is then

$$Y_B = \frac{Qc_B - Qc_{Bf}}{Qc_{Af} - Qc_A} = \frac{c_B - c_{Bf}}{c_{Af} - c_A} \tag{5.4}$$

Answer b:

Equation 5.4 may be used to determine Y_B graphically in $c_A - c_B$ space. A point in this space has a concentration with coordinates of c_A on the *x*-axis and c_B on the *y*-axis.

Figure 5.11(a) provides an illustrative AR with three distinct points, $\mathbf{C}_1$, $\mathbf{C}_2$, and $\mathbf{C}_3$. We can construct a straight line between any point **C** in space and the feed point $\mathbf{C}_f$. Note that the gradient of the line is given by the following relation:

$$m = \frac{c_B - c_{Bf}}{c_A - c_{Af}} = -Y_B$$

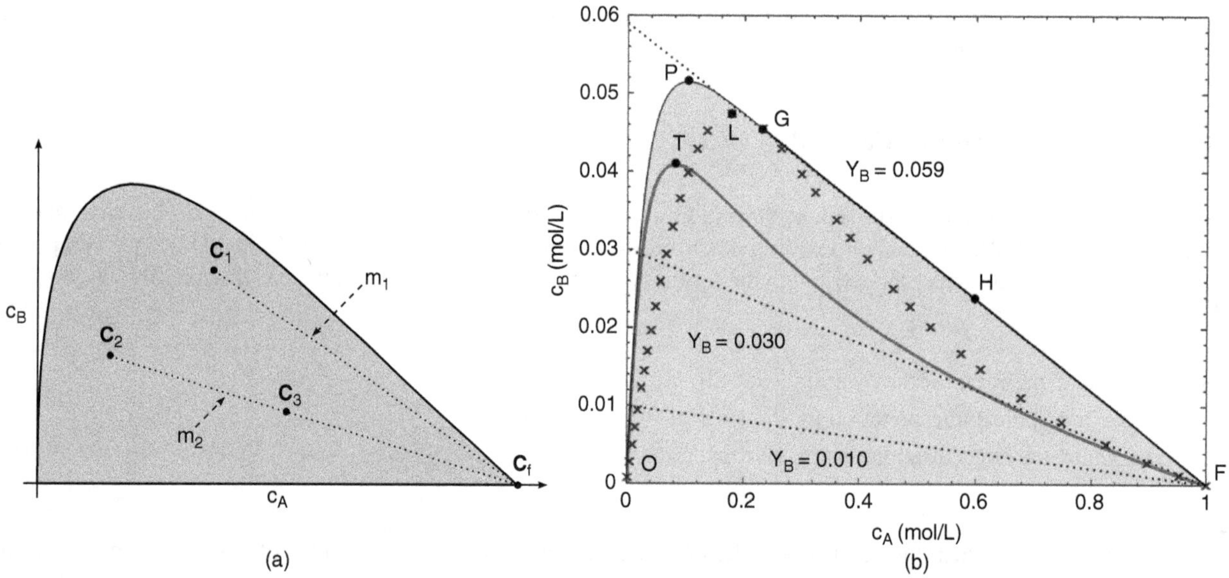

Figure 5.11 (a) Graphical interpretation of yield in $c_A - c_B$ space and (b) yield lines overlaid on the AR for the Van de Vusse system.

The yield of B is related graphically to the gradient of the line segment, m, joining any point in $c_A - c_B$ space with the feed point $\mathbf{C}_f$. In general, the yield of B is greatest when the straight line passing though $\mathbf{C}_f$ has the steepest gradient. A more negative value of m corresponds to a larger value of Y_B in the system.

Hence in Figure 5.11(a), a reactor structure with product concentration of $\mathbf{C}_1$ would have the largest yield of B of all three concentrations. $\mathbf{C}_2$ and $\mathbf{C}_3$ have the same yield because both concentrations lie on the same straight line passing through $\mathbf{C}_f$, although $\mathbf{C}_2$ may still be more desirable since the concentration of B is higher in $\mathbf{C}_2$ compared to $\mathbf{C}_3$.

Answer c:

Y_B is maximized when the straight line passing though $\mathbf{C}_f$ with steepest gradient touches the AR boundary. Using this data, one can plot the yield line for component B on the same axes as Figure 5.6 as shown in Figure 5.11(b). Lines of constant Y_B are also shown for comparison. Note that the steeper the gradient of the line, the higher the yield of component B ($Y_B = -m$). Y_B is hence maximized when it passes through points G and H. The yield at this point is approximately 0.059. Note that this yield can also be achieved with a CSTR with bypass as in Figure 5.9(a). Hence, there are many (in fact, a line) of concentrations that maximize Y_B.

For comparison, Y_B values for other points of interest have also been computed. A summary of the results is given in Table 5.4.

TABLE 5.4 Conversion, Selectivity, and Reactor Yield for the Points in Figure 5.9

Point	Concentration Vector $C=[c_A, c_B]^T$	Fractional Conversion of A[a]	Reactor Yield Y_B[b]
F	$[1, 0]^T$	0.00	N/A
T	$[0.08083, 0.04085]^T$	0.919	0.0444
G	$[0.2166, 0.04636]^T$	0.783	0.0592
H	$[0.604, 0.0235]^T$	0.396	0.0592
L	$[0.1783, 0.04737]^T$	0.822	0.0576
O	$[0, 0]^T$	1.00	0.000
P	$[0.1062, 0.05143]^T$	0.894	0.0575

[a]Calculated using $(c_{Af}-c_A)/c_{Af}$.
[b]Calculated using $(c_B-c_{Bf})/(c_{Af}-c_A)$.

Points H and G are associated with a higher yield than point L (G = 0.0592 compared to L = 0.0576). Note the similar slopes of the PFR trajectories from points G and L. Point P is more desirable for producing species B in terms of maximum concentration, whereas point O has the highest conversion of species A. Note that although point P achieves the largest c_B value, its yield is slightly smaller than point L.

EXAMPLE 3: Instantaneous yield and selectivity

Derive expressions for the following:

(a) The instantaneous yield of component B, Υ_B.
(b) The instantaneous selectivity of component B to D, Σ_{BD}.
(c) The instantaneous selectivity of component B to C, Σ_{BC}.

For each expression, determine the intersection with the AR boundary and comment on the optimal reactor structure required.

Answer:
Instantaneous yield is defined as the rate of formation of the desired product, divided by the rate of reaction of the reactant (Fogler, 2010). For component B, this is

$$\Upsilon_B = r_B / -r_A$$

Substituting Equation 5.2a through 5.2d for the Van de Vusse rate expressions,

$$\Upsilon_B = \frac{k_1c_A - k_2c_B}{k_1c_A + 2k_3c_A^2} \tag{5.5a}$$

Instantaneous selectivity is defined in a similar manner. This gives a measure of the preference of the reaction of a desired product (B) over a competing side-product (D). We define instantaneous selectivity of B to D, Σ_{BD}, as

$$\Sigma_{BD} = \frac{r_B}{r_D}$$

$$= \frac{k_1c_A - k_2c_B}{k_3c_A^2} \tag{5.5b}$$

and similarly for B to C

$$\Sigma_{BC} = \frac{r_B}{r_C} = \frac{k_1c_A - k_2c_B}{k_2c_B} \tag{5.5c}$$

All three expressions are functions of c_A and c_B. For a fixed value of Υ_B and Σ_{BD}, the expressions can be rearranged to express c_B as a quadratic function of c_A. Σ_{BC} results in a linear function of c_B in terms of c_A. These may be plotted over the AR boundary as contours to determine the intersection with the boundary, which is shown in Figure 5.12.

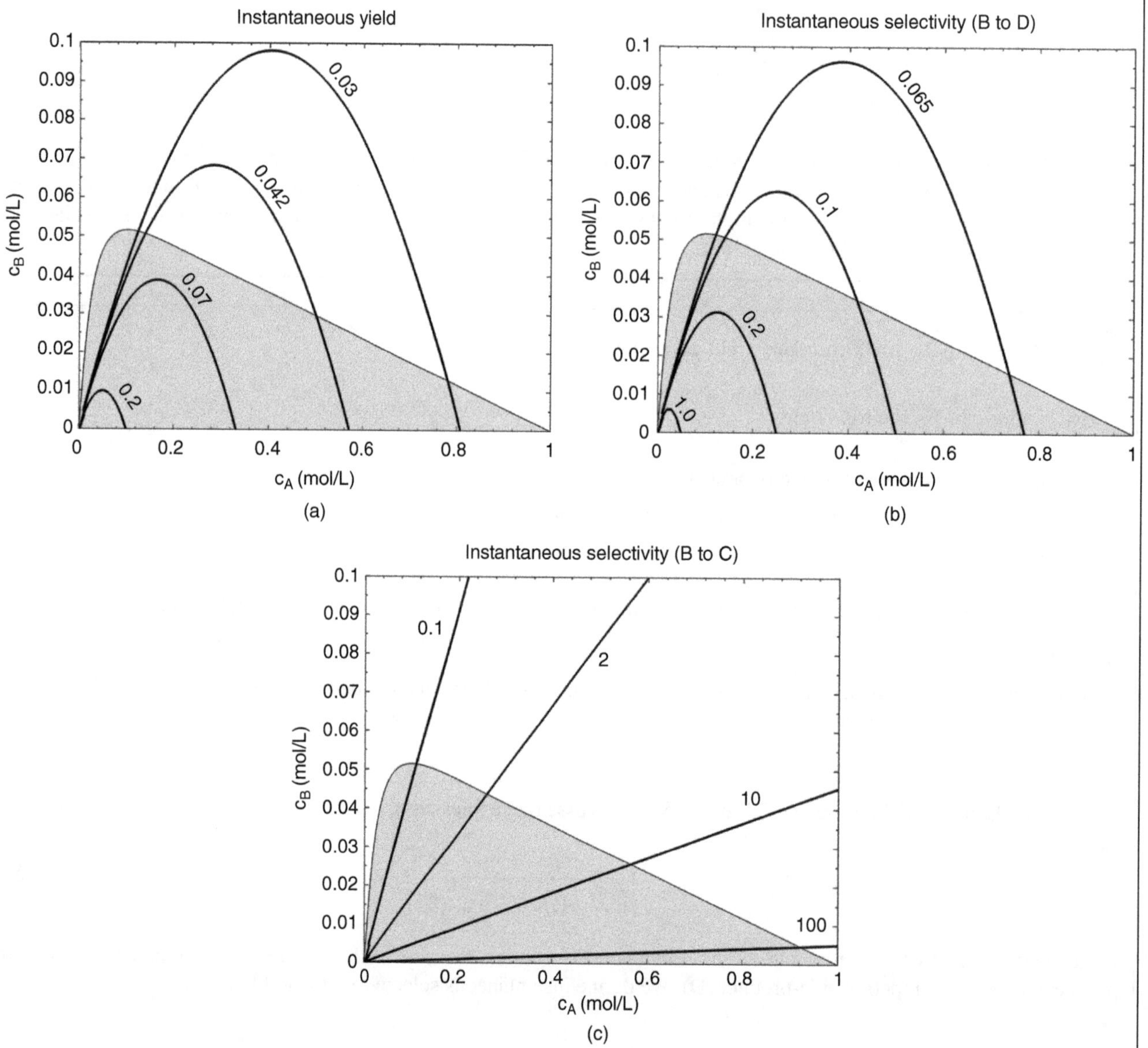

Figure 5.12 (a) Y_B contours for the Van de Vusse system, (b) Σ_{BD} contours, and (c) Σ_{BC} contours.

Figure 5.12 shows a number of contour lines for yield and selectivity. Observe that there are many points of intersection with the AR for these functions. Both Υ_B and Σ_{BD} become larger as the reaction approaches the equilibrium point $[0, 0]^T$ mol/L. This suggests that a reactor operating near the equilibrium point produces the largest instantaneous yield and selectivity. The selectivity of B to C, Σ_{BC}, is slightly different. Σ_{BC} is represented as a linear function in this space, and thus decreases as we move further away from the horizontal line $c_B = 0$ mol/L. Observe that for all expressions of selectivity and yield defined before, all values converge to the equilibrium point $[0, 0]^T$ mol/L.

EXAMPLE 4: Modifications to optimal reactor structure

Suppose that the optimal reactor structure in Figure 5.9(b), for scenario 3, was modified to include another CSTR in series after the PFR (to create a CSTR–PFR–CSTR structure). Using your understanding of the AR boundary, would this structure be able to extend the set of concentrations beyond what is achievable by the region in Figure 5.9(a)?

Answer:

Although it may seem as though the reactor structure in Figure 5.9(b) could be improved by adding new reactors to it, additional reactors will *not* extend the region further. This is because the reactor structure in Figure 5.9(b) generates the true AR boundary, and thus all achievable concentrations are generated by this structure alone. This result implies that all possible reactor configurations, even those that have not yet been devised, will produce concentrations that lie in or on the AR boundary of Figure 5.9(a). Since the optimal structure is known to be a CSTR–PFR only, any modification to the structure in Figure 5.9(b) must produce concentrations that lie in the region specifically and not on the AR boundary for the same feed point. This is the advantage of the AR approach. We find *all* attainable points first, without the need to nominate a new structure, or be influenced by an existing structure.

EXAMPLE 5: Determining optimal reactor sizes

Now that the optimal reactor structure is known for $a_1 > a_2$, it would be useful to compute the volumes of the reactors necessary to achieve point P. Given the information in Table 5.4, calculate the following:

(a) The CSTR volume
(b) The PFR volume
(c) The total reactor volume

Assume that the volumetric flow rate of the feed to the reactor structure is Q = 1.0 L/s.

Answer a:

Conversions of desired reactant in each of the reactors are known, and thus this information may be used to calculate the reactor volumes. Since component A is the principle reactant, volume calculations shall be based with respect to component A. From the CSTR equation, the volume of a CSTR is expressed by

$$V_{CSTR} = \frac{Q(c_A - c_{Af})}{r_A(\mathbf{C})} \tag{5.6a}$$

where V_{CSTR} is the volume of the CSTR (in L), Q is the volumetric flow rate through the CSTR (in L/s), and c_A and c_{Af} are the exiting and entering concentrations of A, respectively. $r_A(\mathbf{C})$ is the rate of reaction of A, evaluated at the *outlet* concentration of the CSTR. In this example, it is stated that Q = 1.0 L/s and also $c_{Af} = 1.0$ mol/L. From the AR construction, the exit concentration of the CSTR must correspond to point G on the AR boundary, and therefore the exit concentration vector is $\mathbf{C} = [0.2166, 0.04636]^T$, as given in Table 5.4. Hence at point G, $c_A = 0.2166$ mol/L. $r_A(\mathbf{C})$ is the species rate expression for component A, which is repeated below for convenience

$$r_A = -k_1 c_A - 2k_3 c_A^2 \tag{5.2a}$$

The rate of reaction for A is only a function of c_A in this instance. (Had it been a function of the other species in the system, it would not make it any more complex seeing as the concentration vector **C** is known.) Therefore, the value for the reaction rate at **C** is known and can be computed, giving

$$\begin{aligned} r_A(\mathbf{C}) &= -(1)(0.2166) - 2(20)(0.2166)^2 \\ &= -2.093\,\text{mol}/(\text{L.s}) \end{aligned}$$

The volume of the CSTR, using Equation 5.6a and 5.6b, may now be calculated as follows:

$$\begin{aligned} V_{CSTR} &= \frac{Q(c_A - c_{Af})}{r_A(\mathbf{C})} \\ &= \frac{(1.0)(0.2166 - 1.0)}{(-2.093)} \\ &= 0.374\,\text{L} \end{aligned}$$

Answer b:

To determine the volume of the PFR, we employ a similar procedure to that given for calculating V_{CSTR}. In this instance, the PFR equation is used. The PFR expression for component A specifically is

$$Q\frac{dc_A}{dV} = r_A(\mathbf{C})$$

where we have used $\tau = V/Q$ to express residence time in terms of PFR volume and volumetric flow rate. This is a variable separable differential equation that can be rearranged with respect to the PFR volume, giving

$$V_{PFR} = Q\int_{c_A}^{c_{A,in}} \frac{dc_A}{-r_A(\mathbf{C})} \tag{5.6b}$$

where $c_{A,in}$ is the feed concentration to the PFR. Unlike the CSTR, determining the PFR volume requires the solution of an integral. Since the particular feed concentration of interest to the PFR is the outlet of the CSTR at point G, $c_{A,in} = 0.2166$ mol/L ($\mathbf{C}_{PFR,in} = \mathbf{C}$ from answer 1). We also wish to terminate the PFR at the highest c_B value. This is given by point P on the AR boundary. The concentration vector corresponding to point P is then $\mathbf{C}_2 = [0.1062, 0.05143]^T$ mol/L, from Table 5.4. Hence, $c_A = 0.1062$ mol/L at P. To calculate the value of the integral in Equation 5.6b, the values of $r_A(\mathbf{C})$ are required over the concentration range of interest. Table 5.5 gives the information necessary to perform this calculation.

TABLE 5.5 Information Necessary to Numerically Evaluate the Integral in Equation 5.6b

c_A (mol/L)	r_A (mol/L/s)	$-1/r_A$ (L.s/mol)	Area of Segment(s)
0.2166	−2.093	0.4777	N/A
0.2043	−1.874	0.5337	0.00622
0.1921	−1.668	0.5994	0.00691
0.1798	−1.473	0.6789	0.00786
0.1675	−1.290	0.7753	0.00894
0.1553	−1.120	0.8928	0.01018
0.1430	−0.9610	1.041	0.01189
0.1307	−0.8140	1.229	0.01396
0.1185	−0.6802	1.470	0.01646
0.1062	−0.5573	1.794	0.02008

A plot of $-1/r_A(\mathbf{C})$ is shown in Figure 5.13 over the desired concentration range. Using the conversion and $1/-r_A(\mathbf{C})$ column in Table 5.5, one can numerically determine the area under the curve between each of the conversion points, as shown in the area under the curve column. This value is determined using the trapezoidal rule

$$\int_a^b f(x)dx \approx (b-a)\left[\frac{f(a)+f(b)}{2}\right]$$

For example, to determine the value under the curve between $c_A = 0.2166$ mol/L and $c_A = 0.2043$ mol/L, the calculation is as follows:

$$(0.2166 - 0.2043)\left[\frac{0.5335 + 0.4777}{2}\right] = 6.219 \times 10^{-3}\ \text{s}$$

Summing each of these sections gives the numerically determined value of the integral (0.102 s) given in Equation 5.6b. Finally, one can multiply by Q to give $V_{PFR} = 0.102$ L.

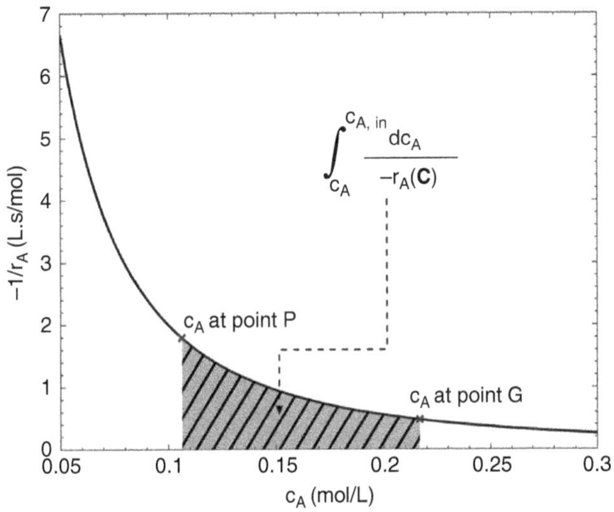

Figure 5.13 Determining the volume of a PFR using optimal AR boundary concentrations.

Answer c:

The total reactor volume is found by summing the individual volumes of the CSTR and PFR from earlier answers (a) and (b), which gives

$$V_{tot} = V_{CSTR} + V_{PFR} = 0.476\text{L}$$

Therefore, to achieve the maximum concentration of $c_B = 0.05143$ mol/L (point P) in the system, for a feed of 1.0 mol/L pure A (point F) and volumetric flow rate of Q = 1.0 L/s, one would first operate a CSTR with volume 0.374 L. Then the product from the CSTR (point G: $c_A = 0.2166$ mol/L and $c_B = 0.04636$ mol/L) is fed to a PFR with a volume of 0.102 L to yield point P.

SIDE NOTE: Solving for V_{CSTR}

Solving for a CSTR residence time τ (or reactor V_{CSTR} given Q) is considerably easier than solving for the CSTR effluent concentration vector given a value for τ.

5.3.5 Review

Let us review the decisions taken in this example and highlight what insights we have gained from them. Note that these insights relate primarily to the case when $a_1 > a_2$; however, similar conclusions may be formed for the other scenarios ($a_1 = a_2$ and $a_1 < a_2$) as well.

- Objective: The goal of the investigation was to determine the maximum amount of species B one can produce from the set of reactions given in Equation 5.1a and 5.1b. Geometrically, this is represented as a horizontal line in $c_A - c_B$ space.
- Optimum amount of species B: The maximum concentration of species B that can ever be produced is 0.05143 mol/L. This is found from constructing the AR for the kinetics and feed point of interest.
- Configuration: The reactor configuration to achieve the optimal amount of species B is a CSTR with a volume of 0.374 L, followed by PFR with volume 0.102 L. This configuration and the equipment size are interpreted directly from construction of the AR boundary. Moreover, if a different objective is required, a different reactor structure appropriate for the new objective function may be found without the need to re-solve a different problem.
- Mixing: The region of products achievable from any combination of CSTR and PFR can be enlarged by utilizing mixing—mixing fills in regions and converts concave regions into convex regions.

The power of AR theory is that the knowledge of all possible states for all possible reactor configurations, even those that have not yet been devised, is obtained. This enables us to define appropriate performance targets for the system of interest. A performance target can be viewed as a potential point of operation, used to make design and feasibility decisions.

If the system of interest is an existing plant (it has already been designed, or is even in operation), this analysis shows the proximity of the existing design in relation to the absolute best performance (the boundary of the AR). This knowledge assists in deciding whether or not to invest additional effort and resources to further improve the current design. Approaching a problem with this knowledge enables one to identify promising pathways and processes to particular products while eliminating costly alternatives that might never be optimal for the desired objectives.

ILLUSTRATION: Nonconvex ARs

In all of the problems that we have examined thus far, the AR for a system could be generated that resulted in a region that was always convex because of mixing. In all of these problems, we have implicitly assumed that there is sufficient flexibility in the reactor network that could allow for mixing between streams. Indeed, in all of the designs that we have examined, our viewpoint has been focused on new (Greenfield) designs, as opposed to existing (Brownfield) designs. Consider then what kind of AR could be generated when we do *not* have free choice in mixing—for instance, where an existing reactor has already been constructed that cannot be modified further. Can we still find the set of achievable concentrations for this reactor?

For example, assume that the Van de Vusse reactions are available. What set of achievable concentrations might be obtained by a single PFR from a feed $\mathbf{C}_f = [c_{Af}, c_{Bf}]^T = [1, 0]^T$ mol/L? In Figure 5.14(a), the PFR trajectory is generated from $\mathbf{C}_f$ and plotted in c_A–c_B space. Since mixing is not allowed (because physically there is no bypass stream connecting the

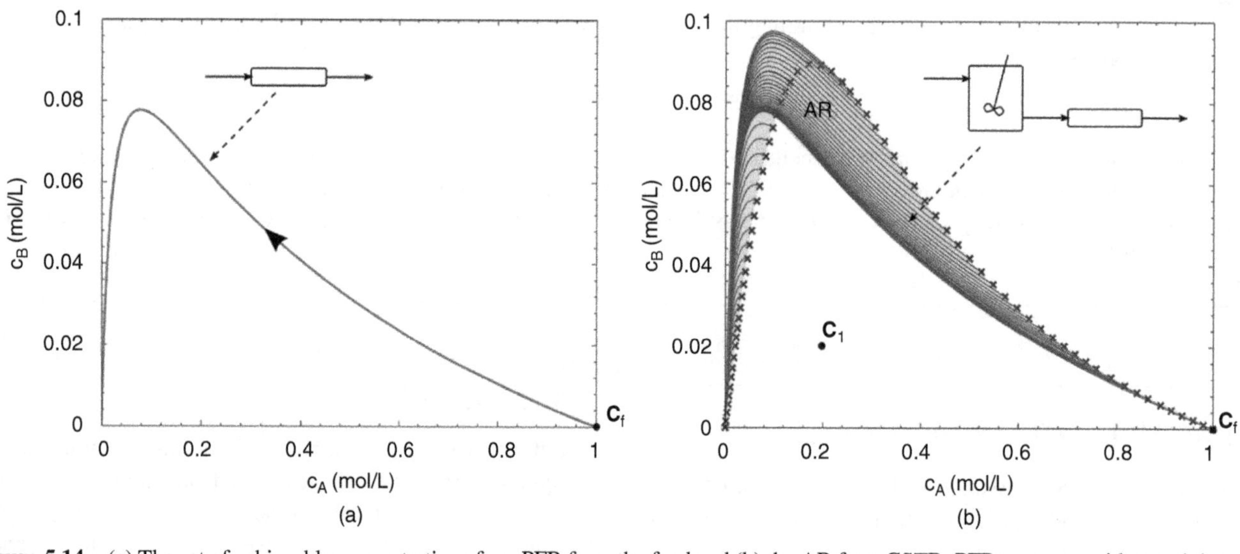

Figure 5.14 (a) The set of achievable concentrations for a PFR from the feed and (b) the AR for a CSTR–PFR structure without mixing. (*See color plate section for the color representation of this figure.*)

feed to the PFR effluent), it is not possible to compute the convex hull of the PFR points to produce a filled region in space. Rather, the set of achievable concentrations remains as a single curve representing the PFR trajectory alone. Hence, the set of achievable concentrations for a PFR from $\mathbf{C}_f$ is simply the concentrations belonging to the PFR solution trajectory only.

Consider now that a PFR in series with a CSTR is available, and that again the reactor structure is already built and there is physically no way to mix any of the streams around the reactors together. What might the set of achievable concentrations for this structure look like?

In Figure 5.14(b), the CSTR locus from the same feed point $\mathbf{C}_f$ is shown. Since the physical reactor structure incorporates a PFR from the CSTR, the CSTR locus points act as feeds to PFR trajectories. In Figure 5.14(b), PFR trajectories emanating from the CSTR locus are also displayed. Yet again, since it is physically not possible to mix between streams, this constraint forbids the computation of the convex hull of the set of points belonging to the CSTR–PFR. Instead, the set of achievable points are those that only belong to the shaded region in Figure 5.14(b). In this instance, the set of achievable points for a CSTR followed by a PFR without mixing between streams represents a *nonconvex* region.

As a result, concentrations lying in the interior of a boundary of achievable points, which would typically be achievable—such as point $\mathbf{C}_1$ in Figure 5.14(b)—are no longer achievable because there is no physical piece of equipment available to achieve these points. An important observation of this result is that for existing plants, it may still be possible to generate a region representing a set of achievable states, but this region may not be convex—the AR for existing plants may contain concavities; it may be *nonconvex* if mixing is not allowed.

5.4 MULTIPLE CSTR STEADY STATES AND ISOLAS

5.4.1 Introduction

The kinetics for the Van de Vusse system is simple enough that the corresponding AR can be determined in a straightforward manner. (Specifically, the full CSTR locus may be computed, with little complication, using an appropriate numerical solver.) In the following example, a system is discussed where the associated kinetics produce *multiple* steady-state CSTR solutions for the same feed point and the same residence time. Although the system to be described is artificial (it has been manufactured to exhibit undesirable properties), these same complexities frequently arise in adiabatic systems, involving the introduction of an energy balance.[7] That is, realistic systems often exhibit this behavior, and hence we must understand how to approach them. We shall again employ the AR construction framework described in Section 5.2.

5.4.2 Step 1: Define the Problem

5.4.2.1 Reactions For this particular example, the system involves a parallel decomposition involving three components A, B, and C, in three reactions (two of the reactions are independent).

$$\mathrm{A} \rightarrow \mathrm{B} \tag{5.7a}$$

$$\mathrm{A} + \mathrm{B} \rightarrow 2\mathrm{B} \tag{5.7b}$$

$$\mathrm{A} \rightarrow \mathrm{C} \tag{5.7c}$$

[7]An adiabatic AR problem will be discussed in Chapter 7.

In Equation 5.7a, component A decomposes to form component B. If B is present, then components A and B may combine, in an autocatalytic reaction, to form additional B by Equation 5.7b. Component A may also decompose in a parallel reaction to form C via Equation 5.7c. For this example, we shall assume that component B is the desired product and seek to determine the optimal reactor structure that maximizes the concentration of B, c_B.

5.4.2.2 Kinetics and State Space The rate expression for this system is assumed to be a complex rate function involving the concentrations of components A and B only, which are given by

$$r_1 = -a_1 c_A - a_2 \frac{c_A^2 c_B^2}{1 + K c_A c_B^2}$$

$$r_2 = -a_3 c_A$$

Here K, a_1, a_2, and a_3 are constants given by $K = 40$, $a_1 = 0.001\ \mathrm{s}^{-1}$, $a_2 = 10\ \mathrm{s}^{-1}$, and $a_3 = 0.1\ \mathrm{s}^{-1}$. Since the system may be expressed entirely in terms of the concentrations of components A and B alone, it is sensible to construct the AR in c_A–c_B space. Once the AR for the system has been determined, the concentration of component C, c_C, is easily determined by mass balance if desired. It follows that the concentration vector and rate vector for the system should be defined in the following way:

$$\mathbf{C} = \begin{bmatrix} c_A \\ c_B \end{bmatrix}$$

$$\mathbf{r}(\mathbf{C}) = \begin{bmatrix} r_A \\ r_B \end{bmatrix} = \begin{bmatrix} r_1 + r_2 \\ -r_1 \end{bmatrix}$$

$$= \begin{bmatrix} -a_1c_A - \frac{a_2c_A^2c_B^2}{1+Kc_Ac_B^2} - a_3c_A \\ a_1c_A + \dfrac{a_2c_A^2c_B^2}{1+Kc_Ac_B^2} \end{bmatrix}$$

5.4.2.3 Feed and Equilibrium Points The feed stream is assumed to be pure in component A so that $\mathbf{C}_f = [1, 0]^T$ mol/L. It is good practice to locate any equilibrium points for the kinetics provided, if possible. This will assist with AR construction later on and will also provide insight into areas of the construction space (c_A–c_B space) where unexpected behavior might occur. Determination of the equilibrium points is performed by setting the component rate expressions to zero and then solving for the set of concentrations that satisfy the system of equations, hence

$$\mathbf{r}(\mathbf{C}^{\#}) = 0$$

As in Chapter 4, $\mathbf{C}^{\#}$ indicates the equilibrium concentration. Writing out the rate expressions explicitly in terms of the individual species gives

$$-a_1c_A - \frac{a_2c_A^2c_B^2}{1+Kc_Ac_B^2} - a_3c_A = 0$$

$$a_1c_A + \frac{a_2c_A^2c_B^2}{1+Kc_Ac_B^2} = 0$$

The objective here is to find the set of concentration vectors (c_A, c_B) that are roots to the given system of equations. Equating the two equations and simplifying the expressions results in the following:

$$2r_1 + r_2 = 0$$

$$(2a_1 + a_3)c_A + 2\frac{a_2c_A^2c_B^2}{1+Kc_Ac_B^2} = 0$$

$$c_A\left[(2a_1 + a_3) + (2a_1 + a_3)Kc_Ac_B^2 + 2a_2c_Ac_B^2\right] = 0$$

This produces a quadratic expression in both c_A and c_B. Observe that the zero vector is obtained whenever $c_A = 0.0$ mol/L. This suggests that there is vertical line at $c_A = 0.0$ mol/L where the system is always in equilibrium.

> **SIDE NOTE: Computing equilibrium points**
>
> Note that it is possible to locate equilibrium points by computing the roots of the rate function in this instance. Depending on the complexity of the kinetics, this approach is not always feasible. Calculating equilibrium points is similar to solving for the CSTR equation in terms of complexity and difficulty, as both methods require the solution of a system of nonlinear equations.

5.4.3 Step 2: AR Construction

5.4.3.1 PFR Trajectory from the Feed Now that the system has been specified in terms of the feed point and reaction kinetics, and equilibrium points have been identified, the AR corresponding to the feed point may be generated. We start by generating the simplest of candidate ARs, which involves finding the PFR trajectory from the feed point. To generate the PFR trajectory, the PFR equation is integrated using $\mathbf{C}_f$ as the initial condition.

$$\frac{d\mathbf{C}}{d\tau} = \mathbf{r}(\mathbf{C}) \qquad \text{At } t = 0\,\text{s} : \mathbf{C} = \mathbf{C}_f$$

The PFR is integrated over a large residence time so as to approximate all states between the feed and the equilibrium state. In this way, the largest possible set of achievable concentrations from the feed PFR may be obtained. Experimentation is required to identify the integration time that adequately approximates the equilibrium state. For the kinetics specified, a residence time range of $0\,\text{s} \leq \tau \leq 80\,\text{s}$ is sufficient—for $\tau > 80\,\text{s}$, no significant change in the concentrations of A and B are observed. Figure 5.15(a) shows the concentration profiles with CSTR residence time, whereas the PFR trajectory in c_A–c_B space is displayed in Figure 5.15(b).

The direction of travel for PFR concentrations is toward an increasing concentration of B. The PFR trajectory appears to be fairly linear (the trajectory does not appear to contain concavities), and thus mixing in this instance will not allow for the achievement of many new concentrations. Note that the equilibrium concentration for c_B is small (~0.011 mol/L) compared to the initial concentration of A (1.0 mol/L), indicating that the majority of A has been converted to C instead of B. Next, the CSTR locus from the feed is determined.

5.4.3.2 CSTR Locus from the Feed To plot the CSTR locus from the feed point, the system of CSTR equations must be solved (simultaneously) over a range of residence times. In comparison to the Van de Vusse system, the kinetics for this system is more difficult to solve analytically. Numerical solution of the CSTR locus is therefore required.

As a first attempt at solving the CSTR locus, a standard numerical root finding routine will be employed. MATLAB's built-in solver routine, `fsolve()`, is used here for convenience, although many other numerical routines may be equally appropriate. A residence time and feed concentration to the CSTR must be supplied in order to

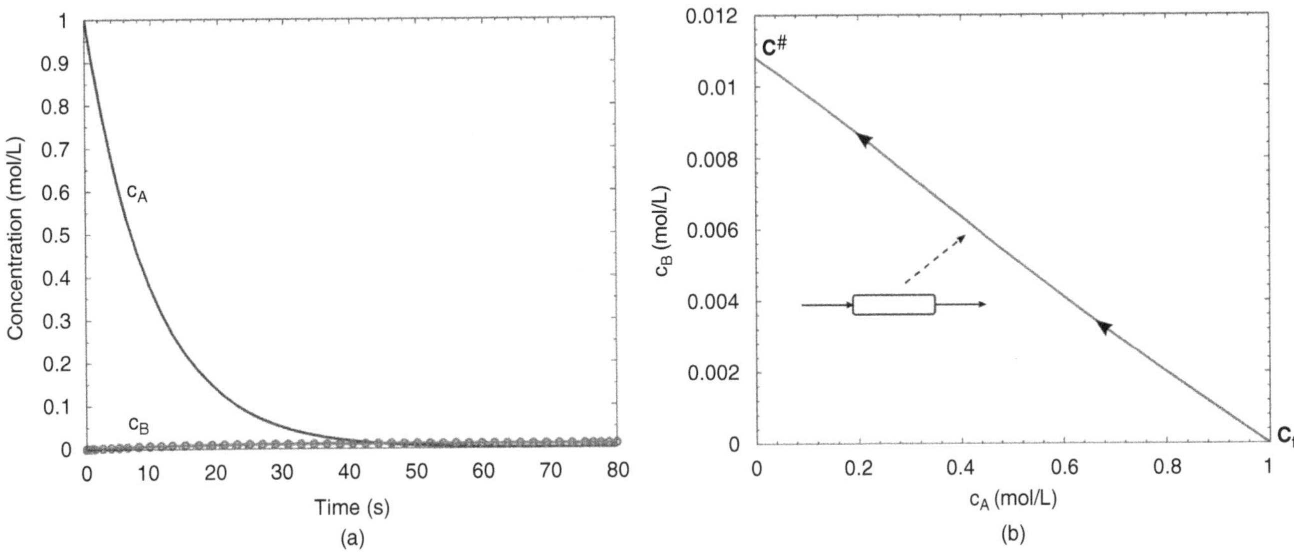

Figure 5.15 (a) Concentration profiles of components A and B in a PFR from the feed. After $\tau \sim 80$ s, little change in the species concentrations is observed. (b) PFR trajectory, corresponding to a PFR from the feed point, plotted in c_A–c_B space.

solve the CSTR system. The `fsolve()` function requires an initial guess of the CSTR effluent concentration before a solution can be obtained. Since no additional information regarding possible CSTR solutions is available, the feed concentration, $\mathbf{C_f}$, is used as the initial guess, and the following system is solved

$$\mathbf{C_f} + \tau_i \mathbf{r}(\mathbf{C}) - \mathbf{C} = 0$$

where τ_i is a CSTR residence time in the range $0 \geq \tau_i \geq \tau_{max}$, and using an initial guess of $[1, 0]^T$ mol/L. A maximum residence time of $\tau_{max} = 1000$ s is found to work well for the problem here. The CSTR locus in Figure 5.16 may then be generated.

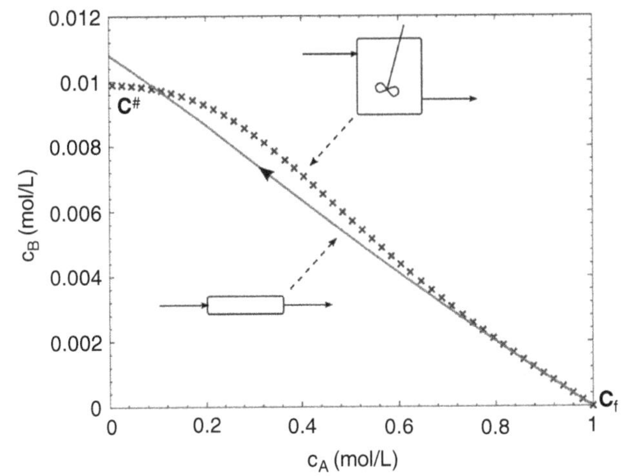

Figure 5.16 CSTR locus and PFR trajectory from the feed operating from the feed concentration.

Again, a smooth locus of points for the feasible CSTR effluent concentrations is generated, which are represented by ×'s in the figure. The results indicate that a higher concentration of B is achieved in a CSTR, in the range $0.85\ \text{mol/L} \leq c_A \leq 0.10\ \text{mol/L}$. For values of c_A outside of this range, larger c_B values are achieved in a PFR.

5.4.3.3 PFR from CSTR Locus Points We observed from the Van de Vusse system, in Section 5.3.4, that a higher concentration in B could be obtained by running a PFR directly after a CSTR from the feed, as a convex hull boundary composed of CSTR points is often not optimal. The convex hull for the entire set of points including CSTRs and PFRs may then be calculated. The appropriate CSTR residence time must be found that maximizes the area of the corresponding convex hull. We find that a residence time of approximately 20.6 s in a CSTR followed by a PFR to equilibrium generates a convex hull with the largest area. The resulting region is shown in Figure 5.17 where point $\mathbf{C_1}$ is the CSTR effluent concentration corresponding to $\tau = 20.6$ s. The shaded region represents a candidate AR for the system specified.

5.4.3.4 CSTR Locus with Different Initial Guesses In Section 5.4.1, we indicated that the kinetics supplied for this system is known to exhibit multiple steady states; however, no such concentrations have been identified in the computation of the current candidate region. A much larger set of achievable concentrations is in fact achievable. To see this, consider solving for the CSTR locus from the feed using a different initial guess.

If, instead, various initial guesses (such as $[0.5, 0.5]^T$ mol/L) are used to solve for the CSTR, then the locus in

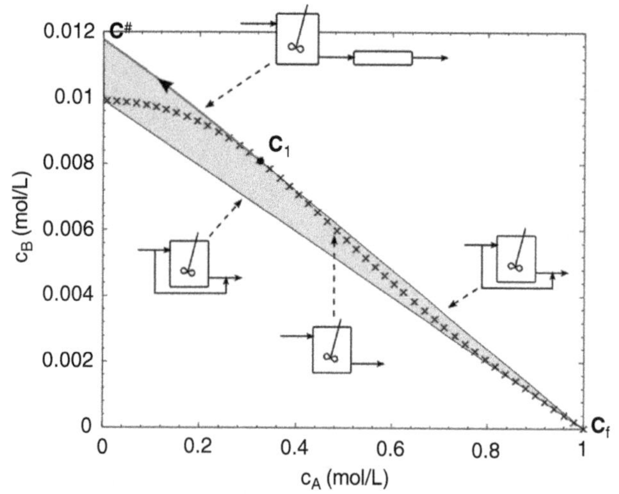

Figure 5.17 Convex hull representing the set of achievable concentrations for the CSTR–PFR structure.

Figure 5.18(a)—for the same residence time range and feed point—is generated.[8] Note that the specific concentrations, used for the initial solution guess, are simply for illustrative purposes. Different guesses may produce different (equally valid) solutions.

Compared to the approximately 0.01 mol/L maximum concentration of B obtained in Section 5.4.3.2, the maximum value of c_B achieved by the new initial guess is much larger ($c_B \sim 0.52$ mol/L). Note that this for *an identical feed point and residence time*.

CONCEPT: Isolas

The closed region of CSTR points obtained in Figure 5.18(a) is termed an *isola* in AR and chemical reactor theory. Isolas result from multiple steady-state solutions present in a CSTR, which are caused mainly from nonlinear terms in the rate expression (Ganapathisubramanian and Showalter, 1984; Gray and Scott, 1983; Scott, 1987).

It is not always straightforward to construct the true AR, particularly when attempting to solve the CSTR equation with nonlinear reaction kinetics. If the existence of multiple CSTR steady states from the kinetics had not been known prior to construction, then only a small candidate region, given by Figure 5.17, would be obtained. By contrast, Figure 5.18(b) shows the resulting region when the newer, expanded CSTR locus is used instead. Note that this region is obtained using the identical CSTR–PFR structure as in Section 5.4.3.3, yet the set of achievable concentrations is very different. Since differences in the size of the region result from the CSTR in particular, great care must be taken to ensure that all CSTR solutions have been identified.

In order to distinguish why the region in Figure 5.18(b) is the true AR, it is appropriate to again refer to some important properties of the AR:

1. Convexity: Both regions are convex.
2. Feed point: It is clear that both regions contain the feed point.
3. Simply connected: It appears as though both regions consist of single subspaces in $\mathbb{R}^2$.
4. Boundary composition: Both regions are composed of straight-line sections (mixing) and PFR trajectories (reaction).
5. Rate vectors on the AR boundary: Figure 5.18(c) shows how rate vectors evaluated on the boundary of the region are either tangent or point into the region. Had rate vectors in the complement region of Figure 5.17 been checked for further expansion of the region, many of the concentrations encompassed by the CSTR locus in Figure 5.18(b) would have been found.[9]

Although multiple steady states do not arise often, their unpredictable behavior may present considerable challenges to the designer. Knowledge of when multiple steady states could occur for a set of kinetics and feed point is therefore desirable when conducting AR constructions on an unfamiliar system for the first time.

EXAMPLE 6: Calculating corresponding CSTR residence times

Suppose that the optimal CSTR effluent concentration, residing on the boundary of the AR in Figure 5.18(b), is $\mathbf{C} = [0.3233, 0.4337]^T$ mol/L. That is, **C** will be used to feed a PFR that traverses along the AR boundary in Figure 5.18(b).

What is the corresponding CSTR residence time required to achieve **C** as an effluent concentration? If it is assumed that a volumetric flow rate of Q = 2.5 L/s is available, what is the volume of the CSTR?

Answer:

The CSTR residence time, τ, may be computed from the CSTR equation if feed and effluent concentrations are

[8]This result is specific to the numerical solver employed. MATLAB's `fsolve()` has been used in this example. The use of a different numerical solver or initial guess may produce different results.

[9]In Chapter 8, we shall discuss an automated AR construction algorithm that uses this property of the complement region to compute candidate regions. This algorithm can then be used to verify this statement.

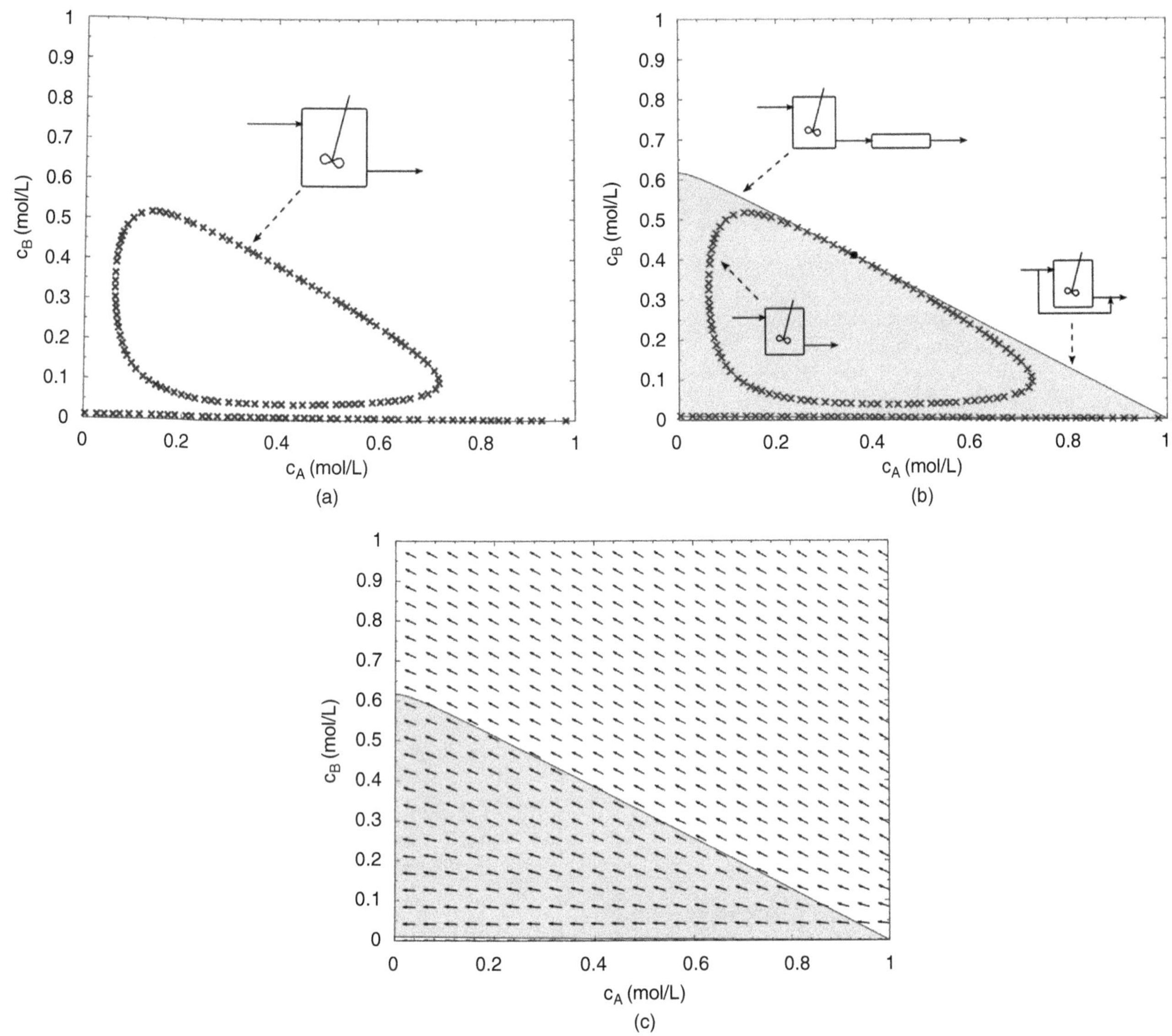

Figure 5.18 (a) The CSTR locus from the feed, produced with various initial guesses. (b) AR obtained using the full set of CSTR points. The optimal reactor structure is still a CSTR from the feed followed by a PFR; however, this region is significantly larger than that obtained initially. (c) Rate field for the isola system.

supplied.

$$\mathbf{C} = \mathbf{C}_\mathrm{f} + \tau\mathbf{r}(\mathbf{C})$$

which may be solved for τ, in component form, as follows:

$$\tau = \frac{c_i - c_{i\mathrm{f}}}{r_i(\mathbf{C})} \tag{5.8}$$

Note that the value of τ must be the same for all components participating in the reaction (there is always only one value of τ for all components); however, Equation 5.8 suggests that the value of τ may be computed using concentration information from many species.

Using information supplied from Section 5.4.2.3, we have

$$\mathbf{C}_\mathbf{f} = [1.0,\ \ 0.0]^\mathrm{T}\,\mathrm{mol/L}$$

and

$$\mathbf{C} = [0.3233,\ \ 0.4337]^\mathrm{T}\,\mathrm{mol/L}$$

The value of $\mathbf{C}$ may be substituted into the rate equation and evaluated to find the rate vector $\mathbf{r}(\mathbf{C})$:

$$\mathbf{r}(\mathbf{C}) = [-0.0899,\ \ 0.0576]^\mathrm{T}\,\mathrm{mol/(L.s)}$$

Substituting species concentrations into Equation 5.8 gives

$$\tau = \frac{c_A - c_{Af}}{r_A} = \frac{0.3233 - 1.0}{-0.0899} = 7.5247\,\text{s}$$

comparing the residence time calculated with concentration information for component B gives

$$\tau = \frac{c_B - c_{Bf}}{r_B} = \frac{0.4337 - 0.0}{0.0576} = 7.5294\,\text{s}$$

which is almost identical to the residence time obtained using concentration information from component A. Hence, the corresponding CSTR residence time to achieve an effluent concentration of $\mathbf{C} = [0.3233, 0.4337]^T$ mol/L is $\tau = 7.52$ s.

Calculating CSTR volume:

The CSTR volume may be found from the residence time relation as follows:

$$V = \tau Q$$

Using a volumetric flow rate to the CSTR of Q = 2.5 L/s, this results in a CSTR volume of V = 18.8 L.

5.4.4 Steps 3–5: Interpretation and Optimization

With the full AR for the isola system now known, the associated optimal reactor structure may be determined. To achieve the entire set of points defined by the region in Figure 5.18(b), a CSTR followed by a PFR is required. This gives the boundary points of the AR from which all interior points may be found via the appropriate mixing operations.

Optimization of the system is easily performed now that the optimal reactor structure used to form the AR boundary is known. This discussion is best left to the following illustration for practice.

ILLUSTRATION: Optimization of the isola system

Consider the following two optimization scenarios. It may be useful to refer to Figure 5.18(b) when working through these examples.

Case 1:

Determine the optimal reactor structure that maximizes the concentration of component B. Let us designate the objective function value by the variable P. The form of the objective function then assumes a simple form given by $P = c_B$.

In Figure 5.19(a), the current objective function is overlaid onto the AR. For this scenario, the shape of the objective function is a horizontal line corresponding to the value of c_B in c_A–c_B space. For larger concentrations of B, the position of the objective function intercepts the

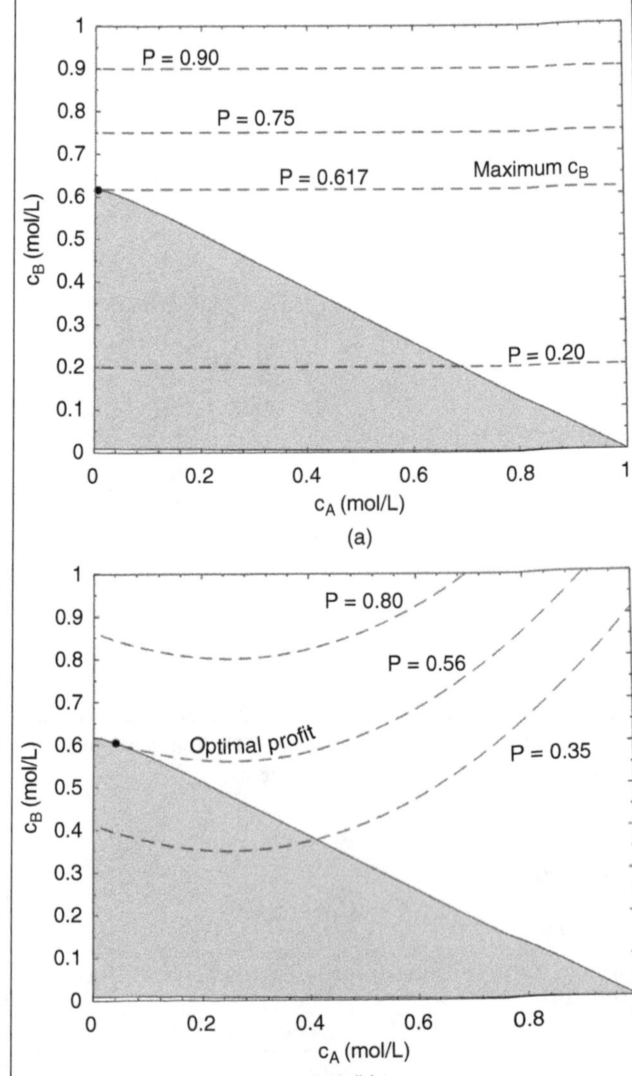

Figure 5.19 (a) AR for complex isola kinetics. The objective function is simply the concentration of component B. (b) A quadratic objective function in c_A representing operating profit from the sale of component B.

y-axis at higher points on the graph. The AR represents the total set of achievable concentrations for the system, and therefore only those values of the objective function that intersect the AR are feasible. In Figure 5.19(a), objective values for P = 0.9 mol/L and P = 0.75 mol/L are shown. These values of P do not intersect the AR, and thus it is not physically possible to achieve these concentrations in practice.

The optimal value of the objective function may be found by slowly decreasing the value of P until it touches the boundary of the AR. This corresponds, graphically, to a vertical (downwards) translation of the straight line associated with P until the line meets the boundary of the AR. For the given system kinetics and AR, this occurs at a value of $c_B = 0.617$ mol/L. As a result, the maximum concentration of B achievable in the system is $c_B = 0.617$ mol/L.

Moreover, since the associated optimal reactor structures are also known for generating all points on the boundary of the AR, the particular reactor arrangement required to achieve P is also known. In this instance, the value of 0.617 mol/L in component B is obtained via a CSTR followed by a PFR.

Although the objective in this instance has been devised in the interest of maximum B, the same function for the objective function might also be used for alternate design scenarios. Suppose that instead of maximizing the concentration of B, we wish to *target* a specific value of the concentration of B. (This might be useful if, for instance, the downstream separations process favors a certain value of B). Hence, in Figure 5.19(a), we also show the objective value for P = 0.2 mol/L. If it is in our interest to target for this specific effluent concentration, then there are an infinite number of reactor structures that could achieve this point—the value of P intersects the interior of the AR along the line intercepting $c_B = 0.2$ mol/L.

In general, intersection points on the AR boundary are the most desirable and convenient operating points, even if the objective function intersects the AR along interior points. The particular value of P = 0.2 mol/L intercepts the AR boundary at a point that suggests the optimal reactor structure for this value of P is obtained by a CSTR with bypass of feed material to the CSTR effluent concentration.

Case 2:

Suppose that component B is to be sold, and component A affects the downstream separation unit operations, which negatively impacts the overall profitability of the process. Assume that the profits obtained from the sale of B (after downstream processing) is given by the following equation:

$$P = c_B - (c_A - 0.25)^2$$

The value of P is in millions of dollars. What is the maximum profit that can be made by the process, and what reactor structure is needed in order to achieve this value?

In Figure 5.19(b), the profit function, corresponding to case 2, is overlaid onto the AR. The shape of the profile is different from that given in case 1; however, the optimization approach is identical. The value of P is again adjusted until the function touches the AR boundary. In this instance, application of this procedure results in an optimal value of P = 0.56, suggesting that the maximum profit derived from this system is \$0.56 million. The corresponding point of intersection with the AR boundary is therefore given by the concentration vector $\mathbf{C} = [0.031, 0.609]^T$ mol/L. Consultation with the optimal reactor structure needed to achieve this point suggests that a CSTR followed by a PFR is required.

Both the optimal value of the objective function and the reactor structure needed to achieve these points have been determined. Moreover, the optimization has been performed on multiple objective functions with little additional effort.

5.5 CONSTRUCTIONS IN RESIDENCE TIME SPACE

5.5.1 Significance of Residence Time Constructions

If a reaction involves the use of a very expensive catalyst, then it is sensible to develop a reactor structure that minimizes *reactor volume* (or catalyst mass), for a desired conversion, rather than objective functions involving concentration. In this way, reactor volume is the primary variable that must be optimized rather than concentration or conversion.

However, all of the AR constructions discussed thus far have originated from viewing data in the phase plane (concentration space) only. Thus, the candidate regions produced from this perspective have catered toward problems involving concentration only.

In keeping with this theme, let us now look at AR constructions involving reactor volume, in the form of residence time. AR constructions involving residence time are easily adapted from concepts in concentration space. In order to maintain the useful geometric properties of mixing for residence time constructions—which are necessary in AR theory—all that must be established is that there is compatibility of residence time with mixing operations (specifically linear mixing) and reaction operations (rate fields).

> **SIDE NOTE: ARs in concentration space**
>
> When one is chiefly concerned with setting performance targets where time is *not* a concern, candidate ARs in concentration space alone serve a valuable purpose.

5.5.2 Mixing in Residence Time Space

Consider Figure 5.20, which shows a simple reactor configuration consisting of two reactors, R_1 and R_2 that are operated in parallel.

Both reactors are physically unique in that each reactor is fed with different feed concentrations ($\mathbf{C}_{f1}$ and $\mathbf{C}_{f2}$), and each are associated with a different reactor volume (V_1 and V_2) and product volumetric flow rate (Q_1 and Q_2). It follows that R_1 and R_2 must have distinct residence times, given by $\tau_1 = V_1/Q_1$ and $\tau_2 = V_2/Q_2$, respectively. Each reactor produces a different effluent concentration as well. The product of the two reactor effluent streams is combined to produce a mixture concentration, $\mathbf{C}^*$; and if density is assumed constant, the total volumetric flow rate is $Q_{tot} = Q_1 + Q_2$.

> **SIDE NOTE: A general reactor structure**
>
> Figure 5.20 has been drawn in a generalized fashion where R_1 and R_2 may represent reactor networks in which V_1 and V_2 would represent the total reactor volume specific to reactor network 1 and 2, respectively.

Computing the residence time of a mixture is equivalent to determining the residence time of the entire system. Let us define the residence time of the system, τ_{mix}, as the total reactor volume (the sum individual reactor volumes) divided by the combined volumetric flow rate:

$$\tau_{mix} = \frac{V_{tot}}{Q_{tot}} \tag{5.9}$$

Each reactor volume may be expressed in terms of the associated residence time and volumetric flow rate as follows:

$$\begin{aligned} V_{tot} &= V_1 + V_2 \\ &= \tau_1 Q_1 + \tau_2 Q_2 \end{aligned}$$

When this expression is substituted into Equation 5.9, we obtain the following:

$$\tau_{mix} = \frac{\tau_1 Q_1 + \tau_2 Q_2}{Q_{tot}}$$

This equation is similar in appearance to the mixing expression for concentrations. If the substitution $\lambda = Q_1/Q_{tot}$

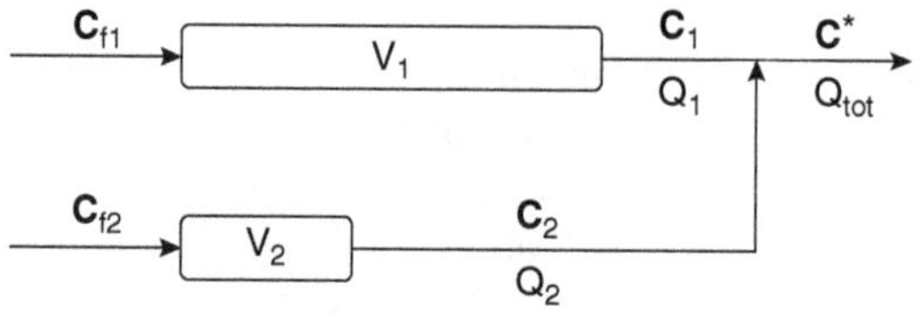

Figure 5.20 Physical interpretation of mixing two different reactors of unique residence times (reactor volumes).

(the ratio of volumetric flow rates to the total mixture flow rate) is made, then a familiar result is obtained as follows:

$$\tau_{mix} = \lambda\tau_1 + (1 - \lambda)\tau_2 \tag{5.10}$$

The form of Equation 5.10 is identical to the equation describing the linear mixing law for concentrations developed in Chapter 2. The process of combining two parallel reactors (or reactor networks) of different residence time results in a linear mixing law as well. This implies that residence time may be used in the construction of candidate ARs in a similar manner to that for concentration.

> **CONCEPT: Building ARs using τ**
>
> Since τ also obeys a linear mixing law, residence time may be incorporated into AR constructions and treated as if it is a component in the concentration vector $\mathbf{C}$. That is, τ may be viewed as a pseudo component in $\mathbf{C}$.

5.5.3 Visualizing Residence Time Data

5.5.3.1 Concentration Vectors

In order to plot data in both residence time and concentration space (c_i–τ space), the state vector $\mathbf{C}$ must be modified accordingly to accommodate τ. Since τ obeys a linear mixing law, it may be included as an element of $\mathbf{C}$ and treated like any other component in the system.

> **ILLUSTRATION: Two-dimensional plot involving τ**
>
> Suppose that a two-dimensional state vector is given of the following form
>
> $$\mathbf{C} = \begin{bmatrix} c \\ \tau \end{bmatrix}$$
>
> where $\mathbf{C}$ represents a system containing a species component c and the residence time of the reactor network τ. By convention, the last component in the state vector $\mathbf{C}$ is typically reserved for the system residence time when residence time constructions are performed.[10]

[10] This appears only for historical convention. In practice, the position of τ as a component in $\mathbf{C}$ does not affect AR constructions in any way.

Consider the case when two unique states, $\mathbf{C}_1$ and $\mathbf{C}_2$, are mixed together. From the theory of mixtures involving concentrations, a mixture of concentrations c_1 and c_2, given by c^*, must lie on a straight line between concentrations c_1 and c_2. But residence time also obeys a linear mixing law. The resulting residence time for the system must therefore also lie on a straight line between the residence times of the individual reactors τ_1 and τ_2, respectively. This interpretation is displayed in Figure 5.21.

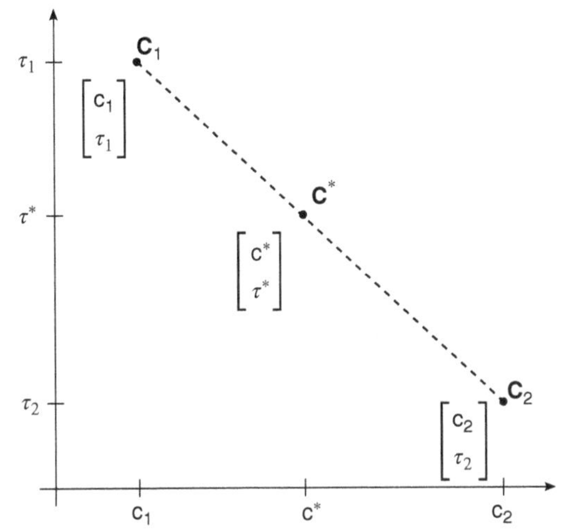

Figure 5.21 Geometric interpretation of mixing in residence time space. Mixtures also lie on a straight line when residence time is used since residence time also obeys a linear mixing law.

Points 1 and 2 are associated with the state vectors $\mathbf{C}_1 = [c_1, \tau_1]^T$ and $\mathbf{C}_2 = [c_2, \tau_2]^T$, respectively. The mixture is hence given by the state vector $\mathbf{C}^* = [c^*, \tau^*]^T$. Geometrically, $\mathbf{C}^*$ must lie on the straight line joining states $\mathbf{C}_1$ and $\mathbf{C}_2$ in c–τ space as both concentration *and* residence time obey a linear mixing law.

5.5.3.2 Rate Vectors The rate of reaction is defined simply in terms of the state vector as follows:

$$\mathbf{r}(\mathbf{C}, \tau) = \frac{d\mathbf{C}}{d\tau} = \begin{bmatrix} \frac{dc_1}{d\tau} & \cdots & \frac{dc_n}{d\tau} & \frac{d\tau}{d\tau} \end{bmatrix}^T$$

The final term in the rate vector is reserved to indicate the system residence time. All other terms present in the vector are simply individual species rate functions. Observe that species rate expressions can be viewed as the instantaneous rate of change with respect to residence time. Thus for rate of reaction for component i, $r_i(\mathbf{C})$, we have

$$\frac{dc_i}{d\tau} = r_i(\mathbf{C})$$

For residence time, the same substitution is applied to obtain the following:

$$\frac{d\tau}{d\tau} = 1$$

Consequently, when rate vectors are to be defined for an AR in residence time space, the corresponding entry for τ is assigned a scalar value of 1:

$$\mathbf{r}(\mathbf{C}, \tau) = \begin{bmatrix} r_1(\mathbf{C}) & \cdots & r_n(\mathbf{C}) & 1 \end{bmatrix}^T$$

where $\mathbf{r}_1(\mathbf{C})$, $\mathbf{r}_2(\mathbf{C})$, ... are the component rate functions participating in the system. This result arises directly as a consequence of the definition of the rate vector.

SIDE NOTE: Value of τ in r(C)

The element corresponding to residence time in the rate vector $\mathbf{r}(\mathbf{C})$ is always unity. This arises as a direct result of the definition of the rate of reaction $r_i(\mathbf{C}) = dc_i/d\tau$ and $d\tau/d\tau = 1$.

5.5.4 Unbounded Regions

Although residence time shares similar geometric traits to concentration, AR constructions in residence time space are inherently different to constructions solely in concentration space. Since there is no limitation on the value of residence time used (we are allowed to make the system residence time as large as desired), the corresponding AR boundaries may then also be arbitrarily large.

Physically, when a state is achieved in the reactor network corresponding to a particular system residence time, it is also achieved for all future residence times as well. Conceptually, one might envision a beaker experiment, similar to that given in Chapter 2, whereby the reaction takes place in the presence of a catalyst that is easily removed to limit the progress of reaction. Then, for any conversion obtained in the beaker, the catalyst may be removed to halt any further reaction from taking place. Concentrations inside the reactor remain fixed indefinitely after the time of catalyst removal. If species concentration profiles were to be plotted over time, the shape of the profiles might take a similar form to that given in Figure 5.22—concentration profiles would appear as horizontal lines at the moment of catalyst removal.

Hence, once a concentration is attained for a specific residence time, the same concentration is available for future residence times as well. From a geometric viewpoint, *AR*

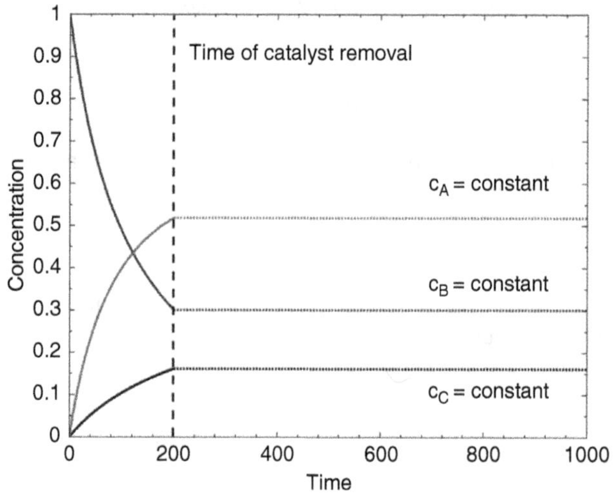

Figure 5.22 Concentration profiles in a beaker after the reaction has been halted.

constructions performed in residence time space are therefore unbounded on the τ axis.

Figure 5.23 shows two equally acceptable representations of candidate regions, generated in residence time space, for an identical feed point and kinetics. The boundary of the AR is defined by the curve AB. In Figure 5.23(b), the boundary is extended to include point C; however, point C is identical in concentration to point B, and only differs to point B in that it is evaluated at a later residence time. (If the time and concentration axes were swapped around in Figure 5.22, the resulting concentration profiles would resemble Figure 5.23(b).) The ARs for both figures (a) and (b) are identical from an achievability point of view as both graphs contain the same set of achievable concentrations.

Since mixing is linear with respect to residence time, the convex hull of a set of points in this space carries the same interpretation. The convex hull represents the region of achievable points obtained via mixing and reaction, which is identical in interpretation to ARs cast only in concentration space.

> **OBSERVATION: Constructions in residence time space**
>
> Candidate ARs in residence time space are necessarily unbounded on the τ axis, owing to the fact that a state achieved at a particular residence time is also achievable for all later residence times.

5.5.5 Example: Optimal Reactor Structure for Minimum Residence Time

5.5.5.1 Reaction and Feed Specification All of the examples described in this chapter possess equivalent versions involving residence time as well. If it is desired to retain the same components as before, and also include residence time in the state vector, then the resulting AR must be of one dimension higher than that originally posed—two-dimensional problems in concentration space are thus three-dimensional problems if residence time is considered as well.[11] With this in mind, consider now the single autocatalytic reaction involving components A and B

$$A + B \rightarrow 2B$$

We wish to determine the optimal reactor structure corresponding to the lowest possible residence time achieved by the system.

[11]It is also possible to simply ignore one of the lesser components and replace the component with residence time if we are comfortable with losing the information associated with the lost component.

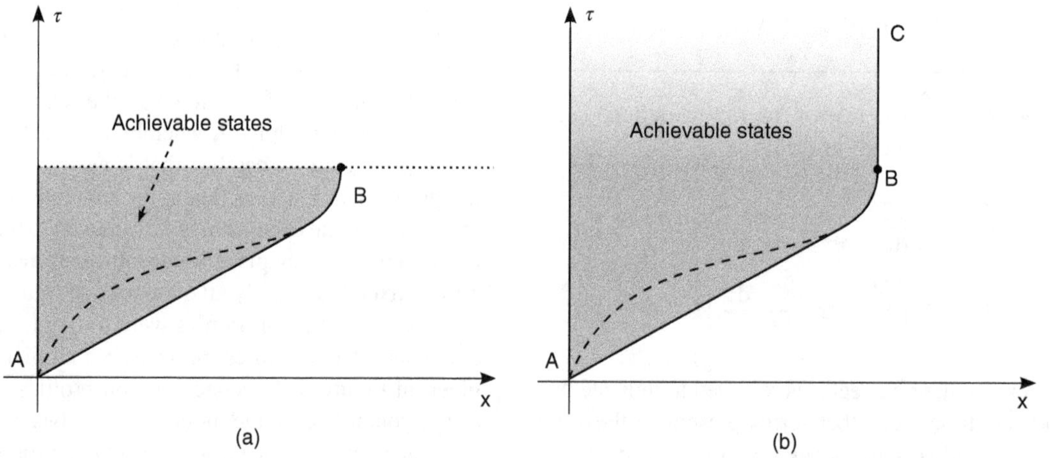

Figure 5.23 Two identical ARs. The region given by figure (b) contains a larger upper bound than that given by figure (a). Both regions contain the exact same concentrations and thus both regions represent the same set of achievable states.

Component A reacts in the presence of component B to form more B, and therefore one mole of A is reacted for every mole of B that is formed overall. It is assumed that B is the desirable component. The associated rate expression for component A is assumed to follow

$$r_A(\mathbf{C}) = -kc_Ac_B$$

where $k = 2 \times 10^{-3}$ mol/(L.s). The rate of production of component B, $r_B(\mathbf{C})$, is therefore also known by stoichiometry. The initial concentrations of A and B are specified as $c_{Af} = 20$ mol/L and $c_{Bf} = 1 \times 10^{-6}$ mol/L, respectively. Assume also that the feed to the reactor network has a volumetric flow rate of $Q = 1.2$ L/s.

5.5.5.2 Concentration and Rate Vectors Since the objective in this example is to determine the reactor structure corresponding to lowest residence time, the state vector for which the AR will be built from must reflect this perspective accordingly. Seeing as component B may be determined by stoichiometry and the value of c_A, one need to only include either component A or B for the other may then be found by mass balance. Hence a two-dimensional state vector may be formed containing the concentration of A and the system residence time. This is given simply as follows:

$$\mathbf{C} = [c_A, \ \tau]^T$$

The feed vector, $\mathbf{C}_f$, is then

$$\mathbf{C}_f = \begin{bmatrix} c_{Af} \\ \tau_f \end{bmatrix} = \begin{bmatrix} 20\,\text{mol/L} \\ 0.0\,\text{s} \end{bmatrix}$$

Note that the initial value for τ in the feed point is always zero, since only the feed point is achievable initially. From Section 5.5.3.2, we can define the rate vector corresponding to the system in a similar fashion:

$$\mathbf{r}(\mathbf{C}, \tau) = \begin{bmatrix} r_A(\mathbf{C}) \\ 1 \end{bmatrix}$$

5.5.5.3 AR Construction

a) PFR Trajectory from the Feed Now that the system is defined, both in terms of a feed point $\mathbf{C}_f$ and state vector $\mathbf{C}$, a candidate region may be constructed in residence time space. Unlike the previous examples, the result of this construction will represent the total set of achievable concentrations in A *and* the corresponding system residence time, which may be converted to reactor volume given a volumetric flow rate Q. To begin, a plot of the PFR solution trajectory from the feed in c_A–τ space is drawn by first integrating the PFR function using the initial condition supplied. This is given in Figure 5.24.

This plot is similar to the BTX data given in the earlier chapters, although the time component is plotted on the

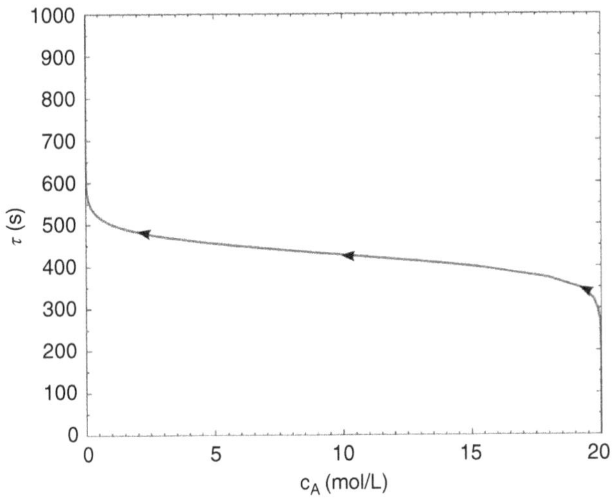

Figure 5.24 PFR trajectory from the feed, plotted now in c_A–τ space.

y-axis instead.[12] Unlike the PFR trajectories plotted solely in concentration space, the trajectory displayed here does not appear to terminate at a definite point. Rather, the concentration of A approaches zero, whereas the τ component increases indefinitely. This is consistent with the physical understanding of the system: after a certain residence time $\tau^{\#}$ (the equilibrium time) in the reactor, the system approaches equilibrium and the value of c_A remains constant for all $\tau \geq \tau^{\#}$.

ILLUSTRATION: Convex hull for the PFR only

Figure 5.25 shows the results of the convex hull computation from the PFR trajectory along with the corresponding reactor structures forming the boundary of the candidate region. Observe that the structures share identical interpretations to those that might be obtained in concentration space alone—the curved portion of the boundary belongs to the PFR from the feed, whereas the straight line from the feed to point X indicates a mixing line. The mixing line, in this instance, includes residence time.

Thus, the result of the candidate region construction assists in revealing that a lower residence time may be achieved by a reactor structure employing a PFR with *larger* residence time and bypass of feed to the product stream. For exit concentrations in A greater than point X, say $c_A = 5$ mol/L, a more efficient reactor structure exhibiting a lower residence time is that of a PFR with

[12] This is the convention used in many AR papers. We adopt the same convention here.

bypass of feed. This result would be difficult to identify if the candidate AR for this system had not been computed.

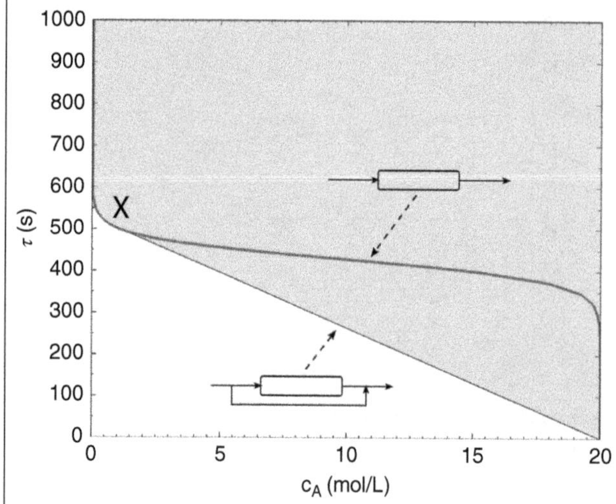

Figure 5.25 Convex hull for the PFR from the feed.

b) CSTR Locus from the Feed Calculating the CSTR locus from the feed is done in the same manner as before. The CSTR equation is solved for different residence times, and a locus of CSTR effluent concentration is thus produced. The following system

$$\mathbf{C}_\mathrm{f} + \tau\mathbf{r}(\mathbf{C},\ \tau) - \mathbf{C} = \mathbf{0}$$

is solved, and the results of the solution are plotted in the same space as the PFR trajectory previously. Figure 5.26(a) is then obtained.

In contrast to the PFR trajectory, the CSTR locus is almost entirely convex, and it achieves a much lower residence time over the majority of the concentration range.

c) Convex Hull Figure 5.26(b) provides the results of the convex hull computation for the CSTR locus and PFR from the feed. The filled area of the plot denotes the set of achievable states obtained for the two reactors when mixing is accounted for. If the objective is to determine the optimal reactor structure that achieves the smallest reactor volume, then focus should be placed on the lower portions of the region as these sections are associated with the reactor structure that exhibits smallest residence time. Hence, the goal of expanding ARs in residence time space relates to extending the lower portion of the region as close to minimum residence time as possible.

5.5.5.4 Optimal Reactor Structure for Minimum Residence Time From Chapter 4, since CSTR points form part of the boundary of the candidate region, rate vectors belonging to this section of the locus must either be tangent to or point out of the region. Points X and Y on the CSTR locus in Figure 5.26(b) are two such points. It is therefore possible to expand the candidate region by using the CSTR effluent concentration as a feed to additional PFRs.

Each point on the CSTR locus corresponds to a different CSTR residence time. To expand the lower portion of

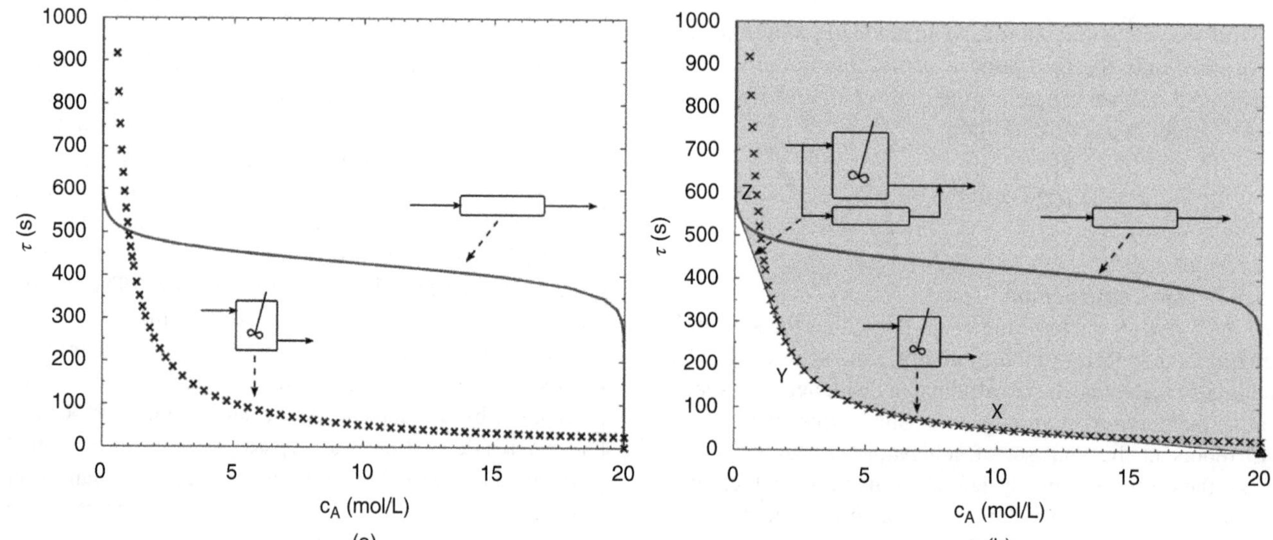

Figure 5.26 (a) CSTR locus from the feed, plotted together with the PFR trajectory obtained previously. The CSTR appears to achieve the majority of concentrations in a much smaller residence time when compared to the PFR. (b) The candidate region produced for a CSTR and PFR from the feed point.

the candidate region in the greatest manner possible, a PFR trajectory must be chosen on the CSTR locus that represents the lowest point of the candidate boundary. In Figure 5.26(b), this condition is given by point X. The value of c_A at X is found to be 10.0 mol/L, and hence this is the optimal feed concentration to the new PFR.

SIDE NOTE: Optimal CSTR residence time

There are potentially many ways that one might approach finding the optimal CSTR residence time that expands the candidate region boundary. One possible method is described below.

Consider that the optimal location on the CSTR locus to initiate a PFR will result in a convex hull with *maximum* area. One method to compute the area is to initiate a PFR from each point on the CSTR locus and then compute the corresponding convex hull from the achievable states generated by the particular CSTR–PFR pair. MATLAB's convhull() and convhulln() functions have the ability to return the convex hull area (volume for hulls greater than two dimensions) along with computing the set of points belonging to the convex hull. A representative plot is shown in Figure 5.27(a). The unbounded nature of residence time plots requires us to place a restriction on the maximum residence time achievable in the system in order to compute a definite value for the AR area (otherwise, the AR area would be infinite). In this example, we use a maximum residence time of 1000 s, as shown in Figure 5.27(a).

Computing the areas for each CSTR–PFR pair at different points on the CSTR locus, and then comparing areas, provides a method for determining the optimal CSTR concentration that maximizes the area of the AR. Traversing along the entire length of the CSTR and computing the convex hull area results in the curve in Figure 5.27(b).

In Figure 5.27(b), the optimal concentration that maximizes area occurs at $c_A = 10$ mol/L. Using Figure 5.26, the corresponding CSTR residence time at $c_A = 10$ mol/L may be determined. When this is done, the optimal CSTR residence time is found to occur at 50 s.

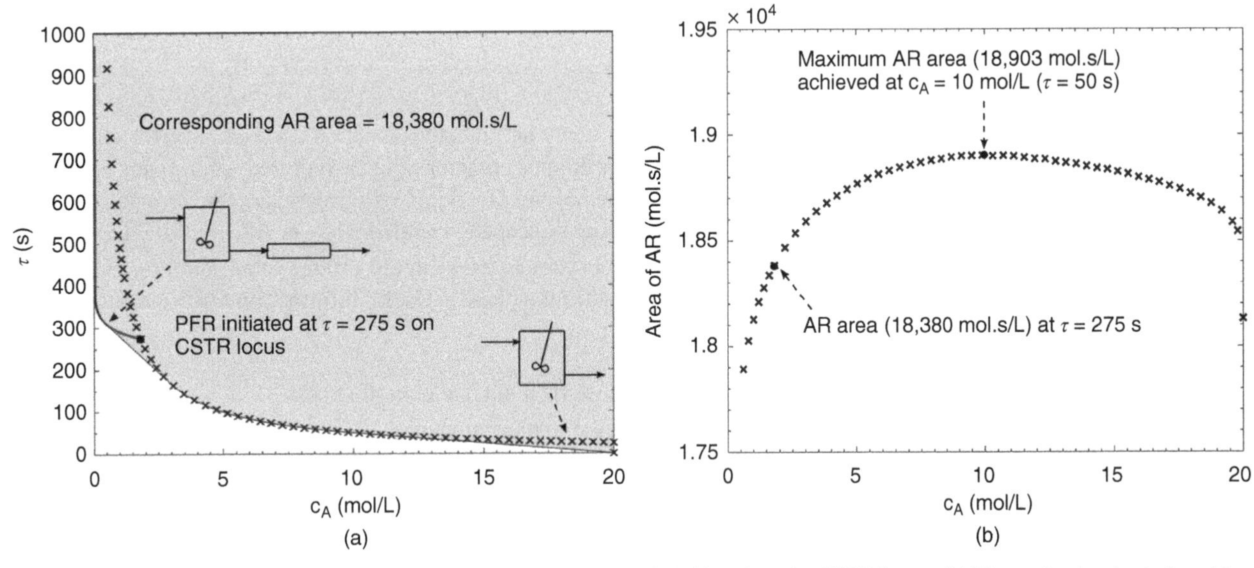

Figure 5.27 (a) The area of the AR varies depending on where a PFR is initiated on the CSTR locus. (b) The optimal point is found by initiating PFRs at each point on the CSTR locus and computing the corresponding AR area.

Figure 5.28 shows the resulting AR and the PFR trajectory taken from point X, the optimal CSTR point.

When the convex hull associated with the improved reactor structure is computed, there are no rate vectors on the boundary that point out of the region. This is clear since the AR boundary is composed entirely out of a PFR trajectory (rate vectors are thus always tangent to this portion of the boundary) and a mixing line to point X, which is obtained from a CSTR effluent concentration that must also be tangent to the boundary. Therefore, there is no possibility of

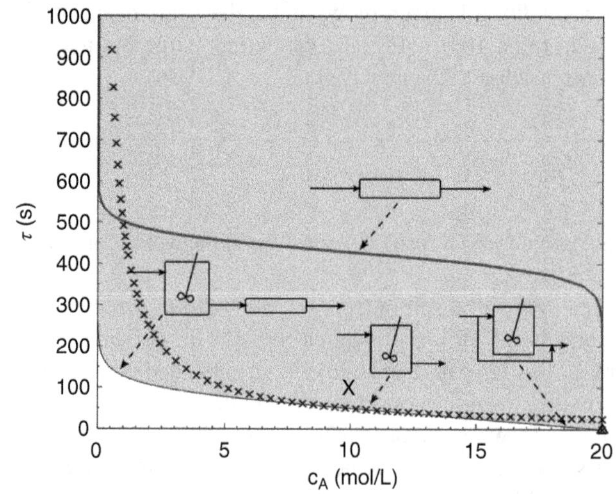

Figure 5.28 Full AR for the residence time problem.

expanding the candidate region further using combinations of CSTRs and PFRs. The region given in Figure 5.28 is therefore the AR for the system in residence time space.

5.5.5.5 Summary of Optimal Reactor Structures The results of the AR construction indicate that the optimal reactor structure for this problem is a CSTR, of residence time equal to 50 s, followed by a PFR to equilibrium. If concentrations greater than 10 mol/L in A are desired, then the optimal reactor for lowest residence time is a CSTR with bypass of feed to the effluent stream. This recommendation is less intuitive, for it suggests that a smaller reactor is achieved by operating *past* $c_A = 10$ mol/L and then diluting the CSTR product stream back to the desired concentration via bypass of feed. Ordinarily, one might presume that simply operating the reactor for small residence time would be the best approach.

In Figure 5.29, a number of different cases for the desired value of c_A, along with associated optimal structures, are shown. The optimal reactor structure varies depending on the final desired product concentration. Although the structure is specific for each objective, all structures are part of a generalized optimal reactor structure related to the AR.

a. For $c_A \geq 10$ mol/L
b. For $c_A = 10$ mol/L
c. For $c_A \leq 10$ mol/L

EXAMPLE 7: Payback period on reactor network

Now that the AR in c_A–τ space has been identified, one can apply an objective function and optimize for a desired scenario. Residence time plots are useful for understanding the minimum reactor volume required for a given output. Since the construction and operation of a reactor is generally an expensive undertaking, requiring a large investment of capital, it would be useful to determine the profitability of the system using concepts generated from an AR in residence time space.

Hence, let us develop a suitable objective function that incorporates the required product concentration and minimum reactor volume, which may be used in profitability calculations for the plant.[13] Useful information for this example is given as follows:

- Costs are split into two categories: the initial capital cost of the plant and operating costs.
  - Capital cost of the plant equipment is $C_{plant} = \$45$ million at present value.
  - Operating costs are assumed to be a function of two factors: total reactor volume V_{tot} and separation costs for component A, given by $C_{op} = 0.05V_{tot} + 0.35c_A^2$ in millions of \$/year).
- Annual revenue from the sale of product (component B) is assumed to be based on the purity of the product, given by $R = 2.4c_B$ (in millions of \$/year).
- Assume that an annual interest rate of 7% is charged on the total capital cost of the plant and compounded annually (the bank charges interest on the money loaned to pay for the initial equipment of the plant).

Using this information, determine the reactor effluent concentration and reactor volume necessary to fulfill payback periods of 2, 5, and 10 years. Recall from Section 5.5.5.1 that the volumetric flow rate of the feed to the reactor network is $Q = 1.2$ L/s.

Answer:

The payback period is the time required to repay the initial capital cost of the plant. Larger profits will allow for shorter payback periods, which are affected by the revenue and operating costs of the plant—a higher purity product will allow for greater revenues, but at the expense of larger reactor volumes and separation costs, which result in higher running costs.

[13]A more sophisticated model could also be employed that may be more realistic, although we provide a version here for demonstration purposes.

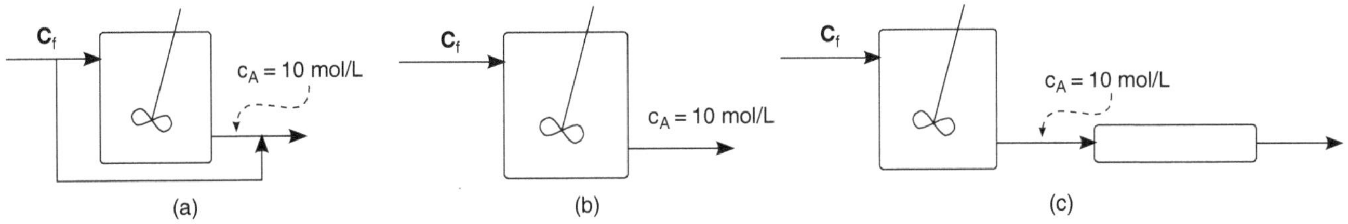

Figure 5.29 Different optimal reactor structures obtained depending on the desired exit concentration. (a) A CSTR with bypass of the feed, (b) a CSTR operated at point X in Figure 5.28 ($C_A = 10$ mol/L), and (c) a CSTR followed by a PFR.

This problem is best discussed by breaking up the solution into subsections, which are explained next.

Annual profit generated

Annual profit, P, is defined as the difference between annual revenues, R, from the sale of reactor product and the annual operating costs, C_{op}, hence

$$P = R\text{-}C_{op}$$

Expressions for R and C_{op} have been supplied previously, thus

$$R = 2.4c_B$$

and

$$C_{op} = 0.05V_{tot} + 0.35c_A^2$$

giving

$$P = 2.4c_B - 0.05V_{tot} - 0.35c_A^2$$

Since P is a function of both c_A and c_B, it will be useful to have expressions that convert c_B to c_A. This may be determined directly from mass balance relations as follows:

$$\begin{aligned} c_A &= c_{Af}(1 - x) \\ c_B &= c_{Bf} + c_{Af}x \\ &= c_{Bf} + c_{Af}\left(1 - \frac{c_A}{c_{Af}}\right) \end{aligned}$$

Furthermore, total reactor volume, V_{tot}, may be expressed in terms of residence time τ and volumetric flow rate Q. Substituting these expressions in for P then gives

$$P = 2.4\left(c_{Bf} + c_{Af}\left[1 - \frac{c_A}{c_{Af}}\right]\right) - 0.05\tau Q - 0.35c_A^2 \tag{5.10a}$$

Total capital cost of plant

Assume for simplicity that the total plant cost is paid as a lump sum amount at a future date, which is specified by the payback period. In other words, if a payback period of 5 years is chosen, the full plant cost will be paid in a single amount 5 years from the present date. Information has been supplied relating the compound interest rate charged on the capital cost of the plant. Hence, in order to determine the future plant cost, the compound interest formula must be employed.

$$C_{tot} = C_{plant}(1 + r)^n \tag{5.10b}$$

Here, $r = 0.07$ corresponds to a 7% annual interest rate and n is the payback period (in years).

Payback period

Now that expressions for the total (future) capital cost and annual profit have been determined, these two expressions may be equated to determine the payback period. Payback period is hence given by the annual profits, generated over n years, needed to recover the total future plant cost:

$$nP = C_{tot}$$

Substituting in the expressions for P and C_{tot} gives

$$n\left[2.4\left(c_{Bf} + c_{Af}\left[1 - \frac{c_A}{c_{Af}}\right]\right) - 0.05\tau Q - 0.35c_A^2\right] = C_{plant}(1 + i)^n \quad (5.10c)$$

Equation 5.10c may be used in conjunction with the AR construction to analyze the profitability of the system. Specification of a value for n results in a function of τ and c_A (a curve in c_A–τ space). These curves may then be overlaid over the AR to find the point of intersection with the AR boundary. Payback periods of 2, 5, and 10 years have been specified; in other words, n = 2, 5, 10 years. These results are shown in Figure 5.30.

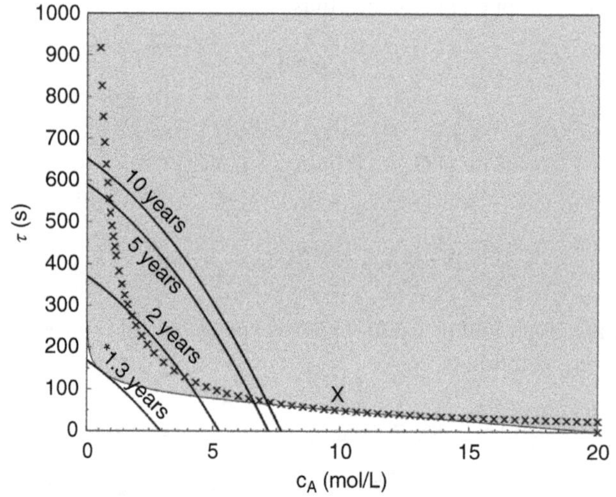

Figure 5.30 Different payback period objective functions overlaid onto the AR to determine the optimal product concentration and reactor volume. The line corresponding to 1.3 years, and marked with an asterisk (*), represents the minimum payback period for the process.

The objective function formulated is particularly useful as it incorporates aspects of both total reactor volume and product purity. Observe that the curve corresponding to the longest payback duration of 10 years is achieved at a point on the AR boundary corresponding to low product purity (low conversion of c_A) and small reactor volume. Following the AR boundary, shorter payback periods are achievable for higher conversions of A (smaller values of c_A).

The payback periods for 2, 5, and 10 years intersect the AR at many points *in* the region as well, indicating that there are multiple operating points (multiple optima) for this system. Since the goal of this problem is to find the reactor structure with smallest reactor volume, the intersection points associated with the lower part of the AR boundary are most favorable. In this example, a higher purity product influences the payback period more than running cost (reactor volume). This behavior is seen by how close the 10- and 5-year payback periods are to each other. It appears that it is more favorable for this example to construct a larger reactor, with larger operating expenses, with the intention of producing a higher quality product. Hence, even if the required reactor system is more complex, the plant is profitable in shorter a period of time.

Furthermore, observe that it is not possible to achieve a payback period of less than 1.3 years irrespective of the reactor configuration employed—smaller payback periods approach the bottom left-hand corner of Figure 5.30, and thus contour lines corresponding to payback periods less than 1.3 years do not intersect the AR boundary at all.

The particular choice of payback period might also influence the optimal reactor structure necessary to achieve it. Longer payback periods will require a CSTR with bypass as the optimal reactor structure (this structure occurs when c_A concentrations are greater than point X = 10.0 mol/L). Table 5.6 provides a summary of the intersection points for the payback periods.

TABLE 5.6 Summary of Required Operating Conditions for Payback Period

Payback Period	Concentration Vector **C**	Required Reactor Volume (L)
10 years	$[7.178, 12.822]^T$	77.676
5 years	$[6.631, 13.369]^T$	81.384
2 years	$[4.391, 15.609]^T$	98.976
[a]Minimum payback (1.297 years)	$[0.5906, 19.4094]^T$	171.120

[a]The minimum payback period entry corresponds to the minimum payback period curve from Figure 5.6.

Table 5.6 also provides the minimum payback period possible. This result is calculated by optimizing Equation 5.10c as an objective function for the optimal payback period n, which is found by minimizing n so that the resulting curve just touches the AR boundary. Reactor volumes in Table 5.6 are calculated from residence time using $V_{tot} = \tau Q$, where Q = 1.2 L/s.

5.6 CHAPTER REVIEW

In this chapter, we applied AR theory to two-dimensional reactor network synthesis problems. Even when the problems are two-dimensional, a great deal of insight into optimal reactor structures may still be gained. AR theory is slightly unconventional in that we wish to find all answers to all possible optimization problems first (by computing the AR), and then introduce an objective function afterward that answers our specific design objectives. Depending on where we wish to operate on the AR boundary, different optimal reactor structures may result, yet it is easier to understand these recommendations once the AR has been generated.

Constructions involving minimum residence time (minimum reactor volume) were also discussed. These constructions are effectively identical to those constructed in concentration space (the phase plane). This is feasible as residence time behaves as a pseudo component in concentration space—τ obeys a linear mixing law and it is easily incorporated into the rate vector **r**(**C**).

SIDE NOTE: End of Section I

This chapter marks the end of Section I of the book. All of the information discussed up to this point provides a firm foundation of the basics of AR theory. In the following chapters, we shall extend on these ideas, and relate them to higher dimensional constructions. Automated AR construction methods and variable density systems will also be discussed, which will allow us to tackle even more realistic problems.

REFERENCES

Chitra, S.P., Govind, R., 1985. Synthesis of optimal serial reactor structures for homogeneous reactions. Part I: Isothermal reactors. AIChE J. 31, 177–184.

Fogler, H.S., 2010. Essentials of Chemical Reaction Engineering, 1st ed. Prentice Hall, Upper Saddle River, NJ.

Ganapathisubramanian, N., Showalter, K., 1984. Bistability, mushrooms, and isolas. J. Chem. Phys. 80, 4177–4184.

Gray, P., Scott, S.K., 1983. Autocatalytic reactions in the isothermal, continuous stirred tank reactor: Isolas and other forms of multistability. Chem. Eng. Sci. 38, 29–43.

Nicol, W., Hernier, M., Hildebrant, D., Glasser, D., 2001. The attainable region and process synthesis: Reaction systems with external cooling and heating. The effect of relative cost of reactor volume to heat exchange area on the optimal process layout. Chem. Eng. Sci. 56, 173–191.

Nicol, W., Hildebrandt, D., Glasser, D., 1997. Process synthesis for reaction systems with cooling via finding the attainable region. Comput. Chem. Eng. 21, S35–S40.

Scott, S., 1987. Isolas, mushrooms and oscillations in isothermal, autocatalytic reaction-diffusion equations. Chem. Eng. Sci. 42, 307–315.

Van de Vusse, J., 1964. Plug-flow type reactor versus tank reactor. Chem. Eng. Sci. 19, 994–996.

SECTION II

EXTENDED TOPICS

6

HIGHER DIMENSIONAL AR THEORY

6.1 INTRODUCTION

Since the problems provided in Chapter 5 are two-dimensional in nature, construction of candidate regions may be achieved by plotting the data in a two-dimensional phase plane and then validating constructions by visual inspection of the plots. The use of DSRs in two-dimensional constructions is also unnecessary. However, for higher dimensional constructions—that is, greater than or equal to three dimensions—the DSR plays an important role in defining the shape of the AR boundary. A framework for handling DSRs must therefore be in place in order to perform higher dimensional constructions appropriately. The difficulty with constructing candidate ARs in higher dimensions is that it is no longer possible to rely solely on visual inspection of the results; viewing higher dimensional constructions is simply more cumbersome when compared to two-dimensional plots.

Furthermore, when AR constructions were carried out in previous chapters, they were described without adequate justification of why the resulting region represented the true AR. In order for these properties to be validated, and also in preparation for higher dimensional examples to be discussed (in Chapter 7), a more detailed AR theory, generalized to n-dimensional spaces, is required. This theory also assists in understanding of the kinds of reactor structures that should be expected when higher dimensional systems are considered in general.

Our goal then, is to provide discussions for a reasonable (but not mathematically rigorous) understanding of the relationships between reaction and mixing, and how these relate to optimal reactor networks in n-dimensional space.[1] In all of the discussions that are to follow, there is a risk of getting lost in the details and forgetting why these concepts are important in the first place. If this occurs, there are two key consequences from this chapter that should be kept in mind:

1. When reaction and mixing are the only two processes available within a system, the AR may be constructed via combinations of CSTRs, PFRs, DSRs and mixing alone. No other reactor types are required to form the AR.
2. It is theoretically possible (but often not practically feasible) to compute concrete conditions (mathematical expressions) that dictate how PFRs, CSTRs, and DSRs should be operated on the AR boundary.

Much of the content in this chapter is taken from important contributions by Martin Feinberg (Feinberg, 1999, 2000a, 2000b; Feinberg and Hildebrandt, 1997). As will be shown, results from four papers by Feinberg, in particular, broadly define the major findings of AR theory in concentration space—at the time of writing, these results have yet to be expanded to wider state spaces, such as mass fraction space. It is for this reason that primary focus will be placed on AR constructions in concentration space alone wherein density is assumed constant.

> **(!) IMPORTANT: Mathematical concepts ahead**
>
> This chapter is highly mathematical in nature. Fundamental AR theory is founded on topics in differential geometry and optimal control theory—in the form of *geometric control theory*. To avoid unnecessary confusion, and to allow the ideas to be understandable to a wider audience, detailed proofs are deliberately omitted from this chapter.

[1] Rather than listing proofs, we prefer to show the final result of the theorems and then attempt to explain them in an understandable manner.

Attainable Region Theory: An Introduction to Choosing an Optimal Reactor, First Edition. David Ming, David Glasser, Diane Hildebrandt, Benjamin Glasser, and Matthew Metzger.

Companion website: http://attainableregions.com

Instead, we shall concentrate only on core results and provide a simplified, potentially inexact, interpretation of the theory. The discussions given are thus explanatory rather than mathematically precise. For those interested in a precise treatment of AR theory, reference is made to the original works wherefrom these discussion are derived: Feinberg (1999, 2000a, 2000b, 2002) and Feinberg and Hildebrandt, (1997).

Readers interested in obtaining practical experience constructing higher dimensional ARs may choose to skip this chapter and move on to Chapter 7 instead, where a number of higher dimensional worked examples are provided. However, it will be useful to at least briefly read over the important results in this chapter before progressing on to Chapter 7.

6.2 DIMENSION AND STOICHIOMETRY

6.2.1 The Stoichiometric Subspace

6.2.1.1 Introduction Given a system of reactions, it is possible to compute bounds in concentration space wherein all feasible concentrations must lie. This space is typically much larger than the space of achievable concentrations (the AR). Consequently, we can use this space as an upper bound on the set of feasible concentrations that the AR must reside in. We call this space the *stoichiometric subspace* and denote it by the set **S**.[2] Determining **S** is also useful for AR construction algorithms, which are discussed in Chapter 8.

6.2.1.2 Geometric Meaning

a) Single Reaction In order to obtain an appreciation for the mathematical concepts used below, it is beneficial to have an idea of the geometric relation between reaction and the stoichiometric subspace. Gaining an understanding of lower dimensional systems in $\mathbb{R}^2$ and $\mathbb{R}^3$, which can be visualized, shall prove useful when higher dimensional spaces are considered. With this in mind, consider the geometric nature of the ammonia synthesis reaction:

$$N_2 + 3H_2 \rightarrow 2NH_3$$

Relative proportions of each component in the reaction are dictated by the reaction stoichiometry—one mole of nitrogen and three moles of hydrogen combine to form two moles of ammonia by the following reaction. Given a feed composition, species concentrations of all components can be expressed in terms of the extent of reaction ε and reaction stoichiometry. Assuming constant density, a mass balance for the ammonia reaction may be written, giving expressions for the instantaneous species concentrations as a function of the initial feed compositions, reaction stoichiometry and extent of reaction as follows:

$$c_{N_2} = c^0_{N_2} - \varepsilon$$
$$c_{H_2} = c^0_{H_2} - 3\varepsilon$$
$$c_{NH_3} = c^0_{NH_3} + 2\varepsilon$$

This system may be written in vector form as

$$\begin{bmatrix} c_{N_2} \\ c_{H_2} \\ c_{NH_3} \end{bmatrix} = \begin{bmatrix} c_{N_2f} \\ c_{H_2f} \\ c_{NH_3f} \end{bmatrix} + \begin{bmatrix} -1 \\ -3 \\ 2 \end{bmatrix} \varepsilon$$

Observe that the left hand side of the equation is simply the concentration vector $\mathbf{C} = [c_{N_2}, c_{H_2}, c_{NH_3}]^T$ mol/L, and the right hand side is a vector sum, consisting of the feed vector $\mathbf{C}_f = [c_{N_2f}, c_{H_2f}, c_{NH_3f}]^T$ mol/L and a vector holding the stoichiometric coefficients of the reaction, $\mathbf{a} = [-1, -3, 2]^T$ mol/L.

$$\mathbf{C} = \mathbf{C}_f + \mathbf{a}\varepsilon \tag{6.1}$$

For any specification of the feed vector $\mathbf{C}_f$ and extent of reaction ε, the set of concentrations compatible with the ammonia reaction must lie on a line in c_{N_2}–c_{H_2}–c_{NH_3} space, for the set of concentrations specified must satisfy the vector equation defined by Equation 6.1. Since Equation 6.1 is in fact a vector relation, we can plot the set of concentrations obeying the ammonia reaction. This plot is shown in Figure 6.1.

Here, a number of feed vectors have been specified together with the corresponding set of concentrations satisfying Equation 6.1. Irrespective of the feed vector used,

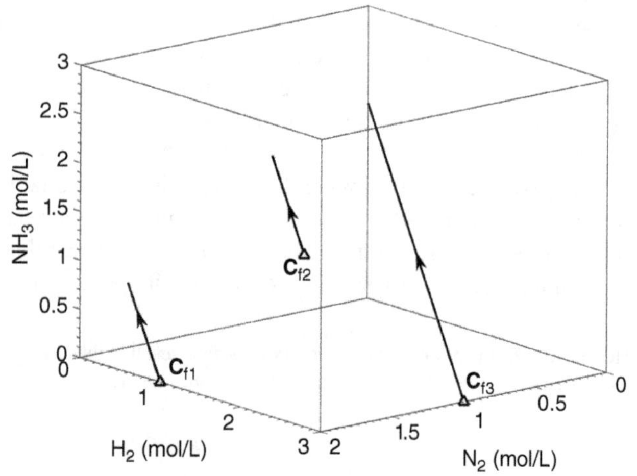

Figure 6.1 Concentrations that are stoichiometrically compatible with the ammonia reaction. The set of concentrations lies on a line in c_{N_2}–c_{H_2} – c_{NH_3} space. The gradient of the line is defined by the vector **a**, whereas its position is determined by the feed vector $\mathbf{C}_f$.

[2]**S** is often also referred to as the *mass balance triangle*.

the set of concentrations satisfying Equation 6.1 lies on a *line* in $\mathbb{R}^3$.

If kinetics for the ammonia system is available, computation of the set of *achievable* concentrations for this system (the AR) would result in concentrations constrained on the line defined by Equation 6.1. The AR for this system would therefore lie in a one-dimensional subspace of $\mathbb{R}^3$. Thus, the set of points achievable by the ammonia reaction, and thus the AR itself, is inherently one-dimensional for the ammonia reaction, even though there are three components present in the system.

EXAMPLE 1: Stoichiometric bounds for the NH_3 reaction

What are the stoichiometric limits of the NH_3 reaction for a feed of $\mathbf{C}_f = [c_{N_2f}, c_{H_2f}, c_{NH_3f}]^T = [1, 1, 1]^T$ mol/L?

Answer:

We know that concentrations achieved in the ammonia reaction follow Equation 6.1 and lie on a one-dimensional line in $\mathbb{R}^3$ so that

$$\begin{bmatrix} c_{N_2} \\ c_{H_2} \\ c_{NH_3} \end{bmatrix} = \begin{bmatrix} 1 \\ 1 \\ 1 \end{bmatrix} + \begin{bmatrix} -1 \\ -3 \\ 2 \end{bmatrix} \varepsilon$$

One bound of the set of achievable concentrations is the feed point $\mathbf{C}_f$ (when $\varepsilon = 0$ mol/L). The other bound must be achieved when 100% conversion of reactants is achieved. Since the feed is *not* in stoichiometric proportion, 100% conversion is based in the limiting reactant. Seeing as the feed is equimolar, the limiting reactant is hydrogen since three moles of H_2 are required for every one mole of N_2 reacted. Dividing the feed concentration of H_2 by the stoichiometric coefficients of H_2 gives

$$\varepsilon_{max} = \frac{1}{3} \text{ mol/L}$$

where ε_{max} is the maximum extent achievable based on the feed and the reaction stoichiometry. Thus the other bound is obtained by substituting ε_{max} into Equation 6.1

$$\begin{bmatrix} c_{N_2} \\ c_{H_2} \\ c_{NH_3} \end{bmatrix} = \begin{bmatrix} 1 \\ 1 \\ 1 \end{bmatrix} + \begin{bmatrix} -1 \\ -3 \\ 2 \end{bmatrix} \varepsilon_{max} = \begin{bmatrix} 1 \\ 1 \\ 1 \end{bmatrix} + \begin{bmatrix} -1 \\ -3 \\ 2 \end{bmatrix} \frac{1}{3} = \begin{bmatrix} 0.667 \\ 0 \\ 1.667 \end{bmatrix} \text{ mol/L}$$

Thus, the two concentrations bounded on the one-dimensional line in $\mathbb{R}^3$ are

$$\left\{ \begin{bmatrix} 1 \\ 1 \\ 1 \end{bmatrix}, \begin{bmatrix} 0.667 \\ 0 \\ 1.667 \end{bmatrix} \right\} \text{ mol/L}$$

b) Multiple Reactions It is straightforward extend the idea of stoichiometrically compatible concentrations to systems involving multiple independent reactions. Consider the system of reactions given by

$$A \rightarrow B$$

$$A + B \rightarrow 2C$$

Here, component A undergoes decomposition to component B and a parallel reaction with B to form C. Since there are now two reactions, two extents of reaction, ε_1 and ε_2, are available. The concentrations of components A, B, and C may again be expressed in terms of material balance relations, now involving two extents of reaction and an initial feed point

$$c_A = c_{Af} - \varepsilon_1 - \varepsilon_2$$

$$c_B = c_{Bf} + \varepsilon_1 - \varepsilon_2$$

$$c_C = c_{Cf} + 2\varepsilon_2$$

which may be expressed in an equivalent matrix form by

$$\begin{bmatrix} c_A \\ c_B \\ c_C \end{bmatrix} = \begin{bmatrix} c_{Af} \\ c_{Bf} \\ c_{Cf} \end{bmatrix} + \begin{bmatrix} -1 \\ 1 \\ 0 \end{bmatrix} \varepsilon_1 + \begin{bmatrix} -1 \\ -1 \\ 2 \end{bmatrix} \varepsilon_2$$

or, more compactly written as

$$\mathbf{C} = \mathbf{C}_f + [\mathbf{a}_1 \ \mathbf{a}_2] \begin{bmatrix} \varepsilon_1 \\ \varepsilon_2 \end{bmatrix} \tag{6.2}$$

where $\mathbf{a}_1 = [-1, 1, 0]^T$ and $\mathbf{a}_2 = [-1, -1, 2]^T$. The set of concentrations stoichiometrically compatible with the system is now governed by the vector equation given in Equation 6.2, parameterized by ε_1 and ε_2.

Since the system now contains two extents reaction (two independent reactions), concentrations compatible with reaction stoichiometry are expected to lie within a *two-dimensional subspace* in $\mathbb{R}^3$. This is more clearly seen from Figure 6.2, which plots the set of stoichiometrically compatible concentrations using a feed of $\mathbf{C}_f = [1, 1, 0]^T$.

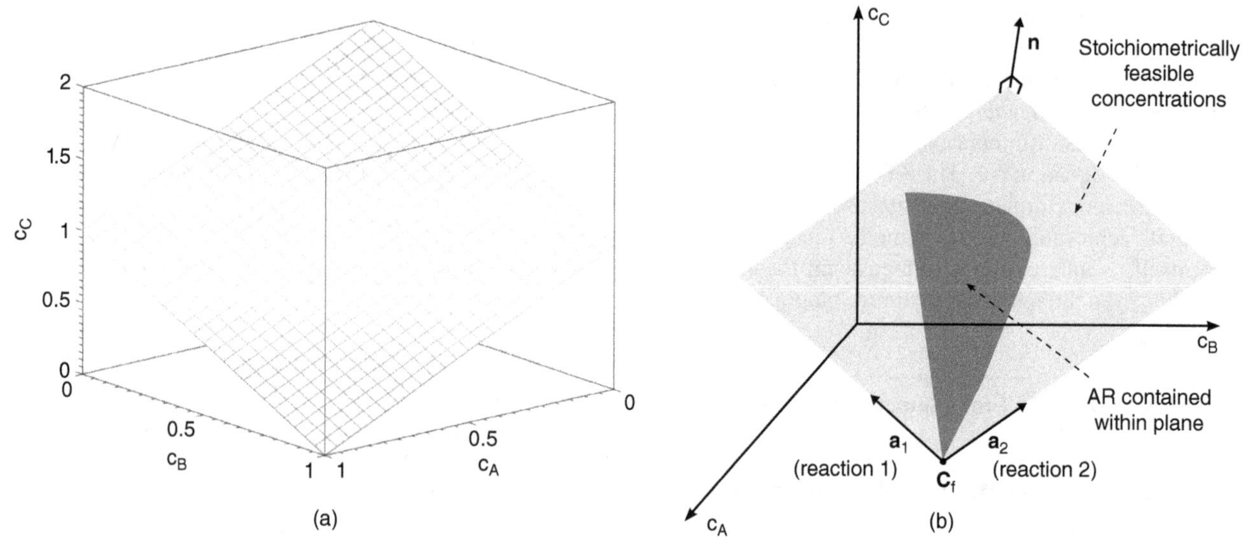

Figure 6.2 (a) The set of concentrations that are stoichiometrically compatible with the feed point $\mathbf{C}_f = [1, 1, 0]^T$ and Equation 6.2. (b) Any AR generated with kinetics using this feed results in a set of compositions that lie on a plane in c_A–c_B–c_C space.

If the plane given in Figure 6.2 represents all concentrations that are compatible with the reaction stoichiometry and feed point, then an AR generated by specification of any reaction kinetics for the same feed point must also lie within this plane. In this instance, the AR is a two-dimensional subspace in c_A–c_B–c_C space residing in the plane spanned by vectors $\mathbf{a}_1$ and $\mathbf{a}_2$.

6.2.1.3 The Stoichiometric Coefficient Matrix Consider a general system involving n components participating in d reactions. The first k species are reactants whereas the remaining $(n - k)$ are products. Expressing all reactions in terms of all species then gives:

$$\nu_{11}c_1 + \cdots + \nu_{1k}c_k \rightleftharpoons \nu_{1(k+1)}c_{k+1} + \cdots + \nu_{1n}c_n \quad \text{(Reaction 1)}$$

$$\vdots$$

$$\nu_{d1}c_1 + \cdots + \nu_{dk}c_k \rightleftharpoons \nu_{d(k+1)}c_{k+1} + \cdots + \nu_{dn}c_n \quad \text{(Reaction } d\text{)}$$

where ν_{ij} is the stoichiometric coefficient of component j participating in reaction i. From this system, a matrix containing the stoichiometric coefficients ν_{ij} can be formed for all components in all reactions. This is termed the *stoichiometric coefficient matrix*, denoted by $\mathbf{A}$ as follows:

$$\mathbf{A} = \begin{bmatrix} \nu_{11} & \cdots & \nu_{d1} \\ \vdots & \ddots & \vdots \\ \nu_{1n} & \cdots & \nu_{dn} \end{bmatrix} \quad (6.3)$$

In general, $\mathbf{A}$ has size $n \times d$. Each column in $\mathbf{A}$ represents a distinct reaction of the system. From the definition of $\mathbf{A}$, the stoichiometric subspace must reside in the space spanned by the columns in $\mathbf{A}$, which correspond to the $\mathbf{a}$ vectors for reactions taking part in the system

$$\mathbf{A} = [\mathbf{a}_1 \ \ldots \ \mathbf{a}_d]$$

The set of concentrations stoichiometrically compatible with the feed point is hence directly related to the stoichiometric coefficient matrix $\mathbf{A}$ and the feed point. The general matrix equation relating this is

$$\mathbf{C} = \mathbf{C}_f + \mathbf{A}\boldsymbol{\varepsilon} \quad (6.4)$$

where $\boldsymbol{\varepsilon}$ is the column vector of extents corresponding to each reaction in the system, $\boldsymbol{\varepsilon} = [\varepsilon_1, \varepsilon_2, \ldots, \varepsilon_d]^T$.

SIDE NOTE: Units of ε

Observe that when ε is written in terms of concentrations—such as in Equation 6.4—ε also has units of [concentration], such as mol/L, kmol/h, etc.

EXAMPLE 2: Stoichiometrically compatible concentrations

Consider the following system involving two independent reactions in three components, A, B, and C

$$A \rightarrow B$$
$$2B \rightarrow 1.5C$$

Species rate expressions are given by the following:

$$\begin{aligned} r_A &= -k_1 c_A \\ r_B &= k_1 c_A - 2k_2 c_B^2 \\ r_C &= 1.5k_2 c_B^2 \end{aligned} \tag{6.5a}$$

where $k_1 = 1\ s^{-1}$ and $k_2 = 1$ L/(mol.s). Notice that the rate expressions in this example are elementary; however, this is done for illustration purposes. Answer the following questions:

(a) Determine the stoichiometric coefficient matrix **A** for this system.
(b) Given a feed vector of $\mathbf{C}_f = [c_{Af}, c_{Bf}, c_{Cf}]^T = [1, 0.2, 0.5]^T$ mol/L, plot the plane of points that are stoichiometrically compatible with the feed in c_A–c_B–c_C space. Plot a PFR from the feed point $\mathbf{C}_f$ and show that the solution trajectory is contained within this plane.
(c) Suppose now that the kinetics is given by

$$\begin{aligned} r_A &= -k_1 c_A c_B \\ r_B &= k_1 c_A c_B - 2k_2 c_B^{1.3} c_C^{0.7} \\ r_C &= 1.5k_2 c_B^{1.3} c_C^{0.7} \end{aligned} \tag{6.5b}$$

Repeat question 2 above for the new set of conditions and compare to the original PFR trajectory.

Answer a:

Two reactions are present involving three components. Letting rows 1–3 correspond to A–C, respectively, the stoichiometric coefficient matrix **A** is therefore a 3 × 2 matrix, given by

$$\mathbf{A} = \begin{bmatrix} -1 & 0 \\ 1 & -2 \\ 0 & 1.5 \end{bmatrix}$$

Observe that component B is an intermediate product; it has a positive stoichiometric coefficient in the first column of **A** (reaction 1), and a negative stoichiometric coefficient in the second column (reaction 2).

Answer b:

Since there are two independent reactions participating in the system, we expect the set of points generated by the system to reside in a two-dimensional subspace in $\mathbb{R}^3$. The matrix equation describing this space is determined by the stoichiometric coefficient matrix **A** and the feed point $\mathbf{C}_f$:

$$\mathbf{C} = \mathbf{C}_f + \mathbf{A}\boldsymbol{\varepsilon}$$

or

$$\mathbf{C} = \begin{bmatrix} 1 \\ 0.2 \\ 0.5 \end{bmatrix} + \begin{bmatrix} -1 & 0 \\ 1 & -2 \\ 0 & 1.5 \end{bmatrix} \begin{bmatrix} \varepsilon_1 \\ \varepsilon_2 \end{bmatrix} \tag{6.5c}$$

ε_1 and ε_2 are the extents of reaction corresponding to reaction 1 and 2, respectively. Equation 6.5c is a vector equation describing concentrations compatible with the reaction stoichiometry and feed point specified. This may be plotted in c_A–c_B–c_C space by specifying a range of value for ε_1 and ε_2 and then computing the corresponding concentration vector $\mathbf{C}$ from Equation 6.5c. All points in the plane are those that satisfy Equation 6.5c.

The PFR equation may be integrated using $\mathbf{C}_f$ as the initial condition and plotted over the plane for comparison. The result is shown in Figure 6.3.

Notice that the PFR trajectory is indeed contained entirely within the plane. This demonstrates that the system is two-dimensional in nature. Had the AR been computed for this system, it would also be contained within the plane. Hence, although the system contains three components, it is actually a two-dimensional system.

Answer c:

Since the feed point and stoichiometry have not changed, the set of concentrations generated by the system is still governed by Equation 6.5c above. However, in this instance, the kinetics is different. The PFR trajectory associated with the new kinetics is expected to be different in shape to the former, although both must be contained within the same plane described by Equation 6.5c. The PFR equation may again be integrated and plotted in c_A–c_B–c_C space. The result of which is shown in Figure 6.3.

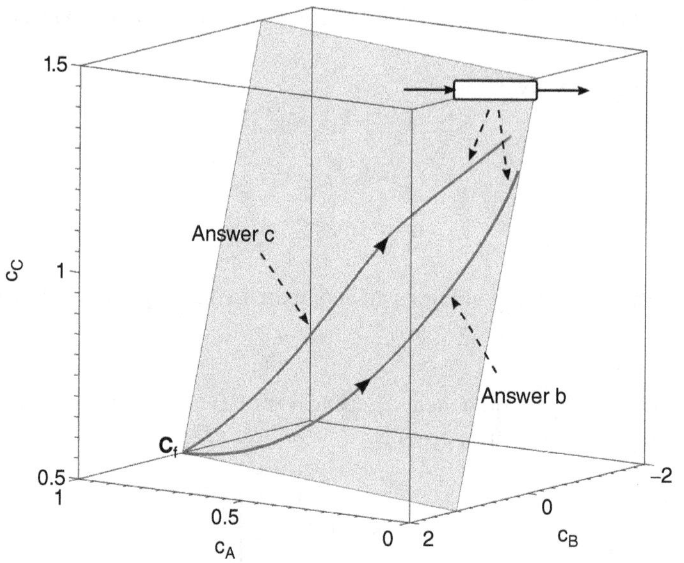

Figure 6.3 PFR trajectory associated with different kinetics, but the same feed point and reaction stoichiometry. (*See color plate section for the color representation of this figure.*)

The previous PFR trajectory has been retained for comparison. As expected, the shape of the PFR trajectories differ, due to the different rate expressions, yet both trajectories reside within the same two-dimensional subspace because they must both obey the reaction stoichiometry and thus both solutions will reside in the same stoichiometric subspace.

Although this discussion has employed PFRs, *any* other reactor type utilizing the same feed point and obeying the same reaction stoichiometry will produce effluent concentrations residing in the plane. It follows that the AR generated for this system will also lie in the plane; it is a two-dimensional subspace in $\mathbb{R}^3$.

CONCEPT: Determining the dimension of the AR

The dimension of the AR is equal to the number of independent reactions participating in the system. This is because the number of independent reactions defines the dimension of the stoichiometric subspace (the rank of the stoichiometric coefficient matrix **A**), and the AR must reside in the stoichiometric subspace.

Hence, the dimension of the AR is governed by the reaction stoichiometry of the system. It is determined by forming the stoichiometric coefficient matrix **A** and then computing its rank (the number of independent rows or columns in **A**).

6.2.1.4 Nonnegativity Constraints Since it is physically impossible to produce a mixture that exhibits negative mass, nonnegativity is an important constraint that further restricts the set of physically achievable concentrations in AR theory. The feasibility of any proposed concentration, achieved in a reactor network, is therefore determined by two important constraints:

1. Mass balance constraints—are all concentrations stoichiometrically compatible with the given reaction stoichiometry?
2. Nonnegativity constraints—are all concentrations in the system positive?

Consideration of nonnegativity constraints, together with the stoichiometric coefficient matrix, can be used to form a set of linear relations that mathematically describe the stoichiometric subspace.

To show this, we use Equation 6.4 to express all species concentrations within a mixture in terms of the stoichiometric coefficient matrix **A** and extent of reaction vector $\boldsymbol{\varepsilon}$

$$\mathbf{C} = \mathbf{C}_f + \mathbf{A}\boldsymbol{\varepsilon} \tag{6.4}$$

If there are n components participating in the reaction, then in addition to Equation 6.4, there are also n inequality relations that describe the nonnegativity restriction of each species concentration

$$c_1 \geq 0$$
$$c_2 \geq 0$$
$$\vdots$$
$$c_n \geq 0$$

These equations can also be written compactly in vector form as

$$\mathbf{C} \geq \mathbf{0}$$

which, when combined with Equation 6.4, gives

$$\mathbf{C}_f + \mathbf{A}\boldsymbol{\varepsilon} \geq \mathbf{0} \tag{6.6}$$

Notice that Equation 6.6 describes n linear *inequality* relations, which are different to standard equations (equality relations): each row in Equation 6.6 describes a linear inequality (a hyperplane) that separates n-dimensional space into two half spaces. The collection of all n inequalities describes a convex region in $\mathbb{R}^n$. Concentrations residing in this region thus satisfy both mass balance and nonnegativity constraints, and hence the region defined by Equation 6.6 describes, mathematically, the stoichiometric subspace **S**.

We shall discuss in greater detail how to numerically compute **S** for a given reaction and feed point in Chapter 8, but for now it is sufficient for us to have a qualitative understanding of the stoichiometric subspace for use in later sections of this chapter.

CONCEPT: The stoichiometric subspace S

The stoichiometric subspace can be defined by the following system of inequalities

$$\mathbf{C}_f + \mathbf{A}\boldsymbol{\varepsilon} \geq \mathbf{0}$$

This system describes a region, or subspace, of $\mathbb{R}^n$ that all feasible concentrations must reside in. A feasible concentration point is hence any *positive* state that obeys the reaction stoichiometry defined by mass balance. The AR must also be contained within **S** (it may constitute a very small or very big part of the space), and thus computing **S** provides an upper bound wherein the AR is known to reside.

CONCEPT: S for multiple feeds

We can extend the concept of the stoichiometric subspace, **S**, to also include *multiple feeds*. Consider the case when two feeds, $\mathbf{C}_{f1}$ and $\mathbf{C}_{f2}$, are available for the same reaction stoichiometry. We may compute the stoichiometric subspace belonging to $\mathbf{C}_{f1}$ and $\mathbf{C}_{f2}$ individually, giving $\mathbf{S}_1$ and $\mathbf{S}_2$, respectively. But since both feeds are achievable, it is also possible to mix between feeds and thus also between points in $\mathbf{S}_1$ and $\mathbf{S}_2$. Thus, we can compute the convex hull of the set of points in both $\mathbf{S}_1$ and $\mathbf{S}_2$:

$$\text{conv}(\{\mathbf{S}_1, \mathbf{S}_2\})$$

The overall stoichiometric subspace $\mathbf{S}_{\text{tot}}$, belonging to multiple feeds may therefore be determined by first computing the stoichiometric subspace $\mathbf{S}_i$ for each feed $\mathbf{C}_{fi}$, and then computing the convex hull of the set of points for all $\mathbf{S}_i$'s, which, for N feeds gives:

$$\mathbf{S}_{\text{tot}} = \text{conv}(\{\mathbf{S}_1, \mathbf{S}_2, \ldots, \mathbf{S}_{\text{N}}\})$$

6.2.2 Concentrations Orthogonal to the Stoichiometric Subspace

6.2.2.1 Null Space For many purposes in AR theory, it is useful to understand the set of concentrations that lie perpendicular (orthogonal) to **S**, which are spanned by the stoichiometric coefficient matrix **A**. For instance, the computation of critical DSR solution trajectories and CSTR effluent compositions that form part of the AR boundary require the computation of this space. It is therefore important that we briefly provide details of this topic here. It is simple to show from linear algebra that all points orthogonal to the space spanned by the columns of **A** are those that obey the following relation:

$$\mathbf{A}^{\text{T}}\mathbf{C} = \mathbf{0} \qquad (6.7)$$

Here, matrix **A** is the stoichiometric coefficient matrix formed from the reaction stoichiometry, and **C** is any concentration vector that satisfies Equation 6.7. Determination of the subspace of concentrations orthogonal to the stoichiometric subspace, is in fact equivalent to computing the *null space* of $\mathbf{A}^{\text{T}}$. We shall denote by **N** the matrix whose columns form a basis for the null space of $\mathbf{A}^{\text{T}}$. Linear combination of the columns in **N** hence generates the set of concentrations orthogonal to the stoichiometric subspace.

$$\mathbf{N} = \text{null}(\mathbf{A}^{\text{T}})$$

Assume that **A** contains n rows (there are n components in the system). If **A** has d linearly independent columns (indicating d linearly independent reactions), then $\text{rank}(\mathbf{A}) = d$. If the rank of **A** is d, then the null space of $\mathbf{A}^{\text{T}}$ must have rank $(n-d)$, or $\text{rank}(\mathbf{N}) = (n-d)$ (Strang, 2003). Thus, **N** is a matrix having n rows and $(n-d)$ columns.

$$\mathbf{N} = [\mathbf{n}_1, \mathbf{n}_2, \ldots, \mathbf{n}_{(n-d)}]^{\text{T}}$$

Vectors $\mathbf{n}_1, \mathbf{n}_2, \ldots, \mathbf{n}_{(n-d)}$ form a basis for the null space of $\mathbf{A}^{\text{T}}$—linear combination of the set of vectors $\{\mathbf{n}_1, \mathbf{n}_2, \ldots, \mathbf{n}_{(n-d)}\}$ result in vectors that satisfy Equation 6.7, which must be perpendicular to **S**.

EXAMPLE 3: Null space practice

Calculate the null space of matrix **A**—null(**A**)—for the following matrices:

(a)

$$\mathbf{A} = \begin{bmatrix} 1 \\ 2 \\ 3 \end{bmatrix}$$

(b)

$$\mathbf{A} = \begin{bmatrix} 4 & 5 & 6 \\ 3 & 5 & 1 \\ 1 & 1 & 1 \end{bmatrix}$$

(c)

$$\mathbf{A} = \begin{bmatrix} 1 & 2 & 3 \end{bmatrix}$$

(d)

$$\mathbf{A} = \begin{bmatrix} 1 & 1 & 1 \\ 3 & 4 & 1 \end{bmatrix}$$

Computing the null space of a matrix is common function in linear algebra. We provide a brief description here for convenience; however, many standard texts (Lay, 2012; Strang, 2003) describe this topic in detail.

Answer a:

A vector **x** is an element of the null space of **A** if it satisfies the relation

$$\mathbf{Ax} = \mathbf{0}$$

A in this instance is a 3×1 matrix. Written as a system of equations, the above relation indicates that there are three equations and one unknown variable x_1:

$$\begin{bmatrix} 1 \\ 2 \\ 3 \end{bmatrix} x_1 = \begin{bmatrix} 0 \\ 0 \\ 0 \end{bmatrix}$$

Expressed explicitly in terms of a system of equations:

$$1x_1 = 0$$
$$2x_1 = 0$$
$$3x_1 = 0$$

Since there are more equations than unknowns, $x_1 = 0$ is the only value that produces the zero vector on the right hand side ($\mathbf{A}x_1 = 0$), and thus the only element in the null space of **A** is $x_1 = 0$. This is the *trivial solution*: zero will always be part of the null space of any matrix, for any matrix **A** multiplied by zero will always result in zero on the right hand side. If we let **N** denote the set of values belonging to the null space of **A**, $\mathbf{N} = \text{null}(\mathbf{A})$, then for this system $\mathbf{N} = \{0\}$ only.

We can validate this result as well using a numerical linear algebra package. Using the `null()` function in MATLAB for the **A** provided produces the result "`Empty matrix: 1-by-0`," indicating that there are no vectors that form part of the null space (excluding the trivial solution).

Answer b:

A is now a 3×3 matrix given by:

$$\mathbf{A} = \begin{bmatrix} 4 & 5 & 6 \\ 3 & 5 & 1 \\ 1 & 1 & 1 \end{bmatrix}$$

If a vector **x** is an element of null(**A**), then the relation $\mathbf{Ax} = \mathbf{0}$ is satisfied.

$$\begin{bmatrix} 4 & 5 & 6 \\ 3 & 5 & 1 \\ 1 & 1 & 1 \end{bmatrix} \mathbf{x} = \begin{bmatrix} 0 \\ 0 \\ 0 \end{bmatrix}$$

x in this case must be a 3×1 vector, $\mathbf{x} = [x_1, x_2, x_3]^T$, or

$$\begin{bmatrix} 4 & 5 & 6 \\ 3 & 5 & 1 \\ 1 & 1 & 1 \end{bmatrix} \begin{bmatrix} x_1 \\ x_2 \\ x_3 \end{bmatrix} = \begin{bmatrix} 0 \\ 0 \\ 0 \end{bmatrix}$$

system of three equations in three unknowns (x_1, x_2, and x_3), which can be solved by standard methods. Let us operations on the above system to make the solution easier to interpret. Hence, performing elementary row ducing the system to only elements on the main diagonal (also known as reduced row echelon form) gives

$$\begin{bmatrix} 1 & 0 & 0 \\ 0 & 1 & 0 \\ 0 & 0 & 1 \end{bmatrix} \begin{bmatrix} x_1 \\ x_2 \\ x_3 \end{bmatrix} = \begin{bmatrix} 0 \\ 0 \\ 0 \end{bmatrix}$$

We emphasize again that the above system is an equivalent matrix that is as a result after elementary row operations have been applied to **A** (the right hand side remains unchanged because it is the zero vector, and it is unaffected by elementary row operations). This system may also be written out as a system of equations:

$$1x_1 + 0x_2 + 0x_3 = 0$$
$$0x_1 + 1x_2 + 0x_3 = 0$$
$$0x_1 + 0x_2 + 1x_3 = 0$$

which indicates that the only values for x_1, x_2, and x_3 that satisfy this relation is when **x** is the zero vector, or

$$\mathbf{x} = \mathbf{0} = \begin{bmatrix} 0 \\ 0 \\ 0 \end{bmatrix}$$

In this example, the null space of **A**, given by **N** = null(**A**), only contains the zero vector **N** = {**0**}. Using the `null()` function in MATLAB for **A** produces the result "`Empty matrix: 3-by-0`," indicating that the zero vector is the only solution.

Answer c:

A is now a 1 × 3 matrix given by:

$$\mathbf{A} = \begin{bmatrix} 1 & 2 & 3 \end{bmatrix}$$

Carrying out a similar procedure as before, we start with the fact that **Ax** = **0** and write out the system of equations belonging to this system

$$\begin{bmatrix} 1 & 2 & 3 \end{bmatrix} \begin{bmatrix} x_1 \\ x_2 \\ x_3 \end{bmatrix} = [0]$$

or

$$1x_1 + 2x_2 + 3x_3 = 0$$

which is a single equation in three unknowns. Hence, there are more unknowns than equations. If the above relation is to be satisfied, then we have free choice over the values of two of the three unknowns (the remaining unknown may then be found by solving for it in the same relation); there are two degrees of freedom.

By example, suppose that $x_2 = 1$ and $x_3 = 0$. Substituting these values into the equation gives

$$1x_1 + 2(1) + 3(0) = 0$$
$$x_1 + 2 = 0$$
$$x_1 = -2$$

Hence, $x_1 = -2$ if $x_2 = 1$ and $x_3 = 0$. We have therefore found a combination of values for x_1, x_2, and x_3 that produce zero when multiplied by **A**. In the language of linear algebra, we have found a *basis for the null space of* **A**. The vector

$$\mathbf{n}_1 = \begin{bmatrix} -2 \\ 1 \\ 0 \end{bmatrix}$$

satisfies $\mathbf{A}\mathbf{n}_1 = \mathbf{0}$ and hence $\mathbf{n}_1$ is one vector that forms part of a basis for the null space of **A**. But, this is not the only vector that satisfies the relation. Since there are two unknowns, we can also set $x_2 = 0$ and $x_3 = 1$ and repeat the process.

Thus

$$1x_1 + 2(0) + 3(1) = 0$$
$$x_1 + 3 = 0$$
$$x_1 = -3$$

Hence, $x_1 = -3$, if $x_2 = 0$ and $x_3 = 1$. Thus, the vector

$$\mathbf{n}_2 = \begin{bmatrix} -3 \\ 0 \\ 1 \end{bmatrix}$$

is also a vector that forms a basis for the null space since $\mathbf{A}\mathbf{n}_2 = \mathbf{0}$. Thus, both $\mathbf{n}_1$ and $\mathbf{n}_2$ produce the zero vector when pre-multiplied by **A**. Furthermore, any scalar combination of $\mathbf{n}_1$ and $\mathbf{n}_2$ are also vectors that will bring **A** to **0**. Vectors $\mathbf{n}_1$ and $\mathbf{n}_2$ are thus two independent vectors that form a basis of the null space of **A**. If **N** denotes the vector space belonging to null(**A**), then **N** will be composed of vectors that are linear combinations of $\mathbf{n}_1$ and $\mathbf{n}_2$:

$$\mathbf{N} = \left\{ \lambda_1 \begin{bmatrix} -2 \\ 1 \\ 0 \end{bmatrix}, \lambda_2 \begin{bmatrix} -3 \\ 0 \\ 1 \end{bmatrix} \right\}$$

To see this, substitute any value for λ_1 and λ_2 and recalculate $\mathbf{A}\lambda_1\mathbf{n}_1$ and $\mathbf{A}\lambda_2\mathbf{n}_2$ (or even the matrix product **AN**). You will find that the result will also be the zero vector. By example if $\lambda_1 = -0.5$ and $\lambda_2 = 3.4$, then $\mathbf{n}_1 = [1, -0.5, 0]^T$ and $\mathbf{n}_2 = [-10.2, 0, 3.4]^T$—prove to yourself that $\mathbf{A}\mathbf{n}_1 = 0$ and $\mathbf{A}\mathbf{n}_2 = 0$ as well.

Geometrically, the columns of **N** (the null space of **A**) span a two-dimensional subspace in $\mathbb{R}^3$ (it is a plane in three-dimensional space). Any points residing in the null space—any point lying on this plane in $\mathbb{R}^3$—produce the zero vector when pre-multiplied by **A**.

In this case, using the `null()` function in MATLAB for **A** produces the result

$$\text{null}(\mathbf{A}) = \begin{bmatrix} -0.5345 & -0.8018 \\ 0.7745 & -0.3382 \\ -0.3384 & 0.4927 \end{bmatrix}$$

The columns of this matrix (the columns space) span the same set of points as $\mathbf{n}_1$ and $\mathbf{n}_2$ (the columns of matrix **N** above).

Answer d:

A is now a 2×3 matrix given by:

$$\mathbf{A} = \begin{bmatrix} 1 & 1 & 1 \\ 3 & 4 & 1 \end{bmatrix}$$

This indicates that there are two equations and three unknowns

$$\begin{bmatrix} 1 & 1 & 1 \\ 3 & 4 & 1 \end{bmatrix} \begin{bmatrix} x_1 \\ x_2 \\ x_3 \end{bmatrix} = \begin{bmatrix} 0 \\ 0 \\ 0 \end{bmatrix}$$

Or

$$1x_1 + 1x_2 + 1x_3 = 0$$
$$3x_1 + 4x_2 + 1x_3 = 0 \tag{6.7a}$$

Inherently, the above system of equations suggests that there is only one degree of freedom (we can only specify one unknown before values for the other two unknowns can be calculated). Performing elementary row operations again we obtain

$$\begin{aligned} 1x_1 + 0x_2 + 3x_3 &= 0 \\ 0x_1 + 1x_2 - 2x_3 &= 0 \end{aligned} \tag{6.7b}$$

Note that Equation 6.7b is equivalent to Equation 6.7a after elementary row operations have been performed on it. Hence, now we can see the effect if we let $x_3 = 1$

$$1x_1 + 3(1) = 0$$

$$1x_2 - 2(1) = 0$$

or

$$x_1 = -3$$

$$x_2 = 2$$

Thus, $x_1 = -3$ and $x_2 = 2$ when $x_3 = 1$. Thus the vector

$$\mathbf{n}_1 = \begin{bmatrix} -3 \\ 2 \\ 1 \end{bmatrix}$$

forms a basis for the null space of **A**. Scalar multiples of $\mathbf{n}_1$ are also vectors in the null space of **A**. Thus matrix **N** is the set of scalar multiples of $\mathbf{n}_1$

$$\mathbf{N} = \{\lambda_1 \mathbf{n}_1\}$$

Geometrically, the columns of **N** (the null space of **A**) span a one-dimensional subspace in $\mathbb{R}^3$ (it is a line in three-dimensional space). Any points residing in the null space of **A**—any point lying on this line in $\mathbb{R}^3$—produce the zero vector when pre-multiplied by **A**.

In this case, using the `null()` function in MATLAB for **A** produces the result

$$\text{null}(\mathbf{A}) = \begin{bmatrix} -0.8018 \\ 0.5345 \\ 0.2673 \end{bmatrix}$$

The values of this matrix are indeed scalar multiples of matrix $\mathbf{n_1}$ (specifically, $\lambda_1 = 0.2673$).

*6.2.2.2 **Nullspace of A^T*** Now that we have a better idea of the nullspace, we can apply the concept to topics in AR theory. This is demonstrated in the following example.

EXAMPLE 4: Computing the null space for reaction stoichiometry

Consider the two sets of reactions:

(a)

$$2A + 3B \rightarrow 4C$$

(b)

$$A + B \rightleftharpoons C$$

$$A + C \rightarrow D$$

For each system:

- Determine the stoichiometric coefficient matrix **A** for the reaction stoichiometry.
- Compute the rank of **A**.
- Determine the matrix **N**—the matrix with columns that form a basis for the null space of $\mathbf{A}^T$.

Answer a:

The stoichiometric coefficient matrix is formed by storing the stoichiometric coefficients for each reaction participating in the system as a column of **A**. Thus for a single reaction involving three components, **A** has size 3 × 1:

$$\mathbf{A} = \begin{bmatrix} -2 \\ -3 \\ 4 \end{bmatrix}$$

There are three components in this system ($n = 3$). This describes a line in $\mathbb{R}^3$ (c_A–c_B–c_C space) for which all possible concentrations must lie in order to be stoichiometrically feasible with the feed. The stoichiometric subspace in this instance is thus a line in $\mathbb{R}^3$ passing through the feed point. Note that computation of the null space does not require us to specify a feed point, and so this has been omitted here.

Since there is only one column in **A** (corresponding to a single reaction), rank(**A**) = 1. To compute the s… concentrations orthogonal to the stoichiometric … we compute the null space of $\mathbf{A}^T$. Hence, since …s one, we expect the rank of the null space … We may compute the null space using … as elementary row operations.[3] It is also often convenient to compute null($\mathbf{A}^T$) numerically with a numerical linear algebra package such as MATLAB if available. We find that the vectors

$$\mathbf{n}_1 = \begin{bmatrix} -1.5 & 1 & 0 \end{bmatrix}^T$$
$$\mathbf{n}_2 = \begin{bmatrix} 2 & 0 & 1 \end{bmatrix}^T$$

(or linear combinations of these vectors) form a basis of the null space of **A**. Thus, matrix **N** may be formed from $\mathbf{n}_1$ and $\mathbf{n}_2$

$$\mathbf{N} = \begin{bmatrix} \mathbf{n}_1 & \mathbf{n}_2 \end{bmatrix} = \begin{bmatrix} -1.5 & 2 \\ 1 & 0 \\ 0 & 1 \end{bmatrix}$$

Matrix **N** describes a two-dimensional subspace in $\mathbb{R}^3$ that is perpendicular to the stoichiometric subspace. Note that $\mathbf{A}^T\mathbf{n}_1 = \mathbf{0}$, $\mathbf{A}^T\mathbf{n}_2 = \mathbf{0}$ demonstrating that $\mathbf{n}_1$ and $\mathbf{n}_2$ are orthogonal to the columns of **A** (the stoichiometric subspace). Linear combination of $\mathbf{n}_1$ and $\mathbf{n}_2$ (such as $-\mathbf{n}_1 + \pi\mathbf{n}_2$) would therefore also be orthogonal to this space.

Answer b:

Let us look at another example. This time, the system of reactions is given by:

$$A + B \rightleftharpoons C$$

$$A + C \rightarrow D$$

There are precisely three reactions in this system, however only two are linearly independent. Two reactions occur in the first line; one corresponding to the forward reaction, the other to the reverse. This is seen by forming the stoichiometric coefficient matrix **A** and then computing its rank:

$$\mathbf{A} = \begin{bmatrix} -1 & 1 & -1 \\ -1 & 1 & 0 \\ 1 & -1 & -1 \\ 0 & 0 & 1 \end{bmatrix}$$

Observe that columns 1 and 2 of **A** are linearly dependent (column 2 is the negative of column 1). Calculation of the rank of **A** gives 2, indicating that only two independent reactions are preset in this system. The columns in **A** thus describe a two-dimensional subspace in $\mathbb{R}^4$ concentration space in this instance.

… form, for instance.

Finally, one may compute the set of vectors orthogonal to the stoichiometric subspace. Again computation gives a set of two vectors that form a basis for the null space. Two representative vectors for this space are given below:

$$\mathbf{n}_1 = \begin{bmatrix}-1 & 2 & 1 & 0\end{bmatrix}^{\mathrm{T}}$$

$$\mathbf{n}_2 = \begin{bmatrix}1 & -1 & 0 & 1\end{bmatrix}^{\mathrm{T}}$$

It follows that

$$\mathbf{N} = \begin{bmatrix}-1 & 1 \\ 2 & -1 \\ 1 & 0 \\ 0 & 1\end{bmatrix}$$

Linear combinations of $\mathbf{n}_1$ and $\mathbf{n}_2$ also describe a two-dimensional subspace in $\mathbb{R}^4$ and in this instance the dimension of the stoichiometric subspace is the same as the dimension of the null space.

6.2.3 Number of Independent Reactor Structures

A useful consequence of the dimension of the AR may be used to relate the maximum number of parallel structures needed to generate the AR, which is achieved by use of Carathéodory's theorem (Carathéodory, 1911; Eckhoff, 1993). Feinberg (2000a) shows that for an AR constructed in $\mathbb{R}^d$, the following limits, in terms of parallel reactor structures, may be enforced:

- The maximum number of parallel reactor structures needed to generate an effluent concentration on the AR boundary is $d-1$.
- The maximum number of parallel reactor structures needed to generate an effluent concentration in the interior of the AR is d.

Here d is a positive integer representing the dimension of **S**, and hence the dimension of the AR. Note also that these results refer to a *maximum* (upper bound) on the number of reactor structures required. A particular system kinetics and feed point may require less than the maximum in order to generate the AR boundary in practice. A convex PFR trajectory in $\mathbb{R}^3$ is an example of such a system. Specific kinetics are described in Chitra and Govind (1985).

It follows that if the AR resides in a d-dimensional subspace in $\mathbb{R}^n$, then one need only consider maximum of d parallel reactor structures in order to generate the AR boundary. Note further these results make no assumption about the complexity of each structure—an AR in $\mathbb{R}^2$ may have as its optimal boundary structure a single PFR for a certain system, whereas an optimal boundary structure of CSTR-PFR-CSTR-PFR for a different set of reactions, which also resides in $\mathbb{R}^2$.

From Section 6.2.1.3, the dimension of the AR is equal to the dimension of **S**, it follows that the maximum number of independent reactor structures is directly related to the number of independent reactions taking part in the system. Moreover, this analysis may be determined in the absence of reaction kinetics and a feed point—the results are a consequence of the system reaction stoichiometry and Carathéodory's theorem only.

CONCEPT: Maximum number of parallel structures

It is possible to determine, beforehand, the maximum number of independent reactor structures needed to generate the AR by computing the rank of the stoichiometric coefficient matrix **A**. The maximum number of parallel structures is related to the number of independent reactions occurring in the system.

EXAMPLE 5: Number of parallel reactor structures

Given the following system of reactions:

$$A \rightleftharpoons B \rightleftharpoons C$$

$$A + C \rightarrow D + E$$

How many parallel reactor structures would be required (in the worst case) to form any achievable concentration, if the feed vector is given by $\mathbf{C}_f = [1, 1, 0, 0, 0]^{\mathrm{T}}$? How many parallel reactor structures would be required if the feed vector was given by $\mathbf{C}_f = [0.5, 1, 0.5, 0.1, 0.3]^{\mathrm{T}}$ instead?

Answer:

The system under consideration involves two reversible reactions and an irreversible reaction involving component A and intermediate component C. Five reactions are thus present in total. To determine the maximum number of parallel reactor structures, we first must determine the dimension of the stoichiometric subspace. The stoichiometric coefficient matrix is then given by:

$$\mathbf{A} = \begin{bmatrix}-1 & 1 & 0 & 0 & -1 \\ 1 & -1 & -1 & 1 & 0 \\ 0 & 0 & 1 & -1 & -1 \\ 0 & 0 & 0 & 0 & 1 \\ 0 & 0 & 0 & 0 & 1\end{bmatrix}$$

Notice that columns 2 and 4 are the reverse reactions corresponding to columns 1 and 3, respectively. The rank of **A** may then be computed. We find that rank(**A**) = 3, demonstrating that there are only three *independent* reactions present in the system.

It follows that the AR is a three-dimensional subspace residing in $\mathbb{R}^5$. The maximum number of parallel reactor structures required to achieve a point on the AR boundary is 2, whereas a maximum of three structures is required to produce a point in the interior of the region.

This analysis is not influenced by specification of the feed point $\mathbf{C}_f$. Computation of the dimension of the stoichiometric subspace involves the reaction stoichiometry alone, and thus the *maximum* number of parallel structures is the same irrespective of the feed point employed—the *actual* number of parallel structures required to achieve the AR will, however, be dependent on the reaction kinetics and feed point.

6.3 THE THREE FUNDAMENTAL REACTOR TYPES USED IN AR THEORY

6.3.1 Introduction

Many fundamental results of the AR, used in everyday practice, can be generalized and extended to higher dimensional systems. Although these ideas are important, they are founded on mathematically sophisticated concepts. In this section, we wish to describe some of these concepts.

A large portion of the paper by Feinberg and Hildebrandt (1997) deals with characterizing the AR boundary in terms of its fundamental features (such as reaction and mixing surfaces). In the following sections, we shall follow three key ideas:

1. The AR is composed of reaction and mixing surfaces only. Reaction surfaces are always convex. This result is based on a simple observation that is highlighted in Section 6.3.2.
2. Points that form convex sections of the AR boundary arise from effluent concentrations specifically from PFR trajectories. This is the topic of theorem 1, which is described Section 6.3.3.1.
3. Points on the AR boundary that initiate these convex PFR trajectories (from point 2 above) arise from specialized CSTRs and DSRs only—no other reactor types are needed to generate these special states. The collection of these points on the boundary is termed a connector. This is the topic of theorem 2, which is described in Section 6.3.3.2.

From these results, we know that the AR is generated from special combinations of PFRs, CSTRs, and DSRs only. Theorems 1 and 2 rely on an important principle (the complement principle), which is briefly discussed in Appendix B.

CONCEPT: Three fundamental reactor types used in AR theory

When reaction and mixing are the only two processes present, the AR may be constructed using reactor combinations involving PFRs, CSTRs, DSRs and mixing only. There is no requirement to devise new (perhaps novel) reactor types that might serve to extend the AR boundary further. Instead, we can focus our attention on arranging combinations of these three fundamental reactor types in an optimal arrangement.

6.3.2 Extreme Points and Reaction

Given a set of points, one is often interested in identifying which points have been generated from reaction and which have been generated from mixing operations. Let us explain how these points can be classified.

Suppose that a convex set of k achievable concentrations in $\mathbb{R}^n$, $\mathbf{X} = \{\mathbf{C}_1, \mathbf{C}_2, \ldots, \mathbf{C}_k\}$, is given. This set is assumed to have been obtained from the operations of a hypothetical reactor network, where many reactors may be linked by recycle, bypass and mixing operations. Suppose we are given a particular concentration $\mathbf{C}^* \in \mathbf{X}$ and are tasked to determine whether:

1. $\mathbf{C}^*$ is achieved from the reactive portion of the network;
2. Or $\mathbf{C}^*$ is obtained from a combination of mixing operations that does *not* employ reaction.

A small amount of reasoning will help to determine the originating point within the reactor network.

Suppose that $\mathbf{C}^*$ is, in fact, obtained from a *mixture* of other points contained in **X**. This situation is shown in Figure 6.4 for a set containing six distinct concentrations $\{\mathbf{C}_1, \mathbf{C}_2, \ldots, \mathbf{C}_6\}$ obtained from the hypothetical network. If $\mathbf{C}^*$ is a mixture of the set, then it must be expressible by:

$$\mathbf{C}^* = \lambda_1\mathbf{C}_1 + \lambda_2\mathbf{C}_2 + \cdots + \lambda_k\mathbf{C}_k \tag{6.8}$$

Concentrations $\mathbf{C}_1, \mathbf{C}_2, \ldots, \mathbf{C}_k$ are members of the set that make up the *convex hull* of **X**, conv(**X**). Recall that points belonging to the convex hull boundary are the unique points of the set. It is clear that the extreme points of the region defined by conv(**X**) are sufficient to generate the entire set of concentrations that lie in the set **X**. This is given

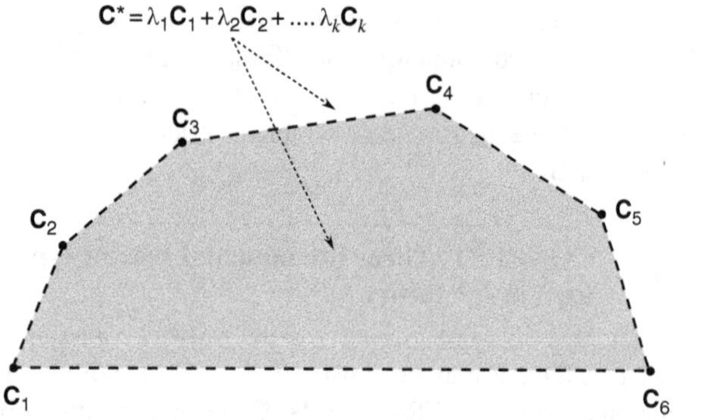

Figure 6.4 A hypothetical set of five points and associated convex hull. The extreme points either belong to the feed set, or else they must be derived from a part of the plant where reaction has taken place.

by the shaded region in Figure 6.4. If $\mathbf{C}^*$ is a mixture of concentrations belonging in $\mathbf{X}$, then it may be written as a linear combination of extreme points in conv($\mathbf{X}$). $\mathbf{C}^*$ must itself lie in the region defined by conv($\mathbf{X}$) as a result.

Since $\mathbf{C}^*$ is expressed by a linear combination of extreme points (it satisfies Eq. 6.8) then it cannot itself be an extreme point—for this would imply that at least one of the λ's in Equation 6.8 would be zero. $\mathbf{C}^*$ must then either lie on a straight line segment joining two extreme points, otherwise $\mathbf{C}^*$ must lie within the region defined by conv($\mathbf{X}$). It follows that *extreme points cannot be expressed as a linear combination of other points in* $\mathbf{X}$. Instead, points in conv($\mathbf{X}$) allow for the generation all other combinations of concentrations in $\mathbf{X}$. Since extreme points cannot be generated from mixing, then they must be generated by some other means.

CONCEPT: Extreme points in a convex set

An extreme point cannot lie between a line segment joining two points on the boundary of the convex hull of a set of points $\mathbf{X}$, conv($\mathbf{X}$). Extreme points (points belonging to a convex hull) are unique and cannot be formed from other points in the convex hull.

Since only reaction and mixing are allowed, extreme points of conv($\mathbf{X}$) must correspond to points that fall into either one of two sets:

1. Points that belong to the set of feed concentrations, denoted here by set $\mathbf{F}$.
2. Points that result from reaction.

It is always possible to compute the convex hull of the set of achievable concentrations derived from a reactor network. Moreover, since the set is derived from reaction and mixing operations alone, the boundary of the AR is therefore composed of surfaces generated from reaction and mixing operations only. The AR boundary, in any dimension, is composed of straight and curved surfaces, which are obtained from reaction and mixing operations. Concentrations obtained via reaction must reside on curved sections of the boundary.

CONCEPT: Exposed points, protrusions and lineations

Exposed points:

An extreme point $\mathbf{x}^* \in \mathbf{X}$ is one that *cannot* be written in the following way:

$$\mathbf{x}^* = \lambda \mathbf{x}_1 + (1 - \lambda)\mathbf{x}_2$$

where $\mathbf{x}_1$ and $\mathbf{x}_2$ are also contained in $\mathbf{X}$ and $0 \leq \lambda \leq 1$. In other words, $\mathbf{x}^*$ is an extreme point if it does not lie on a line between two other points in $\mathbf{X}$. *Exposed points* are a special class of extreme points. We define an exposed point $\mathbf{x}^*$ to be one where a hyperplane touches the set $\mathbf{X}$ at $\mathbf{x}^*$ only once. Consider Figure 6.5, which is an adaption of a figure taken from Feinberg and Hildebrandt (1997).

In Figure 6.5(a), a convex region is shown. A plane is introduced so that it meets the surface joined by extreme points ABCD. In this instance, points contained in the rectangular region bounded by ABCD are not considered as exposed points, for all points in this section are contained within the same hyperplane. (Points A, B, C, and D are all extreme points but they are not exposed points because all four points lie in the hyperplane and thus the hyperplane is not supported at a single point.)

In Figure 6.5(b), a slightly altered convex region is shown. A plane is again introduced; however, in this instance, the plane only touches the arch at $\mathbf{x}^*$. The point $\mathbf{x}^*$ is therefore considered an exposed, for no other point in the convex region touches the plane. Thus, all points on the surface ABCD in Figure 6.5(b) are considered exposed points.

Protrusions:

A *protrusion of dimension m* is defined as a set of exposed points that constitute an m-dimensional "surface." Refer again to Figure 6.5. The surface connecting points A, B, C, and D in Figure 6.5(a) does not constitute a two-dimensional protrusion. This is because points in the plane are not exposed. On the other hand, the arch joining A, B, C, and D in Figure 6.5(b) contains only exposed points, and thus it is considered a two-dimensional protrusion. (If we were to take a little patch of the surface, it would resemble a flat two-dimensional surface of exposed points.)

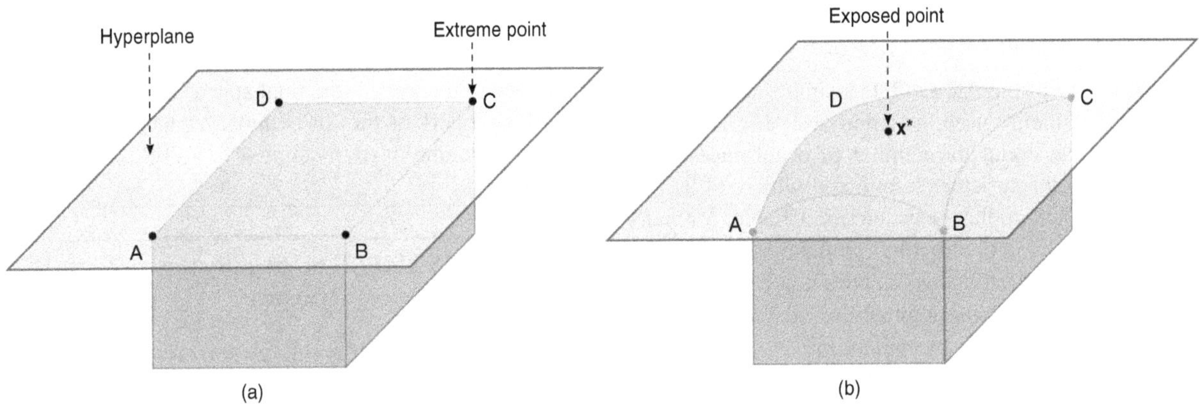

Figure 6.5 Protrusions and nonprotrusions, (a) extreme points and (b) exposed points.

Lineations:

Straight line sections arise in AR theory due to mixing. Feinberg and Hildebrandt (1997) provide a precise definition of these mixing lines specifically for use on the boundary of a closed convex set, which are termed *lineations*. An m-dimensional lineation is a set of points, **L**, on the boundary of convex set **X** with the following properties:

1. All points in **L** are part of an interior line segment (**L** is composed of points from mixing lines).
2. For each point in **L**, the dimension of **X** is m (if we were to stand on points in **L**, the surface would appear to be an m-dimensional "flat" surface on the boundary of the set **X**).

A lineation in **X** is a set of concentrations that form an m-dimensional (linear) surface. Lineations are somewhat the opposite of protrusions—unlike protrusions, where all points contained in a protrusion are extreme, *no* points contained in a lineation are extreme. Regions FPO and FLQ in Figure 6.6 are examples of lineations.

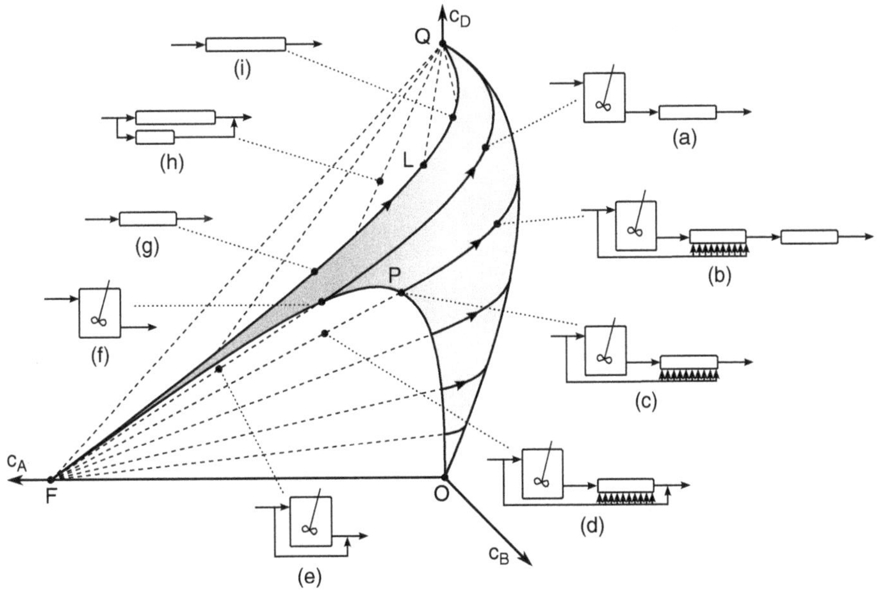

Figure 6.6 The AR boundary for the Van de Vusse system. Feinberg and Hildebrandt (1997). Reproduced with permission of Elsevier. (*See color plate section for the color representation of this figure.*)

6.3.3 Two Important Theorems

6.3.3.1 Theorem 1 Theorem 1 states:

The AR is composed entirely of mixing surfaces and manifolds of PFR trajectories

The boundary of the AR exhibits a simple structure, irrespective of the kinetics used. As a demonstration, refer back to Chapter 5 and count the number of occurrences where the optimal reactor structure terminated with a PFR. (You will find that all examples considered in Chapter 5 resulted in a final optimal reactor structure terminating with a PFR.) Theorem 1 helps prove that this behavior is not a coincidence, for the final approach to *all* optimal reactor structures on the boundary of the AR terminates with a PFR.

From Section 6.3.2 it is understood that extreme points on the AR boundary result from either feed points or reaction surfaces (specifically protrusions). Using theorem 1, this statement may be refined to say that exposed points from reaction originate specifically from PFR trajectories *only*. This result is summarized compactly by the following theorem, which is adapted from Feinberg and Hildebrandt (1997):

Theorem 6.1 (exposed points on the AR boundary are either PFR trajectories or feed points) *Suppose that we have a specified feed set* $\mathbf{F}$ *in* $\mathbb{R}^n$ *and a convex set of achievable points given by* $\mathbf{C}$*, also contained in* $\mathbb{R}^n$*. The rate function* $\mathbf{r}(\mathbf{C})$ *associated with this region is assumed to be continuously differentiable and also defined on* $\mathbb{R}^n$*. Furthermore, the set of concentrations in* $\mathbf{C}$ *is assumed to comply with the complement principle. If it is found that all rate vectors on the boundary of* $\mathbf{C}$ *do not point outward, then any protrusion in* $\mathbf{C}$ *that is separate (disjoint) from* $\mathbf{F}$ *is the union of PFR trajectory segments. The solution trajectories then satisfy the PFR equation* $d\mathbf{C}/d\tau = \mathbf{r}(\mathbf{C})$*.*

Figure 6.6 shows an illustrative example of the AR boundary for three-dimensional Van de Vusse kinetics. The boundary structures have been exaggerated slightly to help emphasize certain characteristics for the discussions below.[4] Elements of the AR boundary are composed of surfaces that are initiated from either mixing or reaction surfaces. Reaction surfaces themselves must produce extreme points (specifically protrusions) that result from PFR solution trajectories alone. Determination of the AR boundary structures is simplified greatly as a result: we know that the final approach to any exposed point on the true AR boundary must be achieved by a PFR trajectory segment—optimal reactor structures must always end with a PFR.

CONCEPT: PFRs as a terminating reactor structure

Due to theorem 1, the final approach to any optimal reactor network on the AR boundary, where reaction and mixing are employed, must involve a PFR.

EXAMPLE 6: Identify the correct AR boundary structure

A number of different reactor structures are given in Figure 6.7. Each structure is composed of different fundamental reactor types. Can you identify which reactor structure would form part of an optimal structure that resides on an *exposed point* of the AR boundary?

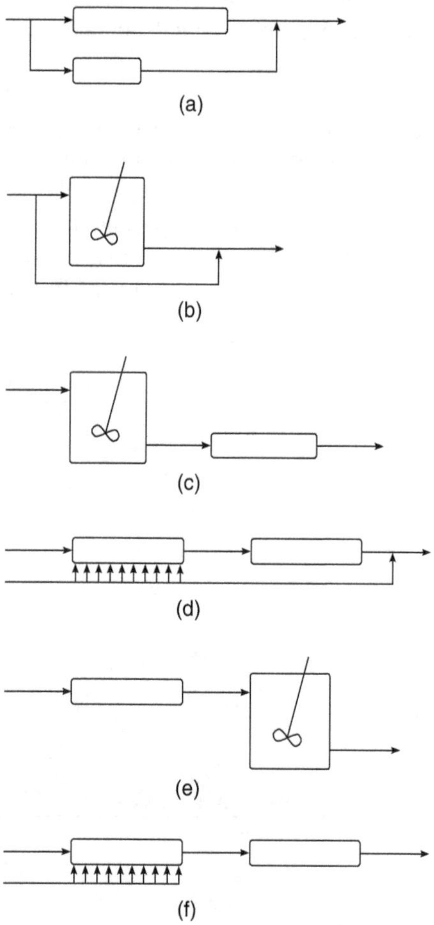

Figure 6.7 Different proposed optimal reactor structures. (a) PFRs in parallel, (b) CSTR with feed bypass, (c) CSTR-PFR, (d) DSR-PFR with feed bypass (e) PFR-CSTR, and (f) DSR-PFR.

[4] In Chapter 7, the actual AR shall be generated for this system using ideas developed from this chapter.

Answer:

Structures corresponding to (a), (b), and (d) in Figure 6.7 involve mixing with the terminal reactor type. This suggests that the final effluent concentration from these structures must lie on a mixing line and not on an exposed point.

Structures (c) and (e) are both similar, as both structures involve combinations of a CSTR and PFR in series. It is known that the final approach to the extreme points of the AR take place as a result of the union of PFR trajectories, and thus we should expect that final fundamental reactor type of any optimal reactor structure on the AR boundary is a PFR. We may conclude that structure (e) does not produce an effluent concentration that is an exposed point on the AR boundary (although the effluent concentration may still lie on the AR boundary, the point will not be exposed). The CSTR feeding the PFR in (c) must therefore produce a concentration that is a point on the AR boundary.

Similarly, structure (f) may also be a valid reactor arrangement that produces an exposed concentration on the AR boundary. The structure terminates with a PFR and there is no mixing with the terminating reactor. The DSR feeding the PFR in (f) must therefore also produce an effluent concentration that is part of the AR boundary.

Note that given the correct kinetics and feed point, it is possible for all three fundamental reactor types to produce an effluent concentration that resides on the AR boundary. However, it is only the PFR that is able to produce exposed points on the AR boundary.

If a reactor network operates at an exposed point on the AR boundary, then the final reactor type of the network is a PFR.[5] Let us again refer to Figure 6.6. Observe that the structure of the AR boundary is consistent with theorem 1. Straight line segments formed by FO and FQ (as well as the triangle section formed by FQO in the c_A–c_D plane) all arise as a result of mixing lines. By comparison, the curved shaded region given in Figure 6.6 is produced *only* by the union of PFR trajectories, which is called a manifold of PFR trajectories. This manifold is also a protrusion of the structure.

6.3.3.2 Theorem 2

a) Introduction Theorem 2 states

CSTRs and DSRs act as connectors to PFRs

The manner by which final PFR trajectories are approached on the AR boundary is highly specialized. We have already noticed how CSTRs may be used as connecting structures to PFRs for two-dimensional constructions in Chapters 4 and 5. Similar kinds of reactor arrangements fulfill the same duty in n-dimensional space as well. These structures are aptly termed *connectors* in AR theory, which are solution trajectories that satisfy the CSTR and DSR equations. Connectors are, in fact, either:

1. Solutions to the DSR equation with specialized α policies.
2. Solutions to the CSTR equation with a very precise requirement on the feed composition and residence time.[6]

Theorem 2 describes how the manifold of PFR trajectories, considered in theorem 1, may be accessed using these two reactors alone. In order for connectors to service the PFR manifold, they must *also* travel on the AR boundary. Furthermore, feed points to either these special CSTRs or DSRs must be taken from a point already situated on the AR boundary.

b) Smooth Connectors A connector may be viewed as an interface point between an exposed point of a PFR trajectory segment and a mixing point (a point that is disjoint from the feed). There is an important characteristic of connectors that we wish to discuss: connectors must meet the manifold of PFR trajectories *smoothly*. The precise definition of what is considered "smooth" is given in Feinberg and Hildebrandt (1997), but we shall give a qualitative description next.

Consider Figure 6.8(a) (a reproduction of a figure from Chapter 4), which is used to describe the space spanned by vectors $\mathbf{r}(\mathbf{C})$ and $(\mathbf{C}^0 - \mathbf{C})$ in a DSR. We can envision that any DSR effluent concentration $\mathbf{C}$ must lie on a point spanned by $\mathbf{r}(\mathbf{C})$ and $(\mathbf{C}^0 - \mathbf{C})$, and so, by the nature of the DSR, all three vectors must lie in the same plane.

Consider also the sketch of a piece of a hypothetical AR boundary, given by Figure 6.8(b). (We want to show that this type of shape cannot be part of the AR boundary.) PFR trajectories are shown to produce a PFR manifold, as is consistent with theorem 1, and this portion of the boundary is denoted by the curved section BCDE. Figure 6.8(b) also shows a surface composed of straight line sections, given by ABCD, representing a family of mixing lines that originate from a common mixing point A to the PFR manifold. The intersection of the two surfaces occurs along a ridge, given in Figure 6.8(b) by the curve BCD. It is clear from the sketch that the meeting point between a PFR trajectory on

[5] Other points that might exist on the AR boundary are critical DSR and CSTR connector points. These will be described by theorem 2.

[6] For now, only the results of the theorem will be discussed. Details of how to compute critical DSR trajectories and critical CSTR solutions are described in Section 6.4.

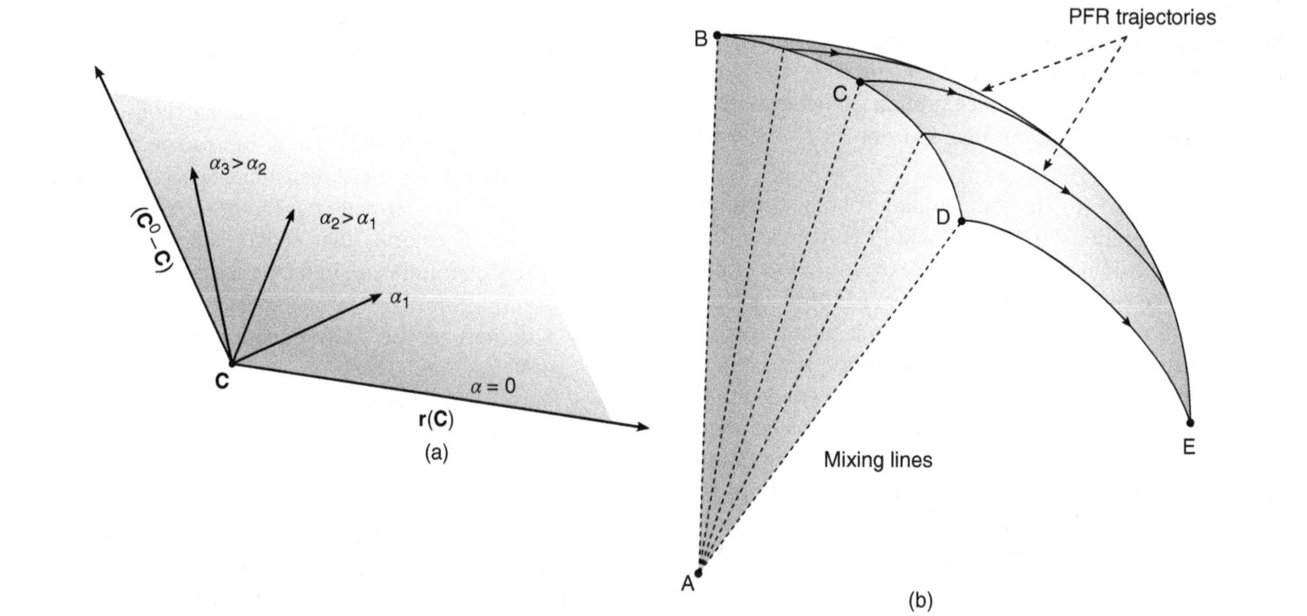

Figure 6.8 (a) The subspace spanned by $\mathbf{r}(\mathbf{C})$ and $(\mathbf{C}^0 - \mathbf{C})$ in a DSR. (b) An inconsistent AR boundary shape. Feinberg and Hildebrandt (1997). Reproduced with permission of Elsevier.

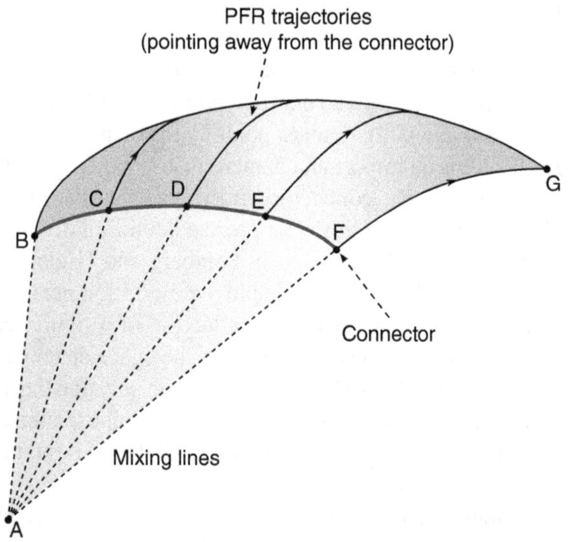

Figure 6.9 The role of connectors on the AR boundary. Mixing lines meet PFR trajectories at a connector transversely. The connector is an initiating point for PFR trajectory segments generated on the AR boundary. Feinberg (2000a). Reproduced with permission of Elsevier.

the manifold and a mixing line with common point A does *not* meet smoothly—there is a distinct change in gradient along the ridge.

But from Figure 6.8(a), we know that the rate vector $\mathbf{r}(\mathbf{C})$ must lie in the same plane as the mixing line (such as mixing line AC in Figure 6.8(b)), yet this cannot occur if there is a ridge as in Figure 6.8(b)—to produce the ridge shape in Figure 6.8(b), rate vectors on the ridge would need to point *out of the plane*, or else they *would* meet the mixing surface smoothly. Hence, the ridge in Figure 6.8(b) cannot exist on the boundary of the true AR.

This result has implications on how the interface point between a manifold of PFR trajectories on the AR boundary may meet mixing lines. As a result, instead of the shape given in Figure 6.8(b), we rather expect a shape such as that given in Figure 6.9. Here, mixing lines and the surface of PFR trajectories meet smoothly (rate vectors and mixing lines lie in the same plane at the connecting point), and rate vectors point away from the connector, because they are initiating points for PFRs.

ILLUSTRATION: Critical reactors for Van de Vusse kinetics

It may be useful to consider again the AR in Figure 6.6 to obtain a sense of the nature of connectors in practice. Specific focus is made on the flat (mixing) surfaces of the region. Three distinct segments are identifiable on the boundary; these are:

1. The triangular region that resides behind the point of view, given by FQO: This is a mixing surface that is formed by mixing points belonging to the PFR and CSTR initiated at the feed point. Both the CSTR and PFR are operated to the equilibrium point (point O). Points on FQO are formed by mixing equilibrium points with the feed point. The interface between this section of the region and the PFR manifold does

not meet smoothly and hence it does not constitute a connector.

2. The region traced out by FLQ: This is a mixing surface formed by two PFRs operated in parallel—one is a PFR from the feed operated to the equilibrium point, the other is a PFR also operated from the feed, but with a smaller residence time. This arrangement is displayed in Figure 6.6 as reactor structure (h). The resulting concentration achieved by mixing these two PFR product streams produces a point on the mixing surface FLQ (dashed lines). The mixing surface meets the PFR at a sharp ridge, given by the curve FLQ (solid line). The ridge itself is composed of concentration vectors that form a protrusion. This protrusion must also exist as a PFR trajectory. This interface also does not constitute a connector because the mixing surface and protrusion do not meet smoothly and rate vectors point towards the interface, not away from it.
3. The region traced out by FPO. The mixing surface meets the manifold of PFR trajectories smoothly and rate vectors point away from the meeting point. The interface defined by FPO is hence considered a connector in this instance.

If a protrusion on the boundary of the AR that is disjoint from the feed meets a mixing line smoothly along a connector, then the connector must contain the following two properties:

1. All rate vectors evaluated at compositions in the connector cannot point out of the protrusion. This ensures that connectors serve as starting points wherefrom PFR trajectories are generated. Rate vectors must be tangent to the boundary.
2. Any end point of a line segment in the lineation of **L** must meet the connector *transversely*. Again this is seen in Figure 6.6 by the mixing lines on the surface defined by FPO. (Mixing lines ending on the curve FPO meet the connector in a diagonal manner.)

c) CSTRs and DSRs as Connectors on the AR Boundary
A fuller description of theorem 2 can now be given, which is viewed as an extension to theorem 1 (Feinberg and Hildebrandt, 1997).

Theorem 6.2 (a connector must satisfy the CSTR or DSR equations) *Suppose that we have a specified feed set* $\mathbf{F}$ *in* $\mathbb{R}^n$ *and a convex set of achievable points given by* $\mathbf{C}$*, also contained in* $\mathbb{R}^n$*. The rate function* $\mathbf{r(C)}$ *associated with this region is assumed to be continuously differentiable and also defined on* $\mathbb{R}^n$*. Furthermore, the set of concentrations in* $\mathbf{C}$ *is assumed to comply of the complement principle. All rate vectors on the boundary of* $\mathbf{C}$ *do not point outward from the boundary. If, on the boundary of* $\mathbf{C}$*, a protrusion disjoint from* $\mathbf{F}$ *meets a lineation* $\mathbf{L}$ *at a connector, and* $\mathbf{C}^*$ *is a point in the connector, then either:*

1. *There is a positive residence time* τ *and a composition* $\mathbf{C}^0$ *in the closure of L such that* $\mathbf{C}^*$ *satisfies the CSTR equation:* $\mathbf{C}^* - \mathbf{C}^0 = \tau\mathbf{r}(\mathbf{C}^*)$
2. *Or* $\mathbf{C}^*$ *belongs to a set of compositions generated by a nontrivial DSR solution trajectory. The solution trajectory must lie entirely in the connector. It follows that the DSR solution trajectory is satisfied by the following equation:* $d\mathbf{C}/d\tau = \alpha(\tau)[\mathbf{C}^0(\tau) - \mathbf{C}(\tau)] + \mathbf{r}(\mathbf{C})$*. Again, the side-stream composition of the DSR given by* $\mathbf{C}^0$ *must be taken from compositions in the closure of L on the boundary of the AR.*

It may be useful to refer to Figure 6.10 for a visual description of these properties. A connector is shown on the boundary of a hypothetical AR segment given by the curve OPQ. It is understood from theorem 2 that the connector is produced by either a specialized CSTR or DSR. In this instance, the connector seems to be formed from a DSR trajectory. Although curve OPQ appears to be generated from a single DSR trajectory, it is common in AR theory for such curves to be formed from a combination of many DSR trajectories. Therefore, curve OP corresponds to one distinct DSR, whereas curve QP corresponds to another separate DSR trajectory, with differing initial condition that is also on the AR boundary. Point P is of special interest, for both DSR trajectories terminate at this point. From Chapter 4 it is known that the equilibrium concentration in a DSR corresponds to a CSTR solution. Hence in addition to being an equilibrium point for curves OP and QP, point P must also be a CSTR effluent concentration. This is shown in Figure 6.10 by the locus of CSTR points given by curve XPW. The majority of points in the CSTR locus lie *within* the AR, but there is single point (point P) that touches the connector OPQ on the AR boundary. Point P is a CSTR solution that forms part of the connector in the AR boundary as well.

Although DSRs do not serve any important function in two-dimensional constructions, theorem 2 shows that they are a critical feature of the AR boundary in higher dimensions—critical DSRs serve as connectors to final PFR extreme points. Theorem 2 also describes the role that mixing and feed points play in relation to connectors on the boundary. Notice that if a connector satisfying theorem 2 is available, then we can conclude that *mixing and feed points to connectors (critical CSTRs or DSRs) must take compositions from mixing lines (lineations) that also reside on the AR boundary.*

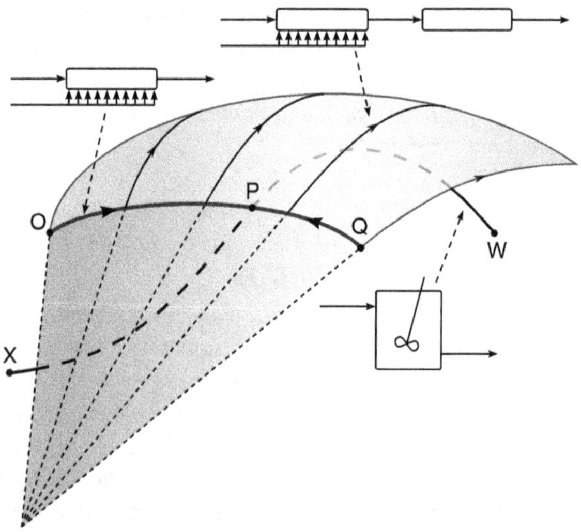

Figure 6.10 Connectors on the AR boundary manifest themselves either as critical CSTR effluent compositions, or critical DSR solution trajectories. (*See color plate section for the color representation of this figure.*)

> **CONCEPT: Critical DSRs and mixing points on the AR boundary**
>
> Mixing points and feed points used by critical CSTRs/DSRs must originate specifically from mixing lines on the AR boundary itself. This implies that there is no need to perform an exhaustive search for potential optimal mixing points for use in critical CSTRs and DSRs. Only a search for points on the AR boundary is required.
>
> Mixing points also *cannot* originate from PFR trajectories (as these points would form part of the AR boundary that is a protrusion and not a lineation). One need only search for points in the closure of the set of concentrations belonging to the mixing lines/surfaces.
>
> Since critical DSRs lie entirely on the AR boundary, these must also be *initiated* on the AR boundary. This places a strict set of constraints on the type of allowable DSRs trajectories that generate the AR that may be used to reduce the search space of potential starting points from which critical DSRs can be operated from.

6.4 CRITICAL DSRs AND CSTRs

6.4.1 Overview

In the previous sections, we described how the boundary of the AR consists of, at most, two distinct surfaces:

1. Ruled surfaces that are obtained from mixing lines (lineations).
2. Protrusions (a surface of exposed points) that are derived specifically from the union of PFR trajectories.

We also discussed how specialized CSTRs and DSRs—termed connectors, or critical reactors—are used in connecting to PFR points on the AR boundary. The underlying structure of the AR boundary is, in fact, fairly simple as a result. Yet we still have not explained how to determine whether a given CSTR or DSR point is critical, or how these specialized reactors are calculated. It will be the goal of this section to understand these topics.

There is a central theme that runs through all discussions related to critical reactors in AR theory—it is the idea of *controllability*. The idea of controllability, and its relation to AR theory, is so important that we must first discuss it in greater detail before critical reactors themselves can be understood. As a result, this section is structured into three main sub-sections:

1. We begin in Section 6.4.2.1 by discussing the basic idea of controllability, in relation to optimal control theory, and linear systems. These discussions are facilitated with reference to a simple example—the controlled motion of a rocket to the moon.
2. In Section 6.4.2.2, we borrow ideas from the controllability of linear systems and discuss how similar concepts apply to nonlinear systems as well. This leads us to determine a condition for critical DSRs in Section 6.4.3.
3. In Section 6.4.4, this theory is extended to include the determination of critical CSTRs.

Again, our approach is to provide the *ideas* behind these concepts, and avoid the discussion of detailed proofs. For readers that wish to gain a deeper understanding of the details the theory, reference is again made to Feinberg (1999, 2000a, 2000b).

6.4.2 Controllability

6.4.2.1 Definition

> **SIDE NOTE: Controlling a rocket to the moon**
>
> The underlying concepts used in this section are based on ideas described by the following (simplified) example. It will be helpful to keep in mind a mental picture of this example when the mathematics is introduced later.
>
> Envision the problem of remotely guiding a rocket to the moon. In reality, the rocket could be controlled by a number of input parameters, such as pitch, yaw, roll and thrust, which all affect the state of the rocket (such as

velocity, altitude, drag and remaining mass of fuel), as depicted in Figure 6.11.

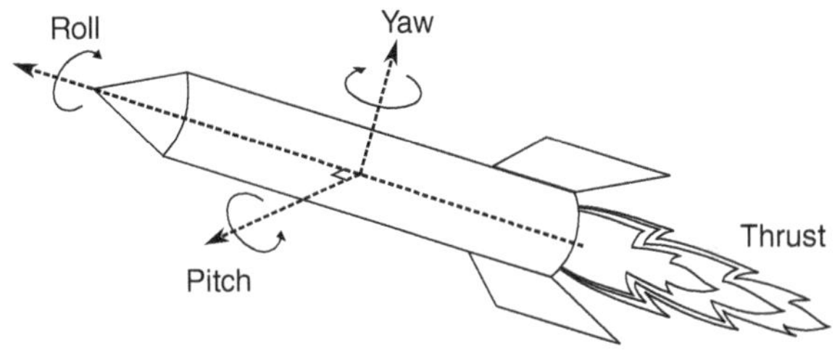

Figure 6.11 Controlling a rocket's movement through space via four input parameters: pitch, yaw, roll, and thrust.

The specific input policies employed in the control of the rocket (how the pitch, roll, yaw and thrust are varied over time), together with initial conditions (such as initial velocity, orientation and current mass), ultimately defines the unique path—or solution trajectory—traversed by the rocket through space. These states may all be modeled by the set of differential equations that describe its motion.

Understanding the controllability of the rocket will allow is to understand how a DSR might be coordinated to produce a trajectory on the boundary of the AR. There are two common viewpoints of controllability, which may be used to interpret a system of differential equations.

a) Controllability from a Perturbation Perspective Suppose that we wish to reach the desired target in as shown in Figure 6.12. Along the rocket's journey, a strong wind is encountered. In the absence of any adjustment to the original control policy, the introduction of the wind (a perturbation) may result in a deviation away from the original flight path of the rocket. The control policy must be adjusted to maintain a path to its desired target.

If the wind is sufficiently strong, so that it is impossible to maintain a feasible trajectory with combination of allowable controls, then the system is not controllable within the control constraints. The manner in which the system responds to the control inputs may then also be viewed as an aspect of controllability.

b) Controllability Intrinsic to the System Whereas controllability can be viewed as an outcome of the current state of system, control theory views controllability as an intrinsic property of the system (without any perturbations). Kirk (2004) defines controllability as *the ability of system to be brought to some desired state (usually the origin) from some different initial state in a finite time using the control parameters*. In the absence of controllability, it is not possible to transfer a system from an initial state to a desired final state. This idea is closely linked to the existence of a

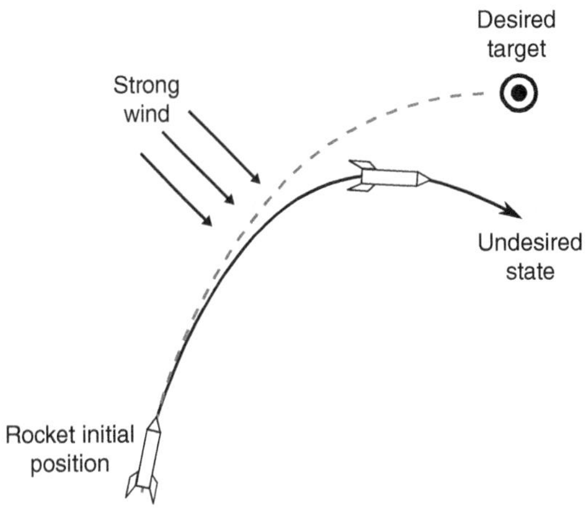

Figure 6.12 Trajectories in the presence of external interferences. The lack of intervening controls in the presence of a strong wind results in an undesired trajectory that does not reach the intended target.

unique solution. In order to obtain a unique solution (one that is required to minimize or maximize a desired objective function) the system must be controlled to do so, and the system must first exhibit *controllability*.

In the rocket example, the system is uncontrollable if there is no combination of inputs (a control policy for pitch, yaw, roll, and thrust) that transfers the initial state of the rocket to a desired final state for the performance objective (minimum fuel consumption, minimum flight time etc.). Computing the uniqueness of a solution can thus be used to determine controllability.

6.4.2.2 Controllability of Linear Systems It is possible to determine if a system of linear differential equations is controllable or not. Although reactive systems found in AR theory are generally nonlinear, the underlying concepts are similar and shall be useful for later discussions. In 1959, Rudolf Kalman showed that specifically for a *linear, time-invariant* system, it is possible to determine whether a system is controllable by computing the rank of a special controllability block matrix, **E** (Kalman, 1959)

$$\mathbf{E} = [\mathbf{B} \vdots \mathbf{AB} \vdots \mathbf{A}^2\mathbf{B} \vdots \cdots \vdots \mathbf{A}^{n-1}\mathbf{B}] \tag{6.9}$$

where **A** and **B** are $n \times n$ and $n \times m$ matrices representing system and control input coefficient matrices, respectively. **E** is a block matrix composed of products of matrix powers of **A** and **B**. The product $\mathbf{A}^{n-1}\mathbf{B}$ is of size $n \times m$, and thus **E** must have size $n \times nm$. In order for the system to be controllable, **E** must contain full row rank (in other words, the rank($\mathbf{E}$) = n) (Chen, 1995; Kirk, 2004). Since the rank of a matrix indicates the number of independent rows or columns, it can be used to determine whether a system is controllable (it admits a unique solution).

EXAMPLE 7: Identifying controllability of a linear system

Consider a system containing n state variables $x_1(t), x_2(t), \ldots, x_n(t)$. The system is assumed to be controlled by m control inputs $u_1(t), u_2(t), \ldots, u_m(t)$, and modeled by n *linear* differential equations, given in the following form:

$$\frac{d\mathbf{x}}{dt} = \mathbf{A}\mathbf{x}(t) + \mathbf{B}\mathbf{u}(t)$$

A and **B** are $n \times n$ and $n \times m$ matrices with entries of coefficients associated with the system. For the purposes of demonstration, assume that $n = 3$ and $m = 2$. Consider the following two specific matrix pairs for **A** and **B**:

Question 1:

A system of three ($n = 3$) linear differential equations is available:

$$\frac{dx_1}{dt} = x_1 + x_2 + x_3 + u_1$$

$$\frac{dx_2}{dt} = 2x_1 + x_2 + 2x_3 + u_2$$

$$\frac{dx_3}{dt} = 3x_1 + x_2 + 3x_3 + u_1 + u_2$$

In other words, **A** and **B** are given by the following matrices:

$$\mathbf{A} = \begin{bmatrix} 1 & 1 & 1 \\ 2 & 1 & 2 \\ 3 & 1 & 3 \end{bmatrix}$$

and

$$\mathbf{B} = \begin{bmatrix} 1 & 0 \\ 0 & 1 \\ 1 & 1 \end{bmatrix}$$

Determine whether system is controllable or not by computing rank(**E**).

Answer 1:

There are three state variables ($x_1(t)$, $x_2(t)$, and $x_3(t)$) and two control inputs ($u_1(t)$ and $u_2(t)$), hence $n = 3$ and $m = 2$.From Equation 6.9, the $n \times nm$ controllability matrix **E** may be formed from matrices **A** and **B** giving:

$$\mathbf{E} = [\mathbf{B} \,\vdots\, \mathbf{AB} \,\vdots\, \mathbf{A}^2\mathbf{B}]$$

$$= \begin{bmatrix} 1 & 0 & 2 & 2 & 12 & 9 \\ 0 & 1 & 4 & 3 & 20 & 15 \\ 1 & 1 & 6 & 4 & 28 & 21 \end{bmatrix}$$

The rank of **E** is computed by standard methods, such as counting the number of independent rows after performing elementary row operations. We prefer here to employ MATLAB. Inputting **E** and computing its rank gives an answer of three. Since $n = 3$, rank(**E**) $= n$. The controllability matrix **E** has full row rank and the system is therefore controllable.

Question 2:

Assume now that **A** is given by the modified matrix:

$$\mathbf{A} = \begin{bmatrix} 1 & 0 & 1 \\ 2 & 1 & 2 \\ 3 & 1 & 3 \end{bmatrix}$$

Assume that matrix **B** remains unchanged as in scenario 1. Determine if the system controllable.

Answer 2:
The controllability matrix **E** may again be computed using **A** and **B** giving the following matrix:

$$\begin{aligned}\mathbf{E} &= [\mathbf{B} \ \vdots \ \mathbf{AB} \ \vdots \ \mathbf{A}^2\mathbf{B}] \\ &= \begin{bmatrix} 1 & 0 & 2 & 1 & 8 & 5 \\ 0 & 1 & 4 & 3 & 20 & 13 \\ 1 & 1 & 6 & 4 & 28 & 18 \end{bmatrix}\end{aligned}$$

The rank of **E** can again be calculated. In this instance rank(**E**) = 2. Thus **E** does not have full rank since rank(**E**) < n. The system given in scenario 2 is therefore not controllable.

SIDE NOTE: Relation to the AR theory

By examination of the system of differential equations for the problem, it is possible to calculate whether the system under investigation is controllable or not. Controllability in itself is important if we wish to obtain a unique solution to the optimal control problem. This is true for the controlled flight of a rocket to the moon, or in our case, the control of the side-stream addition rate (the α policy) in a DSR on the boundary of the AR.

In the next section, we briefly describe how the concept of controllability may be used to find critical DSR trajectories.

6.4.3 Computing Critical DSR Trajectories

6.4.3.1 Nature of a Critical DSR Profile The control of a DSR profile in concentration space is mathematically similar to the control of a rocket to the moon. The dynamics of both systems may be described by a system of ordinary differential equations, affected by the appropriate control inputs:

- Pitch, yaw, roll, and thrust for the rocket example.
- The α policy and side-stream composition, $\mathbf{C}^0$, for the DSR.

Both systems of ODEs may be integrated from an initial condition to obtain a solution trajectory that describes the path of the system in state space along an appropriate time horizon. Whereas in the rocket example, controllability is required everywhere along the solution path, for critical DSRs we will argue that the *opposite* must be enforced—critical DSRs must be completely *uncontrollable* on the AR boundary.

CONCEPT: Controllability of a critical DSR

A critical DSR trajectory cannot be completely controllable for all residence times to travel on the AR boundary.

To motivate why, let us consider the physical significance of traversing on the AR boundary. If a reactor structure achieves effluent concentrations that form part of the true AR boundary, then it is *not* possible for another reactor structure to produce concentrations that could expand the set further—otherwise, these concentrations would form part of a larger region, and the original set would not belong to the true AR.

This is significant from a controllability perspective. Suppose that a critical DSR and associated critical α policy are available, which produce a DSR trajectory that traverses on the AR boundary. Introduction of a perturbation to the α policy would result in a deviation away from the expected optimal solution trajectory. Since the DSR is critical, its trajectory must lie entirely on the AR boundary. Perturbation implies the perturbed DSR trajectory would never result in one that could expand the AR boundary, for if this were satisfied, the original DSR would not be a critical DSR to begin with. Permissible trajectories resulting from the perturbation would only serve to travel *into* the region. This situation is depicted in Figure 6.13.

A critical DSR must hence operate in a highly irregular state. If a perturbation only serves to produce deviations into the region, then not all states can be realized by the system and the system must be uncontrollable. By the nature of the AR boundary, a critical DSR trajectory cannot be completely locally controllable. To determine an α policy that corresponds to a traversal on the AR boundary, the controllability ideas discussed earlier must be related. The same ideas that served as a necessary condition for finding a unique

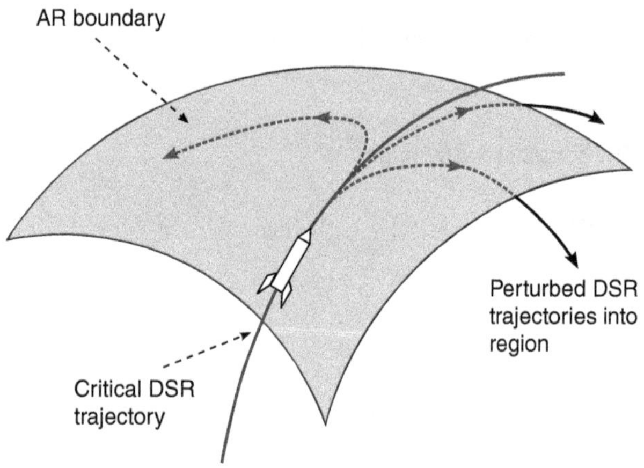

Figure 6.13 DSR trajectories on the AR boundary. All nonoptimal changes in the α policy serve to move the resulting trajectories into the region. The DSR is thus *not* locally controllable on the AR boundary. (*See color plate section for the color representation of this figure.*)

solution for a controllable system might also be used as a necessary condition for *denying* a completely controllable DSR. This in turn provides a condition for computing critical DSR trajectories and α policies.

CONCEPT: A condition for residing on the AR boundary

Local controllability must be denied to achieve a state on the AR boundary. If this were not true, then a perturbation in the α policy would serve to expand the AR boundary, and we would not have been on the true AR boundary initially.

6.4.3.2 Controllability of Nonlinear Systems Section 6.4.2.2 described how it is possible to determine whether a linear system may be controllable by computing the rank of the controllability matrix **E**. For a dynamic system containing n linear ODEs, the system is completely controllable if rank(**E**) = n. Controllability is also directly linked to the uniqueness of solution in a system, and controllability of a system is a necessary condition for uniquely solving the system. Determining the controllability of a linear system is possible because the rank of **E** is computable. For a system of nonlinear ordinary differential equations (the type of systems common in reactor design), these computations are no longer applicable. It is therefore not possible to use the same **E** matrix for a nonlinear system, such as the DSR. However, methods do exist for determining the controllability of nonlinear systems. These techniques no longer utilize linear algebra alone. Instead, *Lie algebra* must be employed.

SIDE NOTE: Lie algebra

Lie algebra is a generalized extension of linear algebra to nonlinear systems, commonly used in physics and differential geometry. Lie algebra is often employed in the theory of special relativity, and also in nonlinear process control.

Since Lie algebra is a generalization of linear (matrix) algebra, it is possible to use Lie algebra in the control of linear systems. But this theory is often unnecessary because matrix algebra is sufficient. In nonlinear systems, Lie algebra replaces matrix algebra, and *Lie derivatives* and *Lie brackets* replace matrix operations.

Lie derivative

Given a scalar field h(**C**) (a surface in $\mathbb{R}^n$) and vector field $\mathbf{f}(\mathbf{C}) = [f_1(\mathbf{C}), f_2(\mathbf{C}), \ldots, f_n(\mathbf{C})]^T$, the Lie derivative of h(**C**) with respect to **f**(**C**), $L_f h(\mathbf{C})$, is defined by

$$L_f h(\mathbf{C}) = [\nabla h(\mathbf{C})]^T \mathbf{f}(\mathbf{C}) = \sum_{i=1}^{n} f_i(\mathbf{C}) \frac{\partial h(\mathbf{C})}{\partial c_i}$$

where $\nabla = \left[\frac{\partial}{\partial c_1}, \frac{\partial}{\partial c_2}, \ldots, \frac{\partial}{\partial c_n}\right]$ is the del operator (or also known as nabla),[7] which is simply the directional derivative of h(**C**) in the direction of vector **f**(**C**). The *Lie derivative operator with respect to* **f**(**C**), L_f, is then defined as

$$L_f = \sum_{i=1}^{n} f_i(\mathbf{C}) \frac{\partial}{\partial c_i}$$

We can apply Lie derivatives repeatedly, such as finding the Lie derivative of h(**C**) first in the direction of vector field **f**(**C**) and then in the direction of vector field **g**(**C**)

$$L_g L_f h(\mathbf{C}) = \sum_{i=1}^{n} g_i(\mathbf{C}) \frac{\partial [L_f h(\mathbf{C})]}{\partial c_i}$$

or even repeatedly in the same direction of **f**(**C**)

$$L_f^2 h(\mathbf{C}) = L_f L_f h(\mathbf{C})$$

[7] By example, if $\mathbf{f}(\mathbf{C}) = [c_1^2 c_2, c_2]^T$, then $\nabla \mathbf{f}(\mathbf{C}) = \left[\frac{\partial(c_1^2 c_2)}{\partial c_1}, \frac{\partial(c_2)}{\partial c_2}\right]^T = [2c_1 c_2, 1]^T$.

Lie bracket

Assume that we have two vector fields, $\mathbf{f}(\mathbf{C})$ and $\mathbf{g}(\mathbf{C})$. We can define a third vector field $[\mathbf{f}, \mathbf{g}](\mathbf{C})$ as:

$$[\mathbf{f},\mathbf{g}](\mathbf{C}) = \begin{bmatrix} \frac{\partial g_1(\mathbf{C})}{\partial c_1} & \cdots & \frac{\partial g_1(\mathbf{C})}{\partial c_n} \\ \vdots & \ddots & \vdots \\ \frac{\partial g_n(\mathbf{C})}{\partial c_1} & \cdots & \frac{\partial g_n(\mathbf{C})}{\partial c_n} \end{bmatrix} \begin{bmatrix} f_1(\mathbf{C}) \\ \vdots \\ f_n(\mathbf{C}) \end{bmatrix} - \begin{bmatrix} \frac{\partial f_1(\mathbf{C})}{\partial c_1} & \cdots & \frac{\partial f_1(\mathbf{C})}{c_n} \\ \vdots & \ddots & \vdots \\ \frac{\partial f_n(\mathbf{C})}{\partial c_1} & \cdots & \frac{\partial f_n(\mathbf{C})}{\partial c_n} \end{bmatrix} \begin{bmatrix} g_1(\mathbf{C}) \\ \vdots \\ g_n(\mathbf{C}) \end{bmatrix}$$

which we call the Lie bracket of $\mathbf{f}(\mathbf{C})$ and $\mathbf{g}(\mathbf{C})$.[8] $[\mathbf{f}, \mathbf{g}](\mathbf{C})$ may be rewritten as

$$L_f L_g - L_g L_f = \sum_{i=1}^{n} [\mathbf{f},\mathbf{g}](\mathbf{C})_i \frac{\partial}{\partial c_i}$$

where $[\mathbf{f}, \mathbf{g}](\mathbf{C})_i$ is the ith element in $[\mathbf{f}, \mathbf{g}](\mathbf{C})$. Lie brackets are composed of Lie derivative operators since

$$L_f L_g - L_g L_f = L_{[f,\, g]}$$

where $L_{[f,\, g]}$ is the Lie derivative operator with respect to the vector field $[\mathbf{f}, \mathbf{g}](\mathbf{C})$. This result shows that the Lie derivative of $[\mathbf{f}, \mathbf{g}](\mathbf{C})$ is equivalent to the difference $L_f L_g - L_g L_f$.

Simply understanding the necessary relationships for a critical DSR is sufficient for one to realize that computing explicit expressions for these reactors are not feasible for many practical applications. These ideas, although powerful, place highly stringent and intricate constraints on the α policy that make it difficult to implement in practice. Nevertheless, these discussions are still important if we wish to understand what *must* be done in the absolute best case, such as when performance targets must be determined.

6.4.3.3 Critical α Policies It is now possible to describe the computation of critical DSR α policies. For nonlinear systems, the use of Lie brackets is required to determine controllability. It will be helpful, nevertheless, to remember the linear, time-invariant version of the controllability problem when going through this section.

Determining whether a DSR is uncontrollable is also dependent on the properties of a specific matrix, and controllability is established by determining if this matrix contains full rank or not. This is achieved with the aid of Lie derivatives and Lie brackets.

ILLUSTRATION: The Lie bracket of a vector field

Let $\mathbf{f}(\mathbf{C})$ and $\mathbf{g}(\mathbf{C})$ be two smooth vector fields on $\mathbb{R}^n$. $\mathbf{C}$ is a vector containing n variables $\mathbf{C} = [c_1, c_2, \ldots, c_n]^T$. The Lie bracket, $\mathbf{z}(\mathbf{C}) = [\mathbf{f}(\mathbf{C}), \mathbf{g}(\mathbf{C})]$, is a third vector field defined by the following

$$[\mathbf{f},\mathbf{g}](\mathbf{C}) = d\mathbf{g}(\mathbf{C})\mathbf{f}(\mathbf{C}) - d\mathbf{f}(\mathbf{C})\mathbf{g}(\mathbf{C}) \tag{6.10}$$

Here, $d\mathbf{f}(\mathbf{C})$ and $d\mathbf{g}(\mathbf{C})$ are shorthand notation used to represent the Jacobian matrices of $\mathbf{f}(\mathbf{C})$ and $\mathbf{g}(\mathbf{C})$, respectively.

$$d\mathbf{f}(\mathbf{C}) = \begin{bmatrix} \frac{\partial f_1(\mathbf{C})}{\partial c_1} & \cdots & \frac{\partial f_1(\mathbf{C})}{c_n} \\ \vdots & \ddots & \vdots \\ \frac{\partial f_n(\mathbf{C})}{\partial c_1} & \cdots & \frac{\partial f_n(\mathbf{C})}{\partial c_n} \end{bmatrix}$$

$$d\mathbf{g}(\mathbf{C}) = \begin{bmatrix} \frac{\partial g_1(\mathbf{C})}{\partial c_1} & \cdots & \frac{\partial g_1(\mathbf{C})}{\partial c_n} \\ \vdots & \ddots & \vdots \\ \frac{\partial g_n(\mathbf{C})}{\partial c_1} & \cdots & \frac{\partial g_n(\mathbf{C})}{\partial c_n} \end{bmatrix}$$

Consider the particular case where $\mathbf{f}(\mathbf{C})$ and $\mathbf{g}(\mathbf{C})$ are given by:

$$\mathbf{f}(\mathbf{C}) = \begin{bmatrix} c_1 & c_2 \end{bmatrix}^T$$
$$\mathbf{g}(\mathbf{C}) = \begin{bmatrix} c_1^2 & c_1 \end{bmatrix}^T$$

We can compute the Jacobians $d\mathbf{f}(\mathbf{C})$ and $d\mathbf{g}(\mathbf{C})$ giving

$$d\mathbf{f}(\mathbf{C}) = \begin{bmatrix} 1 & 0 \\ 0 & 1 \end{bmatrix}$$

$$d\mathbf{g}(\mathbf{C}) = \begin{bmatrix} 2c_1 & 0 \\ 1 & 0 \end{bmatrix}$$

[8] $[\mathbf{f}, \mathbf{g}](\mathbf{C})$ is also sometimes called the commutator or Poisson bracket of $\mathbf{f}(\mathbf{C})$ and $\mathbf{g}(\mathbf{C})$.

The Lie bracket $\mathbf{z}(\mathbf{C})$ can then be found by substituting into Equation 6.10:

$$\begin{aligned}\mathbf{z}(\mathbf{C}) &= [\mathbf{dg}(\mathbf{C})\mathbf{f}(\mathbf{C}) - \mathbf{df}(\mathbf{C})\mathbf{g}(\mathbf{C})] \\ &= \begin{bmatrix} 2c_1 & 0 \\ 1 & 0 \end{bmatrix}\begin{bmatrix} c_1 \\ c_2 \end{bmatrix} - \begin{bmatrix} 1 & 0 \\ 0 & 1 \end{bmatrix}\begin{bmatrix} c_1^2 \\ c_1 \end{bmatrix} \\ &= \begin{bmatrix} c_1^2 \\ 0 \end{bmatrix}\end{aligned}$$

The use of Lie brackets in AR theory arises in the form of *iterated* Lie Brackets. We define the first Lie Bracket, $\mathbf{z}^{[1]}(\mathbf{C})$, as

$$\mathbf{z}^{[1]}(\mathbf{C}) = [\mathbf{dg}(\mathbf{C})\mathbf{f}(\mathbf{C}) - \mathbf{df}(\mathbf{C})\mathbf{g}(\mathbf{C})]$$

It is also possible to form higher order Lie brackets as follows:

$$\begin{aligned}\mathbf{z}^{[2]}(\mathbf{C}) &= [\mathbf{f}(\mathbf{C}),\ \mathbf{z}^{[1]}(\mathbf{C})] \\ &= [\mathbf{f}(\mathbf{C}),\ [\mathbf{dg}(\mathbf{C})\mathbf{f}(\mathbf{C}) - \mathbf{df}(\mathbf{C})\mathbf{g}(\mathbf{C})]]\end{aligned}$$

$\mathbf{z}^{[2]}(\mathbf{C})$ may thus be given as

$$\begin{aligned}\mathbf{z}^{[2]}(\mathbf{C}) &= \mathbf{dz}^{[1]}(\mathbf{C})\mathbf{f}(\mathbf{C}) - \mathbf{df}(\mathbf{C})\mathbf{z}^{[1]}(\mathbf{C}) \\ &= \begin{bmatrix} {c_1}^2 \\ 0 \end{bmatrix}\end{aligned}$$

In this manner, it is possible to compute as many additional Lie brackets by repeated application of Equation 6.10

$$\begin{aligned}\mathbf{z}^{[3]}(\mathbf{C}) &= [\mathbf{f}(\mathbf{C}), \mathbf{z}^{[2]}(\mathbf{C})] \\ &\vdots \\ \mathbf{z}^{[k]}(\mathbf{C}) &= [\mathbf{f}(\mathbf{C}),\ \mathbf{z}^{[k-1]}(\mathbf{C})]\end{aligned}$$

$\mathbf{z}^{[1]}(\mathbf{C})$ is nested within $\mathbf{z}^{[2]}(\mathbf{C})$, $\mathbf{z}^{[2]}(\mathbf{C})$ is nested within $\mathbf{z}^{[3]}(\mathbf{C})$, and so on. It is evident that the computation of Lie brackets involves a large amount of arithmetic, particularly when large systems (large values of n) with many iterated Lie brackets (large values of k) are considered. These computations are thus best managed with the aid of a computer algebra system (CAS) package.

a) Condition for a Critical DSR To determine whether a DSR is uncontrollable, and thus in order for it to be critical on the AR boundary, the following controllability condition must hold:

CONCEPT: Condition for a critical DSR

For a DSR to produce effluent concentrations on the AR boundary, the following set of vectors must be linearly *dependent.*[9]

$$\{\mathbf{f}(\mathbf{C}),\ \mathbf{v}(\mathbf{C}),\ \mathbf{z}^{[1]}(\mathbf{C}),\ \mathbf{z}^{[2]}(\mathbf{C}),\ \ldots,\ \mathbf{z}^{[k]}(\mathbf{C})\}$$

Vectors $\mathbf{f}(\mathbf{C})$ and $\mathbf{v}(\mathbf{C})$ are the DSR expression and mixing point defined as

$$\mathbf{f}(\mathbf{C}) = \mathbf{r}(\mathbf{C}) + \alpha\mathbf{v}(\mathbf{C}) \text{ and } \mathbf{v}(\mathbf{C}) = \mathbf{C} - \mathbf{C}^0$$

$\mathbf{C}^0$ is the DSR side-stream mixing concentration. Vectors $\mathbf{z}^{[1]}$, $\mathbf{z}^{[2]}$, …, $\mathbf{z}^{[k]}$ are iterated Lie brackets as described earlier, given by

$$\begin{aligned}\mathbf{z}^{[1]}(\mathbf{C}) &= [\mathbf{f}(\mathbf{C}),\ \mathbf{v}(\mathbf{C})] = \mathbf{dv}(\mathbf{C})\mathbf{f}(\mathbf{C}) - \mathbf{df}(\mathbf{C})\mathbf{v}(\mathbf{C})\mathbf{z}^{[2]}(\mathbf{C}) \\ \mathbf{z}^{[2]}(\mathbf{C}) &= [\mathbf{f}(\mathbf{C}),\ \mathbf{z}^{[1]}(\mathbf{C})] \\ &\vdots \\ \mathbf{z}^{[k]}(\mathbf{C}) &= [\mathbf{f}(\mathbf{C}), \mathbf{z}^{[k-1]}(\mathbf{C})]\end{aligned}$$

Let us form a controllability matrix $\mathbf{K}$ for a DSR:

$$\mathbf{K} = [\mathbf{f}(\mathbf{C}), \mathbf{v}(\mathbf{C}), \mathbf{z}^{[1]}(\mathbf{C}),\ \mathbf{z}^{[2]}(\mathbf{C}), \ldots, \mathbf{z}^{[k]}(\mathbf{C})] \tag{6.11}$$

If the column vectors in $\mathbf{K}$ are to be linearly dependent, the rank of $\mathbf{K}$ must be *less than* the dimension of the AR (the number of independent reactions):

$$\text{rank}(\mathbf{K}) < n \tag{6.12}$$

where n is the number of linearly independent reactions. Observe that Equation 6.12 does not provide an explicit expression for computing critical DSR trajectories. To circumvent this limitation, first notice that if a matrix is linearly dependent, then the determinant of the matrix will be zero (Lay, 2012). We can exploit of property of indeterminate matrices to express the controllability condition of a critical DSR in terms of determinants. Let us define a new block matrix $\mathbf{E}$[10] composed of two matrices $\mathbf{K}$ and $\mathbf{N}$:

$$\mathbf{E} = [\mathbf{K}, \mathbf{N}]$$

[9]Similar to linear systems, controllability of a system breaks down when the rank of the $\mathbf{E}$ matrix is less than n.

[10]Note that matrix $\mathbf{E}$ used here is technically different to that given for linear systems in Section 6.4.2.2, but it fulfils the same purpose as in a linear system and thus it is displayed by the same symbol.

where **N** is a matrix composed of column vectors that are perpendicular (orthogonal) to the stoichiometric subspace **S**, as described in Section 6.2.1.4.

$$\mathbf{N} = [\mathbf{n}_1, \mathbf{n}_2, \ldots, \mathbf{n}_{n-d}]$$

Then a condition for critical DSRs can then be expressed in terms of a determinant involving matrix **E**:

$$\text{Det}[\mathbf{E}] = 0$$

$$\text{Det}[\mathbf{K}, \mathbf{N}] = 0$$

$$\text{Det}[\mathbf{f}(\mathbf{C}), \mathbf{v}(\mathbf{C}), \mathbf{z}^{[1]}(\mathbf{C}), \mathbf{z}^{[2]}(\mathbf{C}), \ldots, \mathbf{z}^{[k]}(\mathbf{C}), \mathbf{N}] = 0 \quad (6.13)$$

where $\text{Det}[\mathbf{E}] = 0$ indicates that the DSR is not controllable.

EXAMPLE 8: Explaining the role of matrix N in Equation 6.13 with relation to the stoichiometric subspace

Two vectors $\mathbf{x}_1 \in \mathbb{R}^n$ and $\mathbf{x}_2 \in \mathbb{R}^n$ are orthogonal to a vector $\mathbf{y} \in \mathbb{R}^n$

$$\mathbf{x}_1^{\mathrm{T}}\mathbf{y} = 0$$

$$\mathbf{x}_2^{\mathrm{T}}\mathbf{y} = 0$$

or

$$\begin{bmatrix} \mathbf{x}_1^{\mathrm{T}} \\ \mathbf{x}_2^{\mathrm{T}} \end{bmatrix} \mathbf{y} = 0$$

Let

$$\mathbf{X} = [\mathbf{x}_1, \mathbf{x}_2]$$

then

$$\mathbf{X}^{\mathrm{T}}\mathbf{y} = \mathbf{0}$$

Vector **y** must then lie in the null space of $\mathbf{X}^{\mathrm{T}}$ if it satisfies the above relation. Furthermore, notice that $\mathbf{x}_1$ and $\mathbf{x}_2$ span a two-dimensional subspace in $\mathbb{R}^n$. Geometrically, since **X** is composed of $\mathbf{x}_1$ and $\mathbf{x}_2$, the column space of **X** must be orthogonal to **y**—the dimension of the column space of **X** is two (two independent columns in **X**, or $d = 2$), and thus the dimension of the subspace orthogonal to **X** (of which **y** is an element of this subspace) is equal to $n - 2$ (or, $n - d$ in general).

From Equation 6.13, we have that matrix **N** must be orthogonal to the stoichiometric subspace **S**. Recall that the columns of the stoichiometric coefficient matrix **A** span **S**. (Each column in **A** represents a reaction participating in the system, which is represented geometrically as a vector that forms a basis of the stoichiometric subspace.) It follows that if **N** is required to be orthogonal to the **S**, then **N** is found by computing the null space of $\mathbf{A}^{\mathrm{T}}$:

$$\mathbf{N} = \text{null}(\mathbf{A}^{\mathrm{T}})$$

Notice that if we substitute matrix **X** discussed in this example with the stoichiometric coefficient matrix **A**, then the columns in **X** ($\mathbf{x}_1$ and $\mathbf{x}_2$) represent two reactions participating in n-dimensional concentration space, $\mathbb{R}^n$. Hence, to compute **N**, we simply determine the stoichiometric coefficient matrix **A** (as in Section 6.2.1.3), and then compute the null space of $\mathbf{A}^{\mathrm{T}}$. From linear algebra, we can show that if **A** has size $n \times d$ (n components participating in d reactions), the size of matrix **N** will be $n \times (n - d)$.

EXAMPLE 9: Computing N for use in the controllability matrix E

Suppose that the following reaction stoichiometry is given:

$$2\text{A} \rightarrow \text{B}$$

$$3\text{B} \rightarrow 2\text{C} + \text{D}$$

$$\text{D} + \text{A} \rightarrow \text{E}$$

We wish to compute the controllability matrix **E** for this system, which will be used to determine the existence of a critical DSR trajectory. What does the matrix **N** look like for this system?

Answer:

The stoichiometric coefficient matrix **A** is formed from the reaction stoichiometry of the system. There are three reactions ($d = 3$) involving five components ($n = 5$):

$$\mathbf{A} = \begin{bmatrix} -2 & 0 & -1 \\ 1 & -3 & 0 \\ 0 & 2 & 0 \\ 0 & 1 & -1 \\ 0 & 0 & 1 \end{bmatrix}$$

The columns in **A** span the stoichiometric subspace **S** in $\mathbb{R}^5$ (c_A–c_B–c_C–c_D–c_E space). To find **N**—the columns of **N** that form a basis for the nullspace of **S**—we simply compute the null space of $\mathbf{A}^T$.

$$\mathbf{N} = \text{null}(\mathbf{A}^T)$$

or

$$\mathbf{N} = \text{null}\left(\begin{bmatrix} -2 & 1 & 0 & 0 & 0 \\ 0 & -3 & 2 & 1 & 0 \\ -1 & 0 & 0 & -1 & 1 \end{bmatrix}\right)$$

$\mathbf{A}^T$ is a 3×5 matrix and thus the null space of $\mathbf{A}^T$ will be a two-dimensional subspace in c_A–c_B–c_C–c_D–c_E space (the size of matrix $\mathbf{N}$ must be $n\times(n-d)$, or 5×2). To compute the null space of this matrix, we can reduce $\mathbf{A}^T$ to reduced row echelon form by performing elementary row operations on $\mathbf{A}^T$, and determine all of the vectors in the null space (similar in procedure to that shown in Example 3). Hence reducing $\mathbf{A}^T$ to the equivalent matrix gives:

$$\begin{bmatrix} 1 & 0 & 0 & 1 & -1 \\ 0 & 1 & 0 & 2 & -2 \\ 0 & 0 & 1 & 3.5 & -3 \end{bmatrix}$$

which indicates that vectors

$$\mathbf{n}_1 = [-1, -2, -3.5,\ 1,\ 0]^T$$
$$\mathbf{n}_2 = [1,\ 2,\ 3,\ 0,\ 1]^T$$

are vectors that form a basis of the orthogonal subspace to $\mathbf{A}^T$. Thus, the columns of matrix $\mathbf{N}$ are formed from linear combinations of $\mathbf{n}_1$ and $\mathbf{n}_2$.

$$\mathbf{N} = \{\lambda_1\mathbf{n}_1,\ \lambda_2\mathbf{n}_2\} = \left\{\lambda_1\begin{bmatrix} -1 \\ -2 \\ -3.5 \\ 1 \\ 0 \end{bmatrix},\ \lambda_2\begin{bmatrix} 1 \\ 2 \\ 3 \\ 0 \\ 1 \end{bmatrix}\right\}$$

Values of λ_1 and λ_2 are arbitrary. If we let $\lambda_1 = 1$ and $\lambda_2 = 1$, $\mathbf{N}$ takes the form

$$\mathbf{N} = \begin{bmatrix} -1 & 1 \\ -2 & 2 \\ -3.5 & 3 \\ 1 & 0 \\ 0 & 1 \end{bmatrix} \tag{6.14a}$$

Alternatively, we could use a computer linear algebra package and compute $\mathbf{N}$ numerically. Using the `null()` function in MATLAB for $\mathbf{A}^T$ gives:

$$\begin{bmatrix} -0.1203 & 0.2296 \\ -0.2405 & 0.4591 \\ -0.6427 & 0.5085 \\ 0.5638 & 0.3604 \\ 0.4435 & 0.5900 \end{bmatrix} \tag{6.14b}$$

Equation 6.14b differs only to the $\mathbf{N}$ in Equation 6.14a in that the columns of Equation 6.14b are orthonormal (they form an orthonormal basis for null($\mathbf{A}^T$)). However Equation 6.14a and 6.14b span the same subspace.

EXAMPLE 10: Size of E

Using your knowledge of critical DSRs and the DSR controllability matrix $\mathbf{E}$ in Equation 6.13, answer the following questions:

(a) If there are n components present in the system, what is the size of $\mathbf{E}$?

(b) If there are d independent reactions in the system, then what is the value of k? (The number of iterated Lie brackets that must be used to compute the determinant of the DSR controllability matrix.)

Answer a:

The $\mathbf{E}$ matrix in Equation 6.13 must be square since determinants are not defined for nonsquare matrices. Since $\mathbf{E}$ is a block matrix composed of matrices $\mathbf{K}$ and $\mathbf{N}$, then the size of $\mathbf{E}$ is dictated by the number of columns (or rows) in matrices $\mathbf{K}$ and $\mathbf{N}$. The number of rows in $\mathbf{K}$ and $\mathbf{N}$ is equal to the total number of components participating in the system, which we know is n in this example (n components residing in $\mathbb{R}^n$). It follows that the size of $\mathbf{E}$ is also equal to n. That is:

$$\text{size}(\mathbf{E}) = n \times n$$

Answer b:

We know from answer (a) above that $\mathbf{E}$ must be a square $n\times n$ matrix. As a result, $\mathbf{E}$ must contain n columns.

$$\text{size}(\mathbf{E}) = n \times n$$

We also know that the columns in $\mathbf{E}$ are composed of columns in matrices $\mathbf{K}$ and $\mathbf{N}$.

$$\mathbf{E} = [\mathbf{K}, \mathbf{N}]$$

Matrix **N** has columns that form a basis for the nullspace of the stoichiometric subspace **S**, and hence there must be $(n-d)$ columns in **N** if there are n components participating in d independent reactions (the stoichiometric coefficient matrix **A** has size $n \times d$).

$$\text{size}(\mathbf{N}) = n \times (n-d)$$

It follows that the size of **K** must then calculated by taking the total number of columns in **E** (n) and subtracting the number of columns in **N** ($n-d$).

$$\text{size}(\mathbf{K}) = n \times (n-(n-d)) = n \times d$$

Hence, there are d columns in matrix **K**. Finally, it is also known that **K** is composed of vectors two $\mathbf{f}(\mathbf{C})$ and $\mathbf{v}(\mathbf{C})$ and k iterated Lie brackets $\mathbf{z}^{[1]}(\mathbf{C}), \ldots, \mathbf{z}^{[k]}(\mathbf{C})$.

$$\mathbf{K} = [\mathbf{f}(\mathbf{C}), \mathbf{v}(\mathbf{C}), \mathbf{z}^{[1]}(\mathbf{C}),\ \mathbf{z}^{[2]}(\mathbf{C}), \ldots, \mathbf{z}^{[k]}(\mathbf{C})]$$

Therefore, the value of k is equal to the number of columns in **K** (d) minus two (for vectors $\mathbf{f}(\mathbf{C})$ and $\mathbf{v}(\mathbf{C})$):

$$k = d - 2$$

This result also shows that computing the controllability of **E** for critical DSRs is only valid for systems involving greater than two independent reactions (the value of k must greater than or equal one, or $k \geq 1$).

ILLUSTRATION: Demonstrating the nature of E

Example A: Three independent reactions in four components

Consider the following system of three reactions ($d = 3$), involving four components ($n = 4$):

$$A \rightarrow B \rightarrow C \rightarrow D$$

The stoichiometric coefficient matrix **A** may be formed in the usual manner

$$\mathbf{A} = \begin{bmatrix} -1 & 0 & 0 \\ 1 & -1 & 0 \\ 0 & 1 & -1 \\ 0 & 0 & 1 \end{bmatrix}$$

and used to compute **N**—the null space of $\mathbf{A}^T$—in a standard fashion, for use in forming the controllability matrix **E**:

$$\mathbf{N} = \text{null}(\mathbf{A}^T)$$

In this instance, we find that **N** is composed of a single vector $\mathbf{n}_1$:

$$\mathbf{n}_1 = \begin{bmatrix} 1 & 1 & 1 & 1 \end{bmatrix}^T$$

Matrix **N** in Equation 6.13 therefore contains a single vector $\mathbf{N} = \mathbf{n}_1$ in this instance. From this result, the controllability matrix **E** for the DSR may be formed from Equation 6.13:

$$\text{Det}[\mathbf{f}(\mathbf{C}), \mathbf{v}(\mathbf{C}), \mathbf{z}^{[1]}(\mathbf{C}), \ldots, \mathbf{z}^{[k]}(\mathbf{C}), \mathbf{N}] = 0$$

which, for three independent reactions becomes:

$$\text{Det}[\mathbf{f}(\mathbf{C}), \mathbf{v}(\mathbf{C}), \mathbf{z}^{[1]}(\mathbf{C}), \mathbf{n}_1] = 0$$

where $\mathbf{z}^{[1]}(\mathbf{C}) = [\mathbf{f}(\mathbf{C}), \mathbf{v}(\mathbf{C})]$ is the Lie Bracket of $\mathbf{f}(\mathbf{C})$ and $\mathbf{v}(\mathbf{C})$. In the next example, we shall define what $\mathbf{f}(\mathbf{C})$ and $\mathbf{v}(\mathbf{C})$ in the context of critical DSRs.

Example B: Three independent reactions in five components

Consider the following system, now involving five components ($n = 5$) in three reactions ($d = 3$):

$$A + B \rightarrow C$$

$$A + C \rightarrow D$$

$$A + D \rightarrow E$$

The stoichiometric coefficient matrix **A** is now

$$\mathbf{A} = \begin{bmatrix} -1 & -1 & -1 \\ -1 & 0 & 0 \\ 1 & -1 & 0 \\ 0 & 1 & -1 \\ 0 & 0 & 1 \end{bmatrix}$$

We find that rank(**A**) = 3. Hence, exactly two vectors are expected to form the null space of $\mathbf{A}^T$. Computing null($\mathbf{A}^T$) by standard methods, it is possible to show the vectors

$$\mathbf{n}_1 = \begin{bmatrix} -1 & 3 & 2 & 1 & 0 \end{bmatrix}^T$$

$$\mathbf{n}_2 = \begin{bmatrix} 1 & -2 & -1 & 0 & 1 \end{bmatrix}^T$$

form a basis for the null space of $\mathbf{A}^T$. As a result, matrix **N** is given by:

$$\mathbf{N} = \begin{bmatrix} \mathbf{n}_1 & \mathbf{n}_2 \end{bmatrix}$$

The determinant function given by Equation 6.13 can now be formed:

$$\text{Det}[\mathbf{f}(\mathbf{C}), \mathbf{v}(\mathbf{C}), \mathbf{z}^{[1]}(\mathbf{C}), \mathbf{n}_1, \mathbf{n}_2] = 0$$

Notice that in this case, matrix **N** is given by two vectors $\mathbf{n}_1$ and $\mathbf{n}_2$.

Example C: Four independent reactions in six components

As a final example, consider the system

$$\text{A} + \text{B} \rightarrow \text{C}$$
$$\text{A} + \text{C} \rightarrow \text{D}$$
$$\text{A} + \text{D} \rightarrow \text{E}$$
$$\text{A} + \text{E} \rightarrow \text{F}$$

This is a system of four independent reactions ($d = 4$) in six components ($n = 6$). The stoichiometric coefficient matrix is

$$\mathbf{A} = \begin{bmatrix} -1 & -1 & -1 & -1 \\ -1 & 0 & 0 & 0 \\ 1 & -1 & 0 & 0 \\ 0 & 1 & -1 & 0 \\ 0 & 0 & 1 & -1 \\ 0 & 0 & 0 & 1 \end{bmatrix}$$

Calculation of the null space for $\mathbf{A}^{\mathrm{T}}$ gives the following two vectors for the null space:

$$\mathbf{n}_1 = \begin{bmatrix} -1 & 4 & 3 & 2 & 1 & 0 \end{bmatrix}^{\mathrm{T}}$$
$$\mathbf{n}_2 = \begin{bmatrix} 1 & -3 & -2 & -1 & 0 & 1 \end{bmatrix}^{\mathrm{T}}$$

The determinant function is then:

$$\text{Det}[\mathbf{f}(\mathbf{C}), \mathbf{v}(\mathbf{C}), \mathbf{z}^{[1]}(\mathbf{C}), \mathbf{z}^{[2]}(\mathbf{C}), \mathbf{n}_1, \mathbf{n}_2] = 0$$

Observe that this involves two iterated Lie brackets ($\mathbf{z}^{[1]}(\mathbf{C})$ and $\mathbf{z}^{[2]}(\mathbf{C})$) and two normal vectors that are orthogonal to the stoichiometric subspace ($\mathbf{N} = [\mathbf{n}_1, \mathbf{n}_2]$).

OBSERVATION: Observation: E and AR dimension

Notice that **E** is a block matrix composed of matrices **K** and **N**:

$$\mathbf{E} = [\mathbf{K}, \mathbf{N}]$$

For n components participating in d reactions, the size of **E** is $n \times n$ and the size of **K** is $d \times n$. **K** contains information related to the number of reactions in the system, and thus it is related to the dimension of the AR. Since we are generally more concerned with the dimension of the AR (d) and not the total dimension of the system (n)—because we can always find the remaining $n - d$ components by mass balance—it is generally more convenient to express **E** for a general n-component system by

$$\mathbf{E} = [\mathbf{f}(\mathbf{C}), \mathbf{v}(\mathbf{C}), \mathbf{z}^{[1]}(\mathbf{C}), \ldots, \mathbf{z}^{[d-2]}(\mathbf{C}), \mathbf{N}]$$

which specifically describes the dimension of the AR but generalizes the dimension of **N**. The first d columns are related to the dimension of the AR, and the remaining $n - d$ columns are related to **N**.

Recall that for a DSR to be critical, the system must deny controllability, and matrix **E** cannot have full rank as a result. This is similar to how a linear system cannot be controlled if matrix **E** does not contain full rank. The underlying idea behind computing critical DSR policies is thus fairly simple (even if the details are more complex). But the actual process of calculating Det(**E**) involves a considerable amount of algebra and simplification, which is best performed using a CAS package.

EXAMPLE 11: Lie brackets for the DSR equation

Lie brackets are important mathematical objects used for determining critical DSR trajectories. The following examples will utilize properties of iterated Lie brackets. It will be useful to keep these properties in mind when attempting examples later on. In AR theory, $\mathbf{f}(\mathbf{C})$ and $\mathbf{v}(\mathbf{C})$ are vector functions given by the following expressions:

$$\mathbf{f}(\mathbf{C}) = \mathbf{r}(\mathbf{C}) + \alpha\mathbf{v}(\mathbf{C})$$
$$\mathbf{v}(\mathbf{C}) = \mathbf{C} - \mathbf{C}^0$$

Show that:

(a) $d\mathbf{v}(\mathbf{C}) = \mathbf{I}$, where **I** is the identity matrix.
(b) $[\mathbf{f}(\mathbf{C}), \mathbf{v}(\mathbf{C})] = [\mathbf{r}(\mathbf{C}), \mathbf{v}(\mathbf{C})]$

Answer a:

The mixing vector $\mathbf{v}(\mathbf{C}) = [\mathrm{v}_1, \mathrm{v}_2, \ldots, \mathrm{v}_n]^T$ is given by

$$\mathbf{v}(\mathbf{C}) = \mathbf{C} - \mathbf{C}^0$$

$$\begin{bmatrix} \mathrm{v}_1 \\ \mathrm{v}_2 \\ \vdots \\ \mathrm{v}_n \end{bmatrix} = \begin{bmatrix} \mathrm{c}_1 \\ \mathrm{c}_2 \\ \vdots \\ \mathrm{c}_n \end{bmatrix} - \begin{bmatrix} \mathrm{c}_1^0 \\ \mathrm{c}_2^0 \\ \vdots \\ \mathrm{c}_n^0 \end{bmatrix} = \begin{bmatrix} \mathrm{c}_1 - \mathrm{c}_1^0 \\ \mathrm{c}_2 - \mathrm{c}_2^0 \\ \vdots \\ \mathrm{c}_n - \mathrm{c}_n^0 \end{bmatrix}$$

The Jacobian of $\mathbf{v}(\mathbf{C})$, $\mathrm{d}\mathbf{v}(\mathbf{C})$, is found by taking partial derivatives with respect to all n components

$$\mathrm{d}\mathbf{v}(\mathbf{C}) = \begin{bmatrix} \frac{\partial \mathrm{v}_1}{\partial \mathrm{c}_1} & \cdots & \frac{\partial \mathrm{v}_1}{\partial \mathrm{c}_n} \\ \frac{\partial \mathrm{v}_2}{\partial \mathrm{c}_1} & \cdots & \frac{\partial \mathrm{v}_2}{\partial \mathrm{c}_n} \\ \vdots & \ddots & \vdots \\ \frac{\partial \mathrm{v}_n}{\partial \mathrm{c}_1} & \cdots & \frac{\partial \mathrm{v}_n}{\partial \mathrm{c}_n} \end{bmatrix}$$

$$= \begin{bmatrix} \frac{\partial(\mathrm{c}_1 - \mathrm{c}_1^0)}{\partial \mathrm{c}_1} & \cdots & \frac{\partial(\mathrm{c}_1 - \mathrm{c}_1^0)}{\partial \mathrm{c}_n} \\ \frac{\partial(\mathrm{c}_2 - \mathrm{c}_2^0)}{\partial \mathrm{c}_1} & \cdots & \frac{\partial(\mathrm{c}_2 - \mathrm{c}_2^0)}{\partial \mathrm{c}_n} \\ \vdots & \ddots & \vdots \\ \frac{\partial(\mathrm{c}_n - \mathrm{c}_n^0)}{\partial \mathrm{c}_1} & \cdots & \frac{\partial(\mathrm{c}_n - \mathrm{c}_n^0)}{\partial \mathrm{c}_n} \end{bmatrix}$$

$$= \begin{bmatrix} 1-0 & \cdots & 0 \\ 0 & \cdots & 0 \\ \vdots & \ddots & \vdots \\ 0 & \cdots & 1-0 \end{bmatrix}$$

$$= \mathbf{I}$$

Resulting in an $n \times n$ matrix where all entries on the main diagonal are one whilst all other entries are zero, which is the identity matrix $\mathbf{I}$. Therefore, $\mathrm{d}\mathbf{v}(\mathbf{C}) = \mathbf{I}$.

Answer b:

The Lie brackets $[\mathbf{f}(\mathbf{C}), \mathbf{v}(\mathbf{C})]$ and $[\mathbf{r}(\mathbf{C}), \mathbf{v}(\mathbf{C})]$ are defined as

$$[\mathbf{f}(\mathbf{C}), \mathbf{v}(\mathbf{C})] = \mathrm{d}\mathbf{v}(\mathbf{C})\mathbf{f}(\mathbf{C}) - \mathrm{d}\mathbf{f}(\mathbf{C})\mathbf{v}(\mathbf{C}) \qquad (6.15a)$$

$$[\mathbf{r}(\mathbf{C}), \mathbf{v}(\mathbf{C})] = \mathrm{d}\mathbf{v}(\mathbf{C})\mathbf{r}(\mathbf{C}) - \mathrm{d}\mathbf{r}(\mathbf{C})\mathbf{v}(\mathbf{C}) \qquad (6.15b)$$

Since $\mathbf{f}(\mathbf{C}) = \mathbf{r}(\mathbf{C}) + \alpha\mathbf{v}(\mathbf{C})$, this relation may be substituted into Equation 6.15a to give

$$[\mathbf{f}(\mathbf{C}), \mathbf{v}(\mathbf{C})] = \mathrm{d}\mathbf{v}(\mathbf{C})(\mathbf{r}(\mathbf{C}) + \alpha\mathbf{v}(\mathbf{C})) - \mathrm{d}(\mathbf{r}(\mathbf{C}) + \alpha\mathbf{v}(\mathbf{C}))\mathbf{v}(\mathbf{C})$$

From answer (a) above, $\mathrm{d}\mathbf{v}(\mathbf{C}) = \mathbf{I}$. Substitution and simplification gives

$$[\mathbf{f}(\mathbf{C}), \mathbf{v}(\mathbf{C})] = (\mathbf{r}(\mathbf{C}) + \alpha\mathbf{v}(\mathbf{C})) - \mathrm{d}\mathbf{r}(\mathbf{C})\mathbf{v}(\mathbf{C}) - \alpha\mathbf{v}(\mathbf{C})$$
$$= \mathbf{r}(\mathbf{C}) - \mathrm{d}\mathbf{r}(\mathbf{C})\mathbf{v}(\mathbf{C}) \qquad (6.15c)$$

Observe that this is equivalent to Equation 6.15b. Hence $[\mathbf{f}(\mathbf{C}), \mathbf{v}(\mathbf{C})] = [\mathbf{r}(\mathbf{C}), \mathbf{v}(\mathbf{C})]$.

In the next example, a number of results are provided for several higher dimensional systems. The complex form of the α policies is prohibitive for higher dimensions, and hence they are unrealistic to apply in practice. These results are provided simply for demonstration purposes only. Further details are again obtained from Feinberg (2000a, 2000b).

ILLUSTRATION: Results taken from a paper by Feinberg[11]

Now that the condition for controllability of critical DSR has been identified, it is useful to demonstrate the complexity of the results with a number of different dimensions. For the sake of brevity, we shall adopt an abbreviation scheme similar to that given in Feinberg (1999) by not explicitly indicating that functions $\mathbf{f}(\mathbf{C})$, $\mathbf{v}(\mathbf{C})$, and $\mathbf{z}(\mathbf{C})$ are functions of $\mathbf{C}$ (however, we should not forget that $\mathbf{f}$, $\mathbf{v}$, and $\mathbf{z}$ are still functions of $\mathbf{C}$), so that

$$\mathbf{f} := \mathbf{f}(\mathbf{C}) = \alpha\mathbf{r}(\mathbf{C}) + \mathbf{v}(\mathbf{C})$$
$$\mathbf{v} := \mathbf{v}(\mathbf{C}) = (\mathbf{C} - \mathbf{C}^0)$$
$$\mathbf{z}^{[k]} := \mathbf{z}^{[k]}(\mathbf{C}) = [\mathbf{f}(\mathbf{C}), \mathbf{v}(\mathbf{C})]$$

[11] These results are obtained from Feinberg (1999).

A number of results will now be given. The mathematics required to obtain these results is beyond that required here; however, the interested reader is again referred to the original texts found in Feinberg (1999, 2000a).

For $d = 3$:

For the case of three independent reactions ($d = 3$) in $\mathbb{R}^n$, the determinant function takes on the following form:

$$\mathrm{Det}[\mathbf{f}, \mathbf{v}, \mathbf{z}^{[1]}, \mathbf{N}] = 0$$

Since $d = 3$ and $k = d - 2$, there is only one iterated Lie, $\mathbf{z}^{[1]} = [\mathbf{f}, \mathbf{v}] = (\mathrm{d}\mathbf{v})\mathbf{f} - (\mathrm{d}\mathbf{f})\mathbf{v}$. (Recall that $\mathbf{N}$ has size $n \times (n - d)$, but we have not specified the value of n to keep the discussion general to $\mathbb{R}^n$.) It may be shown, after simplification, that the above expression is given by:

$$\mathrm{Det}[\mathbf{r}, \mathbf{v}, [\mathbf{r}, \mathbf{v}], \mathbf{N}] = 0$$

Expanding the Lie bracket expressions and applying the properties of determinants, one may deduce an expression for the equation α in terms of $\mathbf{r}$ and $\mathbf{v}$. Feinberg (1999) shows that although there is no explicit reference to α in the above expression, one may reason that the critical α policy be given by

$$\alpha = -\frac{\mathrm{Det}[\mathbf{r}, \mathbf{v}, [\mathbf{r}, [\mathbf{r}, \mathbf{v}]], \mathbf{N}]}{\mathrm{Det}[\mathbf{r}, \mathbf{v}, [\mathbf{v}, [\mathbf{r}, \mathbf{v}]], \mathbf{N}]}$$

Provided that $\mathrm{Det}[\mathbf{r}, \mathbf{v}, [\mathbf{v}, [\mathbf{r}, \mathbf{v}]], \mathbf{N}] \neq 0$, the above relation describes the condition for which the side-stream feeding rate for a critical DSR in $\mathbb{R}^n$. This result involves the use of two iterated Lie brackets in $\mathbf{r}$ and $\mathbf{v}$.

For $d = 4$:

The procedure for computing a critical α policy for four independent reactions follows that for $d = 3$; however, the determinant function now involves a second iterated Lie bracket $\mathbf{z}^{[2]}$. Thus, application of Equation 6.13 for $d = 4$ gives:

$$\mathrm{Det}[\mathbf{f}, \mathbf{v}, \mathbf{z}^{[1]}, \mathbf{z}^{[2]}, \mathbf{N}] = 0$$

This expression may be expanded and simplified, giving the following equation in α (Feinberg, 2000a):

$$\mathrm{Det}[\mathbf{r}, \mathbf{v}, [\mathbf{r}, \mathbf{v}], [\mathbf{r}, [\mathbf{r}, \mathbf{v}]], \mathbf{N}] + \alpha \mathrm{Det}[\mathbf{r}, \mathbf{v}, [\mathbf{r}, \mathbf{v}], [\mathbf{v}, [\mathbf{r}, \mathbf{v}]], \mathbf{N}] = 0$$

This expression is explicit in α. Rearranging for α thus gives:

$$\alpha = -\frac{\mathrm{Det}[\mathbf{r}, \mathbf{v}, [\mathbf{r}, \mathbf{v}], [\mathbf{r}, [\mathbf{r}, \mathbf{v}]], \mathbf{N}]}{\mathrm{Det}[\mathbf{r}, \mathbf{v}, [\mathbf{r}, \mathbf{v}], [\mathbf{v}, [\mathbf{r}, \mathbf{v}]], \mathbf{N}]}$$

The α expression for when $d = 4$ now involves determinant calculations in terms of $\mathbf{r}$, $\mathbf{v}$, $[\mathbf{r}, \mathbf{v}]$ and $[\mathbf{r}, [\mathbf{r}, \mathbf{v}]]$, provided that the denominator $\mathrm{Det}[\mathbf{r}, \mathbf{v}, [\mathbf{r}, \mathbf{v}], [\mathbf{v}, [\mathbf{r}, \mathbf{v}]], \mathbf{N}] \neq 0$. This is somewhat more complex than the case for three independent reactions; however, it is still possible to solve for α explicitly in this instance.

For $d = 5$:

In the case of $d = 5$, Equation 6.13 takes on the form:

$$\mathrm{Det}[\mathbf{f}, \mathbf{v}, \mathbf{z}^{[1]}, \mathbf{z}^{[2]}, \mathbf{z}^{[3]}, \mathbf{N}] = 0$$

which now involves a third iterated Lie bracket in the form of $\mathbf{z}^{[3]}$. Feinberg (2000a) gives the result for this involving not only higher order powers of α (terms involving α^3 are involved), but also the first derivative of α in terms of the concentration vector. The specific form of the function may be found in Feinberg (2000a).

$$\frac{\mathrm{d}\alpha}{\mathrm{d}\tau} = -\frac{(\alpha^3 y_1 + \alpha^2 y_2 + \alpha y_3 + y_4)}{y_5}$$

where

$$y_1 = \mathrm{Det}[\mathbf{r}, \mathbf{v}, [\mathbf{r}, \mathbf{v}], [\mathbf{v}, [\mathbf{r}, \mathbf{v}]], [\mathbf{v}, [\mathbf{v}, [\mathbf{r}, \mathbf{v}]]], \mathbf{N}]$$

$$y_2 = \mathrm{Det}\, [\mathbf{r}, \mathbf{v}, [\mathbf{r}, \mathbf{v}], [\mathbf{r}, [\mathbf{r}, \mathbf{v}]], [\mathbf{v}, [\mathbf{v}, [\mathbf{r}, \mathbf{v}]]], \mathbf{N}] + 2\mathrm{Det}[\mathbf{r}, \mathbf{v}, [\mathbf{r}, \mathbf{v}], [\mathbf{v}, [\mathbf{r}, \mathbf{v}]], [\mathbf{r}, [\mathbf{v}, [\mathbf{r}, \mathbf{v}]]], \mathbf{N}]$$

$$y_3 = 2\mathrm{Det}[\mathbf{r}, \mathbf{v}, [\mathbf{r}, \mathbf{v}], [\mathbf{r}, [\mathbf{r}, \mathbf{v}]], [\mathbf{r}, [\mathbf{v}, [\mathbf{r}, \mathbf{v}]]], \mathbf{N}] + \mathrm{Det}[\mathbf{r}, \mathbf{v}, [\mathbf{r}, \mathbf{v}], [\mathbf{v}, [\mathbf{r}, \mathbf{v}]], [\mathbf{r}, [\mathbf{r}, [\mathbf{r}, \mathbf{v}]]], \mathbf{N}]$$

$$y_4 = \mathrm{Det}[\mathbf{r}, \mathbf{v}, [\mathbf{r}, \mathbf{v}], [\mathbf{r}, [\mathbf{r}, \mathbf{v}]], [\mathbf{r}, [\mathbf{r}, [\mathbf{r}, \mathbf{v}]]], \mathbf{N}]$$

$$y_5 = \mathrm{Det}[\mathbf{r}, \mathbf{v}, [\mathbf{r}, \mathbf{v}], [\mathbf{r}, [\mathbf{r}, \mathbf{v}]], [\mathbf{v}, [\mathbf{r}, \mathbf{v}]], \mathbf{N}]$$

Thus for higher dimensions, an explicit expression for the critical α policy is no longer available.

For $d = 6$:

The situation becomes increasingly more complex for $d = 6$. In this instance, Equation 6.13 takes the form:

$$\mathrm{Det}[\mathbf{f}, \mathbf{v}, \mathbf{z}^{[1]}, \mathbf{z}^{[2]}, \mathbf{z}^{[3]}, \mathbf{z}^{[4]}, \mathbf{N}] = 0$$

These calculations involve a large number of determinant evaluations on higher order iterated Lie brackets. Feinberg (2000a) shows the resulting expression incorporates second derivatives of α and terms involving α^6. These expressions are impossible to solve for analytically in terms of α explicitly.

Higher dimensions:

The controllability criteria specified in Equation 6.12 is valid for all $d \geq 2$. It is possible, in theory, to apply repeated iterated Lie brackets of the determinant function for as large a system as required. Although for larger systems, application of Equation 6.13 involves higher order evaluations of $\mathbf{f}(\mathbf{C})$, $\mathbf{v}(\mathbf{C})$, and $\mathbf{z}^{[k]}$. The determinant function becomes increasingly more complex in higher dimensions as a result.

The unfortunate truth is that it is difficult to solve for critical α expressions analytically, even with concrete knowledge of the controllability conditions. Furthermore, repeated higher order derivatives of $\mathbf{r}$ bring into question the underlying reliability of the rate expressions themselves (rate functions are typically based on experiment and therefore only valid to within a certain range of accuracy). Nevertheless, these are the conditions for which a critical DSR must abide. It is still valuable to have knowledge of these points, even if they are impractical to achieve in reality, for they serve as targets from which performance benchmarks may be performed.

b) Simplification for d = *3* In the previous sections, we have shown how conditions for critical α policies may be computed by use of iterated Lie brackets. These arise out of a controllability condition that must be enforced in order for a DSR to produce effluent concentrations that lie on the AR boundary. Although these conditions are concrete, the associated calculations are complex. If the dimension of system is three ($d = 3$), then an alternate method for computing critical α policies may be carried out.

To compute a critical α policy in $\mathbb{R}^3$ specifically, the gradient of the DSR trajectory must lie in the plane spanned by $\mathbf{r}(\mathbf{C})$ and $\mathbf{v}(\mathbf{C})$. The Jacobian matrix $d\mathbf{r}(\mathbf{C})$ of the rate function describes the gradient at $\mathbf{C}$, whereas the matrix product $d\mathbf{r}(\mathbf{C})\mathbf{v}$ provides a measure of the gradient of the DSR trajectory in the direction of $\mathbf{v}(\mathbf{C})$. Vectors $\mathbf{v}(\mathbf{C})$, $\mathbf{r}(\mathbf{C})$, and $d\mathbf{r}(\mathbf{C})\mathbf{v}(\mathbf{C})$ must all lie in the same plane (they must be coplanar) on a critical DSR trajectory.

CONCEPT: Condition for critical DSRs in $\mathbb{R}^3$ specifically

Vectors $\mathbf{v}(\mathbf{C})$, $\mathbf{r}(\mathbf{C})$, and $d\mathbf{r}(\mathbf{C})\mathbf{v}$ must all be coplanar on the AR boundary. This provides a method to compute α directly.

A condition involving these three vectors may be formulated using vector products, specific to three dimensions. Orientation of a plane can be defined by its normal vector $\mathbf{n}$ (a vector perpendicular to the plane). Since $\mathbf{r}(\mathbf{C})$ and $\mathbf{v}(\mathbf{C})$ span a two-dimensional plane, a vector cross product involving $\mathbf{r}(\mathbf{C})$ and $\mathbf{v}(\mathbf{C})$ may be applied to find $\mathbf{n}$. Hence, $\mathbf{n} = \mathbf{v}(\mathbf{C}) \times \mathbf{r}(\mathbf{C})$. If the vector $d\mathbf{r}(\mathbf{C})\mathbf{v}(\mathbf{C})$ is to lie in the plane defined by $\mathbf{n}$, then $d\mathbf{r}(\mathbf{C})\mathbf{v}(\mathbf{C})$ must be orthogonal to $\mathbf{n}$. We may express this condition using inner products:

$$[d\mathbf{r}(\mathbf{C})\mathbf{v}(\mathbf{C})]^{\mathrm{T}}\mathbf{n} = 0$$

Next, we define the function $\varphi(\mathbf{C})$ as

$$\varphi(\mathbf{C}) = [d\mathbf{r}(\mathbf{C})\mathbf{v}(\mathbf{C})]^{\mathrm{T}}\mathbf{n}$$

Substitution of $d\mathbf{r}(\mathbf{C})$, $\mathbf{r}(\mathbf{C})$, and $\mathbf{v}(\mathbf{C})$ and simplification of this expression produces the following condition:

$$\varphi(\mathbf{C}) = [d\mathbf{r}(\mathbf{C})(\mathbf{C} - \mathbf{C}^0)]^{\mathrm{T}}[(\mathbf{C} - \mathbf{C}^0) \times \mathbf{r}(\mathbf{C})] = 0$$

This describes a surface in $\mathbb{R}^3$ for which compositions of a critical DSR trajectory must lie on. Observe that all critical DSR effluent compositions must be coordinated in such a way as to maintain $\varphi(\mathbf{C})$ at zero along the entire solution trajectory. This criterion leads to the condition

$$[\nabla\varphi(\mathbf{C})]^{\mathrm{T}}\frac{d\mathbf{C}}{d\tau} = 0$$

$d\mathbf{C}/d\tau$ is the vector of differential equations describing the DSR expression. Substitution of the DSR expression into the above results in the following final form:

$$\alpha(\mathbf{C}) = \frac{[\nabla\varphi(\mathbf{C})]^{\mathrm{T}}\mathbf{r}(\mathbf{C})}{[\nabla\varphi(\mathbf{C})]^{\mathrm{T}}(\mathbf{C} - \mathbf{C}^0)} \tag{6.16}$$

Equation 6.16 expresses the critical α policy for a DSR on the boundary of the AR in terms of the rate function $\mathbf{r}(\mathbf{C})$ and mixing point $\mathbf{C}^0$. This condition is specific to $\mathbb{R}^3$. In the presence of a suitable mixing point and reaction kinetics, Equation 6.16 may be used to compute a critical α policy.

CONCEPT: Critical α policies in three dimensions—the "vDelR" condition

If we wish to compute critical α policies for three-dimensional systems (three independent reactions), we can use the following equations:

$$\alpha(\mathbf{C}) = \frac{[\nabla\varphi(\mathbf{C})]^{\mathrm{T}}\mathbf{r}(\mathbf{C})}{[\nabla\varphi(\mathbf{C})]^{\mathrm{T}}(\mathbf{C} - \mathbf{C}^0)} \tag{6.16}$$

where

$$\varphi(\mathbf{C}) = [d\mathbf{r}(\mathbf{C})(\mathbf{C} - \mathbf{C}^0)]^{\mathrm{T}}[(\mathbf{C} - \mathbf{C}^0) \times \mathbf{r}(\mathbf{C})]$$

which are simpler than the use of determinants and Lie brackets. These equations collectively referred to as the *"vDelR" condition.*

EXAMPLE 12: Critical α policy for three-dimensional Van de Vusse kinetics

In the next chapter, we shall plot critical DSR trajectories for three-dimensional Van de Vusse kinetics. The calculations carried out here will prove useful later on. The reactions are given by:

$$A \rightarrow B \rightarrow C$$

$$2A \rightarrow D$$

with the rate vector:

$$\mathbf{r}(\mathbf{C}) = \begin{bmatrix} r_A(\mathbf{C}) \\ r_B(\mathbf{C}) \\ r_D(\mathbf{C}) \end{bmatrix} = \begin{bmatrix} -c_A - 20c_A^2 \\ c_A - c_B \\ 10c_A^2 \end{bmatrix}$$

Rate constants have already been substituted into the rate expressions for convenience (they are $k_1 = 1\ s^{-1}$, $k_2 = 1\ s^{-1}$, and $k_3 = 10\ L/(mol.s)$). Determine an expression for the critical α policy if the DSR side-stream mixing composition is given by $\mathbf{C}^0 = [1, 0, 0]^T$ mol/L.

Answer:

The Jacobian of the rate vector, $d\mathbf{r}(\mathbf{C})$, and mixing vector $\mathbf{v}(\mathbf{C})$ may be computed giving

$$d\mathbf{r}(\mathbf{C}) = \begin{bmatrix} -40c_A - 1 & 0 & 0 \\ 1 & -1 & 0 \\ 20c_A & 0 & 0 \end{bmatrix}$$

and

$$\mathbf{v}(\mathbf{C}) = \begin{bmatrix} c_A - 1 \\ c_B \\ c_D \end{bmatrix}$$

The normal vector $\mathbf{n}(\mathbf{C})$ may also be computed from the vector cross product of $\mathbf{v}(\mathbf{C})$ and $\mathbf{r}(\mathbf{C})$:

$$\mathbf{n}(\mathbf{C}) = \mathbf{v}(\mathbf{C}) \times \mathbf{r}(\mathbf{C})$$

$$= \begin{bmatrix} 10c_A^2 c_B - c_D\left(c_A - c_B\right) \\ 10c_A^2(1 - c_A) - c_A c_D(20c_A + 1) \\ 20c_A^2 c_B + c_A^2 - c_A + c_B \end{bmatrix}$$

The function $\varphi(\mathbf{C})$ may now be determined from

$$\varphi(\mathbf{C}) = [d\mathbf{r}(\mathbf{C})\mathbf{v}(\mathbf{C})]^T \mathbf{n}(\mathbf{C})$$

upon simplification

$$\varphi(\mathbf{C}) = 10c_A^4 + 20c_A^3 c_D - 20c_A^3 - 20c_A^2 c_B c_D + 20c_A^2 c_B - 20c_A^2 c_D + 10c_A^2 + 40c_A c_B c_D - 20c_A c_B + c_B c_D$$

From this expression, $\nabla\varphi(\mathbf{C})$ can be determined:

$$\nabla\varphi(\mathbf{C}) = \left[y_1\ y_2\ y_3\right]^T$$

where y_1, y_2, and y_3 are functions given by

$$y_1 = 20(2c_A^3 + 3c_A^2c_D - 3c_A^2 - 2c_Ac_Bc_D + 2c_Ac_B - 2c_Ac_D + c_A + 2c_Bc_D - c_B)$$

$$y_2 = -20c_A^2c_D + 20c_A^2 + 40c_Ac_D - 20c_A + c_D$$

$$y_3 = 20c_A^3 - 20c_A^2c_B - 20c_A^2 + 40c_Ac_B + c_B$$

These results may then be used to compute an expression for α in terms of $\nabla\varphi(\mathbf{C})$, $\mathbf{r}(\mathbf{C})$, and $(\mathbf{C}-\mathbf{C}^0)$, given by Equation 6.16

$$\alpha(\mathbf{C}) = \frac{[\nabla\varphi(\mathbf{C})]^T\mathbf{r}(\mathbf{C})}{[\nabla\varphi(\mathbf{C})]^T(\mathbf{C}-\mathbf{C}^0)}$$

The arithmetic is fairly lengthy to perform by hand, and the use of a CAS package is employed to compute the result. After substitution and simplification, $\alpha(\mathbf{C})$ takes the following form

$$\alpha(\mathbf{C}) = \frac{c_A(20c_A^3c_B - 80c_A^2c_B - c_A^2 + 37c_Ac_B + c_A + 2c_B)}{2c_B(c_A - 1)^2}$$

This expression may be substituted directly into the DSR expression along with the mixing vector $\mathbf{C}^0 = [1, 0, 0]^T$. In this instance, observe that $\alpha(\mathbf{C})$ is undefined when $c_B = 0$ mol/L or $c_A = 1$ mol/L, and thus $\alpha(\mathbf{C})$ is undefined at $\mathbf{C}^0$ for the particular kinetics and $\mathbf{C}^0$ used.

The DSR solution trajectory employing the α policy is one that lies entirely on the AR boundary, which is a critical DSR trajectory for the three-dimensional Van de Vusse system.

Figure 6.14 shows the DSR trajectory using the critical α expression supplied above. The AR has also been plotted for comparison. The trajectory lies on the AR boundary at every point along its path. Notice that although there are four components in the system, computation of the critical α policy by Equation 6.16 requires vectors in $\mathbb{R}^3$. The fourth component may be computed by mass balance.

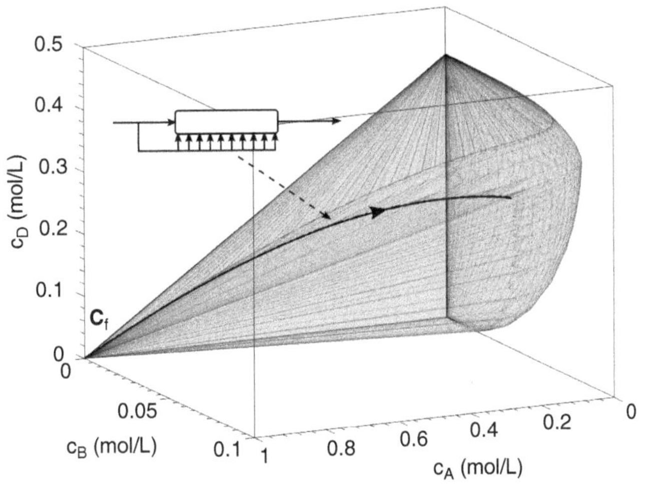

Figure 6.14 Critical DSR trajectory for the Van de Vusse system.

SIDE NOTE: The CSTR tangent vector t

It is sometimes useful to understand the behavior of the CSTR locus in terms of it's derivative with respect to residence time τ. Consider then the standard CSTR expression

$$\mathbf{C} = \mathbf{C}_f + \tau\mathbf{r}(\mathbf{C})$$

For a specified feed concentration $\mathbf{C}_f$, the CSTR locus is parameterized by τ. Specifying a range of values for τ traces out a locus of CSTR steady state solutions in concentration space. The CSTR expression may be differentiated with respect to τ to give a relation between the rate function and the gradient of the CSTR locus as follows

$$\begin{aligned}\frac{d\mathbf{C}}{d\tau} &= \frac{d\mathbf{C}_f}{d\tau} + \frac{d}{d\tau}[\tau\mathbf{r}(\mathbf{C})] \\ &= \frac{d\mathbf{C}_f}{d\tau} + \mathbf{r}(\mathbf{C}) + \tau\frac{d\mathbf{r}}{d\tau}\end{aligned}$$

Recognizing that $\mathbf{C}_f$ is constant and that $d\mathbf{r}/d\tau = (d\mathbf{r}/d\mathbf{C})(d\mathbf{C}/d\tau)$ gives

$$\frac{d\mathbf{C}}{d\tau} = \mathbf{r}(\mathbf{C}) + \tau d\mathbf{r}\frac{d\mathbf{C}}{d\tau}$$

$d\mathbf{r}$ is the familiar shorthand notation for the Jacobian matrix of $\mathbf{r}(\mathbf{C})$.

$$d\mathbf{r} = \frac{d\mathbf{r}}{d\mathbf{C}} = \begin{bmatrix} \frac{\partial r_1(\mathbf{C})}{\partial c_1} & \cdots & \frac{\partial r_1(\mathbf{C})}{\partial c_n} \\ \vdots & \ddots & \vdots \\ \frac{\partial r_n(\mathbf{C})}{\partial c_1} & \cdots & \frac{\partial r_n(\mathbf{C})}{\partial c_n} \end{bmatrix}$$

The derivative $d\mathbf{C}/d\tau$ represents the tangent vector to the CSTR locus. Factorizing out $d\mathbf{C}/d\tau$ gives

$$[\mathrm{I} - \tau d\mathbf{r}]\frac{d\mathbf{C}}{d\tau} = \mathbf{r}(\mathbf{C})$$

or

$$\frac{d\mathbf{C}}{d\tau} = [\mathrm{I} - \tau d\mathbf{r}]^{-1}\mathbf{r}(\mathbf{C})$$

We often reserve the symbol **t** to represent the CSTR tangent vector so that

$$\mathbf{t} = \frac{d\mathbf{C}}{d\tau} = [\mathrm{I} - \tau d\mathbf{r}]^{-1}\mathbf{r}(\mathbf{C}) \quad (6.17)$$

Note that this expression is independent of the feed point and is a function of only $\mathbf{r}(\mathbf{C})$ and τ. Equation 6.17 is valid for any branch of a CSTR locus, even if the particular locus under investigation exhibits multiple steady states. Since multiple effluent concentrations may exist for a single residence time, there is a unique tangent vector for each CSTR steady state effluent concentration **C**.

t will be employed in Chapter 7 to help define a condition for a critical DSR trajectory for a nonisothermal system in $\mathbb{R}^3$.

6.4.4 Computing Critical CSTR Points

A CSTR that produces effluent concentrations on the AR boundary is termed a *critical CSTR*. From Section 6.3.3 it is known that PFRs form the final path to the outermost limits of the AR boundary, whilst DSR trajectories and CSTR points form connectors to these trajectories. Similar to critical DSR trajectories, it is possible to mathematically describe conditions for critical CSTRs to exist. Since both CSTRs and DSRs act as connectors on the AR boundary, the underlying theory for computing critical CSTR solutions is closely related to the theory of critical DSRs.

6.4.4.1 Rocket Example Revisited Consider what might occur if there is now free choice to specify the rocket's initial position and state from Section 6.4.3. If the rocket is launched from a location that coincides with a controllable state (in other words, if the initial state of the rocket is not associated with the **E** matrix having less than full rank), then it is possible to steer the rocket locally for all small inputs. Conversely, if an initial state is chosen that corresponds to an uncontrollable state (where the **E** matrix does not have full rank), the rocket will not be locally controllable with the given control inputs.

Similarly, if the initial condition for a DSR is one that coincides with a critical DSR solution trajectory, then the DSR is uncontrollable for all inputs, and it must lie on the AR boundary. Since CSTRs operate at discrete points in space for a set feed and τ, these might be used as a way of transitioning from an initial feed state to one on the AR boundary. CSTRs thus facilitate "jumping" between different points in space.

Suppose that a CSTR is constructed that produces an effluent concentration $\mathbf{C}^*$ on the AR boundary. If a DSR is connected in series with the CSTR as in Figure 6.15, then the DSR must be critical. Hence, a necessary condition for a critical CSTR is that the corresponding DSR must lie on the AR boundary, and thus the DSR must be critical as defined in Section 6.4.3.3. To determine critical CSTR points for a system, the controllability criteria for a critical *DSR* must also hold.

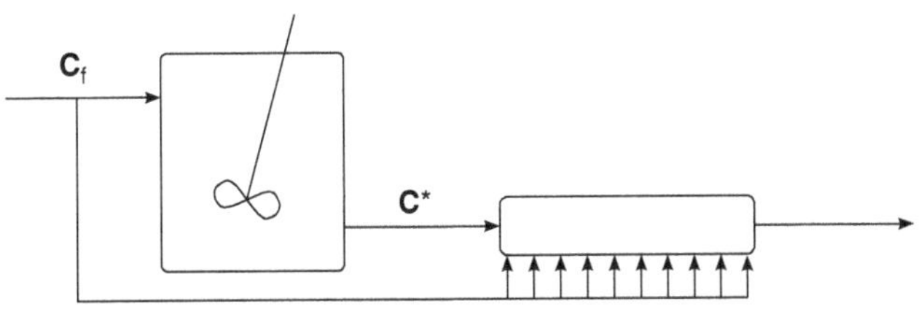

Figure 6.15 A critical CSTR connected to a DSR.

OBSERVATION: Observation: Condition for a critical CSTR

Critical CSTRs are determined by the same condition used to compute critical DSRs. A critical CSTR cannot be locally controllable on the AR boundary.

Feinberg (1999, 2000b) proves that in order for a CSTR to lie on the AR boundary, the following condition must hold:

$$\text{Det}[(\mathbf{C}-\mathbf{C}_f), \mathbf{dr}(\mathbf{C})(\mathbf{C}-\mathbf{C}_f), (\mathbf{dr}(\mathbf{C}))^2(\mathbf{C}-\mathbf{C}_f), \ldots, (\mathbf{dr}(\mathbf{C}))^{d-1}(\mathbf{C}-\mathbf{C}_f), \mathbf{N}] = 0 \quad (6.18)$$

$\mathbf{C}$ is a critical CSTR effluent composition residing on the AR boundary, and $\mathbf{C}_f$ is an attainable feed composition to the CSTR. The columns of the matrix from which the determinant is computed is formed from repeated matrix products between the mixing vector $(\mathbf{C}-\mathbf{C}_f)$ and higher order powers of the Jacobian matrix $\mathbf{dr}(\mathbf{C})$ for the rate function. This is similar to the determinant function for critical DSRs given by Equation 6.13. d is a positive integer corresponding to the dimension of the AR. Thus, if there are five independent reactions taking part in the system, $d=5$. $\mathbf{N}$ is the matrix with columns that are composed of all vectors that are orthogonal to the stoichiometric subspace (the null space), as defined in Section 6.2.1.4.

Since there are, in general, d independent reactions and n components participating in the system, the dimension of the null space is $n-d$. The size of the matrix in Equation 6.18 is therefore $n \times n$ (the first d columns are formed from powers of $\mathbf{dr}(\mathbf{C})$ and $\mathbf{r}(\mathbf{C})$, whereas the final $n-d$ columns are from the columns of $\mathbf{N}$).

SIDE NOTE: Simplifying algebra

We arrive at Equation 6.18 by substituting the CSTR expression into the critical DSR determinant function, and then simplifying the resulting expression.

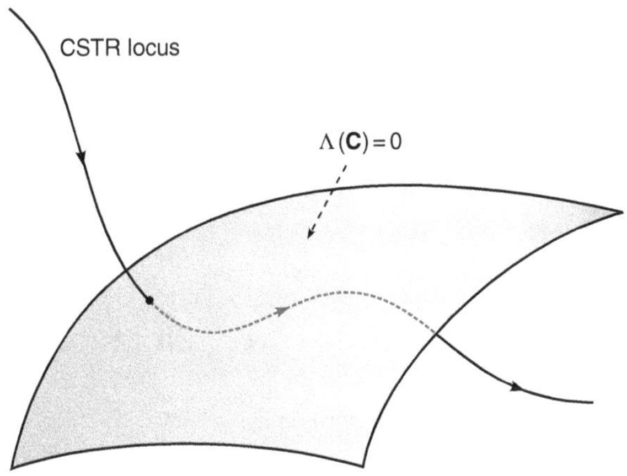

Figure 6.16 Determinant function for a critical CSTR. Only certain points along a residence time range satisfy the condition for a CSTR to be critical.

6.4.4.2 Geometric Significance It is convenient to define the function $\Lambda(\mathbf{C})$ as follows:

$$\Lambda(\mathbf{C}) = \text{Det}[(\mathbf{C}-\mathbf{C}_f), \mathbf{dr}(\mathbf{C})(\mathbf{C}-\mathbf{C}_f), (\mathbf{dr}(\mathbf{C}))^2(\mathbf{C}-\mathbf{C}_f), \ldots, (\mathbf{dr}(\mathbf{C}))^{d-1}(\mathbf{C}-\mathbf{C}_f), \mathbf{N}]$$

The level sets of the $\Lambda(\mathbf{C})$ function described by Equation 6.18 form surfaces in $\mathbb{R}^n$, where n is the dimension of the AR. Note that CSTR effluent compositions on the AR boundary are those that intersect the particular surface defined by $\Lambda(\mathbf{C})=0$. Thus although a CSTR with the correct feed composition may be used to achieve a critical point on the AR boundary, this might only occur at a handful of highly specific residence times, corresponding to points that intersect the surface described by Equation 6.18. This is best described by the sketch given in Figure 6.16.

An equivalent condition for a critical CSTR exists that does not rely on the feed point (Feinberg, 2000b). Substitution of the CSTR expression into Equation 6.18 results in the following equivalent expression after simplification:

$$\Lambda(\mathbf{C}) = \text{Det}[\mathbf{r}(\mathbf{C}), \mathbf{dr}(\mathbf{C})\mathbf{r}(\mathbf{C}), (\mathbf{dr}(\mathbf{C}))^2\mathbf{r}(\mathbf{C}), \ldots, (\mathbf{dr}(\mathbf{C}))^{d-1}\mathbf{r}(\mathbf{C}), \mathbf{N}] = 0 \quad (6.19)$$

This is a useful result for it is possible to compute the critical CSTR condition *without specification of the feed streams*. This comes about as a direct result of the geometric properties of the CSTR—the rate vector evaluated at the CSTR effluent composition is collinear with the mixing point $\mathbf{C}-\mathbf{C}_f$, or

$$\mathbf{C}-\mathbf{C}_f = \tau\mathbf{r}(\mathbf{C})$$

EXAMPLE 13: Proving Equation 6.19

Show that $\Lambda(\mathbf{C})$ written in terms of $(\mathbf{C} - \mathbf{C}_f)$ (Eq. 6.18) is equivalent to $\Lambda(\mathbf{C})$ written in terms of $\mathbf{r}(\mathbf{C})$ (Eq. 6.19). It may be useful to refer to the properties of determinants of matrices, which can be found in Appendix B.

Answer:

Let $\mathbf{M}_1$ be the matrix given by

$$\mathbf{M}_1 = [(\mathbf{C} - \mathbf{C}_f), \mathbf{dr}(\mathbf{C})(\mathbf{C} - \mathbf{C}_f), (\mathbf{dr}(\mathbf{C}))^2(\mathbf{C} - \mathbf{C}_f), \ldots, (\mathbf{dr}(\mathbf{C}))^{d-1}(\mathbf{C} - \mathbf{C}_f), \mathbf{N}]$$

Equation 6.18 may be rewritten as

$$\Lambda(\mathbf{C}) = \mathrm{Det}[\mathbf{M}_1] = 0$$

$\mathbf{M}_1$ has columns composed of powers of $\mathbf{dr}(\mathbf{C})$ and the mixing vector $(\mathbf{C} - \mathbf{C}_f)$. From the CSTR expression

$$\mathbf{C} - \mathbf{C}_f = \tau\mathbf{r}(\mathbf{C})$$

the term $\tau\mathbf{r}(\mathbf{C})$ can be substituted in place of $(\mathbf{C} - \mathbf{C}^0)$ in all of the columns in $\mathbf{M}_1$ where it appears. Upon substitution this results in

$$\mathbf{M}_2 = [\tau\mathbf{r}(\mathbf{C}), \tau\mathbf{dr}(\mathbf{C})\mathbf{r}(\mathbf{C}), \tau(\mathbf{dr}(\mathbf{C}))^2\mathbf{r}(\mathbf{C}), \ldots, \tau(\mathbf{dr}(\mathbf{C}))^{d-1}\mathbf{r}(\mathbf{C}), \mathbf{N}]$$

From the properties of determinants, a scalar multiplied by a column can be factored out of the determinant. Since there are d occurrences of the scalar value τ in the first d columns of the matrix, taking the determinant of $\mathbf{M}_2$ therefore gives

$$\begin{aligned}\mathrm{Det}[\mathbf{M}_2] &= \mathrm{Det}[\tau\mathbf{r}(\mathbf{C}), \tau\mathbf{dr}(\mathbf{C})\mathbf{r}(\mathbf{C}), \tau(\mathbf{dr}(\mathbf{C}))2\mathbf{r}(\mathbf{C}), \ldots, \tau(\mathbf{dr}(\mathbf{C}))^{d-1}\mathbf{r}(\mathbf{C}), \mathbf{N}]\\ &= \tau^d\mathrm{Det}[\mathbf{r}(\mathbf{C}), \mathbf{dr}(\mathbf{C})\mathbf{r}(\mathbf{C}), (\mathbf{dr}(\mathbf{C}))^2\mathbf{r}(\mathbf{C}), \ldots, (\mathbf{dr}(\mathbf{C}))^{d-1}\mathbf{r}(\mathbf{C}), \mathbf{N}]\\ &= \tau^d\mathrm{Det}[\mathbf{M}_1]\end{aligned}$$

since for a critical CSTR

$$\mathrm{Det}[\mathbf{M}_1] = 0$$

setting the determinant of $\mathbf{M}_2$ to zero thus gives

$$\mathrm{Det}[\mathbf{M}_2] = \tau^d\mathrm{Det}[\mathbf{M}_1] = 0$$

Hence since τ^d is a scalar, it can be eliminated from the expression without interfering with the particular level set defined by $\mathrm{Det}[\mathbf{M}_2] = 0$. The condition $\mathrm{Det}[\mathbf{M}_2] = 0$ is equivalent to $\mathrm{Det}[\mathbf{M}_1] = 0$, and hence Equation 6.18 is equivalent to Equation 6.19. Equation 6.19 may therefore also be used to determine the condition for a critical CSTR. Since Equation 6.19 does not depend on a feed point $\mathbf{C}_f$, this is the preferred method for computing critical CSTRs.

EXAMPLE 14: Calculating the surface of critical CSTR points given by $\Lambda(\mathbf{C})$

We shall use the determinant function given by Equation 6.18 to compute critical CSTR residence times in the next chapter. Similar to that given in Section 6.4.3.3, it is helpful to consider an example problem to become comfortable with the calculations. Let us investigate the following system in $\mathbb{R}^3$:

$$\mathrm{X} \rightarrow \mathrm{Y} \rightarrow \mathrm{Z} \rightarrow \mathrm{W}$$

Component X is reacted in a series of three decomposition reactions to form a component W. We shall denote the species concentrations by lower case letters as follows: $c_X = x$, $c_Y = y$, $c_Z = z$, and $c_W = w$. The rate function is then assumed to obey the following simple rate expression:

$$\mathbf{r}(\mathbf{C}) = \begin{bmatrix} -x & x-y & y-z & z \end{bmatrix}^T$$

The feed is assumed to be pure in species X so that $\mathbf{C}_f = [1, 0, 0, 0]^T$. Compute the normal vector to this space and determine the function of the surface for which effluent compositions of a critical CSTR would need to satisfy.

Answer:

Given the system of reactions, it is possible to determine the stoichiometric coefficient matrix for this system.

$$\mathbf{A} = \begin{bmatrix} -1 & 0 & 0 \\ 1 & -1 & 0 \\ 0 & 1 & -1 \\ 0 & 0 & 1 \end{bmatrix}$$

In order to determine the vectors that are orthogonal to the space spanned by $\mathbf{A}$, we can compute null($\mathbf{A}^T$). This gives the desired normal vector $\mathbf{n}$.

$$\mathbf{n} = \begin{bmatrix} 1 & 1 & 0 & 0 \end{bmatrix}^T$$

$\mathbf{n}$ shall be used in the calculation of the determinant function $\Lambda(\mathbf{C})$. To determine the surface for which critical CSTR compositions must intersect, the determinant function must first be formed. $\Lambda(\mathbf{C})$ also requires the mixing vector $\mathbf{v}$ and matrix exponents of the Jacobian matrix for $\mathbf{r}(\mathbf{C})$, $d\mathbf{r}(\mathbf{C})$. The mixing vector is

$$\begin{aligned} \mathbf{v} &= \mathbf{C} - \mathbf{C}_f \\ &= \begin{bmatrix} x-1 & y & z & w \end{bmatrix}^T \end{aligned}$$

Similarly, the Jacobian matrix $d\mathbf{r}(\mathbf{C})$ is computed in a straightforward manner using partial derivatives of $\mathbf{r}(\mathbf{C})$ as follows:

$$d\mathbf{r}(\mathbf{C}) = \begin{bmatrix} -1 & 0 & 0 & 0 \\ 1 & -1 & 0 & 0 \\ 0 & 1 & -1 & 0 \\ 0 & 0 & 1 & 0 \end{bmatrix}$$

Since the system is three-dimensional, we only require one higher power of $d\mathbf{r}(\mathbf{C})$ in order to define $\Lambda(\mathbf{C})$. The matrix exponent $(d\mathbf{r}(\mathbf{C}))^2 = d\mathbf{r}(\mathbf{C})d\mathbf{r}(\mathbf{C})$ may be computed. This is found to be

$$(d\mathbf{r}(\mathbf{C}))^2 = \begin{bmatrix} 1 & 0 & 0 & 0 \\ -2 & 1 & 0 & 0 \\ 1 & -2 & 1 & 0 \\ 0 & 1 & -1 & 0 \end{bmatrix}$$

From $\mathbf{v}$, $\mathbf{r}(\mathbf{C})$, $d\mathbf{r}(\mathbf{C})$, and $(d\mathbf{r}(\mathbf{C}))^2$, we may find $\Lambda(\mathbf{C})$ from Equation 6.18

$$\text{Det}[\mathbf{v}, d\mathbf{r}(\mathbf{C}), (d\mathbf{r}(\mathbf{C}))^2\mathbf{v}, \mathbf{n}] = \text{Det}\begin{bmatrix} x-1 & 1-x & x-1 & 1 \\ y & x-y-1 & 2+y-2x & 1 \\ z & y-z & x-2y+z-1 & 0 \\ w & z & y-z & 0 \end{bmatrix} = 0$$

From this, we may compute the determinant of the matrix. Again, performing the arithmetic by hand is tedious. A CAS package is utilized instead, giving the following result.

$$\begin{aligned}\Lambda(\mathbf{C}) = &-2wx^2 + 2wxy + wxz + 4wx - wy^2 \\ &- 2wy - wz - 2w - x^2z + xy^2 + 3xyz \\ &+ xz^2 + 2xz - y^3 - y^2z - y^2 - 3yz - z^2 - z = 0\end{aligned}$$

The above expression describes a surface in $\mathbb{R}^4$ for which CSTR effluent compositions must intersect in order to be considered critical CSTR points.

EXAMPLE 15: Critical CSTR for three-dimensional Van de Vusse kinetics

Consider again the Van de Vusse system, given by the following reaction scheme:

$$A \rightarrow B \rightarrow C$$
$$2A \rightarrow D$$

and kinetics

$$\begin{aligned}r_A &= -k_1c_A - 2k_3c_A^2 \\ r_B &= k_1c_A - k_2c_B \\ r_C &= k_2c_B \\ r_D &= k_3c_A^2\end{aligned}$$

Answer the following questions regarding critical CSTRs:

(a) From the supplied reaction stoichiometry, compute the stoichiometric coefficient matrix **A** and null space **N**. What is the dimension of the stoichiometric subspace?
(b) Determine the controllability matrix **E** from Equation 6.18.
(c) Compute the equation describing $\Lambda(\mathbf{C})$.
(d) Plot the surface $\Lambda(\mathbf{C}) = 0$ and the CSTR locus from the feed point $\mathbf{C}_f = [c_{Af}, c_{Bf}, c_{Cf}, c_{Df}]^T = [1, 0, 0, 0]^T$ in $c_A - c_B - c_D$ space. Use values of $k_1 = 1$, $k_2 = 1$, and $k_3 = 10$.

Answer a:
The stoichiometric coefficient matrix **A** is found from the reaction stoichiometry

$$\mathbf{A} = \begin{bmatrix} -1 & 0 & -2 \\ 1 & -1 & 0 \\ 0 & 1 & 0 \\ 0 & 0 & 1 \end{bmatrix}$$

N is determined by computing null($\mathbf{A}^T$), giving

$$\mathbf{N} = \begin{bmatrix} 0.5 & 0.5 & 0.5 & 1 \end{bmatrix}^T$$

The dimension of the null space is 1, seeing as the dimension of the stoichiometric subspace is 3 (rank($\mathbf{A}$) = 3), and there are four components in the system.

Answer b:

The controllability matrix $\mathbf{E}$ is found by powers of $\mathbf{dr}(\mathbf{C})$ and $\mathbf{N}$.

$$\mathbf{E} = \left[\mathbf{r}(\mathbf{C})\ \ \mathbf{dr}(\mathbf{C})\mathbf{r}(\mathbf{C})\ \ \mathbf{dr}(\mathbf{C})^2\mathbf{r}(\mathbf{C})\ \ \mathbf{N}\right]$$

Using the component rate expressions given above, the rate vector $\mathbf{r}(\mathbf{C})$ can be found

$$\mathbf{r}(\mathbf{C}) = \begin{bmatrix} -k_1c_A - 2k_3c_A^2 \\ k_1c_A - k_2c_B \\ k_2c_B \\ k_3c_A^2 \end{bmatrix}$$

We can find $\mathbf{dr}(\mathbf{C})$ by computing the Jacobian of $\mathbf{r}(\mathbf{C})$

$$\mathbf{dr}(\mathbf{C}) = \begin{bmatrix} -4c_Ak_3 - k_1 & 0 & 0 & 0 \\ k_1 & -k_2 & 0 & 0 \\ 0 & k_2 & 0 & 0 \\ 2c_Ak_3 & 0 & 0 & 0 \end{bmatrix}$$

the second and third columns of E can be found by multiplication of $\mathbf{r}(\mathbf{C})$ and $\mathbf{dr}(\mathbf{C})$

$$\mathbf{dr}(\mathbf{C})\mathbf{r}(\mathbf{C}) = \begin{bmatrix} c_A\left(2c_Ak_3 + k_1\right)(4c_Ak_3 + k_1) \\ -c_Ak_1(2c_Ak_3 + k_1) - k_2(c_Ak_1 - c_Bk_2) \\ k_2(c_Ak_1 - c_Bk_2) \\ -2c_A^2k_3(2c_Ak_3 + k_1) \end{bmatrix}$$

$$\mathbf{dr}(\mathbf{C})^2\mathbf{r}(\mathbf{C}) = \begin{bmatrix} -c_A\left(2c_Ak_3 + k_1\right)(4c_Ak_3 + k_1)^2 \\ c_Ak_1(2c_Ak_3 + k_1)(4c_Ak_3 + k_1 + k_2) + k_2^2(c_Ak_1 - c_Bk_2) \\ k_2(-c_Ak_1(2c_Ak_3 + k_1) + k_2(-c_Ak_1 + c_Bk_2)) \\ 2c_A^2k_3(2c_Ak_3 + k_1)(4c_Ak_3 + k_1) \end{bmatrix}$$

These expressions are then substituted into $\mathbf{E}$ along with $\mathbf{N}$ to give the controllability matrix for the system resulting in a 4×4 matrix.

Answer c:

The expression $\Lambda(\mathbf{C})$ is found by taking the determinant of the controllability matrix $\mathbf{E}$

$$\Lambda(\mathbf{C}) = \mathrm{Det}(\mathbf{E})$$

Upon simplification, this gives

$$\mathrm{Det}(\mathbf{E}) = -3.5k_1c_A^3k_2k_3(2c_Ak_3 + k_1)(2c_A^2k_1k_3 - 4c_Ac_Bk_2k_3 - c_Ak_1k_2 - c_Bk_1k_2 + c_Bk_2^2)$$

Therefore

$$\Lambda(\mathbf{C}) = -3.5k_1c_A^3k_2k_3(2c_Ak_3 + k_1)(2c_A^2k_1k_3 - 4c_Ac_Bk_2k_3 - c_Ak_1k_2 - c_Bk_1k_2 + c_Bk_2^2) \tag{6.20a}$$

This expression describes the condition for a critical CSTR. Setting $\Lambda(\mathbf{C}) = 0$ describes a surface in concentration space that a critical CSTR locus must intersect. This may be visualized by plotting $\Lambda(\mathbf{C}) = 0$ along with the CSTR locus and seeing where points from both objects meet. Note that this surface is defined without the need for a feed point. Thus all critical CSTR loci, from any feed, must intersect this surface.

Answer d:

The equation given in Answer c can be plotted in c_A–c_B–c_D space. Since Equation 6.20a is only a function of c_A and c_B, it is satisfied for all values of c_D. Using the supplied rate constants, the Equation 6.20a can be simplified to give

$$\Lambda(\mathbf{C}) = 35c_A^4(-400c_A^2 + 800c_Ac_B + 40c_B + 1) \tag{6.20b}$$

Observe that $\Lambda(\mathbf{C}) = 0$ when

$$c_A = 0$$

or

$$c_B = 0.5\left(c_A - \frac{1}{20}\right)$$

These expressions describe two planes in c_A−c_B−c_D space where $\Lambda(\mathbf{C}) = 0$, which are plotted in Figure 6.17.

The CSTR locus from the feed point is also displayed. Notice how the locus intersects the planes at two points. A CSTR locus from the feed is therefore critical at two points for the three-dimensional Van de Vusse system with the kinetics supplied.

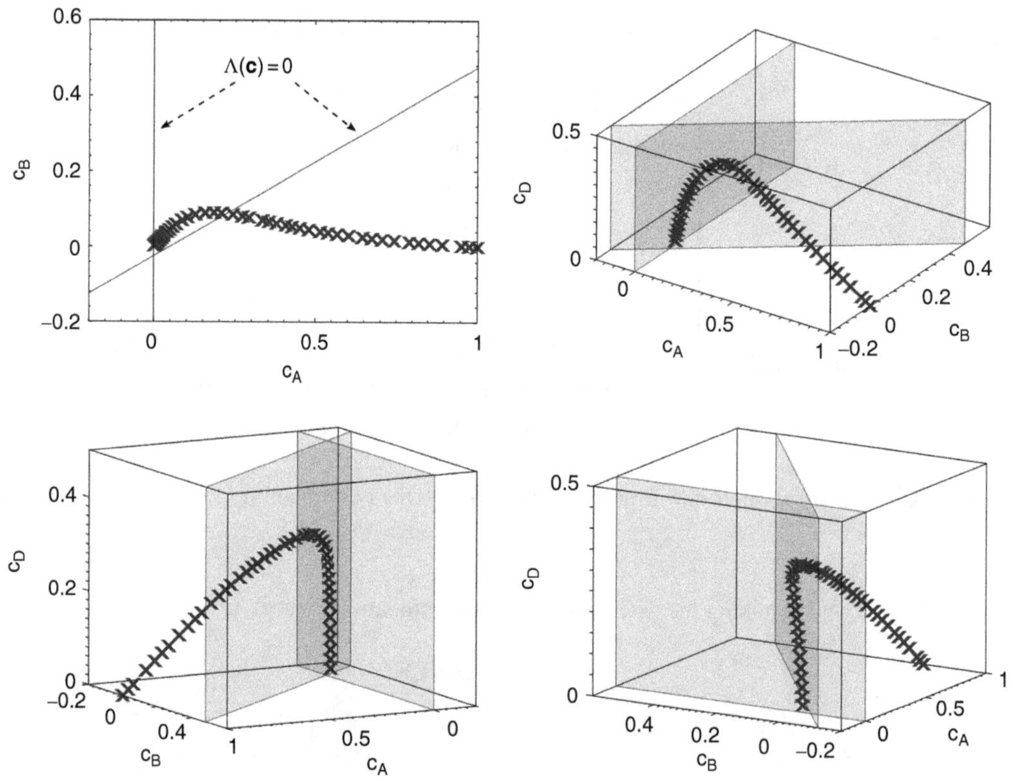

Figure 6.17 Critical CSTR surface ($\Lambda(\mathbf{C}) = 0$) and CSTR locus. The CSTR locus intersects $\Lambda(\mathbf{C}) = 0$ at two points, indicating a CSTR from the feed is critical at two points. (*See color plate section for the color representation of this figure.*)

6.4.4.3 Final Comments Similar to critical DSR policies, the conditions for a critical CSTR may be quite complex, even for very simple kinetics.[12] Nevertheless, one must acknowledge these conditions if a reactor is to lie on the AR boundary. Understanding these conditions guides practical implementation of future designs, for appropriate performance targets can be established, even if the absolute limits are infeasible to achieve for most situations in practice.

6.5 CHAPTER REVIEW

A common misconception concerning the use of AR theory in reactor network synthesis is that it must be performed through visualization of the data, and thus the method is constrained to less than three dimensions. Although visualization of the data greatly assists interpretation and construction of the AR boundary, this is not a requirement. We hope that the results developed in this chapter have helped to motivate this viewpoint.

In this chapter, a number of fundamental topics, essential to our understanding of the AR in higher dimensions have been highlighted. Due to the nonlinear nature of chemical reaction kinetics, and the geometric nature of the AR boundary, higher dimensional AR theory is mathematically complex. We like to see the value of this chapter as assisting to bridge the gap between the very simple and intuitive graphical concepts, found primarily in Section I of this book, and the mathematically rigorous and precise *definition-theorem-proof* style of writing, which is commonly found in published AR literature. We hope that you begin to develop an appreciation from the ingenuity found in these theorems. Although the concepts are mathematically complex, the underlying ideas are basic and intuitive at their core.

It is prudent to understand the nature of the underlying theory so that appropriate application of these properties to difficult reactor optimization problems can be performed. Even if one is not interested in understanding these ideas in detail, it is still powerful to have an awareness of them. Understanding the absolute limits of a system allows us to know *where we stand* in terms of future improvements. Core results in AR theory are often used in practical applications, yet are hardly questioned. Important practical consequences of the theorems are listed as follows:

CONCEPT: Results from this chapter

A number of important results have been described in this chapter. These results are summarized below for ease of reference.

Dimension and complexity of the AR

- We can determine the dimension of the AR based on the number of independent reactions participating in the system. The dimension of the AR is equal to the dimension of the stoichiometric subspace **S**. **S** may be used to form an upper bound on the AR using stoichiometric constraints.
- The stoichiometric subspace is described mathematically as a system of linear inequalities that represent a feasible region in $\mathbb{R}^n$. These inequalities are formed from mass balance and nonnegativity constraints in terms of the stoichiometric coefficient matrix **A** and extent of reaction vector $\boldsymbol{\varepsilon}$.
- As a consequence of Carathéodory's theorem, the maximum number of parallel structures needed to generate the AR is equal to the dimension of the AR (which is computed from rank(**A**)).

Structure of the AR boundary

- The AR is composed entirely of mixing lines (lineations) and manifolds of PFR trajectories (protrusions).
- The AR may be constructed using only combinations of CSTRs, PFRs, and DSRs. PFRs are used to generate the outer extremities of the AR boundary whereas CSTRs and DSRs are used as connectors to these PFRs.

Computation of critical DSR α policies and CSTRs

- Computing critical DSR trajectories and critical CSTR points is based on a lack of controllability.
- Hence, using concepts from geometric control theory, we can state that the condition for a critical DSR is when $\mathrm{Det}(\mathbf{E}) = 0$, where **E** is the controllability matrix for the DSR system.
- For systems in $\mathbb{R}^3$, we can use the "vDelR" condition as a shortcut method to determining critical α policies in a critical DSR.
- For a critical CSTR, the controllability condition is related to the critical DSR, resulting in the final requirement that $\Lambda(\mathbf{C}) = 0$, where $\Lambda(\mathbf{C})$ is the determinant of the controllability matrix specific for the CSTR.
- Computation of $\Lambda(\mathbf{C})$ for a critical CSTR is slightly easier than $\mathrm{Det}(\mathbf{E})$ for a critical DSR because the critical DSR condition involves the use of iterated Lie Brackets (which result in higher derivative functions of α), whereas as the determination of $\Lambda(\mathbf{C})$ only relies on matrix powers of the Jacobian matrix $d\mathbf{r}(\mathbf{C})/d\mathbf{C}$ (which is much easier to solve in practice).

[12]Computation of critical CSTR residence times is slightly less strenuous than that of critical DSR α policies, which require the computation of iterated Lie brackets.

These concepts are crucial to understanding the AR in higher dimensions, when it is no longer easy to visualize the AR boundary in full, ensuring that the correct use of reactor structures is followed.

This discussion effectively completes the theory of attainable regions. (Attainable regions research is still relatively young, and many more results await discovery.) In the next chapter, we will look at applying many of the concepts and ideas discussed here to higher dimensional problems. These examples will mostly be for three-dimensional problems, so that the constructions are still easily visualized.

REFERENCES

Carathéodory, C., 1911. Über den Variabilitätsbereich der Fourier'schen Konstanten von positiven harmonischen Funktionen. Rendiconti Circolo Mat. Palermo 1884–1940 32, 193–217.

Chen, C.-T., 1995. Linear System Theory and Design. Oxford University Press, Inc., Oxford.

Chitra, S.P., Govind, R., 1985. Synthesis of optimal serial reactor structures for homogeneous reactions. Part I: Isothermal reactors. AIChE J. 31, 177–184.

Eckhoff, J., 1993. Helly, Radon, and Carathéodory type theorems. In: P.M. Gruber and J.M. Wills (eds), Handbook of Convex Geometry, Vol A, 389–448.

Feinberg, M., 1999. Recent results in optimal reactor synthesis via attainable region theory. Chem. Eng. Sci. 54, 2535–2543.

Feinberg, M., 2000a. Optimal reactor design from a geometric viewpoint. Part II. Critical sidestream reactors. Chem. Eng. Sci. 55, 2455–2479.

Feinberg, M., 2000b. Optimal reactor design from a geometric viewpoint. Part III. Critical CFSTRs. Chem. Eng. Sci. 55, 3553–3565.

Feinberg, M., 2002. Toward a theory of process synthesis. Ind. Eng. Chem. Res. 41, 3751–3761.

Feinberg, M., Hildebrandt, D., 1997. Optimal reactor design from a geometric viewpoint—I. Universal properties of the attainable region. Chem. Eng. Sci. 52, 1637–1665.

Kalman, R., 1959. On the general theory of control systems. IRE Trans. Autom. Control 4, 110–110.

Kirk, D.E., 2004. Optimal Control Theory: An Introduction. Dover-Publications.com.

Lay, D.C., 2012. Linear Algebra and Its Applications. Addison-Wesley, Boston.

Strang, G., 2003. Introduction to Linear Algebra. SIAM, Philadelphia.

7

APPLICATIONS OF AR THEORY

7.1 INTRODUCTION

In this chapter, we wish to apply aspects of higher dimensional AR theory to a number of problems. These problems are organized into the following three sections:

1. Higher dimensional constructions in concentration space
2. Residence time constructions involving temperature
3. Application of AR theory to batch reactors

These examples are more complex than those described in Chapter 5. In this chapter, we also (finally) answer the BTX problem, which was originally posed in Chapter 1.

7.2 HIGHER DIMENSIONAL CONSTRUCTIONS

7.2.1 Three-Dimensional Van de Vusse Kinetics

Many aspects of the Van de Vusse system have already been discussed in previous chapters—particularly with respect to two-dimensional constructions in Chapter 5 and critical reactors in Chapter 6—but we have not yet given a full description of how the AR for the three-dimensional system can be generated. We now wish to describe the AR construction for this system for two noteworthy reasons:

1. The three-dimensional Van de Vusse system is considered a complete system in AR theory. The AR for the system is well understood and can be generated with confidence. This construction allows us to demonstrate (geometrically) the role that all three fundamental reactor types play in the formation of the AR boundary. It is not always easy to identify, beforehand, whether the true AR has been found or not, and thus this example is convenient because the true AR is guaranteed by theory.
2. Three-dimensional Van de Vusse kinetics has been used extensively in AR research papers in the past. Since the system is well understood, AR practitioners often use the system as an "acid test" for many AR construction algorithms and hypotheses. Understanding this system therefore assists in understanding many research investigations that employ the system, and future research in the field of AR theory is easier to undertake if we are able to understand past work.

7.2.1.1 Problem Statement and State Space Specification

We wish to compute the AR and the corresponding optimal reactor structure for the set of reactions belonging to the Van de Vusse system. The full system involves three independent reactions, assumed to obey mass action kinetics, and is given as follows:

$$\mathrm{A} \xrightarrow{k_1} \mathrm{B} \xrightarrow{k_2} \mathrm{C}$$

$$2\mathrm{A} \underset{k_3}{\rightarrow} \mathrm{D}$$

EXAMPLE 1: Dimension of the Van de Vusse System

In Chapter 6, we described how the dimension of the AR depends on the number of independent reactions participating in the system. Using results from Chapter 6, determine the stoichiometric coefficient matrix **A** and the dimension of the AR for the Van de Vusse system.

Answer:

The particular matrix for the Van de Vusse system is as follows:

$$\mathbf{A} = \begin{bmatrix} -1 & 0 & -2 \\ 1 & -1 & 0 \\ 0 & 1 & 0 \\ 0 & 0 & 1 \end{bmatrix}$$

Attainable Region Theory: An Introduction to Choosing an Optimal Reactor, First Edition. David Ming, David Glasser, Diane Hildebrandt, Benjamin Glasser, and Matthew Metzger.

Companion website: http://attainableregions.com

A in this instance is a 4 × 3 matrix (four components and three reactions), with row 1 corresponding to component A, and row 4 corresponding to component D. Computation of the rank of **A** gives a value of 3, which validates that there are three independent reactions occurring in the system, and that the stoichiometric subspace (and hence also the AR) is a three-dimensional subspace in $\mathbb{R}^4$.

Isothermal operation is assumed. AR constructions for the system frequently occur in c_A–c_B–c_D space. We will hence adopt this convention and define the concentration vector as $\mathbf{C} = [c_A, c_B, c_D]^T$ mol/L. The corresponding rate vector, $\mathbf{r(C)}$, is then defined by the following set of component rate expressions:

$$\mathbf{r(C)} = \left[r_A(\mathbf{C}) \quad r_B(\mathbf{C}) \quad r_D(\mathbf{C})\right]^T = \begin{bmatrix} -k_1c_A - 2k_3c_A^2 \\ k_1c_A - k_2c_B \\ k_3c_A^2 \end{bmatrix}$$

The specific values for the rate constants are taken as those found in Feinberg and Hildebrandt (1997): $k_1 = 1.0\,s^{-1}$, $k_2 = 1.0\,s^{-1}$, and $k_3 = 10.0$ L/(mol.s). A feed concentration must also be specified in order to construct the AR. In keeping with what has traditionally been accomplished for the system, a single feed stream containing pure A is assumed available. The feed vector for construction may then be written as $\mathbf{C_f} = [1, 0, 0]^T$ mol/L.

7.2.1.2 *Initial Construction and First Candidate Region*

It is easiest to begin construction by considering single reactor structures from the feed. The simplest method of achieving this is by computing the PFR trajectory and CSTR locus from $\mathbf{C_f}$. This is equivalent to the procedure performed for the two-dimensional Van de Vusse system in Chapter 5.

a) PFR from the Feed Upon integration of the PFR equation using the feed point as the initial condition, the convex region in Figure 7.1 is obtained.

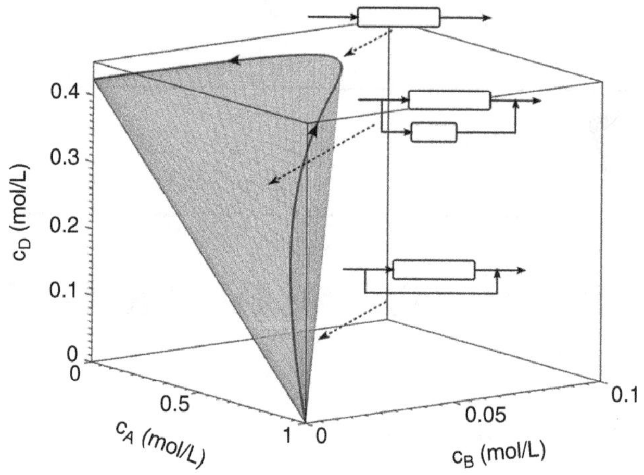

Figure 7.1 Convex hull of PFR trajectory from the feed.

Note that the convex hull belonging to the PFR is narrow. Unlike the two-dimensional Van de Vusse system, where the PFR trajectory separated the space into two regions, the PFR trajectory exists as a one-dimensional curve in $\mathbb{R}^3$ (and also higher dimensions), which no longer divides space.

The region associated with the PFR from the feed is composed of straight-line sections, indicating mixing lines, connected along different sections on the PFR trajectory. Rate vectors are tangent along all points on the PFR solution trajectory, and hence there is no need to check for rate vectors that point out from the curve.

b) CSTR from the Feed Next, the CSTR locus from the feed is generated to identify whether the region may be expanded further. The kinetics associated with this system is fairly easy to solve in that Van de Vusse kinetics does not exhibit multiple steady states. The residence time range for the CSTR locus is made large enough so that the entire range of CSTR effluent concentrations may be captured and plotted. For each residence time specified, a single corresponding CSTR effluent composition is computed. When plotted in concentration space, the set of CSTR effluent concentrations is represented by a single, unbroken locus.

The collection of achievable points obtained from both the PFR trajectory and CSTR locus can be combined and the convex hull for the entire set is then determined, which is shown in Figure 7.2(a). It is evident that the candidate region is enlarged by the inclusion of the CSTR locus.

7.2.1.3 *PFR Trajectories from CSTR Locus Points*

As a further step, it is useful to consider PFRs in series from the CSTR locus. As discussed in Chapter 6, PFRs make up the final approach to the extremities of the AR boundary, and thus it is helpful to include PFR segments as the final reactor type to all structures. From Figure 7.2(a) we can also see that the entire CSTR locus lies on the AR boundary. This suggests that there are rate vectors on the current candidate region that are either tangent or point out of the boundary. In essence, we would like to generate the convex hull for the structure displayed in Figure 7.3.

This is achieved algorithmically by taking the set of CSTR effluent points generated previously as feed points for the PFR expression. For N CSTR points calculated previously, N PFR trajectories may be computed. These points may be appended to the existing set of achievable concentrations and the convex hull for the larger set may again be computed, the results of which are shown in Figure 7.2(b).

The region is again enlarged by the inclusion of this structure. A PFR from the feed generated previously is effectively identical to considering Figure 7.3 with a hypothetical CSTR of zero volume. Observe that the shape of the AR is again composed of PFR trajectories and straight-line

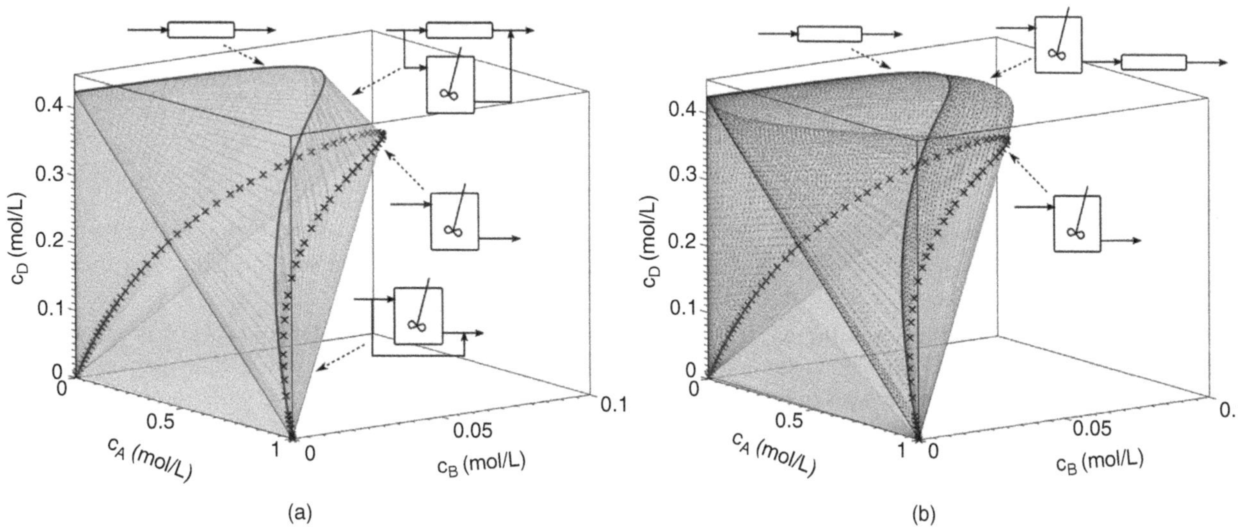

Figure 7.2 (a) Convex hull of a PFR and CSTR from the feed and (b) convex hull of the CSTR–PFR reactor structure from the feed. Note that this is not the full AR for the three-dimensional Van de Vusse system. (*See color plate section for the color representation of this figure.*)

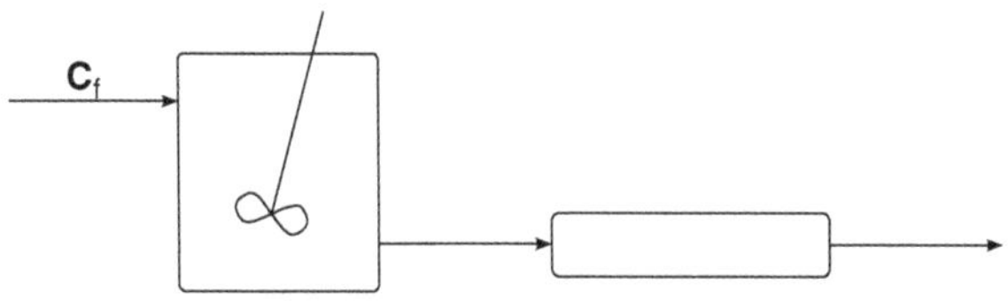

Figure 7.3 A CSTR in series with a PFR from the feed point.

sections, although there is more "curvature" to the boundary points as a result of the manifold of PFR trajectories.

> **SIDE NOTE: Partial Candidate Region for Optimization**
>
> It is acceptable to terminate construction of the AR here, as a preliminary optimization, if we are content with the set of points described by Figure 7.2(b). Since the reactor structure is simple, the implementation and construction of the corresponding reactor network is equally straightforward (the initial reactor arrangement would correspond to Figure 7.3).

Let us consider the role of critical reactors in the formation of the AR boundary. The aim of this example is to provide a complete set of all achievable points, and so we shall continue investigating further expansion of the region. In order to do this, critical reactors must be introduced, which will require us to use ideas and theory developed previously in Chapter 6.

7.2.1.4 Critical CSTRs It is conceptually (and also computationally) easier to begin the expansion of the region by considering critical CSTRs first. This is initiated by computing the controllability matrix **E** for a critical CSTR, previously discussed in Chapter 6.

Note that determining the behavior of critical CSTRs and DSRs requires that the full dimension of the system (i.e., all components as opposed to a subset) be considered. To construct the controllability matrix **E** and compute the determinant function, $\Lambda(\mathbf{C})$, for this system, a rate vector containing all components present in the system must be employed. Therefore, a four-component variant of the rate vector given in Section 7.2.1.1 is utilized, rather than the typical three-component rate vector, which corresponds to the number of independent reactions present in the system. The original rate vector is hence augmented with the rate expression corresponding to component C in the system, so that the following rate vector is obtained:

$$\mathbf{r}'(\mathbf{C}) = \begin{bmatrix} r_A(\mathbf{C}) & r_B(\mathbf{C}) & r_C(\mathbf{C}) & r_D(\mathbf{C}) \end{bmatrix}^T = \begin{bmatrix} -k_1c_A - 2k_3c_A^2 \\ k_1c_A - k_2c_B \\ k_2c_B \\ k_3c_A^2 \end{bmatrix}$$

Vector $\mathbf{r}'(\mathbf{C})$ is different from the standard rate vector, $\mathbf{r}(\mathbf{C})$, in that $\mathbf{r}'(\mathbf{C})$ contains all four components. Similarly, the Jacobian matrix, **J**, corresponding to $\mathbf{r}'(\mathbf{C})$ may be computed as follows:

$$\mathbf{J}(\mathbf{C}) = \begin{bmatrix} -k_1 - 4k_3c_A & 0 & 0 & 0 \\ k_1 & -k_2 & 0 & 0 \\ 0 & k_2 & 0 & 0 \\ 2k_3c_A & 0 & 0 & 0 \end{bmatrix}$$

EXAMPLE 2: Calculating $\Lambda(\mathbf{C})$ for the Van de Vusse System

Determine the expression for the critical CSTR surface $\Lambda(\mathbf{C})$.

Answer:

$\Lambda(\mathbf{C})$ is found from the controllability matrix **E** for the CSTR. To construct **E**, the set of vectors orthogonal to the stoichiometric subspace must be known. This is done by finding a basis for the null space of the vectors spanned by the reaction system given by the stoichiometric coefficient matrix **A**.

From Section 7.2.1.1, the dimension of the AR is three ($d = 3$) and there are four components ($n = 4$). It is expected that the dimension of the subspace orthogonal to the stoichiometric subspace is $(4 - 3) = 1$. Therefore, for the three-dimensional Van de Vusse system, the null space is given by a one-dimensional subspace (in other words, the basis for the null space is composed of a single, linearly independent vector). This is confirmed when the null space of $\mathbf{A}^{\mathrm{T}}$ is computed, giving

$$\mathrm{Null}(\mathbf{A}^{\mathrm{T}}) = \mathbf{n} = \begin{bmatrix}1 & 1 & 1 & 2\end{bmatrix}^{\mathrm{T}}$$

Linear combinations of **n** hence form the set of vectors that are orthogonal to the stoichiometric subspace. From Chapter 6, it is known that the condition for a CSTR to lie on the AR boundary occurs when the controllability matrix **E** does not contain full rank. An expression may be determined for this by computing the determinant of **E** and setting it equal to zero.

$$\mathrm{Det}(\mathbf{E}) = 0$$

$$\mathrm{Det}[\mathbf{r}(\mathbf{C}), \mathrm{d}\mathbf{r}(\mathbf{C}), (\mathrm{d}\mathbf{r}(\mathbf{C}))^2\mathbf{r}(\mathbf{C}), \ldots, (\mathrm{d}\mathbf{r}(\mathbf{C}))^{d-1}\mathbf{r}(\mathbf{C}), \mathbf{N}] = 0$$

Thus for the Van de Vusse system, the determinant function is given by

$$\mathrm{Det}[\mathbf{r}(\mathbf{C}),\ \ \mathrm{d}\mathbf{r}(\mathbf{C}),\ \ (\mathrm{d}\mathbf{r}(\mathbf{C}))^2\mathbf{r}(\mathbf{C}),\ \ \mathbf{n}] = 0$$

Upon expanding this expression and simplifying,[1] the following expression is generated:

$$\Lambda(\mathbf{C}) = -3.5c_A^3 k_2 k_3 (2c_A k_3 + k_1)(2c_A^2 k_1 k_3 - 4c_A c_B k_2 k_3 - c_A k_1 k_2 - c_B k_1 k_2 + c_B k_2^2) \quad (7.1)$$

The level sets of function $\Lambda(\mathbf{C})$ describe surfaces in $\mathbb{R}^3$. From this discussion, critical CSTRs must satisfy the condition $\Lambda(\mathbf{C}) = 0$.

For a CSTR effluent concentration to exist on the AR boundary, the associated CSTR locus must intersect the surface described by $\Lambda(\mathbf{C}) = 0$. It follows that CSTR effluent concentrations may be substituted into $\Lambda(\mathbf{C})$ to determine if the point is a solution to the function. If it is found that a CSTR effluent concentration **C** is a root of $\Lambda(\mathbf{C})$, then **C** must exist on the AR boundary.

Since the CSTR locus at the feed point has been solved in a previous step, straightforward substitution of the locus values into $\Lambda(\mathbf{C})$ may be carried out. Each CSTR solution is associated with a unique residence time. We may therefore graph the value of $\Lambda(\mathbf{C})$ as a function of τ to identify the unique values of τ that produce $\Lambda(\mathbf{C}) = 0$. Thus, instead of computing $\Lambda(\mathbf{C})$ as a function of **C** to obtain a surface in $\mathbb{R}^3$, we plot $\Lambda(\mathbf{C}(\tau))$ versus τ, where τ is the corresponding CSTR residence time associated with **C**. This produces the two-dimensional plot, given in Figure 7.4.

A root of $\Lambda(\mathbf{C})$ exists near $\tau \sim 36.7$ s. The value of $\Lambda(\mathbf{C})$ also appears to approach zero as the value of τ approaches large values. This suggests that the equilibrium CSTR point also acts as a critical CSTR point to the Van de Vusse system. The curve in Figure 7.4 implies that there are exactly two concentrations that lie on the AR boundary. The remaining concentrations do not lie on the boundary and are thus *not* associated with an optimal reactor structure from Chapter 6. PFR trajectories initiated from the CSTR locus not associated with the two critical CSTR solutions therefore also do not form part of the manifold of PFR trajectories

[1] It is easier to use a CAS program, such as the Sympy package in Python, to perform the arithmetic as these controllability expressions are often complicated to derive by hand.

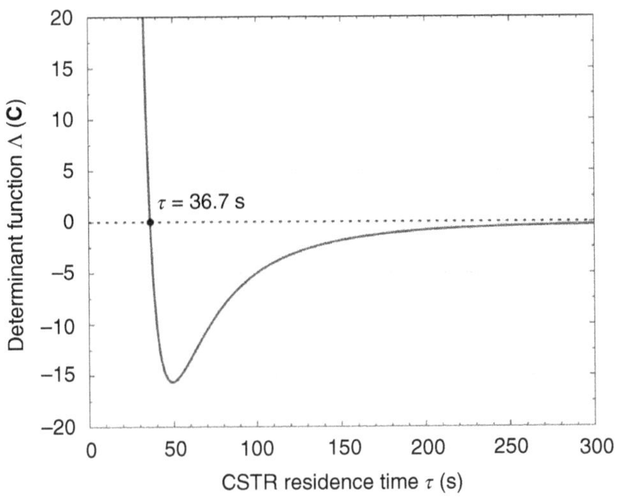

Figure 7.4 The determinant of the controllability matrix E as a function of CSTR residence time from the feed point. Two roots exist: one at $\tau \sim 36.7$ s and the other at the CSTR equilibrium point.

on the true AR boundary. Other structures must therefore exist that constitute the AR boundary. Fortunately, the only remaining fundamental reactor type available is the DSR.

7.2.1.5 ***Critical DSRs*** Let us now investigate the role of DSRs in the formation of the AR boundary. The form of the critical α policy for the DSR is generalized, and therefore potentially various α policies may be computed that all conform to the controllability criteria for critical DSRs.[2] As noted in Chapter 6, sidestream concentrations $\mathbf{C}^0$ used in a critical DSR must originate from points on the AR boundary. This assists in refining the set of compatible concentrations in a DSR. Since the feed point is always available, we often set $\mathbf{C}^0 = \mathbf{C}_f$ for convenience.

A critical DSR profile for the system may now be computed. Since the system under investigation is a three-dimensional problem, the following two possibilities are available for computing the critical α policy for the system:

1. The controllability criterion for the system may be applied by setting the determinant of the **E** matrix to zero and solving this together with the DSR equation.
2. The "vDelR" condition, specific to three-dimensional problems described in Chapter 6, may also be employed.

Often for three-dimensional problems, the second approach is quicker, and so this will be done here. The familiar "vDelR" condition is reproduced as follows:

$$\varphi(\mathbf{C}) = [\mathbf{J}(\mathbf{C})(\mathbf{C} - \mathbf{C}^0)]^{\mathrm{T}}[(\mathbf{C} - \mathbf{C}^0) \times \mathbf{r}(\mathbf{C})]$$

and

$$\alpha(\mathbf{C}) = \frac{[\nabla\varphi(\mathbf{C})]^{\mathrm{T}}\mathbf{r}(\mathbf{C})}{[\nabla\varphi(\mathbf{C})]^{\mathrm{T}}(\mathbf{C} - \mathbf{C}^0)} \tag{7.2}$$

The expression produced is too large to display in terms of the system variables alone. However, Equation 7.2 may be simplified slightly by substituting numerical values for the rate constants k_i and sidestream concentration $\mathbf{C}^0$. The expression given in Feinberg (2000) is then obtained as follows:

$$\alpha(\mathbf{C}) = \frac{c_A(20c_A^3c_B - 80c_A^2c_B - c_A^2 + 37c_Ac_B + c_A + 2c_B)}{c_B(c_A^2 - 2c_A + 1)} \tag{7.3}$$

> **SIDE NOTE: The Specific form of Equation 7.3**
>
> Note that Equation 7.3 is specific to the feed point $\mathbf{C}^0 = \mathbf{C}_f = [1,\ 0,\ 0]^{\mathrm{T}}$ mol/L and the values of the rate constants k that are given in Section 7.2.1.1. If different conditions are used, these values must again be substituted into Equation 7.2 and simplified for the new α expression.

Initial conditions for the DSR are considered next. It is intuitive to initiate a DSR from the feed point, and so we shall start our construction with this approach here. The DSR expression is therefore integrated, together with the critical α policy provided in Equation 7.3, to produce a DSR trajectory. From Equation 7.2, the critical α policy expression is undefined at $\mathbf{C} = \mathbf{C}^0$, and hence the initial concentration for the DSR must be offset slightly from $\mathbf{C}^0$ to ensure that α a positive real number at $\tau = 0$. We can approximate the feed concentration to the DSR that is very near to $\mathbf{C}_f$ using points obtained from either the CSTR or PFR from the feed. (In practice, this is achieved by selecting a point on the PFR trajectory from the feed that is sufficiently close to $\mathbf{C}_f$, such as the second PFR concentration just after the PFR integration has advanced off the feed point.) Hence, an initial DSR concentration of $\mathbf{C}_f = [0.9999,\ 5.0235 \times 10^{-6},\ 5.0235 \times 10^{-5}]^{\mathrm{T}}$ mol/L is supplied as opposed to $[1,\ 0,\ 0]^{\mathrm{T}}$ mol/L. The DSR equation is then integrated, together with the critical α policy, over a sufficiently long integration time to approximate all concentrations between the feed and equilibrium points. The resulting profile is given in Figure 7.5(a). The convex hull corresponding to this candidate is not shown here so that the underlying boundary structures are more easily identified.

[2] In fact the sidestream concentration may even be catered to vary along the length of the reactor, although we never consider this as it further complicates a problem that is already quite complicated.

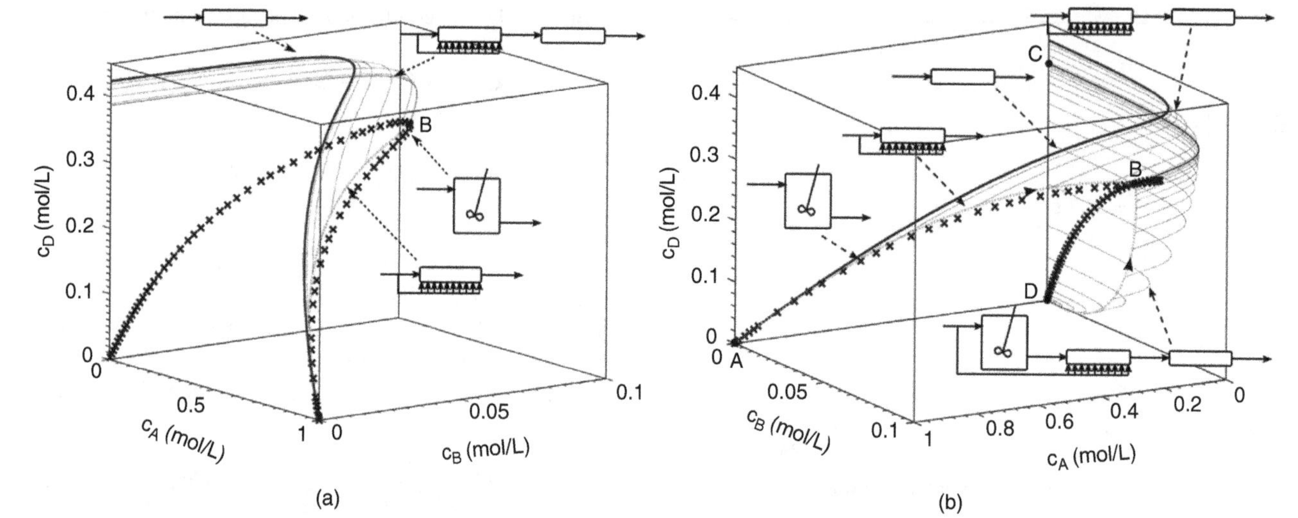

Figure 7.5 (a) Unfilled candidate region for the three-dimensional Van de Vusse kinetics including a critical DSR trajectory from the feed point and (b) the full AR for the three-dimensional Van de Vusse system in c_A–c_B–c_D space. Mixing lines have been removed from the plot to make interpretation of the AR boundary structures easier to identify. (*See color plate section for the color representation of this figure.*)

All of the points on the critical DSR trajectory lie on the AR boundary. PFR trajectories in series with the DSR will therefore serve to fill out a manifold of extreme points on the AR boundary, which are also displayed in Figure 7.5(a).

It is interesting to note that the critical DSR trajectory obtained in Figure 7.5(a) terminates at the critical CSTR point obtained in Section 7.2.1.4. This is labeled as point B in Figure 7.5(a). The second critical CSTR solution, located at the CSTR equilibrium point, does not appear to intersect the DSR trajectory. Since the AR boundary is composed of PFR trajectories connected by critical CSTRs and DSRs, a portion of the AR boundary is missing that joins together the two critical CSTR points obtained previously. Additional DSR structures must exist that connect the two critical points together that are not currently associated with the proposed structure.

Since the feed point $\mathbf{C}_f$ has already been used as an initial condition to the critical DSR that passes through one of two critical CSTR points (the DSR equilibrium concentration is at the critical CSTR concentration), the only critical CSTR concentration available is that obtained at the CSTR equilibrium concentration.

Accordingly, to complete construction of the AR boundary, another DSR trajectory is *initiated at the CSTR equilibrium point.* (Similar to the critical DSR initiated at $\mathbf{C}_f$, this critical DSR trajectory is approximated in practice by choosing a CSTR concentration that is sufficiently close to the CSTR equilibrium point—which may be obtained by solving the CSTR equation for a very large residence time and using the result of the solution as the initial concentration to the critical DSR integration.) When this is performed, the resulting critical DSR trajectory is different from that initiated at $\mathbf{C}_f$: the trajectory appears to connect the two critical CSTR concentrations together. This second DSR trajectory is an additional connector on the AR boundary. PFR trajectories are again initiated from critical DSR trajectory to fill out the remaining underside of the AR boundary. This serves to complete construction of the AR boundary for the three-dimensional Van de Vusse system. The corresponding convex hull of points is given in Figure 7.5(b).

A summary of the optimal structures is given as follows:

1. Structure 1
   A DSR followed by a PFR with sidestream composition given at the feed point $\mathbf{C}_f$. This structure fills out the first part of the AR boundary.
2. Structure 2
   A CSTR operating at the equilibrium point, followed by a DSR and then a PFR. This structure fills out the underside of the AR boundary.

These structures are also shown in Figure 7.6.

Combination of these two reactor structures alone allows for the synthesis of *every possible concentration for*

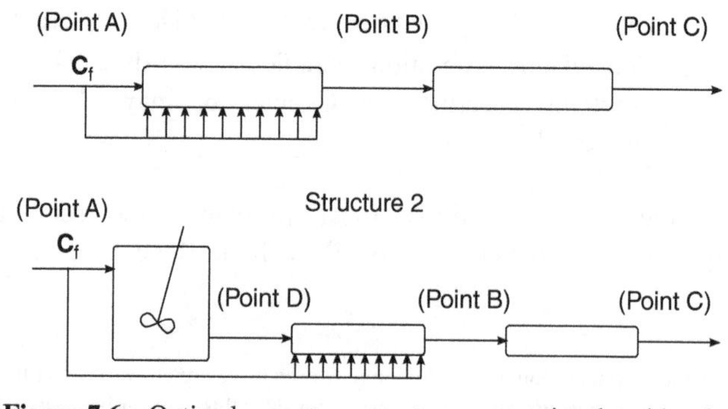

Figure 7.6 Optimal reactor structures associated with the three-dimensional Van de Vusse system. Note that both structures terminate with a PFR.

every possible reactor configuration imaginable. Different optimizations may now be carried out to identify points where the AR boundary and objective function intersect. Once these intersection points have been established, the appropriate optimal reactor structures may be employed to achieve the points of interest.

Note that these structures are consistent with what is expected from the theory developed in Chapter 6:

1. The AR is convex.
2. DSR trajectories and CSTR critical points form connectors to a manifold of PFR extreme points.
3. The boundary of the AR is composed of straight-line sections (indicating mixing) and PFR trajectories only. The final approach to all reactive portions of the boundary is achieved via a PFR.

This concludes the AR construction for the three-dimensional Van de Vusse kinetics. Note that the inclusion of critical CSTRs and DSRs complicates construction, but these structures are required in order to generate the true AR.

EXAMPLE 3: The effect of Different Objective Functions

Now that the AR and the associated optimal structures have been determined, optimization for a number of design scenarios may be investigated. Suppose that it is desired to achieve the maximum amount of intermediate component B. From the previous discussion, both structures 1 and 2, given in Figure 7.6, pass through point B, and therefore both may be used to achieve maximum concentration of component B. The AR for the three-dimensional Van de Vusse system is composed of at most two unique structures, and as a result, all concentrations on the boundary may be achieved by these two structures together with mixing.

(a) Suppose that we wish to limit the formation of component D to a maximum value of 0.3 mol/L whilst still achieving maximum B. What is the optimal reactor structure needed?

(b) Assume that the concentration of component D required is $c_D = 0.4$ mol/L instead. What reactor structure should be used in this scenario if it is desired to still maximize B?

Answer a: the case for $c_D = 0.3$ mol/L

The optimal operating point must lie on the intersection of the AR boundary and the plane described by $c_D = 0.3$ mol/L. Note that the concentration at point B (the critical CSTR point) is approximately $[0.223, 0.08686, 0.316]^T$ mol/L. Although there are a number of points that intersect this plane with the AR, we wish to determine the point that maximizes the concentration of component B. This occurs on the AR boundary at point I. Note that this point corresponds the point of maximum c_B when $c_D = 0.3$ mol/L, and it is different from point B in Figure 7.5(b). Point I is located below point B (for point I contains less c_D than point B), and therefore it is obtained by structure 2 (given by Figure 7.6) and not by structure 1. A CSTR operated at the equilibrium point at point D from the feed, followed by a critical DSR to point H, terminated by a PFR to point I is therefore the required reactor structure. The path ADHI given in Figure 7.7(a) must therefore be followed in order to achieve the desired concentration.

In fact, if it is desired to only achieve the concentration given at point I, then there is no need to generate the entire AR boundary structure and thus structure 1 in Figure 7.6 need not be utilized. Nevertheless, it is still important to understand the optimal reactor structures that form the AR boundary as the optimization of several objective functions may require the use of more than one optimal structure.

Answer b: the case for $c_D = 0.4$ mol/L

In this scenario, the AR intersects the plane at $c_D = 0.4$ mol/L at a number of points. However, we seek to find the unique point on the plane that maximizes c_B. This occurs on the AR boundary at point G in Figure 7.7(b). In this instance, the desired concentration of component D is larger than that obtained at the critical point B. The reactor structure needed to achieve point G is therefore different when $c_D = 0.3$ mol/L described previously as structure 1 (given by Figure 7.6) is now required. This is true even though scenarios (a) and (b) share similar objective functions—0.4 mol/L in component D in scenario (b) compared to 0.3 mol/L in scenario (a). A critical DSR fed at the feed point operated to point F, followed by a PFR to point G is now required. The reduced path AFG shown in Figure 7.7(b) is needed in this instance. Again, if it is required to only achieve point G, then structure 2 need not be considered.

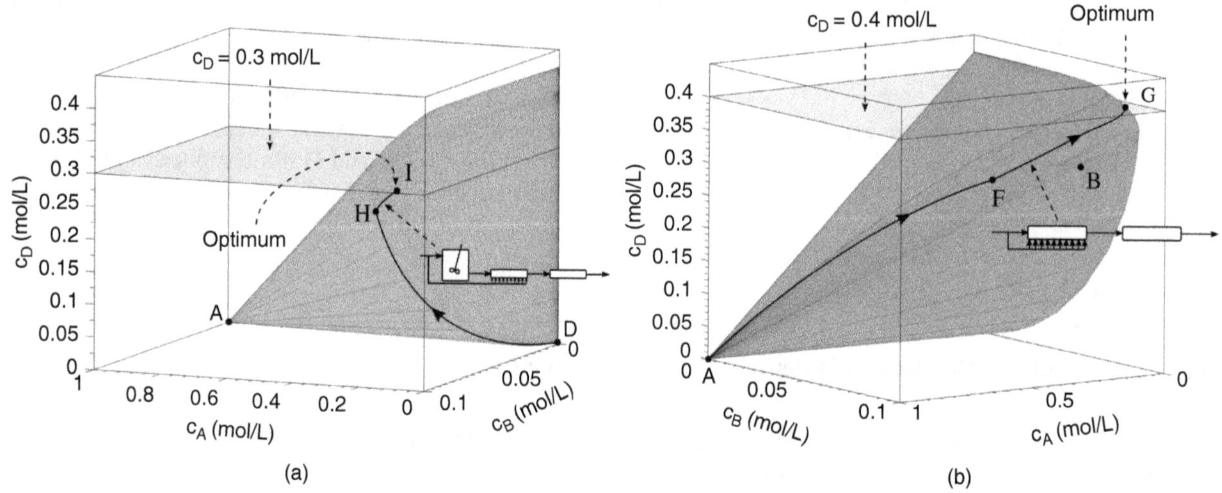

Figure 7.7 The three-dimensional Van de Vusse system for different terminating objective functions: (a) $c_D = 0.3$ mol/L and (b) $c_D = 0.4$ mol/L. (*See color plate section for the color representation of this figure.*)

Nevertheless, since two optimization scenarios have been investigated that each required different optimal structures, an understanding of the AR boundary structures that form the entire region has been beneficial. Under a traditional optimization scenario, two separate optimizations would need to be performed in order to obtain an answer (and also provided that the correct optimal reactor structure is identified in the optimization as well). Hence, not only have we determined the limits of achievability, but we also have an understanding of what reactor structures are required for different optimization scenarios.

7.2.1.6 ***Closing Remarks*** The three-dimensional Van de Vusse system has held an important position in AR research as a model example whereupon many fundamental properties, as well as new ideas in AR research, have been tested against. The system contains all the necessary features from classical AR theory discussed in Chapter 6, which makes it an ideal example for instructional purposes. Understanding how to construct the AR associated with this system also assists with the understanding of more advanced topics, discussed in published scientific literature. Many variants of the Van de Vusse system exist depending on specification of the feed point, rate constants, and the reversibility of reactions. However, all variants share a common construction methodology—the construction methodology that has been detailed in these sections.

7.2.2 BTX Kinetics

7.2.2.1 ***Introduction: An Old Problem Revisited*** In Chapter 1, we introduced the BTX system and posed the following question:

> How do we maximize the amount of toluene produced from a feed of 0.5 moles of ethylene and 1.0 moles of benzene?

Now that we have a full understanding of AR theory, we are in a better position to answer this question. Let us determine the point of maximum toluene concentration, and provide the optimal reactor structure required to achieve the point.

We have already generated the AR in $\mathbb{R}^2$ for this system in Chapter 3, where the procedure shown involved repeated use of batch (PFR) reactions followed by partial mixing with fresh feed for a subsequent batch. This process is somewhat lengthy and impractical in practice.

For the full three-dimensional BTX system, we find that the AR construction procedure is similar to that for Van de Vusse kinetics. Hence, we can summarize the key steps as follows:

1. Specify reaction kinetics and feed point.
2. Compute the PFR trajectory and CSTR locus from the feed.
3. Compute critical CSTR points.
4. Determine the existence of critical DSR trajectories.

7.2.2.2 ***System Specification*** The BTX system is repeated here for convenience. Benzene (B), ethylene (E), and toluene

(T) react to form xylene (X), diphenyl, (D) and hydrogen (H) in the following three reactions:

$$\mathrm{B} + \frac{1}{2}\mathrm{E} \overset{k_1}{\rightarrow} \mathrm{T}$$
$$\mathrm{T} + \frac{1}{2}\mathrm{E} \overset{k_2}{\rightarrow} \mathrm{X}$$
$$2\mathrm{B} \overset{k_3}{\rightarrow} \mathrm{D} + \mathrm{H} \tag{7.4}$$

The molar rates of reaction for all components are assumed to obey the following rate expressions:

$$\begin{aligned}
r_B &= -k_1 c_B c_E^{0.5} - 2k_3 c_B^2 \\
r_E &= -0.5k_1 c_B c_E^{0.5} - 0.5k_2 c_T c_E^{0.5} \\
r_T &= k_1 c_B c_E^{0.5} - k_2 c_T c_E^{0.5} \\
r_X &= k_2 c_T c_E^{0.5} \\
r_D &= k_3 c_B^2 \\
r_H &= k_3 c_B^2
\end{aligned} \tag{7.5}$$

Here, c_i is the molar concentration of component i in the system. Rate constants are specified as $k_1 = 1\,\mathrm{L}^{0.5}/\mathrm{mol}^{0.5}.\mathrm{h}$, $k_2 = 1\,\mathrm{L}^{0.5}/\mathrm{mol}^{0.5}.\mathrm{h}$, and $k_3 = 10\,\mathrm{L/mol.h}$. Since the system is composed of three reactions, it is sufficient to specify information—such as the concentrations and associated rate expressions—for three components only. The remaining components may then be computed by mass balance.

Our objective is to determine the largest toluene concentration using the feed components. As a result, it is sensible to generate the AR in c_B–c_E–c_T space, which provides information related to all significant reactants and products. The concentration vector is then given as $\mathbf{C} = [c_B, c_E, c_T]^T$ mol/L. The AR is therefore a three-dimensional subspace in $\mathbb{R}^6$, and the state of the system is specified by these three concentrations alone.[3] From the original problem statement, a concentration of $c_B = 1.0$ mol/L and $c_E = 0.5$ mol/L is available in the feed. It follows that the feed vector for this system is then $\mathbf{C}_f = [c_{Bf}, c_{Ef}, c_{Tf}]^T = [1, 0.5, 0]^T$ mol/L.

*7.2.2.3 **PFR from the Feed*** To begin, the PFR equation is integrated over the residence time range $0 \le \tau \le 10$ h, using $\mathbf{C}_f$ as the feed. The resulting trajectory is displayed in Figure 7.8.

Although the BTX system has been considered in detail in previous chapters, this is the first time that the full profile has been displayed. As the reaction proceeds in the PFR, the concentration of toluene approaches a maximum value (~0.062 mol/L), and then slowly decreases as toluene is converted into xylene by Equation 7.4.

[3]This result can also be verified by computing the rank of the stoichiometric coefficient matrix **A**. We find that for the BTX system, rank(**A**) = 3.

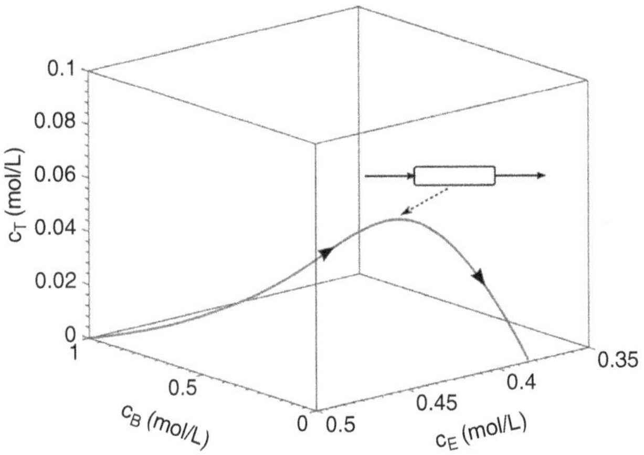

Figure 7.8 PFR trajectory from the feed for the BTX system.

*7.2.2.4 **CSTR from the Feed*** Next, the CSTR equation is solved, using $\mathbf{C}_f$ as the CSTR feed, over the residence time range $0\,\mathrm{h} \le \tau \le 20\ 000\,\mathrm{h}$. When a plot of c_T versus τ is generated, a maximum toluene concentration of approximately 0.074 mol/L is achieved near a CSTR residence time of approximately 1.49 h. Once the resulting CSTR locus has been determined, it may be plotted in conjunction with the PFR trajectory from Section 7.2.2.3. This is given in Figure 7.9(a). Observe that the terminating points for the CSTR locus and PFR trajectory are different.

Guided by the knowledge from Chapter 6 that the extreme points of the AR boundary are composed of PFR trajectories, we can use concentrations belonging to the CSTR locus as feed points for PFRs. The required reactor structure in this instance is then a CSTR followed by a PFR, similar to that given in Figure 7.3 for the three-dimensional Van de Vusse kinetics. The resulting set of achievable concentrations, after the convex hull has been computed for the points, is thus given in Figure 7.9(b). Figure 7.9(b) serves as a candidate AR for the BTX system in $\mathbb{R}^3$.

*7.2.2.5 **Critical CSTRs*** Let us now investigate the existence of any critical CSTR points in the BTX system. The determination of critical CSTR points follows the procedure given in Section 7.2.1.4 for Van de Vusse kinetics. We shall use the CSTR locus from the feed point in this analysis.

ILLUSTRATION: Calculating Λ(C) for BTX Kinetics

To determine whether a CSTR effluent concentration is a critical CSTR point, $\Lambda(\mathbf{C}) = 0$, which depends on the controllability matrix **E**. To find **E**, vectors forming the subspace orthogonal to the stoichiometric subspace

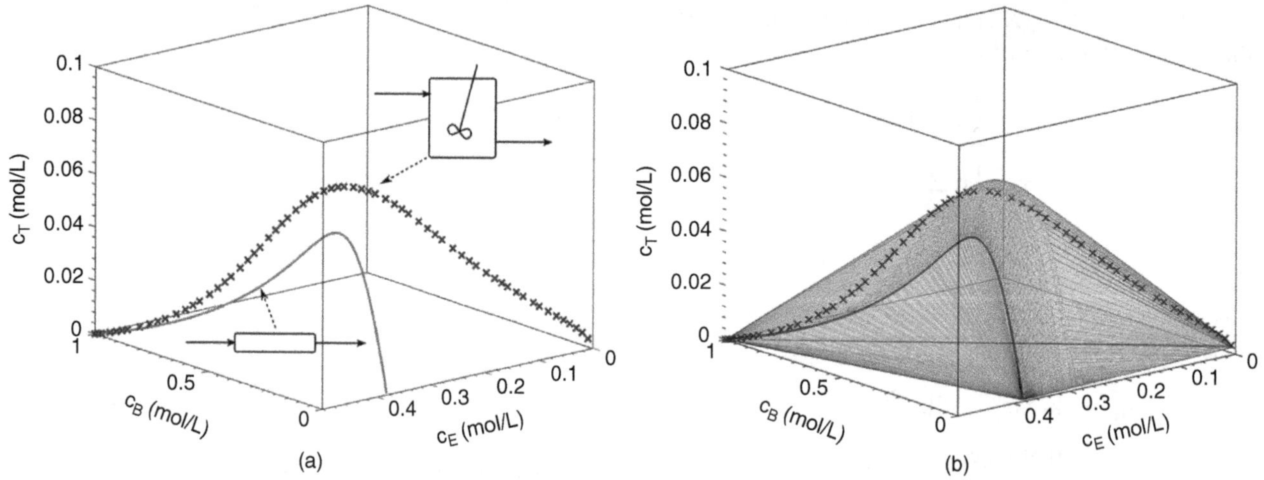

Figure 7.9 (a) CSTR from the feed for the BTX system and (b) candidate AR for a CSTR followed PFR from the feed.

(given by the matrix **N**) must be determined. To achieve this, the stoichiometric coefficient matrix **A** must first be determined. The BTX system involves three reactions in six components. The resulting stoichiometric matrix is therefore a 6 × 3 matrix of the following form:

$$\mathbf{A}=\begin{bmatrix}1 & 0 & -2\\ -0.5 & -0.5 & 0\\ 1 & -1 & 0\\ 0 & 1 & 0\\ 0 & 0 & 1\\ 0 & 0 & 1\end{bmatrix}$$

Each row in **A** represents a component in the BTX system in the following order: (i) benzene, (ii) ethylene, (iii) toluene, (iv) xylene, (v) diphenyl, and (vi) hydrogen. Computing rank(**A**) results in an answer of three, which validates that all three reactions participating in the BTX system are independent. The stoichiometric subspace is hence a three-dimensional subspace residing in $\mathbb{R}^6$. Matrix **N** may be computed from null($\mathbf{A}^T$) (the null space to the stoichiometric subspace). We expect the size of **N** to be 6 × 3, since the total number of species is 6 and rank(**A**) = 3. Performing the null space computation, we find that the following three vectors

$$\mathbf{n}_1=\begin{bmatrix}0 & 1 & 0.5 & 1 & 0 & 0\end{bmatrix}^T$$
$$\mathbf{n}_2=\begin{bmatrix}0.5 & -0.5 & 0.25 & 0 & 1 & 0\end{bmatrix}^T$$
$$\mathbf{n}_3=\begin{bmatrix}0.5 & -0.5 & 0.25 & 0 & 0 & 1\end{bmatrix}^T$$

form a basis for the null space of $\mathbf{A}^T$. It follows that matrix **N** is given by $\mathbf{N}=[\mathbf{n}_1, \mathbf{n}_2, \mathbf{n}_3]^T$. Now that **N** is known, the determinant function for the CSTR locus, $\Lambda(\mathbf{C})$, may be formed. $\Lambda(\mathbf{C})$ for this system then assumes the following form:

$$\Lambda(\mathbf{C}) = \text{Det}[\mathbf{r}(\mathbf{C}),\ \mathbf{dr}(\mathbf{C}),\ (\mathbf{dr}(\mathbf{C}))^2\mathbf{r}(\mathbf{C}),\ \mathbf{n}_1,\ \mathbf{n}_2,\ \mathbf{n}_3] \tag{7.6}$$

Here, $\mathbf{dr}(\mathbf{C})$ represents the Jacobian matrix of the rate vector *for all six components* (refer to Chapter 6 if in doubt) as follows:

$$\mathbf{dr}(\mathbf{C})=\begin{bmatrix}\dfrac{\partial r_B}{\partial c_B} & \cdots & \dfrac{\partial r_B}{\partial c_H}\\ \vdots & \ddots & \vdots\\ \dfrac{\partial r_H}{\partial c_B} & \cdots & \dfrac{\partial r_H}{\partial c_H}\end{bmatrix}$$

CSTR effluent concentrations are considered to lie on the AR boundary (they are considered to be critical points) when $\Lambda(\mathbf{C}) = 0$. Note that we have expressed the controllability expression in terms of $\mathbf{r}(\mathbf{C})$ instead of the mixing vector $\mathbf{v}(\mathbf{C})$. This is the preferred method as it does not involve the specification of an appropriate mixing point $\mathbf{C}^*$ on the AR boundary, which is unknown at present.

Equation 7.6 describes a set of surfaces in $\mathbb{R}^6$. We wish to determine the CSTR effluent concentrations that intersect the particular surface at $\Lambda(\mathbf{C}) = 0$. Using the CSTR locus generated in Section 7.2.2.4, CSTR concentrations may be substituted into the expression given by Equation 7.6, and plotted against their corresponding CSTR residence times τ. Equation 7.6 requires that the determinant of a 6 × 6 matrix be found, which is too large to reproduce here. Instead, the critical CSTR plot is found by directly substituting

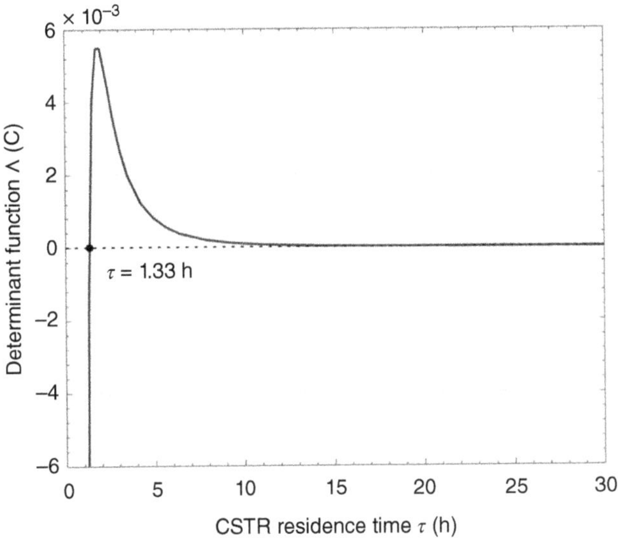

Figure 7.10 The AR for the BTX system (a) computed by an automated AR construction algorithm. The region obtained is in agreement with the theoretical prediction given by (b), although there is still a moderate difference between the two regions.

CSTR concentrations into Equation 7.6 and evaluating the determinant of the 6×6 matrix numerically.

The roots of Equation 7.6 are therefore CSTR residence times belonging to effluent concentrations that fall on the true AR boundary. A plot of $\Lambda(\mathbf{C})$ is shown in Figure 7.10.

To adequately display the features of the plot, the CSTR locus is only shown up to a maximum residence time of 30 h (the trend observed at 30 h remains unchanged for longer residence times). A distinct root is seen on the graph at $\tau \sim 1.33$ h, signifying a critical CSTR point. However, for CSTR residence times greater than approximately 10 h, the value of $\Lambda(\mathbf{C})$ lies very close to zero (and only truly reaches zero at the CSTR equilibrium point for extremely long residence times).

We can use the candidate region generated by an automated AR construction technique as a way to check if there are points that are still missing from the true AR. Hence, the candidate AR given in Figure 7.11(a) is obtained from a numerical AR construction algorithm.[4] The region generated in Figure 7.11(b) represents the candidate region obtained for the BTX system using identical kinetics and feed point. Note that the shape of the numerically computed region is slightly different from the candidate region generated by analytical solution of the CSTR–PFR structure, and further investigation is hence required to consolidate these differences.

7.2.2.6 Critical DSRs

a) Computing the Critical α Policy As a final analysis, the existence of critical DSRs shall be investigated. Since the BTX system is three-dimensional, the "vDelR" condition may be used to compute an analytical expression for a critical α policy. To compute the critical α policy using the "vDelR" condition, $\varphi(\mathbf{C})$ is first formed from the Jacobian of the rate vector $\mathbf{dr}(\mathbf{C})$ and the sidestream mixing vector $\mathbf{C}^0$.

$$\varphi(\mathbf{C}) = [\mathbf{dr}(\mathbf{C})(\mathbf{C} - \mathbf{C}^0)]^{\mathrm{T}}[(\mathbf{C} - \mathbf{C}^0) \times \mathbf{r}(\mathbf{C})]$$

Since $\varphi(\mathbf{C})$ operates specifically on vectors in $\mathbb{R}^3$ (due to the use of the vector cross product), $\mathbf{r}(\mathbf{C})$ and $(\mathbf{C} - \mathbf{C}^0)$ are vectors composed of only c_B, c_E, and c_T—this is unlike the $\mathbf{r}(\mathbf{C})$ and $\mathbf{v}(\mathbf{C})$ vectors used by the standard approach using Lie brackets, or the computation of the determinant function $\Lambda(\mathbf{C})$ for critical CSTR concentrations, which are methods that both require vectors of

[4] Specifically, the parallel construction method, discussed in Chapter 8, has been used to generate the region.

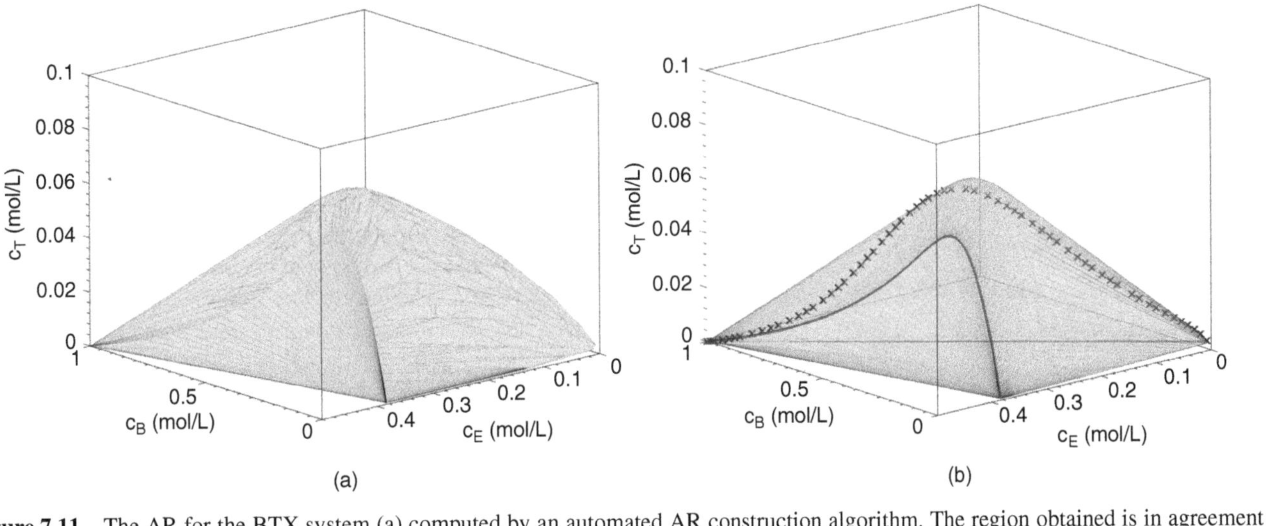

Figure 7.11 The AR for the BTX system (a) computed by an automated AR construction algorithm. The region obtained is in agreement with the theoretical prediction given by (b), although there is still a moderate difference between the two regions.

all six species present in the system. $\mathbf{r}(\mathbf{C})$ and $(\mathbf{C}-\mathbf{C}^0)$ are thus given by $\mathbf{r}(\mathbf{C})=[r_B(\mathbf{C}),\ r_E(\mathbf{C}),\ r_T(\mathbf{C})]^T$ and $(\mathbf{C}-\mathbf{C}^0)=[(c_B-c_B^0),\ (c_E-c_E^0),\ (c_T-c_T^0)]^T$. The mixing point $\mathbf{C}^0$ shall be taken as the feed concentration $\mathbf{C}_f$, so that $\mathbf{C}^0=\mathbf{C}_f=[1.0, 0.5, 0.0]^T$ mol/L. This is done because it is known that critical DSRs must be fed with points on the true AR boundary, and $\mathbf{C}_f$ is one such point that fulfills this criterion. The Jacobian $\mathbf{dr}(\mathbf{C})$ is given by

$$\mathbf{dr}(\mathbf{C})=\begin{bmatrix} -40c_B-c_E^{0.5} & -\dfrac{c_B}{2c_E^{0.5}} & 0 \\ -\dfrac{c_E^{0.5}}{2} & \dfrac{-(c_B+c_T)}{4c_E^{0.5}} & -\dfrac{c_E^{0.5}}{2} \\ c_E^{0.5} & \dfrac{c_B-c_T}{2c_E^{0.5}} & -c_E^{0.5} \end{bmatrix}$$

Substituting $\mathbf{dr}(\mathbf{C})$ and $(\mathbf{C}-\mathbf{C}^0)$ into the expression for $\varphi(\mathbf{C})$ and simplifying gives[5]

$$\begin{aligned}\varphi(\mathbf{C}) =& -10c_B^3c_E^{1.5}-5c_B^3c_E^{0.5}c_T-2.5c_B^3c_E^{-0.5}c_T+2.5c_B^3c_E^{-0.5} \\ &+10c_B^2c_E^{1.5}c_T+20c_B^2c_E^{1.5}-5c_B^2c_E^{0.5}c_T^2 \\ &+10c_B^2c_E^{0.5}c_T-10c_B^2c_E^{0.5}-2.5c_B^2c_E^{-0.5}c_T^2 \\ &-2.5c_B^2c_E^{-0.5}c_T-40c_Bc_E^{1.5}c_T+20c_Bc_E^{0.5}c_T^2 \\ &+20c_Bc_E^{0.5}c_T+c_Bc_Ec_T-c_E^2c_T+0.5c_Ec_T^2-0.5c_Ec_T\end{aligned} \tag{7.7}$$

The critical α policy may then be found from

$$\alpha(\mathbf{C})=\frac{[\nabla\varphi(\mathbf{C})]^T\mathbf{r}(\mathbf{C})}{[\nabla\varphi(\mathbf{C})]^T(\mathbf{C}-\mathbf{C}^0)} \tag{7.8}$$

In this instance, we cannot simplify the α expression into a simpler form (i.e., what could be done with the Van de Vusse system, by expressing one of the components in terms of the other two), and thus we arrive at the following irrational function:

$$\alpha(\mathbf{C})=\frac{c_E(c_B^4y_1+c_B^3y_2+c_B^2y_3+c_By_4+y_5)}{c_B^3u_1+c_B^2u_2+c_Bu_3+u_4} \tag{7.9}$$

Here,

$$\begin{aligned}y_1 =& 2.5c_E^{2.5}+1.25c_E^{1.5}c_T-2.5c_E^{1.5}-0.625c_E^{0.5}c_T \\ &+0.625c_E^{0.5}+600c_E^3+300c_E^2c_T+150c_Ec_T-150c_E \\ y_2 =& 40c_E^{3.5}+10c_E^{2.5}c_T-5c_E^{2.5}+2.5c_E^{1.5}c_T^2+2.5c_E^{1.5}c_T \\ &-7.5c_E^{1.5}-1.25c_E^{0.5}c_T^2-400c_E^3c_T \\ &-800c_E^3+200c_E^2c_T^2-400c_E^2c_T+400c_E^2+100c_Ec_T^2 \\ &+100c_Ec_T\end{aligned}$$

$$\begin{aligned}y_3 =& -30c_E^{3.5}c_T-80c_E^{3.5}+12.5c_E^{2.4}c_T^2+5c_E^{2.5}c_T+40c_E^{2.5} \\ &+1.25c_E^{1.5}c_T^3+2.5c_E^{1.5}c_T^2 \\ &+5c_E^{1.5}c_T-0.625c_E^{0.5}c_T^3-0.625c_E^{0.5}c_T^2+800c_E^3c_T \\ &+c_E^3-400c_E^2c_T^2-400c_E^2c_T \\ y_4 =& 80c_E^{3.5}c_T-30c_E^{2.5}c_T^2-40c_E^{2.5}c_T-5c_E^{1.5}c_T^3-5c_E^{1.5}c_T^2 \\ &-c_E^4-0.5c_E^3-0.75c_E^2c_T^2+0.25c_E^2c_T \\ y_5 =& c_E^4c_T+0.5c_E^3c_T-0.25c_E^2c_T^3+0.25c_E^2c_T^2\end{aligned}$$

and

$$\begin{aligned}u_1 =& 45c_E^4+22.5c_E^3c_T-7.5c_E^3+7.5c_E^2c_T-6.25c_E^2 \\ &+0.625c_Ec_T-0.625c_E \\ u_2 =& -45c_E^4c_T-100c_E^4+22.5c_E^3c_T^2-42.5c_E^3c_T+40c_E^3 \\ &+7.5c_E^2c_T^2+1.25c_E^2c_T+5c_E^2 \\ &+0.625c_Ec_T^2+0.625c_Ec_T \\ u_3 =& -3c_E^{3.5}c_T+0.5c_E^{2.5}c_T+160c_E^4c_T+40c_E^4-80c_E^3c_T^2 \\ &-60c_E^3c_T-20c_E^3 \\ u_4 =& 3c_E^{4.5}c_T-1.5c_E^{3.5}c_T^2+c_E^{3.5}c_T+0.25c_E^{2.5}c_T^2 \\ &-0.25c_E^{2.5}c_T-40c_E^4c_T+20c_E^3c_T^2+20c_E^3c_T\end{aligned}$$

Equation 7.9 is the critical α policy for the BTX system when $\mathbf{C}^0=\mathbf{C}_f=[1.0, 0.5, 0.0]^T$ mol/L is employed. This can be substituted into the DSR equation to compute critical DSR trajectories that serve to further expand the region.

b) DSR from the Feed Point Equation 7.9 may be substituted into the DSR expression and integrated with the appropriate feed condition. From Chapter 6, we understand that critical DSR trajectories must lie entirely on the AR boundary, and thus critical DSRs must be initiated on the AR boundary. A common point for initiating a DSR is from $\mathbf{C}_f$. When this is carried out, we obtain the DSR trajectory AB in Figure 7.12(a).

The CSTR locus and PFR trajectory from the feed point have also been overlaid for comparison. The DSR trajectory appears to terminate at point B in the figure. This is sensible because point B on the DSR trajectory also corresponds to the CSTR effluent concentration (of $[0.1623, 0.4005, 0.0742]^T$ mol/L) corresponding to a CSTR residence time of $\tau=1.33$ h. The intersection of the critical DSR trajectory with the CSTR locus confirms that point B is in fact an equilibrium point for the critical DSR trajectory displayed in Figure 7.12(a). Although this trajectory serves to expand the set of achievable points, comparison with the region in Figure 7.11(a) suggests that there are still elements of the AR boundary that are missing.

c) DSR from the CSTR Equilibrium Point From Figure 7.10, a critical CSTR concentration also exists

[5] For convenience, the Python package Sympy is used to simplify the algebra for finding $\alpha(\mathbf{C})$.

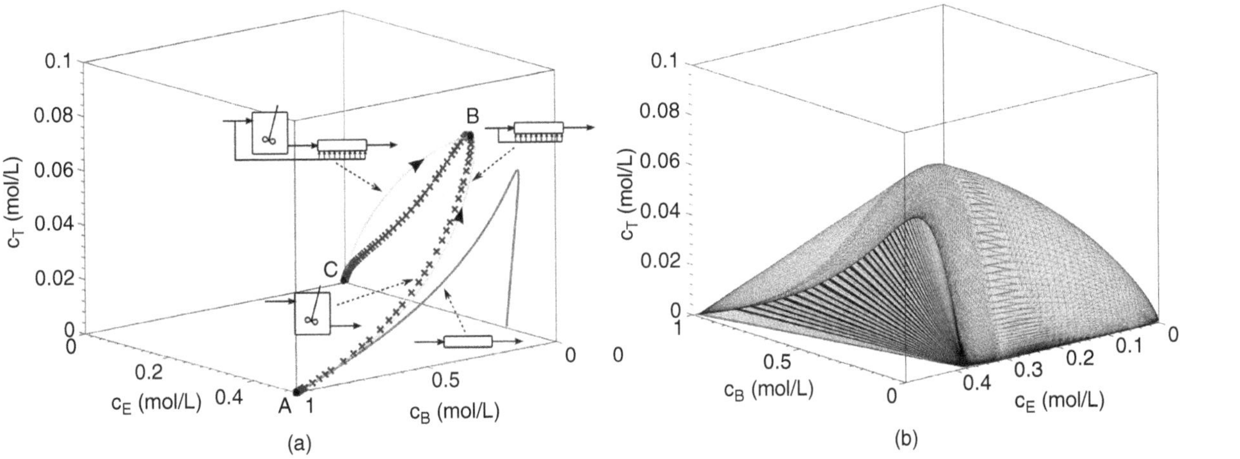

Figure 7.12 (a) Critical DSR trajectories from the feed (curve AB) and from the CSTR equilibrium point (curve CB). (*See color plate section for the color representation of this figure.*) (b) Full AR for the BTX system.

at the CSTR equilibrium point (point C). Guided by the AR construction results for the automated method, as well as the methodology used in the three-dimensional Van de Vusse system, a DSR initiated from the CSTR equilibrium point may then also expand the region.

To generate this critical DSR trajectory, the DSR expression is integrated together with the critical α policy, given by Equation 7.9, using the CSTR equilibrium point as the initial condition to the DSR. From point C, the DSR trajectory may be computed and plotted, which is also displayed in Figure 7.12(a).

The DSR trajectory from point C travels in a direction opposite to the CSTR locus and terminates at the first DSR equilibrium point from Section "DSR from the feed point" (point B). The path taken from the feed point to the second DSR equilibrium points is then A to C (in a CSTR) and then C to B (in a critical DSR). This DSR trajectory further expands the set of achievable points.

PFR trajectories form the extremities of the AR boundary, and critical DSR trajectories provide access to these trajectories. It is sensible to therefore finalize construction by initiating PFR trajectories from the concentrations generated by both critical DSR trajectories. The expanded set of points, including PFR trajectories, is given in Figure 7.12(b).

Comparison of the analytically constructed region (Figure 7.12(b)) to the region obtained by the automated AR method (Figure 7.11(b)) is again used as a check, and the two regions appear to closely match each other. This suggests that the AR in Figure 7.12(b) is the full AR for the BTX system, and construction for the specified feed is thus complete.

The physical reactor structures corresponding to the region in Figure 7.12(b) are then given by Figure 7.13. Note that these structures are identical to the optimal reactor structures given in Figure 7.6, which are for the Van de Vusse system, even though the kinetics and feed point are different.

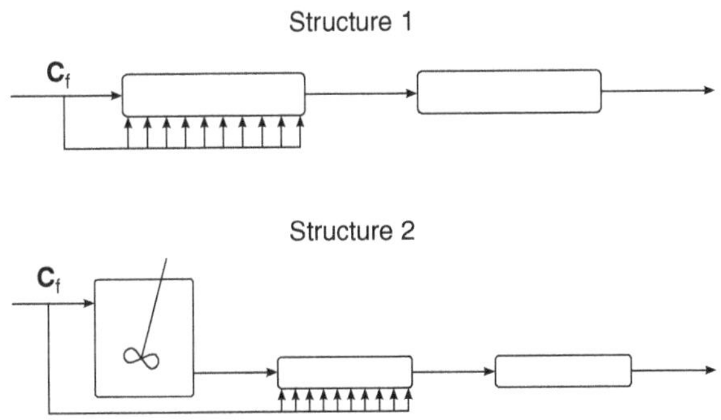

Figure 7.13 Optimal reactor structures for the BTX system.

ILLUSTRATION: Maximum toluene concentration in the BTX system

Since the AR for the BTX system has already been computed, finding the maximum toluene concentration is straightforward. Our objective function in this instance is given by a hyperplane in the c_B–c_E plane that just touches the AR at the point of maximum toluene, as shown in Figure 7.14. We can employ a standard optimization method to determine the point of intersection of the plane with the AR boundary. When this is carried out, the point $\mathbf{C} = [0.0804, 0.4032, 0.0807]^T$ mol/L is achieved. Hence, this is the operating point in the BTX system with the highest toluene concentration. There are no other points that achieve a higher toluene concentration for the feed point specified.

Note also that the point of maximum toluene concentration corresponds to a position on the AR boundary that requires a critical DSR followed by a PFR in series to reach it. This is hence the optimal reactor structure needed to obtain the desired toluene concentration.

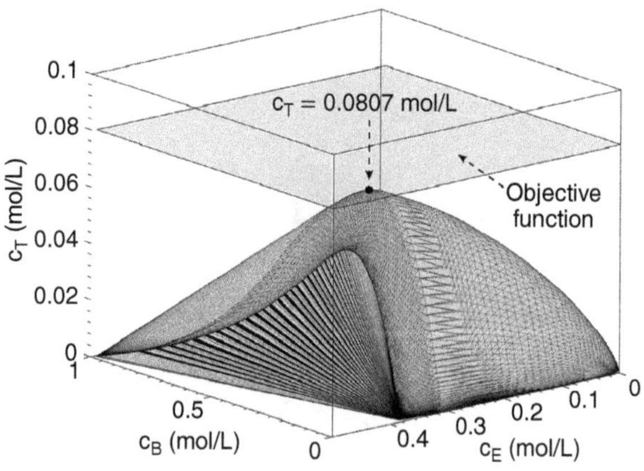

Figure 7.14 Maximum toluene surface for the BTX system.

Had Sam, Adam, and Donald used AR theory in their analysis, they would have needed to use this structure in their recommendation to the company.

ILLUSTRATION: BTX AR in H_2 Space

Suppose we wish to determine the AR for the BTX system involving hydrogen (i.e., in Chapter 1, when Sam Alex and Donald's boss requests that they investigate minimum hydrogen production). Since the AR has already been generated in c_B–c_E–c_T space, the corresponding boundary concentrations can be found in terms of other species participating in the BTX reactions by back-calculation using the reaction stoichiometry and mass balance.

Hence, any concentration belonging to the AR is given by the concentration vector $\mathbf{C} = [c_B, c_E, c_T]$T mol/L. We may express any concentration by

$$\begin{bmatrix} c_B \\ c_E \\ c_T \end{bmatrix} = \begin{bmatrix} c_{Bf} \\ c_{Ef} \\ c_{Tf} \end{bmatrix} + \begin{bmatrix} -1 & 0 & -2 \\ -0.5 & -0.5 & 0 \\ 1 & -1 & 0 \end{bmatrix} \begin{bmatrix} \varepsilon_1 \\ \varepsilon_2 \\ \varepsilon_3 \end{bmatrix}$$

or

$$\mathbf{C} = \mathbf{C}_f + \hat{\mathbf{A}}\boldsymbol{\varepsilon} \tag{7.10}$$

where $\hat{\mathbf{A}}$ is a matrix containing rows of the stoichiometric coefficient matrix A corresponding to benzene, ethylene, and toluene (the components represented by the AR) and $\boldsymbol{\varepsilon}$ is a vector containing the extents of reaction in the BTX system. Hence, for each value of $\mathbf{C}$, Equation 7.10 represents a system of linear equations for which the extent vector $\boldsymbol{\varepsilon}$ may be solved for

$$\boldsymbol{\varepsilon} = \hat{\mathbf{A}}^{-1}(\mathbf{C} - \mathbf{C}_f)$$

Thus, once the value of $\boldsymbol{\varepsilon}$ is known, the remaining concentrations (which in this instance is xylene, diphenyl, and hydrogen) can be found by substituting $\boldsymbol{\varepsilon}$ into the mass balance expression for all components as follows:

$$\begin{bmatrix} c_B \\ c_E \\ c_T \\ c_X \\ c_D \\ c_H \end{bmatrix} = \begin{bmatrix} c_{Bf} \\ c_{Ef} \\ c_{Tf} \\ c_{Xf} \\ c_{Df} \\ c_{Hf} \end{bmatrix} + \begin{bmatrix} -1 & 0 & -2 \\ -0.5 & -0.5 & 0 \\ 1 & -1 & 0 \\ 0 & 1 & 0 \\ 0 & 0 & 1 \\ 0 & 0 & 1 \end{bmatrix} \begin{bmatrix} \varepsilon_1 \\ \varepsilon_2 \\ \varepsilon_3 \end{bmatrix}$$

In Figure 7.15, a plot of the optimal reactor structures forming the BTX AR boundary in benzene, toluene, hydrogen space is given. Had Sam, Alex, and Donald constructed the AR for the BTX system initially, they would have been able to determine the limits of hydrogen production as well as understood the optimal reactor structures needed to achieve these limits.

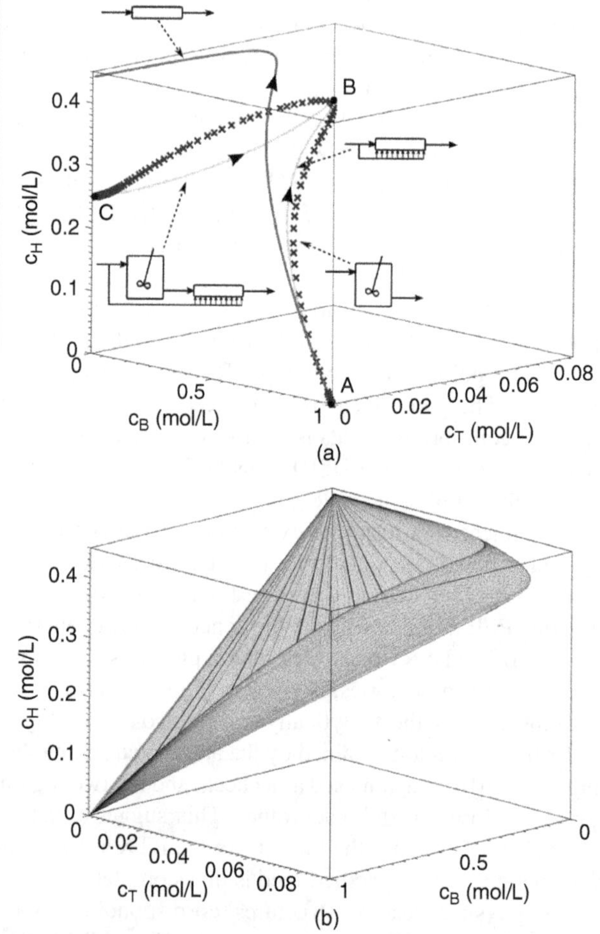

Figure 7.15 (a) Optimal reactor structures in c_B–c_T–c_H space and (b) filled AR for the BTX system in c_B–c_T–c_H space. (*See color plate section for the color representation of this figure.*)

7.3 NONISOTHERMAL CONSTRUCTIONS AND REACTOR TYPE CONSTRAINTS

7.3.1 Adiabatic Reaction

7.3.1.1 ***Motivation*** In this example, we wish to propose a potential solution for generating a lower dimensional AR, from a higher dimensional problem, when certain constraints on the design are imposed. In the following discussion, a series of subproblems of increasing difficulty will be provided where the following two common design constraints are enforced:

1. An adiabatic constraint: An AR will be constructed when an adiabatic energy balance is introduced. The implications of how this constraint impacts AR construction will provide for an interesting discussion. Temperature will be an important consideration in this instance, and hence it is important to understand how temperature may be accommodated in AR constructions.
2. A reactor type constraint: In this scenario, the construction of the AR is carried out by using PFRs alone. This approach is similar to that discussed in Chapters 2 and 3. This is done because situations might arise when only access or knowledge to a specific reactor type is available. We will be interested in how this constraint impacts construction of the AR, and ultimately how this influences what states are achieved.

Constraints such as these are significant because many practical problems in reality might arise that manifest themselves as a combination of these two types. For example, in ammonia synthesis, a series of adiabatic reactors with intermediate cooling between stages is employed to increase the overall conversion of reactants past what would be achieved in a single adiabatic reactor (Denbigh and Turner, 1984; Howard, 1977). Thus, despite what is known about the true AR and its optimal structures, practical considerations may restrict us from implementing these recommendations, and alternate ways of advancement must be found.

7.3.1.2 ***Problem Statement*** In Chapter 5, a system involving multiple CSTR steady states is shown (the isola example). Multiple steady states often arise in isothermal constructions when the kinetics is nonlinear. Nonlinearity is also often introduced in nonisothermal systems (i.e., in adiabatic systems, for example), and thus multiple steady states appear in these systems as well; this is true even with simple kinetics. Consider an adiabatic reaction of the following form (Hildebrandt et al. 1990):

$$A \xrightarrow{k_1} B$$

$$A \underset{k_2}{\rightarrow} C$$

with kinetics

$$r_A = -k_1 c_A^2 - k_2 c_A$$

$$r_B = k_1 c_A^2$$

$$r_C = k_2 c_A$$

Two independent reactions are present, and thus the AR must reside in $\mathbb{R}^2$. We aim to compute the AR for this system in concentration space. This will allow for the determination of all feasible concentrations for components A and B (component C can then be found by mass balance). Before we proceed, we can simplify the notation of the system slightly by combining certain terms in the rate expression. The ratio of r_A and r_B is given by

$$\frac{r_B}{r_A} = -\frac{k_1 c_A}{k_1 c_A + k_2} = -\frac{a}{a+1}$$

where

$$a = \frac{k_1}{k_2} c_A \tag{7.11}$$

Substitutions $x = (c_A/c_{Af})$ and $y = (c_B/c_{Af})$ may be made to further simplify the expression. The PFR expression is then written directly in terms of new variables x and y as follows:

$$\frac{dy}{dx} = \frac{-a}{a+1}$$

Rate constants k_1 and k_2 are assumed to follow an Arrhenius form

$$k_1 = \exp(11.5)$$

$$k_2 = \exp\left(\frac{5000}{T}\right)$$

which, upon substitution into Equation 7.11, gives

$$a = \exp\left(11.5 - \frac{5000}{T}\right)(x c_{Af}) \tag{7.12}$$

We will assume that a feed containing pure A is available, so that $\mathbf{C}_f = [x, y]^T = [1, 0]^T$. Isothermal operation is no longer assumed. Rather, we will use an expression to describe an adiabatic energy balance around the reactor network.

7.3.1.3 ***The Energy Balance*** Since the system is assumed to operate adiabatically, an adiabatic energy balance may be written over the reactor system, and an expression for temperature in terms of species concentrations x and y may then be obtained. This expression may be substituted into the temperature dependent rate constants, found in Equation 7.12. Due to the nonlinear nature of both the kinetics and energy balance terms, the final kinetic expression is typically a nonlinear function. For the purposes of demonstration, we shall assume that the resulting energy balance expression, after simplification, is of the following form:

$$T = T_{in} + (540 - 100)y + 100(1 - x) \qquad (7.13)$$

Here, T_{in} is the inlet temperature to the system, which is specified at $T_{in} = 290$ K for the current problem. The expression for T, given by Equation 7.13, may be substituted into Equation 7.12 and the rate expressions. The resulting rate field after substitution now incorporates the adiabatic constraint imposed on the system, and it is necessarily different compared to an equivalent isothermal system.

> **SIDE NOTE: Simplified Energy Balance and Temperature Expression**
>
> The temperature expression obtained from the energy balance for this system is artificial, although this is done only to simplify the discussions slightly. A more realistic energy balance could also be utilized, without any loss of generality, to the following discussions.

7.3.1.4 AR Construction Since the reactor temperature has been expressed explicitly in terms of components x and y, construction of the AR for the adiabatic system from this point onward follows the same approach as that for an isothermal system. The system is two-dimensional as there are two independent reactions participating in the system, and therefore we need only consider combinations of PFRs and CSTRs in the construction of the candidate region. It is customary to begin by generating the PFR trajectory and CSTR locus from the feed. Due to the nonlinear nature of the kinetics introduced by the adiabatic constraint, the system exhibits multiple CSTR solutions from the feed point. We find that the construction process is similar to that of the isola example described in Chapter 5.

Figure 7.16 shows the PFR trajectory and CSTR locus from the feed for the adiabatic system. We find that the PFR trajectory from the feed lies close to the *x*-axis.

The CSTR locus from the feed point contains two separate solution branches, which are similar in shape to the isola example from Chapter 5. One of the branches lies close to the *x*-axis. The other branch is an isola. The set of effluent concentrations associated by this branch serve to significantly expand the entire set of concentrations achievable by the system. If the existence of the second branch is not considered, the corresponding AR for the system is small in comparison. A PFR may be initiated from point X in Figure 7.16(a) and integrated to complete construction. The resulting region obtained is then given by Figure 7.16(b).

The reactor structures that make up the AR boundary are also provided in Figure 7.16. The inclusion of the second CSTR branch enlarges the region significantly. Indeed, it can be shown that the stoichiometric subspace is given by the triangular region transcribed by points ABC. The set of achievable points for the adiabatic system thus represents a large portion of the total space achievable by mass balance. The optimal reactor structure for the AR is given in Figure 7.17.

Hence, the optimal reactor structure for the adiabatic system is a CSTR from the feed point followed in series by a PFR. All concentrations represented by the shaded region in

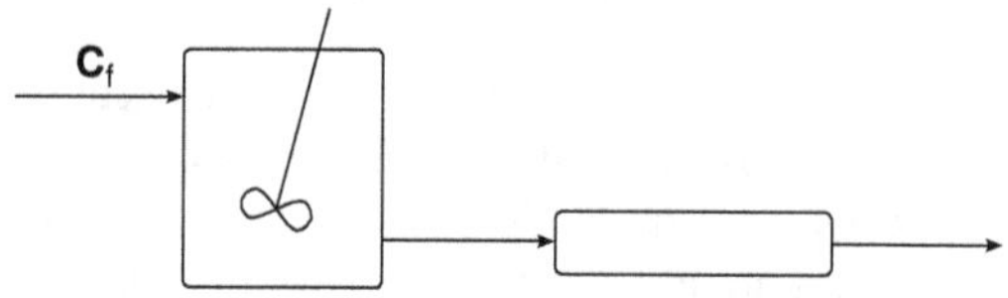

Figure 7.17 Optimal reactor structure for the adiabatic system.

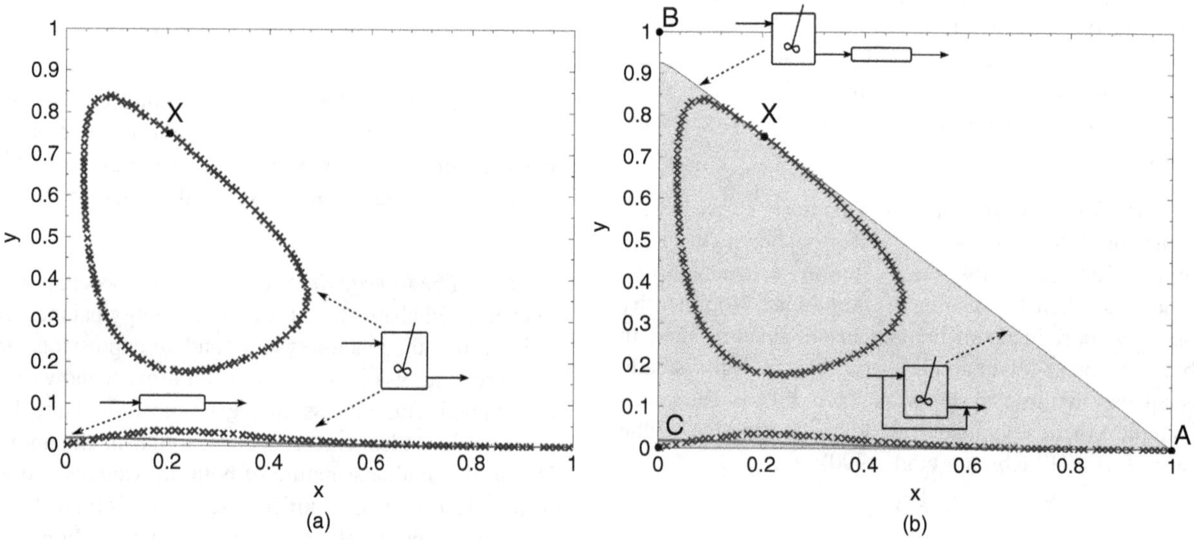

Figure 7.16 (a) PFR trajectory and CSTR locus from the feed for the adiabatic example and (b) AR for the adiabatic example.

Figure 7.16(b) may be obtained by this adiabatic structure with mixing.

This example demonstrates how multiple CSTR steady states may arise in nonisothermal systems, even when the associated kinetics are simple, and temperature is assumed to be linear in terms of concentration. The inherent nonlinear nature of rate expressions in general thus often leads to complex behavior even when the energy balance is of a simple form. Multiple steady states must be included in the AR in order to understand the true bounds of achievability. Omission of these states may have important implications on subsequent optimizations, such as if we wish to maximize the concentration of component B.

SIDE NOTE: The Absence of an Explicit Temperature Expression

In the preceding example, we have shown the AR for a nonisothermal system. The particular example given employs an adiabatic system wherein the energy balance could be reduced to a simple, linear, form. This allows for temperature to be expressed explicitly in terms of the component concentrations, which may then be substituted into the temperature-dependent rate expressions.

For more complex systems, we may not be as fortunate with the energy balance, and hence it may not be as simple (or even possible) to express temperature in terms of concentration explicitly. For these systems, one cannot proceed with computing the AR in the manner discussed here.

There are methods that exist that address these constraints; however, these require additional theory to understand. In the absence of an explicit temperature expression, it is often easier to tackle the problem numerically, with the aid of an automated AR construction scheme. In Chapter 8, a number of AR construction methods are discussed that may be used for nonisothermal systems. Although these methods often do not suggest an optimal reactor structure, knowledge of the limits of achievability for a nonisothermal is often sufficient for setting design targets.

ILLUSTRATION: The Isothermal Case

For comparison, the AR associated with an isothermal reactor network shall be inspected briefly. This situation may arise if heating or cooling equipment is available to maintain a constant temperature throughout reaction.

Since the system is now isothermal, the effect of the adiabatic energy balance developed previously shall be ignored. Rather, we shall use a constant value for the system temperature, which can be substituted into the rate expression. The particular value for temperature is arbitrary; however since the reaction is known to be exothermic, this value must lie between the inlet temperature of 290 K and the maximum adiabatic temperature that is associated with maximum conversion. The maximum temperature associated with this concentration is then found to be 830 K. For the purposes of illustration, a median temperature of 560 K will be utilized for the computations.[6]

The AR may again be constructed for the isothermal case, which is similar to that provided in Section 7.3.1.2. When this is done, the region given in Figure 7.18 is obtained.

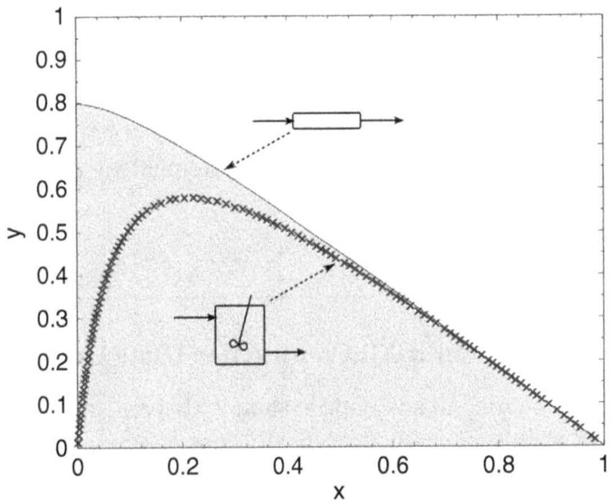

Figure 7.18 AR obtained for isothermal operation (T = 560 K).

Note that the region obtained in this instance is smaller when compared to the adiabatic case. It is not possible to achieve the same maximum concentration in component B as the adiabatic case when isothermal conditions are employed. The CSTR locus corresponding to the feed point has also been plotted for comparison. Observe that only a single branch is present, and thus the system no longer exhibits multiple steady states.

Furthermore, observe that the CSTR locus is completely contained *within* the PFR trajectory. An isothermal PFR operating at 560 K is entirely convex for the kinetics in this case, and hence a PFR from the feed is the optimal reactor structure for the isothermal AR. This is notably different to the CSTR–PFR structure required

[6] Any value between 290 and 830 K may be adopted; however, the associated AR shall be different from the region shown here.

for the adiabatic case. Thus, the following two different design scenarios may be generated that may evaluated against each other:

1. In the adiabatic case, a higher maximum concentration of B is achievable. It is also possible to build the reactor network without the need to install complicated temperature control, which simplifies the design. However, the optimal reactor structure involves two reactors, and thus the capital cost of installing and operating two reactors might be high.
2. In the isothermal case, a lower concentration in B is achievable, and an appropriate control system must be in place to maintain isothermal operation. However, a benefit of this approach is that the optimal reactor structure is a single PFR from the feed. This is notably easier to construct and operate than the CSTR–PFR structure expected by the adiabatic case. Furthermore, the maximum concentration of B achieved is only approximately 10% lower than the adiabatic case. The benefit realized by slightly purer concentrations in B may not warrant the requirement of operating two separate reactors.

7.3.2 Constrained AR Construction Using Only PFRs

In this scenario, an adiabatic system with a reactor type constraint is investigated. Specifically, the construction of the AR shall be carried out using PFRs only. The techniques developed in this section will be utilized in later examples, when interstage and cold-shot cooling in an adiabatic system is employed. The AR will be constructed in residence time space to find a reactor configuration that minimizes the total reactor volume for a given output concentration.

7.3.2.1 Problem Statement We shall assume that a single reaction is present as follows:

$$A \leftrightarrow B$$

The rate of reaction is given by

$$r_A = -(5 \times 10^5)\exp(-4000/T)c_A + (5 \times 10^8)\exp(-8000/T)(c_{Af} - c_A) \quad (7.14)$$

and the reactor is assumed to operate adiabatically. Similar to Section 7.3.1, an adiabatic energy balance can be written over the system and used to express system temperature T in terms of concentration. We assume that the particular temperature for our system is given by

$$T = T_{in} + T_{ad}(c_{Af} - c_A) \quad (7.15)$$

T_{ad} is the adiabatic temperature rise over the reactor, specified here as $T_{ad} = 200$ K and T_{in} is the feed temperature to the reactor. In this work, we assume that $T_{in} = 300$ K.

Since we wish to generate the AR in c_A–τ space, the concentration vector is given by $\mathbf{C} = [c_A, \tau]^T$ mol/L. The feed to the reactor system is assumed to be pure in A so that $c_{Af} = 1.0$ mol/L and $\mathbf{C}_f = [1, 0]^T$ mol/L. The rate vector is thus given by $\mathbf{r}(\mathbf{C}) = [r_A, 1]^T$ mol/(L.h). Equation 7.15 may be substituted into Equation 7.14 and used for the AR construction.

7.3.2.2 Construction Using Only PFRs

a) Motivation Since construction of the AR has been constrained to the use of PFRs only, our construction procedure must be modified accordingly. The method of construction follows an iterative procedure that is similar to the successive batch experiments performed for the BTX system in Chapter 3.

Consider a PFR integrated from the feed point. The trajectory and associated convex hull of the initial set of achievable points is shown in Figure 7.19. Note that the time axis refers to the residence time of the system, which in this case is generated by PFRs only. Observe that the PFR trajectory from the feed, given by curve ABCD, is concave with respect to residence time, which may be filled by joining points A and C with a mixing line. The straight-line AC therefore represents a PFR from the feed with bypass of feed material to the PFR exit. Thus, although the initial PFR trajectory shows that a certain residence time is needed to achieve a value of c_A on the PFR, we can actually do better between points A and C—this is achieved by operating the PFR up to point C and then mixing feed material with this stream to obtain the desired overall concentration. Since the PFR trajectory lies higher up on the graph than the mixing line AC, bypassing feed results in a smaller system residence time overall

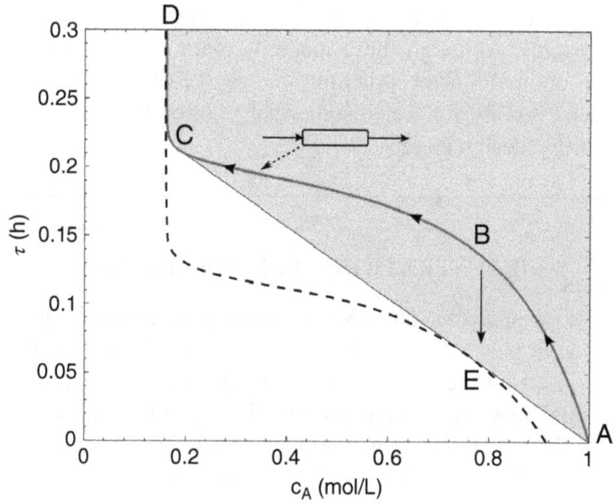

Figure 7.19 Sliding a PFR trajectory downward.

for the same exit concentration. Note that this is only possible because of the concavity in the original PFR trajectory, and hence regions of low reaction rate in the PFR are to be bypassed by the use of mixing.

Observe how this procedure might be used to expand the total set of achievable concentrations in c_A–τ space. Concavities in the PFR solution trajectory allow for bypass and mixing, which in turn result in a smaller overall system residence time. Furthermore, since Equation 7.14 relies only on c_A, the PFR trajectory given in Figure 7.19 is identical for all residence times. We shall call the original PFR trajectory ABCD from the feed the *base trajectory*.

The base trajectory may therefore be used as a means to decrease the overall system residence time by moving down curve ABCD, until the curve just touches the mixing line given by AC, which is displayed in Figure 7.19 by point E. Point E represents the lowest residence time on the boundary of the current candidate region wherefrom an additional PFR may be run to expand the region.

SIDE NOTE: Rate Vector on the AR Boundary

Recall from discussion of the properties of the AR that all rate vectors on the boundary of the AR must be tangent or point into the region. Since PFR trajectories are being used in this example, we know that the gradient of the trajectory is equivalent to the rate vector at that point. By moving the base trajectory ABCD down and tracking how it intersects the mixing line AC, we have a visual indication of how rate vectors on the candidate region boundary point out of the region. The point where the rate vector is tangent to the current boundary is therefore given by point E. This process is shown in Figure 7.20.

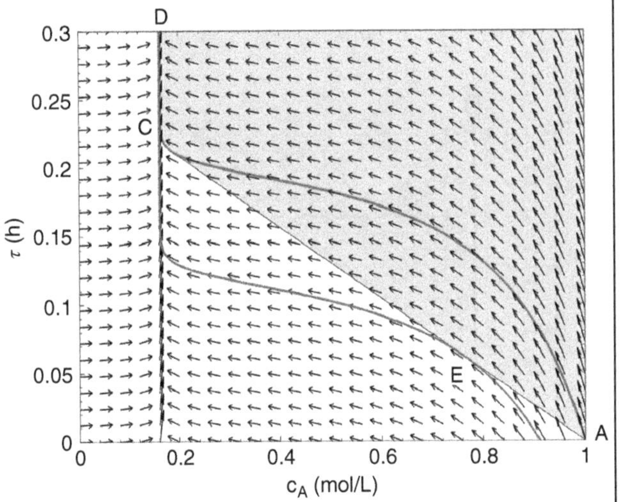

Figure 7.20 Rate field for the current candidate region.

Thus, when the base trajectory is translated downward, we are checking for condition 6 of the AR properties described in Chapter 4. Evidence that we can identify rate vectors that point out of the region indicates that the region given in Figure 7.19 is not the true AR, and thus further expansion of the region is possible.

b) Two PFRs In Figure 7.21(a), the PFR trajectory initiated from point E is shown. Note that the shape of the resulting trajectory, EFD, is identical to the base trajectory ABCD, which is simply shifted downward. The rate vector at point E is tangent to the shaded boundary. This produces a trajectory with a lower residence time than the PFR with bypass that was suggested previously. Hence, by running a PFR from the bypass point, we may achieve an even lower overall residence time for the system. This structure is still not optimal, for the resulting trajectory EFD still appears to contain a concavity. Thus, mixing may be used to achieve a lower residence time.

c) Many More PFRs One may begin to see how the method of improvement follows that of the batch experiments discussed in Chapter 3. For each subsequent mixing operation, a PFR trajectory may be appended to the structure, and mixed to fill in any concavities. With each successive PFR, the potential for additional improvements decreases—because the size of the concave region reduces in size—and we slowly approach a state whereby no additional improvements are possible. This state is only reached in the limit of an infinite number of PFRs, however.

In Figure 7.21(b), the PFR trajectory, and the shape of candidate region, is shown for an additional PFR. The point at which the base trajectory touches the current boundary is now given by point G. In order to achieve point G, two PFRs in series are required with interstage bypass of feed to the exit of each reactor. The PFR trajectory GHD achieves a lower overall residence time than before. Again, the inclusion of an additional PFR increases the complexity of the structure; however, a lower system residence time is obtained.

In Figure 7.21(c), we show a final trajectory for a third PFR. The point of initiation is given by point I. Note that with each successive PFR, the vertical distance between the current and previous PFR trajectories decreases, suggesting a smaller improvement in system residence time is achieved with each additional PFR segment that is added.

Observe also that the tangency points (E, G, and I) move toward the left of the plot. Each additional stage produces a smaller change in the boundary. In the limit of many stages, we reach the lower boundary of the true AR for the system.

In Figure 7.22(a), we show the final region obtained after sufficiently many PFR trajectories have been run. The lower part of the boundary, given by AJD, thus represents a close approximation of the true AR boundary for this system using

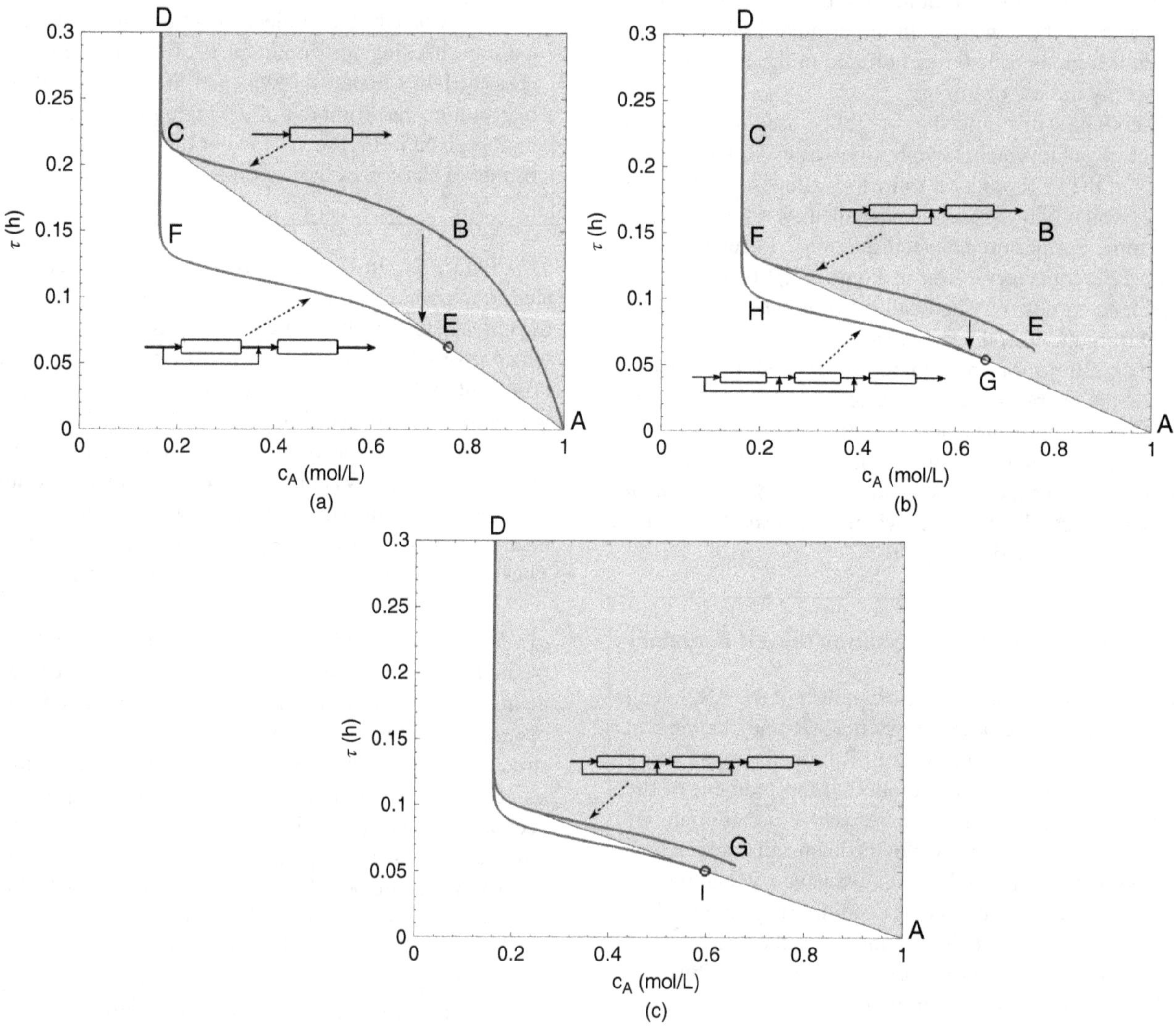

Figure 7.21 Generating a candidate AR using only PFRs and a base trajectory. Extension via (a) two PFRs (b) three PFRs, and (c) four PFRs in series, respectively.

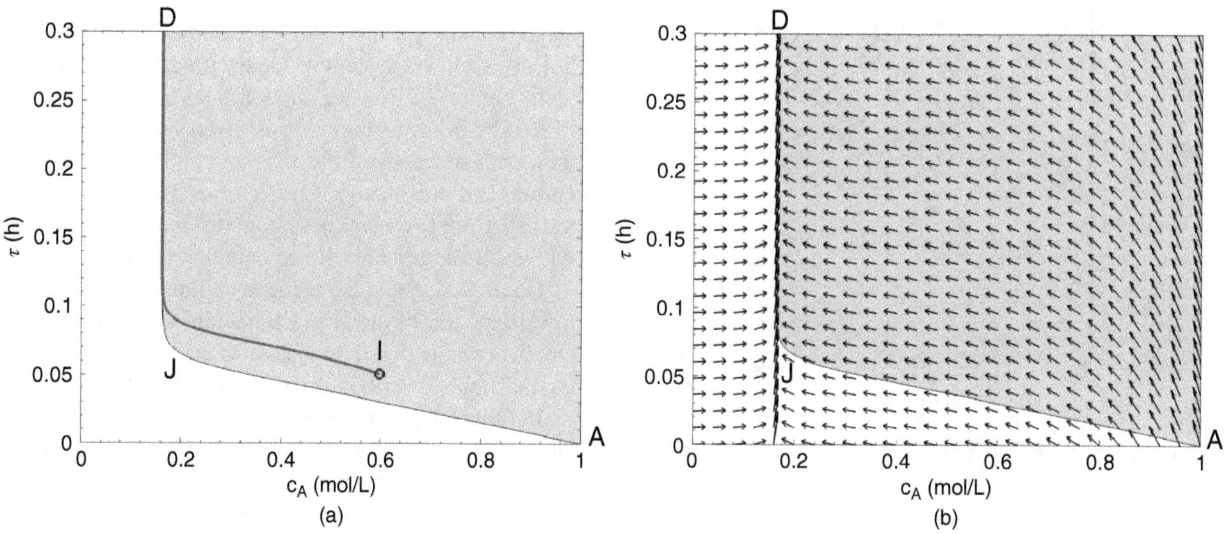

Figure 7.22 (a) AR generated using only PFRs and (b) the rate field showing that there are no rate vectors pointing out of the boundary.

combinations of PFRs only. (The true AR boundary for this system would require an infinite number of PFRs.) For comparison, the PFR trajectory after three PFRs is also shown, which is given by curve ID. Although the AR boundary in Figure 7.22(a) has been expanded, the difference between the two regions is small in comparison to the number PFRs used—that is, many PFR stages are needed to produce curve AJD, whereas boundaries in Figure 7.21 require less than four PFRs in total. To demonstrate that the true AR boundary in Figure 7.22(a) has been approximated, the rate field is overlaid in Figure 7.22(b). Note that it is difficult to identify rate vectors that point out of the boundary, which indicates that no further significant expansion of the region is possible.

SIDE NOTE: What if we were not Constrained by PFRs Only?

We have shown how the AR can be generated using only PFRs. Figure 7.22 is the full AR for the system, and thus it does not change if other reactor types are used in its construction.

Observe that tangency points (i.e., points E, G, and I) are those that lie on a mixing point with the feed concentration. These points are also tangent to the PFR base trajectory, for this is how lower overall system residence times were achieved. But rate vectors evaluated at points E, G, and I are also tangent to the boundary, and thus they must be collinear with the vectors $(\mathbf{C}_E - \mathbf{C}_f)$, $(\mathbf{C}_H - \mathbf{C}_f)$, and $(\mathbf{C}_I - \mathbf{C}_f)$. It is not difficult to see then that points E, G, and I must also satisfy the condition of a CSTR from the feed point. Hence, points E, H, and I are CSTR effluent concentrations corresponding to different residence times.

If we were not constrained by PFRs only, the resulting optimal reactor structure would therefore be a CSTR from the feed followed by a PFR. This is easily verified by plotting the CSTR locus from the feed point and observing how the CSTR locus intersects the base trajectory ABCD as it is moved downward (verify this behavior for yourself.) Points E, H, and I are thus intersection points between the base trajectory and different points along the CSTR locus from the feed.

7.3.3 Insights into Interstage and Cold-Shot Cooling Operation

In the previous two examples, we looked at how to construct the AR for an exothermic adiabatic reaction, when construction is constrained to PFRs only. These two constraints are fairly common when dealing with exothermic equilibrium limited reactions, such as in ammonia synthesis—one may need to operate a reactor at a high temperature, in favor of a high rate of reaction, but at the expense of a lower overall (equilibrium) conversion. In order to circumvent this limitation, the reaction may be divided into a number of smaller adiabatic reactors in series, and cooled between adiabatic reactors. Let us now investigate how these two common approaches (interstage and cold-shot cooling) may be explained from a geometric perspective using AR theory, and how they may be improved from the insights gained.

7.3.3.1 Minimum Residence Time for Interstage Cooling

a) A Distinct Number of Reactors In this example, we wish to briefly discuss the operation and optimization of the interstage cooling arrangement, such as that given in Figure 7.23.

Heat exchangers are placed before the feed to each new reactor to control the feed temperature. In this situation, the reactor structure is fixed (it is already known how the adiabatic reactors should be arranged), and we wish to identify the minimum residence time achieved for this arrangement (specifically, by understanding what cooling duty is required between reactors, and the residence times of reach reactor).

For simplicity, it is assumed that the same reaction and kinetics are available as in Section 7.3.2 so that each reactor obeys Equations 7.14 and 7.15. The characteristic shape of each PFR trajectory is then uniquely defined when an inlet temperature T_{in} is specified. Identifying the reactor structure with minimum residence time is then found by generating a candidate AR in concentration–residence time space. In Figure 7.24, three PFR trajectories from the feed are plotted corresponding to inlet temperatures of 300, 320, and 350 K, respectively.

A balance between a high inlet temperature and the final equilibrium limited concentration can be visualized—the PFR trajectory corresponding to 350 K produces the fastest reaction rate (the profile is situated the lowest to the τ axis), yet it is also the profile corresponding to the lowest overall conversion. By comparison, the profile belonging to 300 K produces the highest equilibrium conversion, but it is also associated with the largest residence time. In order to expand the region of achievability in an optimal manner, a *falling temperature profile* is required. (One where a reaction arrangement that combines for both a high initial reaction rate with a gradual cooling of the product stream as the reaction proceeds is employed.)

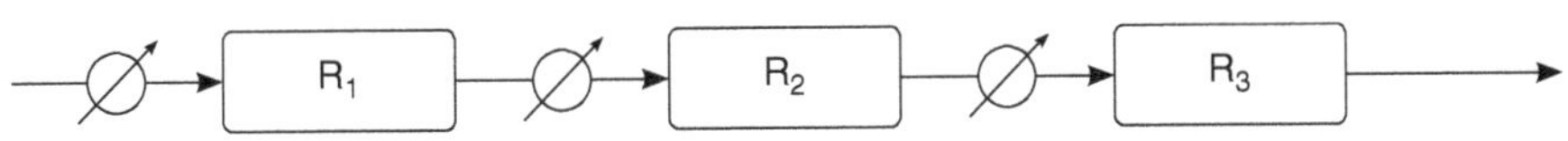

Figure 7.23 Interstage cooling without bypass.

EXAMPLE 4: Transitioning between Adiabatic Profiles

Consider again Figure 7.24. If an inlet temperature of $T_{in} = 350\,K$ is specified to the first reactor, then the profile ABX is obtained. Assume that the PFR is sized so that the exit concentration from the first reactor is that given by point X. If the concentration vector at point X is $\mathbf{C}_X = [0.2508, 0.1057]^T$, answer the following questions:

(a) What are the values of c_A and τ at point X, and what does this state physically represent?

(b) What is the temperature at the **exit** of the reactor if the **inlet** temperature is 350 K?

(c) If point X were to be achieved by using an inlet temperature of 320 K instead, then what would the exit temperature of this reactor be? How could this be used to formulate an interstage cooling strategy involving two adiabatic reactors?

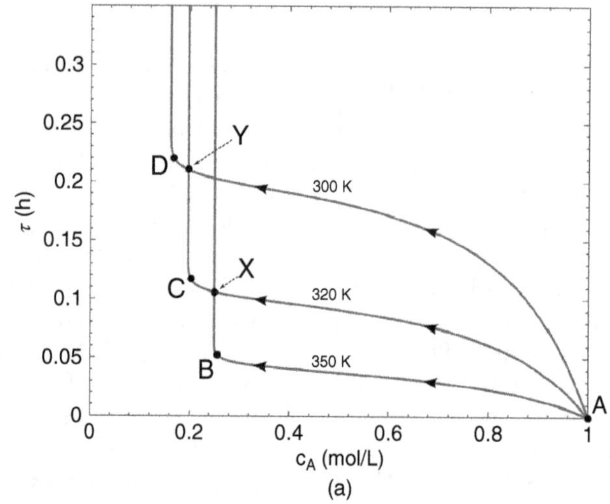

Figure 7.24 Adiabatic PFR trajectories for three different inlet temperatures.

Answer a:

Since the state vector is given by $\mathbf{C}_X = [0.2508, 0.1057]^T$, and the profiles are plotted in c_A–τ space, then $c_A = 0.2508$ mol/L and $\tau = 0.1057$ h at point X. These values represent the exit concentration of A and the corresponding residence time for PFR trajectories with feed temperatures of 350 and 320 K.

Answer b:

The temperature expression given by Equation 7.15 may be used to determine the exit temperature from the first reactor. Hence, we can solve directly for the exit temperature of the reactor with a 350 K feed as follows:

$$\begin{aligned} T_1 &= T_{in} + T_{ad}(c_{Af} - c_A) \\ &= 350 + 200(1 - 0.2508) \\ &= 499.84\text{ K} \end{aligned}$$

Answer c:

The procedure to calculate the exit temperature in this case is the same as the previous example; however, the inlet temperature used is now 320 K instead of 350 K

$$\begin{aligned} T_2 &= T_{in} + T_{ad}(c_{Af} - c_A) \\ &= 320 + 200(1 - 0.2508) \\ &= 469.84\text{ K} \end{aligned}$$

Point X is also an intersection point with the PFR profile with an inlet temperature of $T_{in} = 320$ K. Since the shape of the adiabatic profiles is fixed for an inlet temperature, we may traverse along the second PFR trajectory XCY if the inlet temperature to the second reactor is equal to T_2. In this way, we transition from the adiabatic curve corresponding to 350 K to the curve corresponding to 320 K and achieve a slightly higher conversion in the process. Since the exit temperature from the first reactor is $T_1 = 499.84$ K, one must cool this stream down to a temperature that corresponds to the temperature at X for the second reactor with an inlet temperature of $T_2 = 469.84$ K.

A transition between trajectories, by using interstage cooling strategy, may then be achieved by coordinating the feed conditions to the second reactor as if these conditions are same as running a single PFR with an inlet temperature of 320 K.

This process may be repeated for a third reactor. Assume that the second adiabatic PFR is operated up to point Y in Figure 7.24, which is given by the concentration vector $\mathbf{C}_Y = [0.196, 0.2099]^T$. Again, using an inlet temperature of 320 K and exit concentration of $c_A = 0.196$ mol/L, Equation 7.15 may be used to find the exit temperature of the second reactor. We find that this temperature is $T_2^{\#} = 480.8$ K. In order to traverse along the profile associated with 300 K, the reactor product stream must be cooled from $T_2^{\#} = 480.8$ K to the equivalent exit temperature for the reactor associated with the 300 K curve. Using Equation 7.15, the value for this exit temperature may be solved for, giving $T_3 = 460.8$ K. A transition from curve XCY to curve AYD may again be achieved to obtain a slightly larger overall conversion. Figure 7.25 shows a summary of the reactor arrangement associated with this process.

b) Extension to Many Adiabatic Reactors In Figure 7.26, we extend this concept to include a number of different PFR trajectories, at many more inlet feed temperatures. Recall that a unique PFR trajectory is obtained for a fixed feed temperature.

The family of PFR trajectories is given by the solid curves, where each curve corresponds to an adiabatic PFR trajectory with a unique inlet temperature, T_{in}. If we examine the PFR trajectories with inlet temperatures of 310, 320, and 330 K,

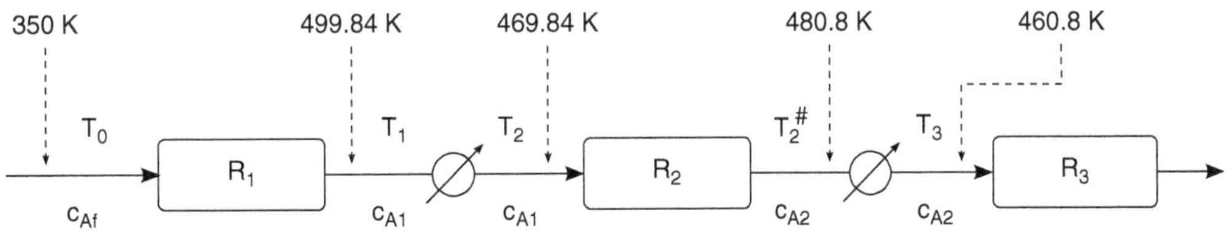

Figure 7.25 Recommended arrangement for three adiabatic PFRs.

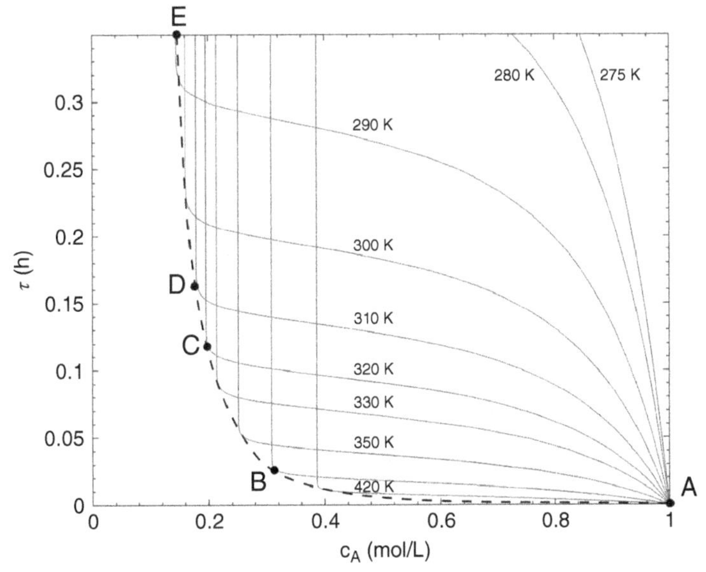

Figure 7.26 A family of PFR trajectories for different starting inlet temperatures T^0.

it is clear that the PFR trajectory with an inlet temperature of 320 K lies on the boundary for exit conditions between points C and D. For concentrations lower than D (i.e., a larger conversion of c_A), the trajectory corresponding to inlet temperatures smaller than 310 K is needed.

The region produced by the family of PFR trajectories hence serves as a candidate AR for the reactor structure involving interstage cooling. We can find the convex hull of the trajectories by integrating the PFR equation for various inlet temperatures and then computing the convex hull belonging to the entire set of PFR trajectories. When this is done, the boundary of the convex hull is given by the dotted line ABCDE. A vertical line of achievable points extended from the feed point $\mathbf{C}_f = [1, 0]^T$ is also included in the set. (Recall that in residence time plots, concentrations achieved at time $\tau = \tau_0$ are also concentrations that are achievable at $\tau \geq \tau_0$.)

From the figure, it is reasonable to expect that the ideal arrangement for many adiabatic reactors connected in series thus follows a falling temperature profile—by first preheating the feed to the first PFR to a high temperature, and then gradually reducing the feed temperature for subsequent reactors.

If the reaction is split into many small pieces, with cooling between each piece, then each reactor would traverse along a small segment on the boundary ABCDE in Figure 7.26. As these sections are made smaller, we approach the structure of a differentially cooled reactor (DCR), and thus the candidate region boundary associated with the DCR is therefore given by an *envelope of adiabatic PFR trajectories* with different feed temperatures.

It is possible to show that the optimal temperature profile for traversing along the boundary ABCDE is given by the following condition:

$$\left(\frac{\partial \tau}{\partial T^0}\right)_{c_A} = \frac{\partial}{\partial T^0}\left[\int_{c_{Af}}^{c_A^*} \frac{1}{r_A} dc_A\right]$$
$$= -\left(\int_{c_{Af}}^{c_A^*} \frac{1}{{r_A}^2}\left(\frac{\partial r_A}{\partial T}\right)_{c_A}\left(\frac{\partial T}{\partial T^0}\right)_{c_A} dc_A\right) = 0 \tag{7.16}$$

ILLUSTRATION: Deriving Equation 7.16

Although the form of Equation 7.16 may appear complex, its derivation simply involves a small amount of calculus. The reaction investigated is exothermic, and therefore a balance in the operation of the reactor must be established between high temperature (favoring high reaction rate), and low temperature (favoring high conversion). In order to obtain the boundary given in Figure 7.26, a reactor that is differentially cooled along its length must be employed. Hence, we wish to determine how the reactor residence time may be minimized with respect to the *feed temperature*, T^0, of a differential slice along the reactor. If residence time must be *minimized* with respect to T^0, then we may express this minimum mathematically by setting the derivative of τ with respect to T^0 to zero along the reactor length as follows:

$$\left(\frac{\partial \tau}{\partial T^0}\right)_{c_A} = 0 \tag{7.17a}$$

In order to relate τ to the inlet temperature T^0, we start with the definition of reaction rate,

$$\frac{dc_A}{d\tau} = r_A(c_A, T)$$

We shall neglect the parenthesis for r_A for brevity. The definition of the rate of reaction is a variable separable differential equation. In general, τ may be computed from the following integral with appropriate initial conditions on c_A:

$$\tau(c_A) = \int_{c_{Af}}^{c_A} \frac{dc_A}{r_A} \tag{7.17b}$$

Substituting Equation 7.17b into Equation 7.17a gives

$$\left(\frac{\partial \tau}{\partial T^0}\right)_{c_A} = \frac{\partial}{\partial T^0}\left(\int_{c_{Af}}^{c_A} \frac{dc_A}{r_A}\right)_{c_A}$$
$$= \int_{c_{Af}}^{c_A} \frac{\partial}{\partial T^0}\left(\frac{1}{r_A}\right)_{c_A} dc_A \tag{7.17c}$$

The integrand may be expanded giving

$$\frac{\partial}{\partial T^0}\left(\frac{1}{r_A}\right)_{c_A} = -\frac{1}{{r_A}^2}\left(\frac{\partial r_A}{\partial T^0}\right)_{c_A}$$

and by the chain rule for partial derivatives

$$\left(\frac{\partial r_A}{\partial T^0}\right)_{c_A} = \left(\frac{\partial r_A}{\partial T}\right)_{c_A}\left(\frac{\partial T}{\partial T^0}\right)_{c_A}$$

which gives

$$\frac{\partial}{\partial T^0}\left(\frac{1}{r_A}\right)_{c_A} = -\frac{1}{{r_A}^2}\left(\frac{\partial r_A}{\partial T}\right)_{c_A}\left(\frac{\partial T}{\partial T^0}\right)_{c_A} \tag{7.17d}$$

Substitution of Equation 7.17d into Equation 7.17c results in the desired expression

$$\left(\frac{\partial \tau}{\partial T^0}\right)_{c_A} = -\int_{c_{Af}}^{c_A} \frac{1}{{r_A}^2}\left(\frac{\partial r_A}{\partial T}\right)_{c_A}\left(\frac{\partial T}{\partial T^0}\right)_{c_A} dc_A = 0$$

This expression is general, and it may be used for any rate expression that is a function of c_A and T (and hence T^0). Evaluation of the partial derivatives in the integrand is straightforward given the appropriate rate and temperature expression. The integration of this equation provides a relation between T^0 and c_A for when residence time is minimized. This relation may be used in conjunction with the rate expression and Equation 7.17b to give the concentration profile in c_A–τ space that traverses the boundary ABCDE given in Figure 7.26.

EXAMPLE 5: Numerical Determination of the DCR Boundary

Using the results obtained for the adiabatic system, compute the optimal temperature profile required by a DCR if it is to traverse along the candidate region boundary given by ABCDE in Figure 7.26.

Answer:

The rate expression for the system is given in Section 7.3.2.1, and is repeated here for ease of reference.

$$r_A = -k_1 \exp\left(-\frac{E_1}{T}\right) c_A + k_2 \exp\left(-\frac{E_2}{T}\right)(c_{Af} - c_A) \tag{7.14}$$

Here, the reactor temperature is given by

$$T = T_{in} + T_{ad}(c_{Af} - c_A) \tag{7.15}$$

Rate constants and activation energies are given as $k_1 = 5 \times 10^5\ h^{-1}$, $k_2 = 5 \times 10^8\ h^{-1}$, $E_1 = 4000$ K, and $E_2 = 8000$ K. Similarly, T_{ad} is the adiabatic temperature rise, specified as $T_{ad} = 200$ K.

The condition for traversing along the boundary ABCDE is given by Equation 7.16. This involves the solution of a nonlinear function, which itself is defined by the integral of products of partial derivatives involving r_A and T. Analytical solution of Equation 7.16 is difficult for this reason, and we prefer, instead, to evaluate this function numerically. See the following text.

The partial derivatives $(\partial r_A/\partial T)_{c_A}$ and $(\partial T/\partial T^0)_{c_A}$ are first computed. For the rate and temperature expressions defined, these derivatives are given as

$$\left(\frac{\partial r_A}{\partial T}\right)_{c_A} = -k_1 \exp\left(\frac{-E_1}{T}\right) c_A \left(\frac{E_1}{T^2}\right) + k_2 \exp\left(\frac{-E_1}{T}\right)(c_{Af} - c_A)\left(\frac{E_2}{T^2}\right)$$

and

$$\left(\frac{\partial T}{\partial T^0}\right)_{c_A} = 1$$

It is also possible to compute the partial derivatives numerically, although the expressions given here are simple enough that they can be found analytically. The simplicity of the temperature expression in this instance allows for a straightforward relation for the $(\partial T/\partial T^0)_{c_A}$ term. Substitution of these partial derivatives into Equation 7.16 produces a function in terms of c_A, T and T_{in}. Noting that T_{in} is a function of c_A and T, let us define a function $f(c_A, T)$ as follows:

$$f(c_A, T_{in}) = \left(\frac{\partial \tau}{\partial T^0}\right)_{c_A}$$

$$= -\int_{c_{Af}}^{c_A} \frac{1}{{r_A}^2}\left(\frac{\partial r_A}{\partial T}\right)_{c_A}\left(\frac{\partial T}{\partial T^0}\right)_{c_A} dc_A$$

To traverse along boundary ABCDE, we wish to find the specific pair of values, c_A^* and T^*, that satisfies the following relation:

$$f(c_A^*, T_{in}^*) = 0$$

One approach to solving this equation is to specify a reactor temperature T and then solve for the corresponding exit concentration c_A that satisfies $f(c_A, T_{in}) = 0$.[7] Solution of f, given a value for T, involves the solution of a nonlinear function in a single variable (c_A in this instance). This may be achieved with any suitable numerical nonlinear solver.[8] Figure 7.27(a) shows a representative plot of f versus c_A, when T is specified at 450 K.

The solution may be repeated for a range of T values. The corresponding (c_A, T) pairs that solve f = 0 may then be used to generate a profile.

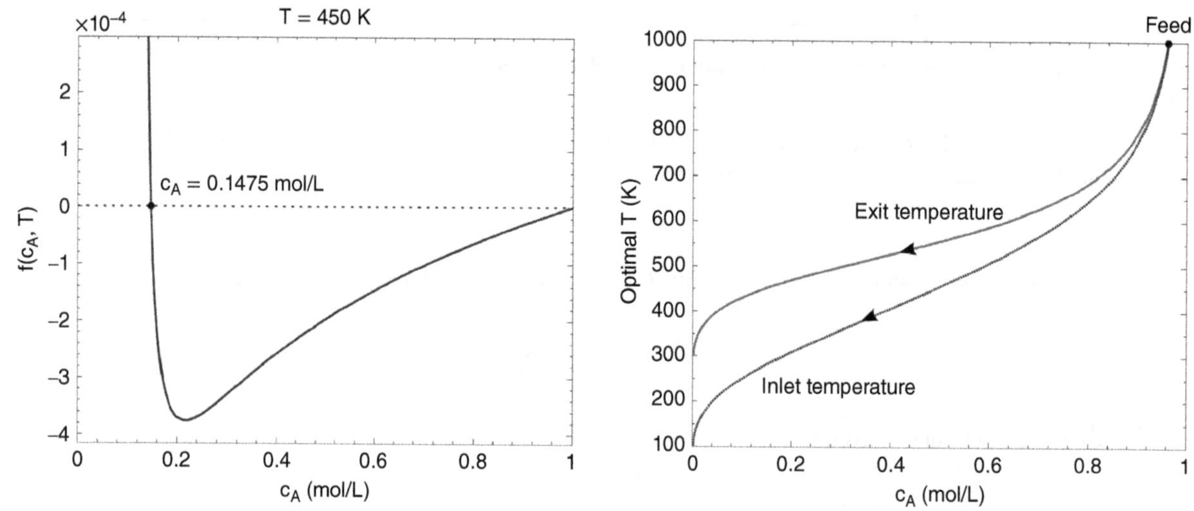

Figure 7.27 (a) The roots of function f when T = 450 K and (b) optimal temperature profiles for a differentially cooled reactor for traversal along boundary ABCDE.

We should be mindful of the T that we are attempting to solve for in f—multiple temperatures exist at a given c_A. T in this case is the reactor *exit* temperature (the temperature corresponding to the reactor exit *before* it is cooled to the next differential slice of reactor). However, Equation 7.15 may also be used to calculate the inlet temperature, T_{in}, to the next differential slice of reactor. Thus, two profiles may be plotted. The results of this process are given in Figure 7.27(b).

[7] We might also choose to specify a c_A and then solve for the corresponding T, although this involves a highly nonlinear function in T.

[8] In this example, MATLAB's `fzero()` function has been used to compute the corresponding T.

This is the required temperature profile that a DCR must employ in order to traverse along the boundary ABCDE in Figure 7.26. Observe that a falling temperature profile is required as expected. The profile given by the top curve is that of the current reactor temperature (before cooling), whereas the latter is the "inlet" temperature to the next differential slice of reactor (after cooling). The difference in these two values for a fixed value of c_A provides an indication of the cooling duty at any particular exit concentration. A larger cooling duty, which is indicated by the vertical distance between inlet and outlet temperatures for a given value of c_A, is required as the reaction proceeds.

7.3.3.2 Minimum Residence Time for Cold-Shot Cooling

a) Introduction In Section 7.3.3.1, we showed how a falling temperature profile could be achieved in an interstage cooling reactor network. Cold-shot cooling is another popular technique for achieving intermediate cooling as it does not require additional cooling equipment. In this example, we wish to determine the candidate AR for a reactor system involving cold-shot cooling. Cold-shot cooling derives its name from the manner in which reactor effluent streams are cooled between reaction stages. The cooling in this instance is achieved via a "cold-shot" of feed material, as displayed in Figure 7.28.

This arrangement is similar to that of the interstage cooling structure provided earlier. However in this scenario, a portion of the feed is bypassed and mixed with reactor product after each reaction stage, and there is no need to use dedicated cooling equipment, which lowers the costs of operation. The mixture temperature must lie at an intermediate value between the feed and product temperatures. It follows that an expression for how reactor feed temperatures vary with mixing is thus necessary.

EXAMPLE 6: Demonstrating Cold-Shot Cooling

Assume that two adiabatic reactors are available with the same kinetics given in Section 7.3.3.1. Using an inlet temperature of 300 K, plot the PFR trajectory for the first reactor as well as the possible PFR trajectories for the second reactor if cold-shot cooling is employed. Use mixing fractions of $\lambda = 0.25$, 0.5, and 0.75. Assume that the feed vector is given by $\mathbf{C}_f = [c_{Af}, \tau^0]^T = [1, 0]^T$, and that the exit concentration from the first reactor is $c_A = 0.2$ mol/L.

For this example, a relation to describe how temperature changes with mixing is required. For simplicity, you may assume that temperature obeys a linear mixing law, so that $T^* = \lambda T_1 + (1 - \lambda)T_2$.

Answer:

Since is it assumed that temperature obeys a linear mixing law, we can use the values of λ provided to compute mixture concentrations as well as temperatures that act as feed conditions to the second reactor. Therefore, using the exit concentration of the first reactor, the corresponding residence time can be found. This information, in conjunction with the feed, then gives the following mixing relations:

$$\begin{aligned} c_A^* &= \lambda c_{Af} + (1 - \lambda)c_A \\ \tau^* &= \lambda \tau^0 + (1 - \lambda)\tau \\ T^* &= \lambda T^0 + (1 - \lambda)T \end{aligned} \tag{7.18}$$

Here

$$T = T_{in} + T_{ad}(c_{Af} - c_A) \tag{7.15}$$

$c_{Af} = 1$ mol/L, $\tau^0 = 0$ h, and $T^0 = 300$ K. In this instance, the exit concentration from the first reactor is known ($c_A = 0.2$ mol/L). Integrating the PFR equation for an inlet temperature of 300 K, we can find the corresponding residence time for $c_A = 0.2$ mol/L. This is $\tau = 0.2091$ h. The exit conditions from the first reactor may then be used with the feed conditions along with Equation 7.18 and Equation 7.15 to find the mixture state used in the feed to the second reactor. Table 7.1 provides a summary of the mixing results.

Plots of the PFR trajectories using the data from Table 7.1 are shown in Figure 7.29.

TABLE 7.1 Mixing Results for Different Values of λ

λ	c_A^* (mol/L)	τ^* (h)	T^* (K)
0.2	0.36	0.1673	428
0.4	0.52	0.1255	396
0.6	0.68	0.0836	364
0.8	0.84	0.0418	322

Note that for values up to $\lambda \sim 0.75$, the PFR trajectory associated with the second reactor after cold-shot cooling achieves a lower overall residence time. Using this approach, we could compute the optimal mixing fraction for the second reactor that gives the lowest residence time overall for the two reactor cold-shot cooling structure.

In Figure 7.29(b), we show the candidate region for the collection of PFR trajectories that is each associated with

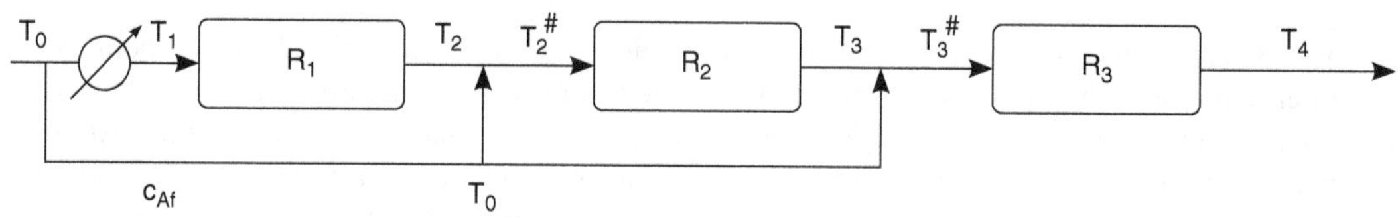

Figure 7.28 Cold-shot cooling for an adiabatic exothermic reaction.

a different value of λ. Note that a concavity still exists in the region, and thus it is possible to improve the system further (by decreasing the overall residence time of the network) if a third reactor is employed. The constructive nature of this exercise is similar to the sliding process of PFR construction from Section 7.3.2, although, in this instance, one must also account for the effect of mixtures changing both inlet temperature and composition to the next reactor segment.

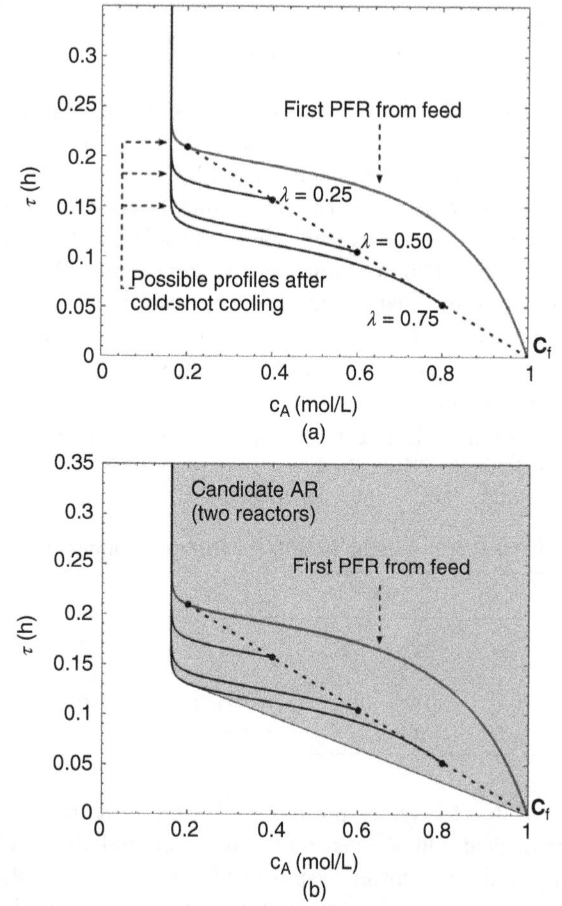

Figure 7.29 (a) Cold-shot cooling for different mixing fractions using two reactors and (b) candidate region obtained for all mixing fractions for the two reactor arrangement.

b) Expanding the Differentially Cooled Reactor Boundary
In the previous example, we showed how the candidate AR for a two reactor cold-shot cooling arrangement could be achieved. The particular region obtained is dependent on the feed temperature and exit concentration from the first reactor. Note that although any exit concentration on the PFR trajectory can be used as a basis for constructing a mixing line as a starting point for the next reactor, this point should be chosen in a way that maximizes expansion of the region.

In Section 7.3.3.1, we showed the candidate AR for a DCR boundary. But a DCR might be viewed as a unique reactor type in itself, and combinations of DCRs with other reactor types could also be used to expand the region further. For example, a PFR could be operated from any point inside the candidate region of the DCR in Figure 7.26. Physically, this arrangement would represent a PFR in series with the DCR with bypass of feed. The analysis is made slightly more complex because now temperature is involved and thus each point in c–τ space is associated with many rate vectors, since each point is associated with multiple temperatures.

Consider now Figure 7.30(a). The DCR boundary is shown together with two representative PFR trajectories. Feed temperatures of 300 and 350 K have been used for demonstration. Although these trajectories have no physical significance in the context of the DCR arrangement, their corresponding profiles are related to the boundary. For each PFR from the feed point, we can locate the intersection point with the current boundary. In Figure 7.30(a), this is given by points B and C for inlet temperatures of 300 and 350 K, respectively.

Assume that a DCR is employed from the feed point and operated up to point B on the boundary of Figure 7.30(a). Point B is also a point on the adiabatic PFR trajectory from the feed with an inlet temperature of 300 K. Assume that a sidestream of cold-shot material is also available at the feed temperature before preheating to the first reactor. If this stream is mixed with the DCR effluent stream at point B, this process would be represented graphically by joining together points B and A with a mixing line. Mixing lines may then be constructed joining the feed point $\mathbf{C}_f$ with all points on the boundary, as shown in Figure 7.30(b) by straight lines AB and AC. Any point on these lines represents a mixture point from cold-shot cooling, using the feed and product from a

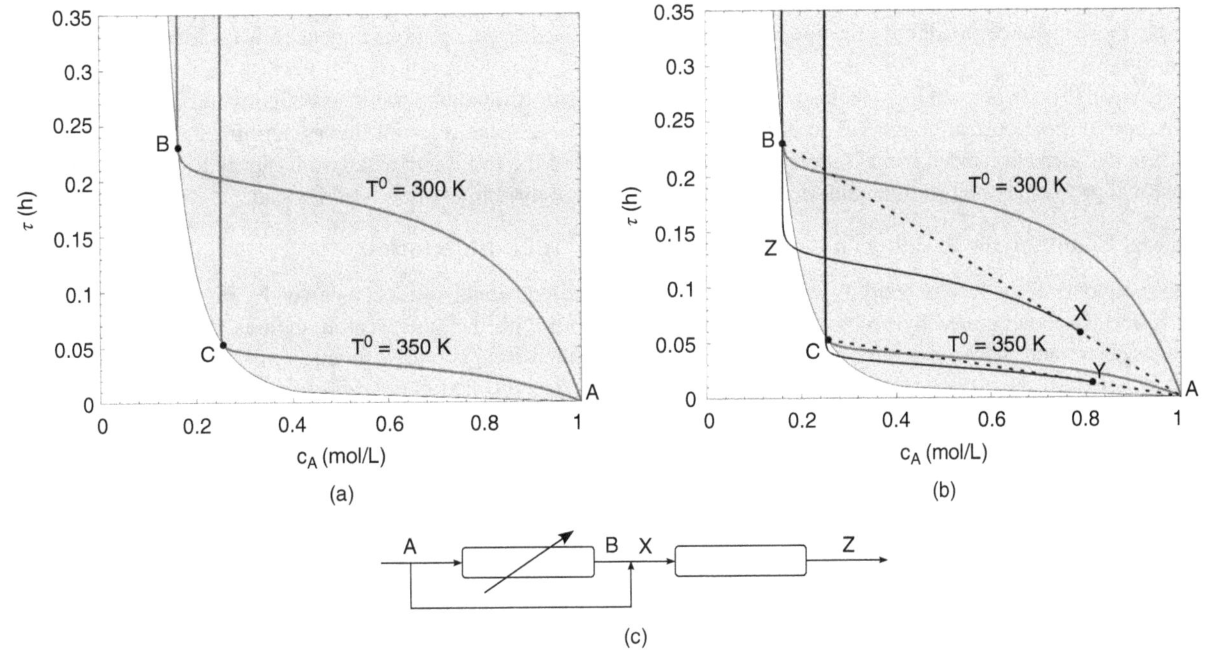

Figure 7.30 (a) Points on the DCR AR boundary, (b) mixing lines representing cold-shot cooling between DCR effluent with the feed, and (c) a DCR in series in a PFR corresponding to the PFR trajectories in (b).

DCR. For all points along these lines, the same procedure as in Example 6 can be carried out, generating a family of PFR trajectories. Points X and Y in Figure 7.30(b) are two representative mixing points that may be used to generate PFR trajectories. The PFR trajectories resulting from point X and Y are also shown in Figure 7.30(b).

Note that portions of both trajectories emanating from points X and Y manage to extend the boundary of the DCR given in Section 7.3.3.1. Point Z in Figure 7.30(b) is thus a PFR effluent concentration that extends the candidate AR boundary. Hence, we are able to expand the candidate region further by use of cold-shot cooling in conjunction with the DCR. The reactor structure belonging to this region in general is then given by Figure 7.30(c). This structure produces a second envelope of extrema that may be used to further extend the boundary. (The letters in Figure 7.30(c) correspond to the particular points required to achieve point Z in Figure 7.30(c).)

For more than two reactors, the same procedure may be carried out. However, successive application of additional reactors produces a smaller improvement each time. This approach highlights an important point. The reactor structure in this investigation has been fixed, yet we are interested in understanding how best to improve the system given these constraints. Although these restrictions result in constructions that do not represent the full AR, the geometric interpretation afforded by viewing candidate regions in this way allows one to still gain insight into the maximum performance *for the constrained system.*

ILLUSTRATION: When Temperature does not Obey Linear Mixing

In this chapter, a number of simplified examples have been outlined to demonstrate how AR theory may be used to answer common reactor synthesis problems related to adiabatic systems and minimum residence time. A number of natural extensions to these discussions may be carried out that enhance the use of AR theory to nonisothermal systems. For the interested reader, two notable papers are available that extend on the ideas discussed here. Nicol et al. (1997) show how the AR for an exothermic reaction may be generated that incorporates external heating or cooling, whereas Glasser et al. (1992) extend the two-dimensional preheating examples, shown in this chapter, to $\mathbb{R}^3$ involving x–T–τ space.

Realistic systems often are not operated solely under adiabatic or isothermal conditions however. Many systems of interest rely on a controlled variation of temperature in an optimal manner. Although methods, such as optimal control theory, may be used in these instances to identify the temperature profiles required, these methods may not always identify discontinuities in the profile.

ARs may also be generated for systems where a control parameter (often temperature) is employed. This theory is not part of the scope of this book, although we do provide a brief discussion of the relevant theory here for interest. The following discussion has ideas borrowed from Godorr et al. (1999), and all of the following discussions apply to $\mathbb{R}^3$ alone. For higher dimensional problems, the use of an automated AR construction algorithm is often used instead.

A General Condition for Residing on the AR Boundary in $\mathbb{R}^3$ (PFR trajectories)

The AR boundary for a system residing in $\mathbb{R}^3$ is given by a two-dimensional surface, as given by Figure 7.31. It is assumed that this surface is generated by process vectors that change both with the state vector, such as the concentration vector $\mathbf{C} = [x_1, x_2, x_3]^T$ and the reactor temperature T. The rate vector, therefore, is both a function of **C** and T, or **r**(**C**, T). In this scenario, there is no assumption that T obeys a linear mixing law. As a result, T is not represented as a component in the state vector **C**, but rather as a parameter in the component rate expressions. (Note that residence time, τ, may still be included in **C**, as it always obeys a linear mixing law.) The AR boundary is assumed to be parameterized by two scalar quantities: τ (the reactor residence time) and s. τ represents the arclength of a PFR solution trajectory satisfying

$$\frac{d\mathbf{C}}{d\tau} = \mathbf{r}(\mathbf{C}, \mathrm{T})$$

whereas scalar s is constant along any given PFR trajectory that varies smoothly from one trajectory to the next. In this sense, it is more correct to define the rate vector by partial derivatives involving τ and s as follows:

$$\mathbf{r}(\mathbf{C}, \mathrm{T}) = \left(\frac{\partial \mathbf{C}}{\partial \tau}\right)_{\mathrm{s}}$$

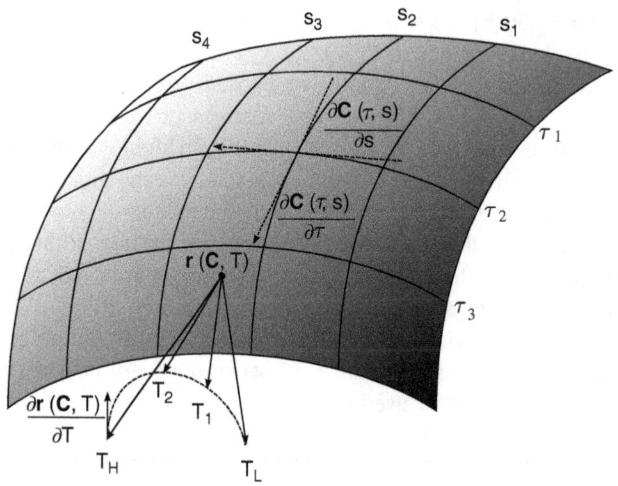

Figure 7.31 A segment of the AR boundary in $\mathbb{R}^3$ when temperature is a control parameter. Adapted from Godorr et al. (1999).

At any particular point on the AR boundary, the rate vector may now point in *multiple directions*, depending on the value of T. A locus of potential points may be traced out at any value of **C** on the AR boundary by specifying a range of temperatures. The tangent vector to this locus is given by the partial derivative of **r**(**C**, T) with respect to T for a given **C**, or $\partial\mathbf{r}/\partial\mathrm{T}$. In order for this locus to reside on the AR boundary, this vector must be tangent to the surface. A tangent plane to the AR boundary may be formed by spanning vectors in terms of the two surface parameters τ and s. Hence, a tangent plane is uniquely defined at a point **C** on the AR boundary by the vectors $\partial\mathbf{C}(\tau, \mathrm{s})/\partial\tau = \mathbf{r}(\mathbf{C}, \mathrm{T})$ and $\partial\mathbf{C}(\tau, \mathrm{s})/\partial\mathrm{s}$. From this, the locus of potential points, generated by T, must be coplanar with the tangent plane defined by $\partial\mathbf{C}(\tau, \mathrm{s})/\partial\tau$ and $\partial\mathbf{C}(\tau, \mathrm{s})/\partial\mathrm{s}$ on the AR boundary, and thus it must also be orthogonal to the normal vector belonging to the plane defined by $\partial\mathbf{C}(\tau, \mathrm{s})/\partial\tau$ and

$\partial \mathbf{C}(\tau, s)/\partial s$. Mathematically, this is given by

$$\xi_1 = \left[\mathbf{r}(\mathbf{C}, T) \times \frac{\partial \mathbf{C}}{\partial s}\right]^T \frac{\partial \mathbf{r}}{\partial T} = 0$$

where $\times$ is the standard vector cross product in $\mathbb{R}^3$. This condition must hold true for a PFR trajectory to lie on the AR boundary. Hence, derivatives of ξ_1 must therefore also be zero so that $d\xi_1/d\tau = 0$. From this, it may be shown that in order for a control parameter, such as temperature, to lie on the AR boundary, the temperature change with respect to residence time be given by the following expression:

$$\frac{\partial T}{\partial \tau} = -\frac{\left[\mathbf{r}(\mathbf{C}, T) \times d\mathbf{r}(\mathbf{C}, T)\frac{\partial \mathbf{r}}{\partial T}\right]^T \frac{\partial \mathbf{r}(\mathbf{C}, T)}{\partial T} + \left[\mathbf{r}(\mathbf{C}, T) \times \frac{\partial \mathbf{r}}{\partial T}\right]^T \frac{\partial}{\partial \mathbf{C}}\left(\frac{\partial \mathbf{r}}{\partial T}\right)\mathbf{r}(\mathbf{C}, T)}{\left[\mathbf{r}(\mathbf{C}, T) \times \frac{\partial \mathbf{r}}{\partial T}\right]^T \frac{\partial^2 \mathbf{r}}{\partial T^2}}$$

This condition applies when temperature is bounded between the limits $T_{min} \le T \le T_{max}$.

Nonisothermal Critical DSRs in $\mathbb{R}^3$

For DSRs, the "vDelR" condition for critical α policies must still apply, although the conditions must be amended to include the temperature parameter as well.

$$\varphi_1(\mathbf{C}, T) = [\mathbf{r}(\mathbf{C}, T) \times \mathbf{v}]^T d\mathbf{r}(\mathbf{C}, T)\mathbf{v} = 0 \tag{7.19a}$$

An additional condition is required to fully specify a critical DSR with temperature variation

$$\varphi_2(\mathbf{C}, T) = [\mathbf{r}(\mathbf{C}, T) \times \mathbf{v}]^T \frac{\partial \mathbf{r}}{\partial T} = 0 \tag{7.19b}$$

A critical DSR trajectory with a varying temperature profile is one that is satisfied by Equation 7.19a and 7.19b. For a DSR to be critical, φ_1 *and* φ_2 must be held at zero throughout the entire integration range. Differentiating Equation 7.19a and 7.19b results in an expression in terms of the α policy as well as derivatives of T with respect to τ. Hence, by the total derivative of φ_1, we have the following:

$$d\varphi_1 = \frac{\partial \varphi_1}{\partial x_1}dx_1 + \frac{\partial \varphi_1}{\partial x_2}dx_2 + \frac{\partial \varphi_1}{\partial x_3}dx_3 + \frac{\partial \varphi_1}{\partial T}dT$$

Dividing by $d\tau$, and since $\varphi_1 = 0$, gives

$$0 = \left[\frac{\partial \varphi_1}{\partial x_1}, \frac{\partial \varphi_1}{\partial x_2}, \frac{\partial \varphi_1}{\partial x_3}\right]^T \left[\frac{\partial x_1}{\partial \tau}, \frac{\partial x_2}{\partial \tau}, \frac{\partial x_3}{\partial \tau}\right] + \frac{\partial \varphi_1}{\partial T}\frac{\partial T}{\partial \tau}$$

$$0 = [\nabla \varphi_1]^T[\mathbf{r}(\mathbf{C}, T) + \alpha \mathbf{v}] + \frac{\partial \varphi_1}{\partial T}\frac{\partial T}{\partial \tau} \tag{7.19c}$$

The same may be performed for φ_2 to give

$$0 = [\nabla \varphi_2]^T[\mathbf{r}(\mathbf{C}, T) + \alpha \mathbf{v}] + \frac{\partial \varphi_2}{\partial T}\frac{\partial T}{\partial \tau} \tag{7.19d}$$

The α policy and $\partial T/\partial \tau$ are the two unknown quantities present in the expressions. Simultaneous solution of Equation 7.19c and 7.19d provides a method for solving for both. When this is done, the following expressions are obtained:

$$\alpha = -\frac{\left[\frac{\partial \varphi_2}{\partial T}\nabla \varphi_1 - \frac{\partial \varphi_1}{\partial T}\nabla \varphi_2\right]^T \mathbf{r}(\mathbf{C}, T)}{\left[\frac{\partial \varphi_2}{\partial T}\nabla \varphi_1 - \frac{\partial \varphi_1}{\partial T}\nabla \varphi_2\right]^T \mathbf{v}} \tag{7.19e}$$

and for temperature

$$\frac{\partial T}{\partial \tau} = \frac{[\nabla\varphi_2]^T[\mathbf{r}+\alpha\mathbf{v}] - [\nabla\varphi_1]^T[\mathbf{r}+\alpha\mathbf{v}]}{\left[\frac{\partial\varphi_1}{\partial T} - \frac{\partial\varphi_2}{\partial T}\right]} \tag{7.19f}$$

Note that the policies defined by Equation 7.19e and 7.19f apply for an *unconstrained* temperature profile. In reality, we must enforce bounds on T, which are governed by our specific practical application. Thus, a maximum temperature, T_{max}, and minimum temperature, T_{min}, constraint must also be included so that value of T lies in the range $T_{min} \leq T \leq T_{max}$. Traversal on the AR boundary follows Equation 7.19e and 7.19f provided that T lies within T_{min} and T_{max}. If one of the temperature bounds is activated, then Equation 7.19b no longer applies and T is held at the temperature of the bounds.

DSR Equilibrium Points

Degenerate solutions to the critical DSR conditions given by Equation 7.19a and 7.19b occur when a DSR equilibrium point is reached (when this occurs, $\mathbf{r}(\mathbf{C}, T)$ and $\mathbf{v}$ are collinear and Equation 7.19a and 7.19b produce the null vector). These points represent solutions to a nonisothermal CSTR equation. However, one must also account for CSTR point as starting points for PFR and critical DSR trajectories. Additional conditions must be satisfied in this situation. In order for a CSTR to be a connector on the AR boundary for a nonisothermal system in $\mathbb{R}^3$, it may be shown that the tangent vector of the CSTR locus be coplanar with either the mixing or rate vector as follows:

$$\varphi_2(\mathbf{C}, T) = [\mathbf{r}(\mathbf{C}, T) \times \mathbf{t}]^T \frac{\partial \mathbf{r}}{\partial T} = 0$$

Here, $\mathbf{t}$ is the vector tangent to the CSTR locus, as defined in Chapter 6.

$$\mathbf{t} = [\mathbf{I} - \tau d\mathbf{r}(\mathbf{C}, T)]^{-1}\mathbf{r}(\mathbf{C}, T)$$

We note again that these conditions are specific to $\mathbb{R}^3$ only. Although similar results may be derived for higher dimensions, one often employs the use of an automated AR construction method instead.

7.4 AR THEORY FOR BATCH REACTORS

7.4.1 Introduction

Up to now, the entire development of the AR, including theory and all supporting examples, have been based on the idea of continuous operation. (This is because classic AR theory was developed with continuous reaction in mind.) With the expansion of chemical engineering into biological, pharmaceutical, and other specialist industries, the modern process engineer is often confronted with problems related to batch operation. As a result, the ideas discussed so far are not useful unless they can be converted for use with batch reaction.

In this section, we wish to briefly describe how a candidate AR, originally generated from continuous operation, may be transformed into an equivalent batch reactor structure, using simple conversion guidelines. This allows the use of the AR for both continuous and batch reaction. It is then possible to employ a single procedure to potentially optimize for two separate design scenarios. This approach allows for flexibility in design and analysis.

IMPORTANT: Continuous and Batch Operation

In the following sections, we will want to show how the states realized in PFRs, CSTRs, and DSRs can be achieved using batch reactors. Moreover, we would like to demonstrate that the necessary transformations from continuous to batch operation can be used in conjunction with the AR to help guide the design of optimal batch structures. Before discussion of the similarities between batch and continuous reactors may begin however, a clear difference between the two operating regimes should be highlighted.

Even though we wish to show that it is possible to attain steady state *concentrations* in a batch, it is not assumed that steady state itself can ever be attained. Batch processes, by their very nature, cannot achieve this. Hence, the outlook here will not be of the view that batch reactors, specifically fed-batch reactors, can ever be operated in any particular way that is able to reach steady state. We would, however, like to demonstrate that it is still possible to achieve the specific concentration associated with that steady state. Certainly, this will require a special operating regime to achieve (which shall be detailed in the following sections), but this will always be with the intention that the reactor is operated under batch conditions—that is, with a distinct cycle time where the state variables of the batch reactor (volume, concentration, etc.) do, in fact, vary for the duration of this period.

7.4.2 Similarities Between Batch and Continuous Reactive Equipment

7.4.2.1 ***The Standard Batch*** By standard batch, we have in mind a batch reactor where there is no inflow or outflow of material during the reaction period. The reactor contents are well mixed, although the concentration of species evolves with respect to time. Given an initial concentration of species, the concentration profile within a standard batch reactor may be determined by the integration of the appropriate rate expression. Thus, we have for the standard batch

$$\frac{d\mathbf{C}}{dt} = \mathbf{r}(\mathbf{C}) \tag{7.20}$$

The solution trajectory obtained by the integration of Equation 7.20 is identical to that obtained by the integration of the PFR equation for the same initial conditions. Reaction time of the batch replaces residence time of the PFR as the integration parameter in this regard. As a result, optimal reactor structures where PFRs form part of the AR boundary may be achieved in batch processes by running a standard batch reactor with the appropriate reaction time.

7.4.2.2 ***Fed-Batch Reaction*** Fed-batch reactors, often also referred to as semi-batch reactors, are those that allow the feed of additional material into the reaction vessel during the reaction period, and may be used to bias the selectivity of components in competing reactions. The general form of the fed-batch reactor is given as follows:

$$\begin{aligned}\frac{d\mathbf{C}}{dt} &= \mathbf{r}(\mathbf{C}) + \left[\frac{F(t)}{V(t)}\right](\mathbf{C}^0 - \mathbf{C}) \\ &= \mathbf{r}(\mathbf{C}) + \alpha(\mathbf{C}^0 - \mathbf{C}) \end{aligned} \tag{7.21}$$

Here, $\alpha = F(t)/V(t)$, F(t) is the rate of material addition into the reactor, and V(t) the reactor volume. It is assumed here that the sidestream feed concentration, $\mathbf{C}^0$, is constant in the derivation of the expression (Eq. 7.21). From Chapter 6, we know that this assumption does not affect the final result however.

EXAMPLE 7: Deriving the Fed-Batch Equation

Derive a differential equation that expresses the concentration change in a fed-batch reactor with respect to reaction time t. A *constant* sidestream concentration $\mathbf{C}^0$ is available and is fed into the fed-batch at a time-dependent volumetric rate of F(t). Constant density may be assumed.

Answer:

We may express the change in concentration of component *i* in the fed-batch batch by beginning with a molar balance of component *i* with respect to fed-batch reaction time t.

$$n_i^{t+\Delta t} = n_i^t + V(t)r_i(\mathbf{C})\Delta t + Fc_i^0\Delta t$$

Here, V(t) is the time-dependent total volume of the batch and $r_i(\mathbf{C})$ is the rate of reaction for component *i*. Dividing by Δt and taking the limit as $t \rightarrow 0$,

$$\frac{dn_i}{dt} = V(t)r_i(\mathbf{C}) + F(t)c_i^0$$

The moles of component *i* in the fed-batch as a function of time, $n_i(t)$, may be expressed in terms of concentration $c_i(t)$ and reactor volume V(t),

$$n_i(t) = c_i(t)V(t)$$

and when substituted into the differential equation gives

$$\frac{d[c_i(t)V(t)]}{dt} = V(t)r_i(\mathbf{C}) + F(t)c_i^0$$

$$V(t)\frac{dc_i}{dt} + c_i(t)\frac{dV}{dt} = V(t)r_i(\mathbf{C}) + F(t)c_i^0$$

Since density is assume constant, we may express the total volume change of the fed-batch over time in terms of the sidestream feeding rate F(t),

$$\frac{dV}{dt} = F(t)$$

which may be substituted into the previous expression and simplified to give

$$\frac{dc_i}{dt} = r_i(\mathbf{C}) + \left[\frac{F(t)}{V(t)}\right](c_i^0 - c_i)$$

Note that this expression is generalized for component i participating in the system. The vector equivalent version of this expression is therefore given by

$$\frac{d\mathbf{C}}{dt} = \mathbf{r}(\mathbf{C}) + \left[\frac{F(t)}{V(t)}\right](\mathbf{C}^0 - \mathbf{C})$$

The properties of fed-batch reactors are such that they may assume several functions, corresponding to the continuous fundamental reactor types discussed in Chapter 4. This is easily understood by way of an illustrative example.

Consider then Figure 7.32, which represents a hypothetical scenario for a number of fed-batch trajectories in concentration space. We assume that each trajectory is obtained by the integration of Equation 7.21 for a unique α policy. The feed conditions for each fed-batch trajectory are the same, and are given here by point A, that is, point A represents $\mathbf{C}_f$. In the following text, we wish to describe the behavior of a fed-batch trajectory under three specific cases for α.

a) Case 1: $\alpha = 0$ The solid curve AO is that of a solution trajectory corresponding to a fed-batch reactor when $\alpha = 0$. Point O is therefore the equilibrium point of the fed-batch. Clearly, the form of the fed-batch for this situation reduces to that of Equation 7.20 for the standard batch, and thus curve AO is also a solution for a PFR with feed composition given by point A. The analogy between standard batch reaction and PFRs has already been discussed above, and therefore this case is not considered further.

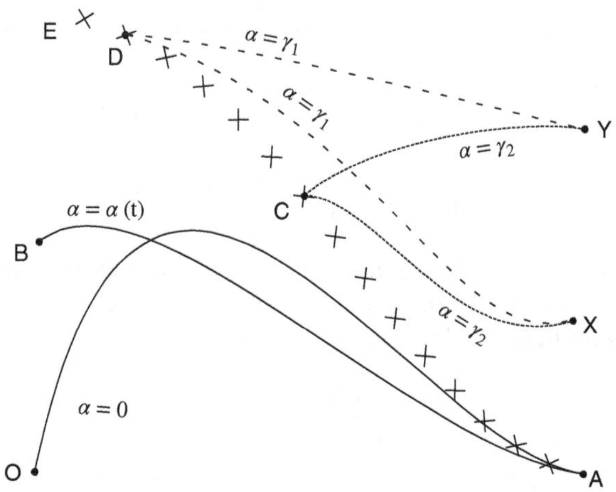

Figure 7.32 Graphical interpretation for various fed-batch reactor trajectories. Each trajectory corresponds to a different α policy. The sidestream composition of all trajectories is given by point O. Adapted from Ming et al. (2013).

OBSERVATION: Achieving PFR Concentrations in a Fed-Batch Reactor

When $\alpha = 0$ in a fed-batch reactor, Equation 7.21 reduces to the standard batch equation, which is identical to the PFR equation under continuous operation.

b) Case 2: $\alpha = \alpha(C(t))$ The most familiar method of operation is one whereby the feed addition policy is varied throughout the reaction period. Again, Equation 7.21 may be integrated if an initial condition $\mathbf{C}_f$, sidestream composition $\mathbf{C}^0$, and α policy are supplied. The curve AB is representative of such a solution, where point B is the equilibrium point for this trajectory. Observe once more that the form of the fed-batch equation is identical to the DSR equation, and identical behavior may be achieved in the batch by appropriate interpretation of the α variable as a result.

The set of concentrations belonging to the trajectory AB is thus equivalent to that obtained in a DSR with feed composition equal to point A, and sidestream composition given by $\mathbf{C}^0$ (this point is not shown in the figure). With similar reasoning to the standard batch analogy, residence time in the DSR is replaced by reaction time in the fed-batch. The value of α in the DSR represents the feed rate with respect to DSR length, whereas in the batch, $\alpha = F/V$ (the ratio of sidestream addition feed rate to reactor volume). F(t) and V(t) are determined by the integration of the α policy with respect to time and the differential expression for the volume change, usually given by $dV/dt = F(t)$. Nevertheless, note that there is a direct analogue between fed-batch reactors with varying α policies and DSRs. AR boundary structures that contain DSR segments are thus easily transformed to an equivalent fed-batch system as a result.

EXAMPLE 8: Converting α Expressions to Feeding Rate Profiles

Suppose that the following α policy

$$\alpha(t) = 2t$$

is to be used in an equivalent constant density fed-batch reactor. What are the corresponding expressions for the feeding rate profile, F(t), and fed-batch volume V(t)?

Answer:

We can find expressions for F(t) and V(t) from the definition of $\alpha(t)$ and dV/dt

$$\alpha(t) = \frac{F(t)}{V(t)}$$

$$F(t) = V(t)\alpha(t)$$

and

$$\frac{dV}{dt} = F(t)$$

$$= V(t)\alpha(t)$$

Substituting in the expression for $\alpha(t)$ gives

$$\frac{dV}{dt} = V(t)2t$$

which is a variable separable differential equation

$$\int_{V_0}^{V} \frac{dV}{V} = \int_0^t 2tdt$$

where V_0 is the initial volume of the fed-batch at $t=0$. This expression may be integrated to give

$$V(t) = V_0 \exp(t^2)$$

The expression for V(t) may then be used to find F(t) as follows:

$$F(t) = V(t)\alpha(t)$$

$$= 2tV_0 \exp(t^2)$$

OBSERVATION: Achieving DSR Concentrations in a Fed-Batch Reactor

When the value of α varies continuously throughout the reaction period in a fed-batch reactor ($\alpha = \alpha(\mathbf{C}(t))$), Equation 7.21 is identical to the DSR equation under continuous operation.

c) Case 3: α = Constant Consider now the case when α is maintained at a *constant* value throughout the entire reaction period. Again, if an initial condition and fed-batch addition composition are specified, Equation 7.21 may be integrated for any arbitrary reaction time. The pair of curves XD and YD are representative solutions for when α is kept at a constant value of $\alpha = \gamma_1$, whereas the pair XC and YC is that which is obtained when α is kept at a different constant value given by $\alpha = \gamma_2$. Clearly, XC and XD share the same initial point, given by point X, and points YC and YD share the same initial condition given by Y. For $\alpha = \gamma_1$, the fed-batch trajectories both terminate at point D, even though the starting conditions differ. The same behavior is observed for $\alpha = \gamma_2$. All trajectories, however (XC, YC, XD, and YD), are fed with material of concentration given by point A.

For a sufficiently long reaction time, the concentration of species within the vessel approaches equilibrium, and Equation 7.21 reduces to

$$\mathbf{0} = \mathbf{r}(\mathbf{C}) + \alpha(\mathbf{C}^0 - \mathbf{C}) \tag{7.22}$$

End points C and D therefore represent *equilibrium points in the fed-batch for a constant α policy*, and are thus also solutions to Equation 7.22. For all values of α in the range $0 \le \alpha \le \infty$, the integration of Equation 7.21 to equilibrium may be carried out with the fed-batch feed concentration $\mathbf{C}^0$ set at point O. The locus of ×'s, given by the curve ACDE, therefore represents the set of equilibrium concentrations achieved in the system.

Equilibrium concentrations for the fed-batch are equivalent to CSTR concentrations when the CSTR feed composition, $\mathbf{C}_f$, is equal to point A. This may be more easily seen by comparing Equation 7.22 with the general CSTR expression.

$$\mathbf{C} = \mathbf{C}_f + \tau_{CSTR}\mathbf{r}(\mathbf{C})$$

The value of α *at equilibrium* is thus related to the CSTR residence time by

$$\alpha = \frac{1}{\tau_{CSTR}} \quad \text{at equilibrium}$$

It follows that for all values of α in the range $0 \le \alpha \le \infty$ for fed-batch reactors operating at equilibrium correspond to the set of CSTRs with residence times between the range $0 \le 1/\tau_{CSTR} \le \infty$.

The curve ACDE in Figure 7.32 thus also corresponds to solutions for a CSTR (with feed point A), and therefore the curve ACDE is representative of a CSTR locus. CSTR compositions may therefore be obtained in a batch by operating a fed-batch reactor at its equilibrium concentration. The particular location of the equilibrium point is defined by the specific constant α value chosen, which also corresponds to a CSTR effluent concentration with residence time given by $\tau = 1/\alpha$.

OBSERVATION: Achieving CSTR Concentrations in a Fed-Batch Reactor

When the value of α is constant throughout the reaction period in a fed-batch reactor, Equation 7.21 resembles a constant α DSR. We understand that equilibrium points of constant α DSRs correspond to CSTR solutions, and thus we can use a fed-batch reactor with a constant value of F/V to achieve CSTR concentrations.

It is remarkable that we can achieve CSTR concentrations in a batch at all, for the CSTR is in many ways the opposite of a batch. Yet, using the geometric behavior of CSTRs and their connection with DSRs, we now have an understanding for how to obtain these states.

Since CSTRs can be problematic, several observations should be made for constant α fed-batch reactors, when their purpose is to obtain CSTR concentrations.

INITIAL CONDITIONS Note that the equilibrium point in the fed-batch is defined by the value of α and the fed-batch feed concentration, and is not a function of the initial condition at

$t = 0$. It follows that equivalent CSTR behavior may only be achieved in the fed-batch at equilibrium; therefore, in order to obtain CSTR concentrations in a batch system, the reactor must be initiated *at the equilibrium conditions*. In practice, this may be achieved by several methods as follows:

1. Initially run a small CSTR at the appropriate residence time to produce the desired equilibrium composition. Then transfer the contents of the CSTR into a fed-batch.
2. Begin at a condition sufficiently close to the desired CSTR concentration so that equilibrium may be approached during the reaction period.
3. Artificially form the desired CSTR concentration by combining pure chemicals in the appropriate amounts to achieve the correct equilibrium composition.

Irrespective of the method employed, this procedure need only be carried out once at the beginning of the first batch cycle. Once product of the correct CSTR composition is produced, a portion may be used to seed further fed-batch cycles, and the sequence is self-sustaining as a result.

INITIAL VOLUME At equilibrium, the concentration of species within the vessel is maintained at a constant value of **C** throughout the entire reaction period. Reactor volume therefore varies without a change in concentration in this instance.

In order to achieve a production rate comparable to that of an equivalent CSTR, the correct initial volume of material must be chosen. The final batch volume must allow for an additional small volume V_{seed} for seeding of subsequent batches. Therefore, the final volume produced in the batch at $t = \tau_{CSTR}$ must be equal to the equivalent CSTR volume (V_{CSTR}) and the initial volume required for successive batches (V_{seed}). We then have

$$\frac{dV}{dt} = \alpha V \tag{7.23}$$

with boundary conditions given by $V(t_0) = V_{seed}$ and $V(\tau_{CSTR}) = V_{seed} + V_{CSTR}$

MULTIPLE STEADY STATES We have seen that multiple steady states are often observed in CSTRs, and fed-batch reactors may exhibit the same behavior for an equivalent fixed α policy and sidestream feed composition. Thus, care must be taken in the choice of equilibrium concentration applied, as rather different outcomes in performance may arise, even when the correct optimal structure belonging to the AR is enforced. Additional information on multiple steady states in CSTRs can be found in (Gray and Scott, 1984, 1983; Schlosser and Feinberg, 1994).

DYNAMIC BEHAVIOR OF DSRS AND FED-BATCH REACTORS Observe that the system of differential equations described by Equation 7.21 contains equilibrium points associated with solutions to Equation 7.22. These points, often termed "fixed points," are time-independent solutions that describe the location of stable and unstable states in concentration space. Understanding the position and nature of these points assists in both describing how concentrations evolve over time and predicting the final state of the system from an initial condition (i.e., the equilibrium concentration in a fed-batch reactor from a known feed concentration). The stability of fixed points is often distinguished by an examination of the eigenvalues of the Jacobian matrix of the system (Strogatz, 2001). These are classified as either stable, if all real parts of the eigenvalues are negative, or unstable, if at least one eigenvalue has a positive real part. The stability and position of the fixed points in state space influence the dynamical behavior of the system, and thus also influence the choice of initial condition—nodes associated with unstable equilibrium points may exhibit chaotic behavior and unstable operation. Furthermore, depending upon the feed condition and kinetics specified, multiple steady states may arise, and thus the stability of each equilibrium point should be considered in turn. Again, the initial condition of the fed-batch should be considered carefully to ensure that the desired equivalent CSTR behavior is maintained.

ILLUSTRATION: Multiple Steady States in a Fed-Batch Reactor

To demonstrate the influence of different starting concentrations on a fed-batch reactor, Figure 7.33 shows a plot of many (constant α) fed-batch solution trajectories for a system known to exhibit multiple steady states in $\mathbb{R}^2$. The trajectories are shown as solid lines and are obtained by supplying different starting concentrations

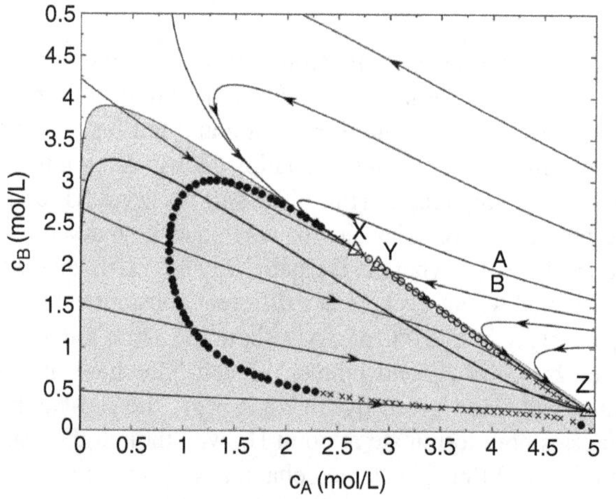

Figure 7.33 A fed-batch trajectory map for complex kinetics containing multiple nodes (multiple CSTR solutions). This map is generated by plotting many constant α trajectories for different feed point and the same sidestream composition $\mathbf{C}^0$. Adapted from Ming et al. (2013).

for the fed-batch for the same value of α (the particular value of α is not important for this discussion.) Also shown in the plot is a CSTR locus. Different sections of the locus are shown by different symbols, indicating the type of solution obtained in the fed-batch. Filled circles indicate a stable focus (complex eigenvalues, negative real part), crosses indicate stable nodes (real eigenvalues both negative in sign), and unfilled circles correspond to saddle points (real eigenvalues that differ in sign).

Three equilibrium points are associated with the chosen α value. These are given by triangles at points X, Y, and Z. Points X and Z lie on sections of the CSTR locus corresponding to stable solutions, and therefore they are achievable in a constant α fed-batch. Point Y, on the other hand, lies on a section of the locus associated with an unstable solution, and thus is not achievable even though it is a concentration that satisfies the CSTR equation and a fed-batch equilibrium point. Hence, if a starting concentration at point A is used, the corresponding CSTR solution obtained is point X, and the path followed by the fed-batch trajectory is then AX. If a starting point at point B is used instead, the resulting fed-batch trajectory approaches point Y, and then quickly diverges to point Z. The path traveled in this case is BYZ.

Observe that points A and B are situated closely to one another in c_A–c_B space; however, their corresponding equilibrium solutions are significantly different. Thus, we must be cautious when choosing starting points for fed-batch reactors if CSTR solutions are required, and, if possible, an analysis similar to that of Figure 7.33 should be performed, to understand how concentrations evolve in the fed-batch reactor over time.

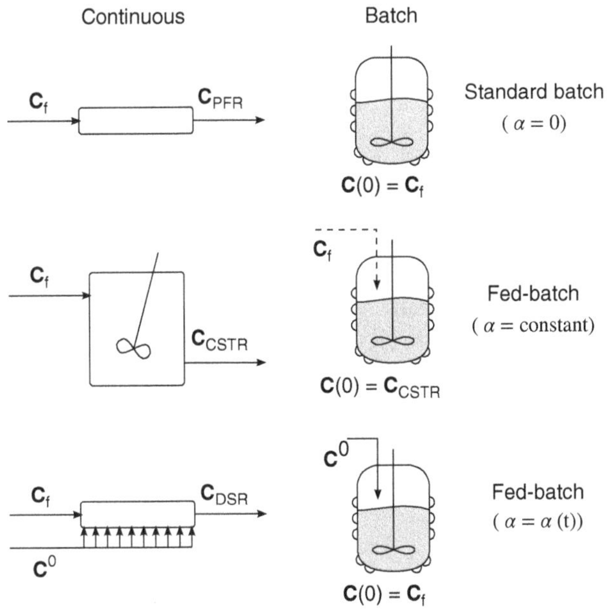

Figure 7.34 Comparison between continuous and batch reactive equipment. All three continuous reactors required to form the AR share equivalent batch structure. Adapted from Ming et al. (2013).

A comparison between continuous and batch equipment is given in Figure 7.34, with defining parameters required by the fed-batch reactor summarized in Table 7.2. These results provide for a convenient point of reference when conversions between batch and continuous reactors are carried out.

EXAMPLE 9: Converting Continuous Reactor Structures to Batch

Consider the following two-dimensional reaction:

$$A \rightarrow B \rightarrow C$$

The kinetics for this are assumed to be given by

$$r_A = -k_1 c_A{}^2$$
$$r_B = k_1 c_A{}^2 - k_2 c_B$$
$$r_C = k_2 c_B$$

where $k_1 = 1$ L/(mol.h) and $k_2 = 1\ h^{-1}$. The feed concentration is given by $\mathbf{C}_f = [1, 0, 0]^T$ mol/L.

Question 1:

A PFR feed at the feed concentration $\mathbf{C}_f$ is available. The volumetric flow rate to a PFR is Q = 2.5 L/min, and the volume of the PFR vessel is $V_{PFR} = 1.0\ m^3$. What is the initial concentration of the equivalent batch, and how long must the batch to run to give the same concentration as in the PFR? What is the required volume of the batch?

Answer 1:

There is a one-to-one relation between the concentration profile in a PFR to that in a standard batch for the same initial condition and residence time. Residence time in the PFR is equivalent to batch reaction time in a standard batch. The reaction time of the batch may therefore be calculated from the residence time of the PFR. The PFR residence time is given as

$$\tau = \frac{V_{PFR}}{Q}$$

where $V_{PFR} = 1.0\ m^3 = 1000$ L and Q = 2.5 L/min. The residence time of the PFR is then 400 min, or approximately 6.67 h. It follows that in order to achieve the same concentration in an equivalent batch as to that in

TABLE 7.2 Summary of Fed-Batch Operating Parameters

Continuous Structure	Initial Condition	α Policy	Sidestream Composition	Volume	Reaction Time
PFR	$C(0) = \mathbf{C}_f$	$\alpha = 0$	$\mathbf{0}$	V = constant	$t_f = \tau_{PFR}$
DSR	$C(0) = \mathbf{C}_f$	$\alpha = \alpha(t)$	$\mathbf{C}^0$	$V(t) = F(t)/\alpha(t)$	$t_f = \tau_{DSR}$
CSTR	$C(0) = \mathbf{C}_{CSTR}$	α = constant	$\mathbf{C}_f$	$V(t) = F(t)/\alpha$	t_f = free

the PFR, a standard batch must be initiated at the feed concentration $\mathbf{C}_f = [1, 0, 0]^T$ mol/L and reacted for a time of $t_{batch} = \sim 6.67$ h. Note the same concentration profile will be produced in the batch irrespective of the size of the batch vessel—only the amount of product produced will change. Thus, the volume of the batch may be as large or small as desired.

Question 2:

A DSR operating at the feed point $\mathbf{C}_f$ is assumed to have the following α policy, parameterized in terms of τ as follows:

$$\alpha(t) = \exp(-t)$$

Determine the expressions for the corresponding batch volume and feeding rate in a batch if the batch is initiated at an initial volume of $V_0 = 1.5$ L and the sidestream composition to the DSR is given by $\mathbf{C}^0 = [0.5, 0.5, 0.5]^T$ mol/L.

Answer 2:

A fed-batch reactor with a varying α policy may be used to mimic the concentrations achieved in a DSR. The initial conditions of the fed-batch must be equal to the feed concentration to the DSR. The sidestream concentration used in the DSR must be that provided to the fed-batch. Hence, the DSR initial conditions are at t = 0.

$$V = V_0 = 1.5 \text{ L and } \mathbf{C} = \mathbf{C}_f = [1, 0, 0]^T \text{ mol/L}.$$

Also, the concentration of the feed stream to the fed-batch is given by the sidestream concentration, $\mathbf{C}^0$, in the DSR $\mathbf{C}^0 = [0.5, 0.5, 0.5]^T$ mol/L. Similar to the PFR, the fed-batch cycle time is also equivalent to the DSR residence time τ_{DSR}.

Since the α policy for the DSR has already been parameterized in terms of DSR residence time, we may compute the feeding rate F(t) and batch volume V(t) directly from $\alpha(t)$. Hence, from Equation 7.23, the volume of the fed-batch is given by the following differential equation:

$$\frac{dV}{dt} = F(t)$$

which from the definition of α for fed-batch systems gives

$$\frac{dV}{dt} = \alpha V$$

Substitution of the α policy into this expression thus gives

$$\frac{dV}{dt} = \exp(-t)V$$

This is a variable separable differential equation. Integration with the appropriate boundary conditions gives

$$\int_{V_0}^{V} \frac{dV}{V} = \int_0^t \exp(-t)dt$$

Or

$$V(t) = V_0 \exp(-\exp(-t))$$

This gives the volume of the fed-batch as a function of batch reaction time t. The feeding rate policy F(t) may then be found from the definition of α and V(t)

$$\begin{aligned} F(t) &= \alpha(t)V(t) \\ &= V_0 \exp(-t)\exp(-\exp(-t)) \end{aligned}$$

Plots of $\alpha(t)$, V(t), and F(t) for the fed-batch are provided in Figure 7.35.

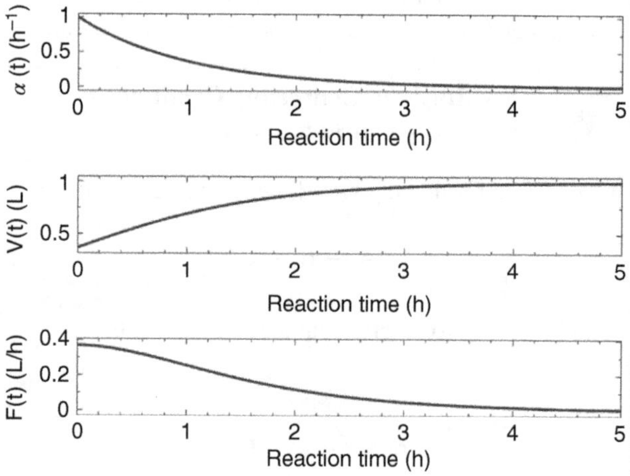

Figure 7.35 Feeding rate and volume profiles for an equivalent fed-batch mimicking a DSR.

Note also that the concentration of the sidestream feed in the fed-batch does not influence the expressions for $\alpha(t)$, V(t), or F(t). Only the concentration profile will be affected by $\mathbf{C}^0$.

Question 3:

A CSTR is operated with a feed concentration given at the feed point $\mathbf{C}_f$. The volumetric flow rate and CSTR volume are given as Q = 2.0 L/h and $V_{CSTR} = 1.3\,m^3$. If it is desired to achieve the same CSTR effluent concentration in a batch, what type of batch reactor would be required? It shall be assumed that a seed concentration $\mathbf{C}_{seed} = [0.6, 0.2, 0.0]^T$ mol/L is available, which will be used to seed the initial concentration of the fed-batch at start up. Plot the concentration profile in the equivalent batch as a function of batch reaction time t.

Answer 3:

In order to achieve the same concentrations of a CSTR in a batch setting, a fed-batch reactor operated with a *constant* α policy is required. Furthermore, the fed-batch must be initiated and operated, ideally, at the CSTR steady-state concentration.

It has already been established that, at equilibrium, the fed-batch equation reduces to the CSTR equation. Thus, the value of α in the fed-batch may be computed from the CSTR residence time τ_{CSTR} by

$$\alpha = \frac{1}{\tau_{CSTR}}$$

Hence, with the CSTR residence time given by $\tau_{CSTR} = V_{CSTR}/Q = 0.65$ h, α is given as $1.538\,h^{-1}$.

The volume V(t) and feeding rate F(t) of the fed-batch may be computed in a manner identical to that for the DSR. In fact, calculation of V(t) is simplified by the fact that α for a CSTR is now constant. Hence,

$$\int_{V_0}^{V} \frac{dV}{V} = \alpha \int_0^t dt$$

$$V(t) = V_0 \exp(\alpha t)$$

and

$$\begin{aligned} F(t) &= \alpha V(t) \\ &= \alpha V_0 \exp(\alpha t) \end{aligned}$$

Using the CSTR residence time provided, it is suitable to compute the CSTR effluent concentration in the continuous case. We find that $\tau_{CSTR} = 0.65$ h with the feed concentration $\mathbf{C}_f$ gives an exit concentration of $\mathbf{C}_{CSTR} = [0.6903,\ 0.1877]^T$ mol/L. By mass balance, the corresponding concentration for component C is approximately 0.1219 mol/L. When one compares this to the seeding concentration provided, we see that the two concentrations are similar, but they are not identical. We may conclude from this that the fed-batch will need to run for an initial period to approach the correct CSTR concentration, which is associated with a CSTR residence time of $\tau_{CSTR} = 0.65$ mol/L. The integration of the fed-batch expression

$$\frac{d\mathbf{C}}{dt} = \mathbf{r}(\mathbf{C}) + \alpha(\mathbf{C}_f - \mathbf{C})$$

along with the initial conditions at t = 0 h: $\mathbf{C} = \mathbf{C}_{seed}$ results in concentration profiles for components A, B, and C, which are displayed in Figure 7.36(a).

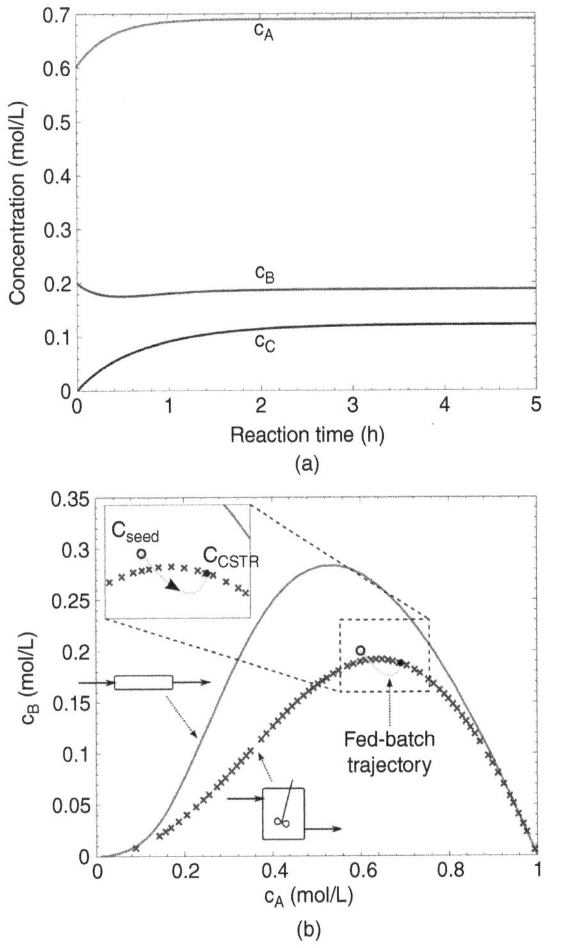

Figure 7.36 (a) Concentration profiles for components A, B, and C, achieved in a constant α fed-batch initiated at the seed concentration. (b) Trajectories and locus achieved in the equivalent fed-batch. (*See color plate section for the color representation of this figure.*)

Observe that the concentrations start at the CSTR seed concentration, $\mathbf{C}_{seed}$, and slowly approach the CSTR concentration associated with $\tau_{CSTR} = 0.65$ h. It appears that a fed-batch reaction time of approximately 3.0 h is sufficient to approximate the steady-state CSTR concentration. Once this concentration has been achieved, a fraction of the product volume may be spared and used for seeding of subsequent fed-batch cycles.

As further validation, in Figure 7.36(b) the steady-state curves in c_A–c_B space for a PFR from the feed, the CSTR locus from the feed, and a DSR initiated at $\mathbf{C}_{seed}$ are plotted. The DSR trajectory must then also correspond to a constant α fed-batch trajectory for the same operating conditions. This profile corresponds to the concentration profiles supplied in Figure 7.36(a). Observe that the fed-batch trajectory initially begins at $\mathbf{C}_{seed}$, and then approaches $\mathbf{C}_{CSTR}$, corresponding to the steady-state CSTR concentration associated with $\tau_{CSTR} = 0.65$ h, which lies on the CSTR locus as expected.

7.4.3 Example: Three-Dimensional Van de Vusse Kinetics Revisited

7.4.3.1 Optimal Continuous Structure In Section 7.2.1, we examined the nature of the optimal continuous reactor structure for the three-dimensional Van de Vusse reaction scheme. Let us now investigate the associated optimal batch structure for the same problem. This exercise will demonstrate the role that DSRs, and by extension fed-batch reactors with varying α policies, play in the formation the AR boundary.

The system under investigation is identical to the system established in Section 7.2.1. As a result, the AR for the batch system will be constructed in the same space and feed point ($\mathbf{C}_f = [c_{Af}, c_{Bf}, c_{Df}]^T = [1, 0, 0]^T$ mol/L). We are also able to utilize the optimal reactor structures, developed in Section 7.2.1, and convert them to batch structures without the need to perform additional analysis. The critical α policy for the DSR, given in Section 7.2.1.5, will be used for the F/V policies in the batch system.

By consideration of Equation 7.3, together with the rate expressions and feed point provided earlier, the AR may be constructed in the standard fashion given in Section 7.2.1. That is, construction of the AR occurs with the assumption that continuous operation is available.

Recall from Section 7.2.1.5 that the optimal continuous structure is composed of two parallel configurations, which are traversed by paths ABC and ADBC on the AR boundary in Figure 7.5(b). Refer back to Figure 7.6 for a reminder of the optimal continuous reactor structures for the three-dimensional Van de Vusse system.

1. Path ABC: The first structure is a DSR from the feed (point A) to the critical DSR equilibrium (point B), followed by a PFR to equilibrium (point C).
2. Path ADBC: The second structure forms the underside of the AR, and its corresponding structure is slightly more subtle. Here, a CSTR–DSR–PFR structure is required.

OBSERVATION: Conversion Approach for Batch Systems

Observe that we first determine a candidate region with continuous operation in mind. Only after the continuous AR structures have been determined are they converted to equivalent batch reactor structures.

7.4.3.2 Conversion to Batch Since the AR is defined by two parallel continuous structures, it follows that, at most, two batch structures are required to achieve all points on the AR boundary as well. Figure 7.37 provides the necessary conversion from continuous to batch for the Van de Vusse system. A one-to-one correspondence between batch and continuous equipment is observed. For both optimal batch structures, a fed-batch reactor with a varying α policy is seen to contribute to the overall set of achievable concentrations. (Recall that the varying α policies correspond to the critical DSR trajectories needed to form the AR boundary.)

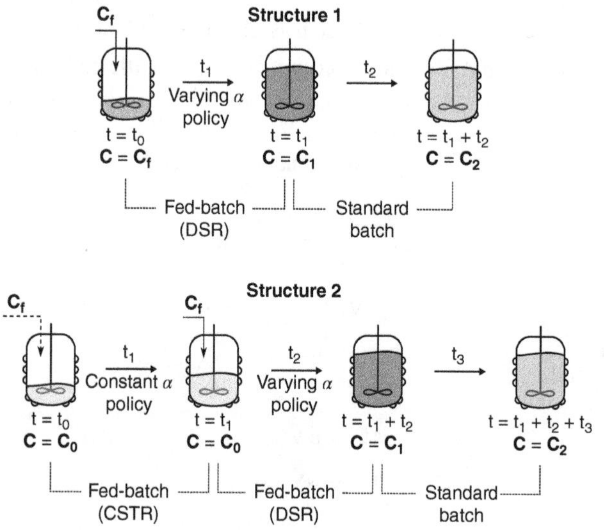

Figure 7.37 Optimal batch reactor structures for the Van de Vusse system. Adapted from Ming et al. (2013), (*See color plate section for the color representation of this figure.*)

The two following batch structures are thus required to achieve all achievable points:

1. For structure 1, a fed-batch reactor followed by a period of standard batch reaction is needed. This is provided by path ABC in Figure 7.5(b).
2. For structure 2, three distinct reaction intervals are needed, two of which are fed-batch operations. The cycle is initiated with a constant α interval, followed by a nonconstant α interval, and completed with a period of standard batch reaction. This is given in Figure 7.5(b) by the path ADBC.

If one is interested in achieving concentrations along the unique PFR trajectory BC, reduced segments belonging to structures 1 and 2, such as those of AFG and DHI, can be used. For concentrations occurring within the region however, combinations of these structures, along with intermediate mixing, may be used.

7.4.3.3 ***Different Objective Functions Revisited*** Let us investigate the possible reactor structures needed if we are to minimize c_B for fixed values of $c_D = 0.3$ mol/L and $c_D = 0.4$ mol/L. Geometrically, this is equivalent to inserting two hyperplanes, spanned in the $c_A - c_B$ plane, at levels $c_D = 0.3$ mol/L and $c_D = 0.4$ mol/L, and checking for points of intersection with the AR boundary that *minimize* the concentration of component B. (This objective function is different from the example given in Example 3, where we sought to find the point of intersection with the AR boundary that *maximized* component B for $c_D = 0.3$ mol/L and $c_D = 0.4$ mol/L.)

Two such profiles satisfying these objectives are shown in Figure 7.38: for the objective when $c_D = 0.4$ mol/L, a

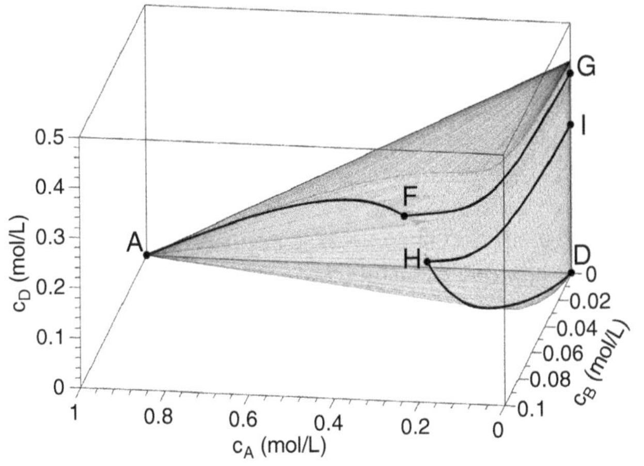

Figure 7.38 Potential paths traversed on the AR boundary that satisfy the objectives functions for a batch Van de Vusse system. Adapted from Ming et al. (2013).

critical DSR from the feed (point A) to point F, followed by a PFR to equilibrium (point G) is used. For the case where $c_D = 0.3$ mol/L, a CSTR–DSR–PFR, which traverses the path ADHI on the AR boundary, is used. (Note that points I and G are not extreme, and thus we could also achieve these by operating a PFR at equilibrium from the feed and a CSTR at equilibrium from the feed in parallel and then mixing between these two structures. Mixture compositions would then lie on the straight line CD in Figure 7.5(b).)

The associated α policies and concentration profiles for component D for a single, multipurpose, fed-batch reactor for both structures 1 and 2 are provided by Figure 7.39. In particular, paths AFG belonging to the objective function $c_D = 0.4$ mol/L and paths DHI belonging to $c_D = 0.3$ mol/L in Figure 7.38 are shown. The value of α follows Equation 7.3, which is the critical α expression for the Van de Vusse system developed earlier.

$$\alpha(\mathbf{C}) = \frac{c_A(20c_A^3c_B - 80c_A^2c_B - c_A^2 + 37c_Ac_B + c_A + 2c_B)}{c_B(c_A^2 - 2c_A + 1)} \tag{7.3}$$

From Chapter 6, critical α policies depend on the controllability matrix $\mathbf{E}$ of the system, which is a function of the concentration vector $\mathbf{C}$. Hence, α is expressed in terms of $\mathbf{C}$ and not residence time τ. Nevertheless, an explicit function for α in terms of τ may be found once the optimal DSR concentration profile is determined, which is accomplished by substituting DSR concentrations, for a certain value of τ, into Equation 7.3 and determining the value of α corresponding to the $\mathbf{C} - \tau$ pair.

From this information, we can also determine the feeding rate, F(t), and volume, V(t), for the fed-batch using Equation 7.23 with the appropriate boundary conditions for the fed-batch volume at t = 0.

$$V(t) = V_0 \exp(\alpha t)$$

and

$$F(t) = \alpha V(t)$$
$$= \alpha V_0 \exp(\alpha t)$$

Again, these equations often result in impractical requirements for the operation of the fed-batch reactor in practice that are difficult to accept, yet these conditions *must* be enforced if the fed-batch to operate on the true AR boundary.

For a target value of $c_D = 0.4$ mol/L, structure 1 is required. The α policy begins with an exponential decline in sidestream addition to a time of approximately 1.536 s, followed by a period of standard batch reaction to termination. The total batch cycle time for this structure is thus 3.5 s. Theoretically, the correct α policy for this structure

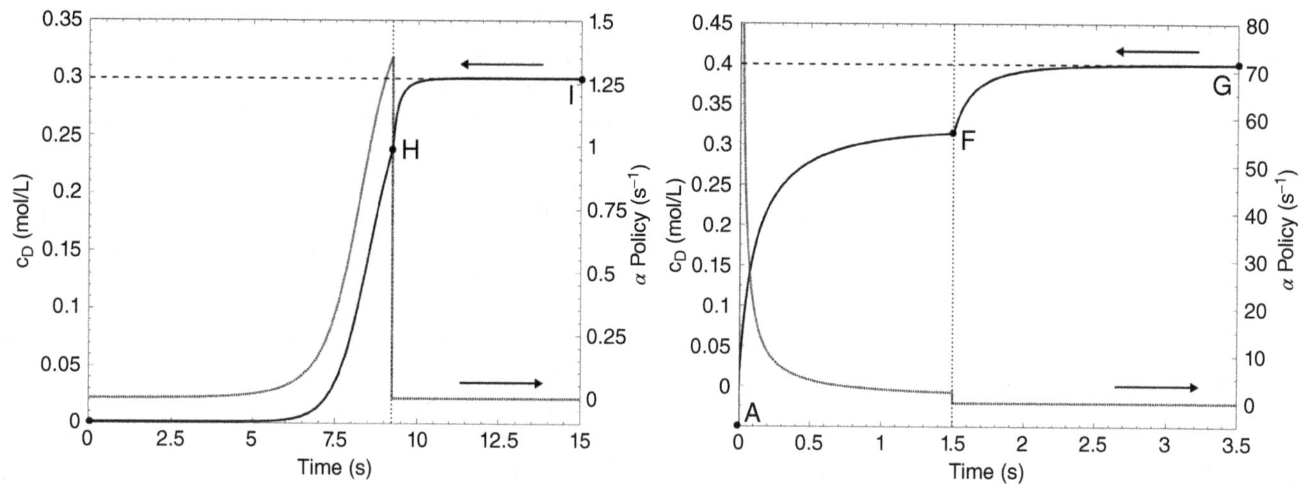

Figure 7.39 α (F(t)/V(t)) policies required to achieve the desired objective function values. (a) Feeding policy for when $c_D = 0.3$ mol/L is the objective function. Structure 2 is required. (b) Feeding policy for when $c_D = 0.4$ mol/L is the objective function. Structure 1 is required.

starts at a value of ∞, however this has been approximated by utilizing a starting concentration that gives an initial α value of approximately $200\,s^{-1}$ ($\mathbf{C} = [0.9885, 0.0006, 0.0055]^T$ mol/L).

For $c_D = 0.3$ mol/L, the recommended operating policy is structure 2. This structure is different compared to structure 1 in that three reaction periods are required. The batch cycle begins with a fed-batch period of constant α in accordance with the equilibrium CSTR concentration, given by point D. The cycle is then brought into a period of varying α, where the sidestream addition is observed to increase sharply to a maximum value of approximately $1.356\,s^{-1}$, and then completed with a standard batch period lasting approximately 9.24 s; the total batch cycle time for this structure is thus roughly 15 s.

EXAMPLE 10: Optimal Batch Structure for BTX Kinetics

Had Sam, Alex, and Donald known about how to convert continuous reactors to batch, what optimal batch structures would be required to generate the BTX AR with batch equipment?

Answer:

The optimal continuous reactors for the BTX system are described in Section 7.2.2. Hence, an equivalent optimal batch structure is then

1. A critical fed-batch reactor with α policy given by Equation 7.9 followed by a standard batch.
2. A constant α fed-batch operating at the CSTR equilibrium point, followed by a critical fed-batch with α policy also given by Equation 7.9 followed by a standard batch.

The feeding rate policies for the fed-batch reactors may then be found from the critical DSR α policies, whereas the batch cycle times are found from the equivalent reactor residence times under continuous operation.

SIDE NOTE: Performing Reactions in Space Versus Time

If the reaction is to be performed in a continuous setting, two separate sets of equipment are required and must be operated simultaneously. Compositions are then obtained by mixing the necessary streams belonging to each structure in the appropriate proportions. By comparison, if it is desired to perform the same task under batch conditions, an identical outcome may be achieved by two serial operations together with intermediate storage of material used in mixing processes. Hence, it may turn out that a single set of reactive equipment is sufficient to achieve the same result.

Whereas in continuous systems one is concerned with the arrangement of equipment through space, in batch systems one may also employ time. The time during which batch operations are carried out is important in developing an efficient batch reactive structure.

7.5 CHAPTER REVIEW

In Chapter 1, we introduced the BTX system, and described it in relation to the story of Sam, Alex, and Donald. Our goal at the time was to find a reactor, or set of reactors, that could maximize the toluene concentration in the system. Although we succeeded in improving the toluene concentration, there was always uncertainty in our recommendations, for it was not known at the time if better solutions existed.

On reflection, we have come a long way to understanding the nature of the reactor network synthesis problem, and by extension the problem that Sam, Alex, and Donald originally faced. In this chapter, we have finally answered the BTX problem initially posed in Chapter 1, by computing the true AR for this system.

Along with finding the AR for the BTX system, we also calculated the AR for the three-dimensional Van de Vusse system. It should be clearer now how one might approach construction of a candidate region, using fundamental AR theory developed in Chapter 6. Knowledge of this system also allows for further study and understanding of AR research papers that have often been written using the Van de Vusse system in mind.

A number of small examples, centered on the optimization of exothermic adiabatic reactions, were also covered. These types of reactions are commonly found in industry. We find that operating procedures such as interstage and cold-shot cooling are appropriate for minimum reactor volume problems, and may be validated by viewing the problem from an AR perspective. These discussions also demonstrate how a candidate region may be generated when we are restricted to a certain reactor type, such as a PFR.

The conversion of continuous reactor structures to batch structures was also described in this chapter, which is useful because the AR need only be determined once. In this way, determination of a candidate region for a given reaction system and feed point may potentially guide the construction of both continuous and batch reactor structures, via a single construction step.

The constructions discussed in this chapter provide a small sample of the type of problems now solvable, using AR concepts. We hope that these descriptions will spark many more independent investigations and analyses of your own.

Yet, a number of complex, real-life problems are beyond the scope of traditional analytical methods, such as those methods discussed here. For this reason, we are often reliant on the results obtained by automated AR construction algorithms. In Chapter 8, we shall seek to understand some of these techniques.

REFERENCES

Denbigh, K.G., Turner, J.C.R., 1984. Chemical Reactor Theory: An Introduction. Cambridge University Press, New York.

Feinberg, M., 2000. Optimal reactor design from a geometric viewpoint. Part II. Critical sidestream reactors. Chem. Eng. Sci. 55, 2455–2479.

Feinberg, M., Hildebrandt, D., 1997. Optimal reactor design from a geometric viewpoint- I. Universal properties of the attainable region. Chem. Eng. Sci. 52, 1637–1665.

Glasser, B., Hildebrandt, D., Glasser, D., 1992. Optimal mixing for exothermic reversible reactions. Ind. Eng. Chem. Res. 31, 1541–1549.

Godorr, S., Hildebrandt, D., Glasser, D., McGregor, C., 1999. Choosing optimal control policies using the attainable region approach. Ind. Eng. Chem. Res. 38, 639–651.

Gray, P., Scott, S.K., 1983. Autocatalytic reactions in the isothermal, continuous stirred tank reactor': Isolas and other forms of multistability. Chem. Eng. Sci. 38, 29–43.

Gray, P., Scott, S.K., 1984. Autocatalytic reactions in the isothermal, continuous stirred tank reactor': Oscillations and instabilities in the system A + 2B -> 3B; B -> C. Chem. Eng. Sci. 39, 1087–1097.

Hildebrandt, D., Glasser, D., Crowe, C.M., 1990. Geometry of the attainable region generated by reaction and mixing: With and without constraints. Ind. Eng. Chem. Res. 29, 49–58.

Howard, F., 1977. Chemical reactor design for process plants. Case Study 106 Ammonia Synthesis. John Wiley & Sons, Inc., New York.

Ming, D., Glasser, D., Hildebrandt, D., 2013. Application of attainable region theory to batch reactors. Chem. Eng. Sci. 99, 203–214.

Nicol, W., Hildebrandt, D., Glasser, D., 1997. Process synthesis for reaction systems with cooling via finding the attainable region. Comput. Chem. Eng. 21, S35–S40.

Schlosser, P.M., Feinberg, M., 1994. A theory of multiple steady states in isothermal homogeneous CFSTRs with many reactions. Chem. Eng. Sci. 49, 1749–1767.

Strogatz, S.H., 2001. Nonlinear Dynamics and Chaos: With Applications to Physics, Biology, Chemistry, and Engineering. Westview Press, Cambridge.

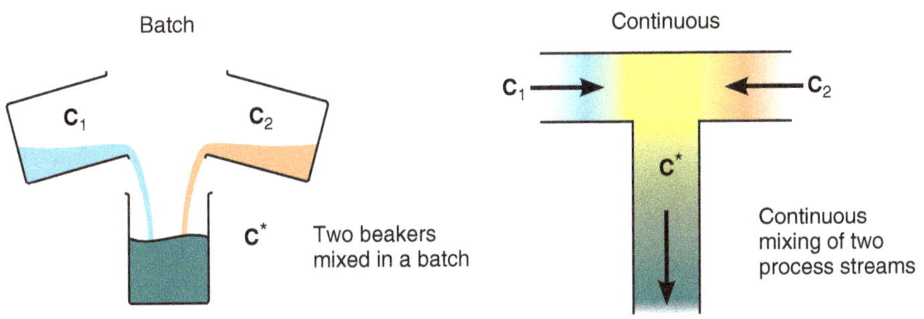

Figure 2.10 Physical interpretation of mixing for batch and continuous processes.

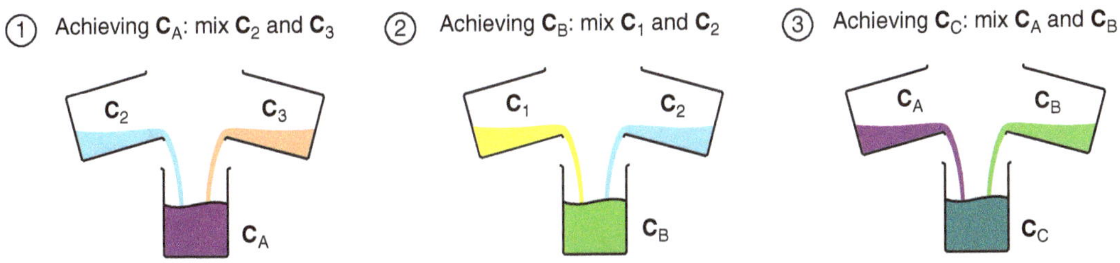

Figure 2.16 Forming a mixture concentration C_C as a series of three mixing steps.

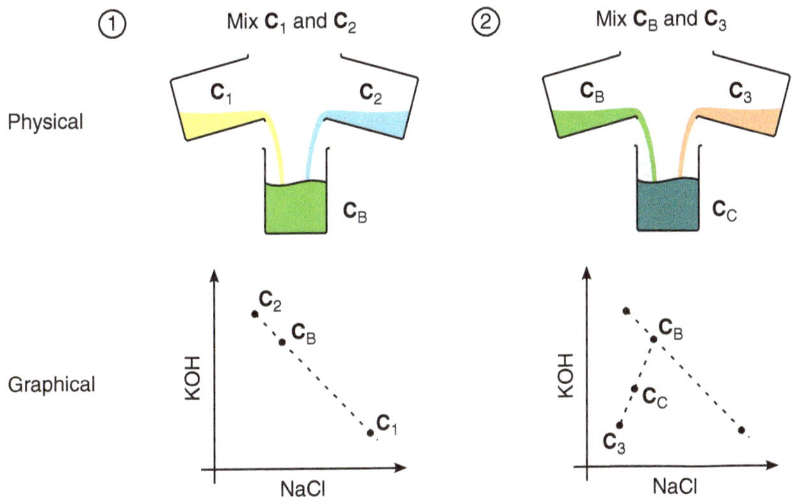

Figure 2.18 Physical steps needed to achieve C_C as well as the corresponding geometric interpretation in NaCl–KOH space.

Attainable Region Theory: An Introduction to Choosing an Optimal Reactor, First Edition. David Ming, David Glasser, Diane Hildebrandt, Benjamin Glasser, and Matthew Metzger.

Companion website: http://attainableregions.com

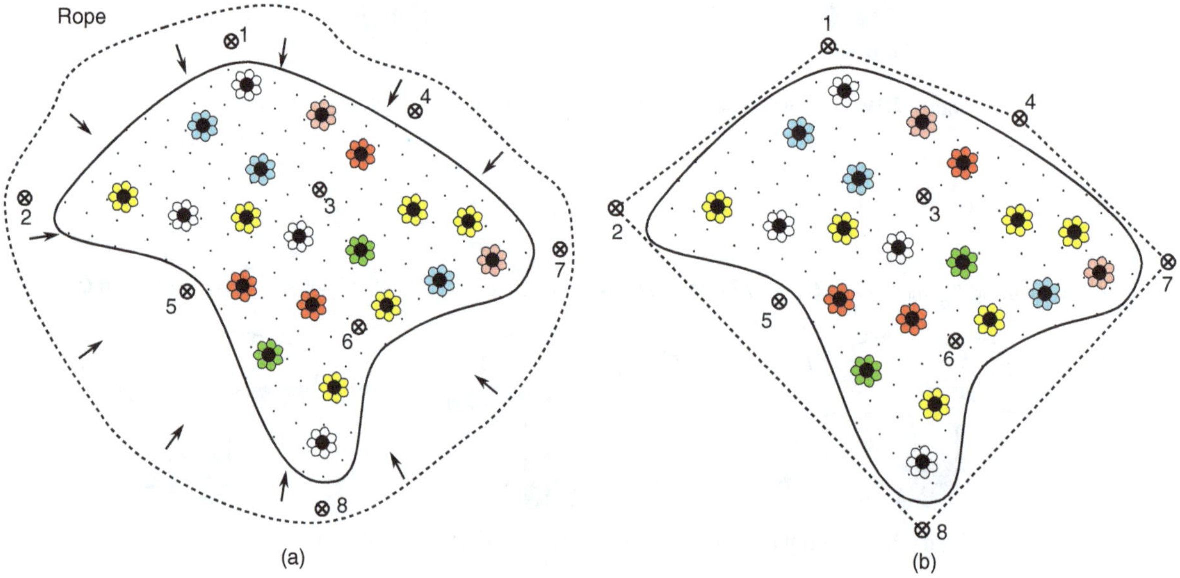

Figure 2.27 Top-down view of a garden. (a) Loose rope enclosing the garden and (b) tightened rope representing a convex hull that encloses all eight lights.

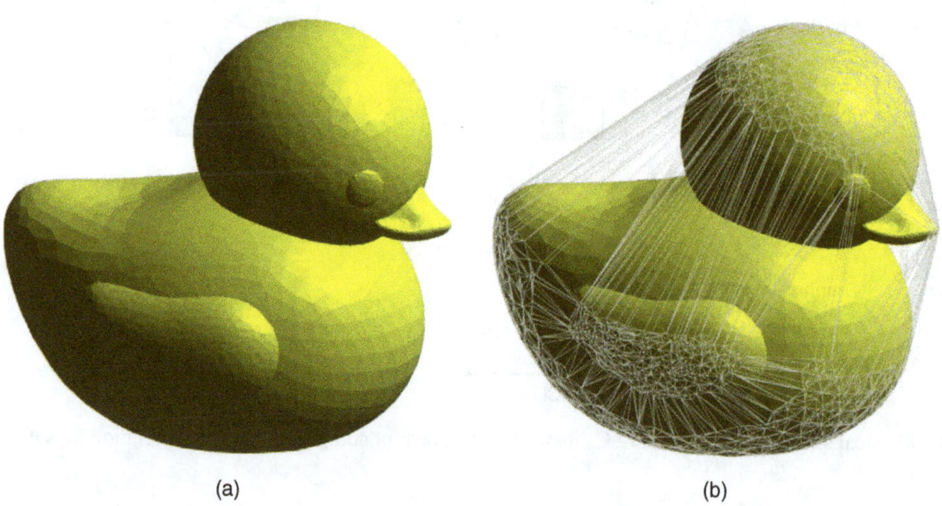

Figure 2.28 (a) A three-dimensional CAD drawing of a rubber duck and (b) the associated convex hull, given in a wireframe representation.

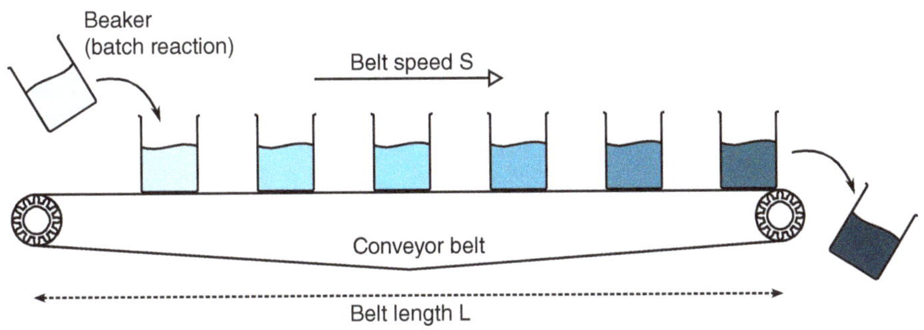

Figure 4.7 Beakers (batch reactors) on a conveyor belt, which approximate PFR behavior.

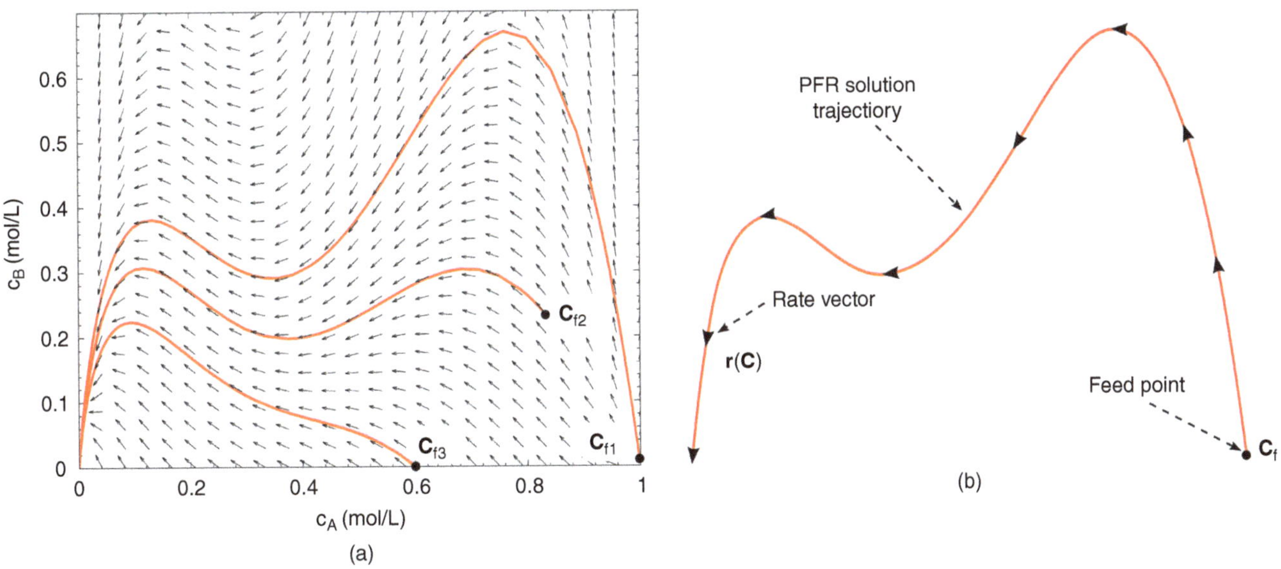

Figure 4.11 (a) Geometric interpretation of the PFR trajectory and (b) rate vectors are tangent to the PFR trajectory.

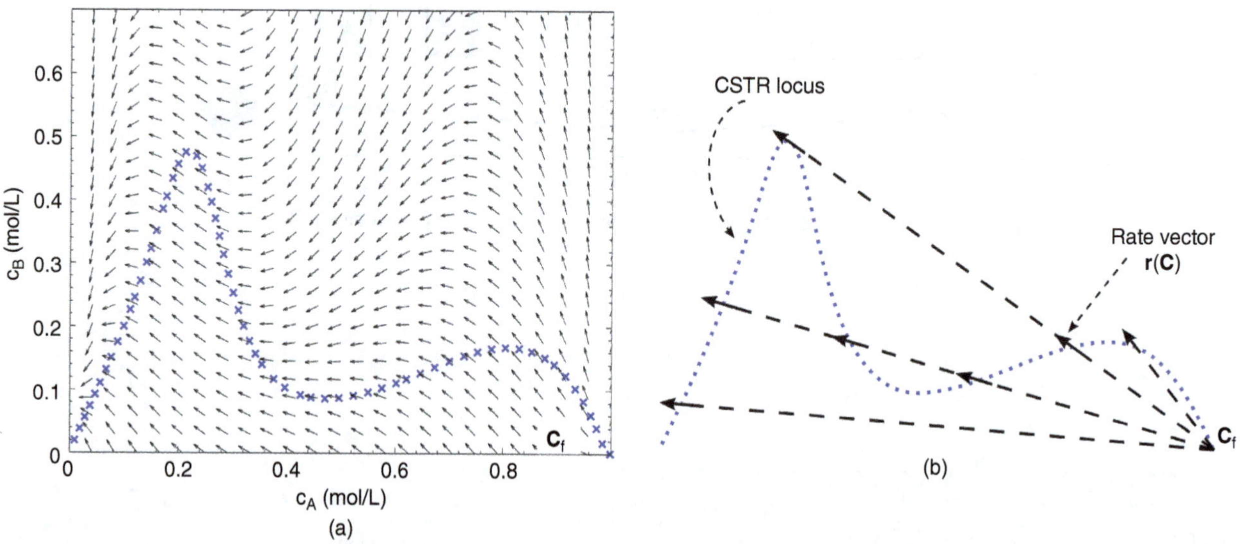

Figure 4.18 (a) Geometric interpretation of the CSTR. (b) CSTR solutions are collinear with the rate vector evaluated at that point and the feed.

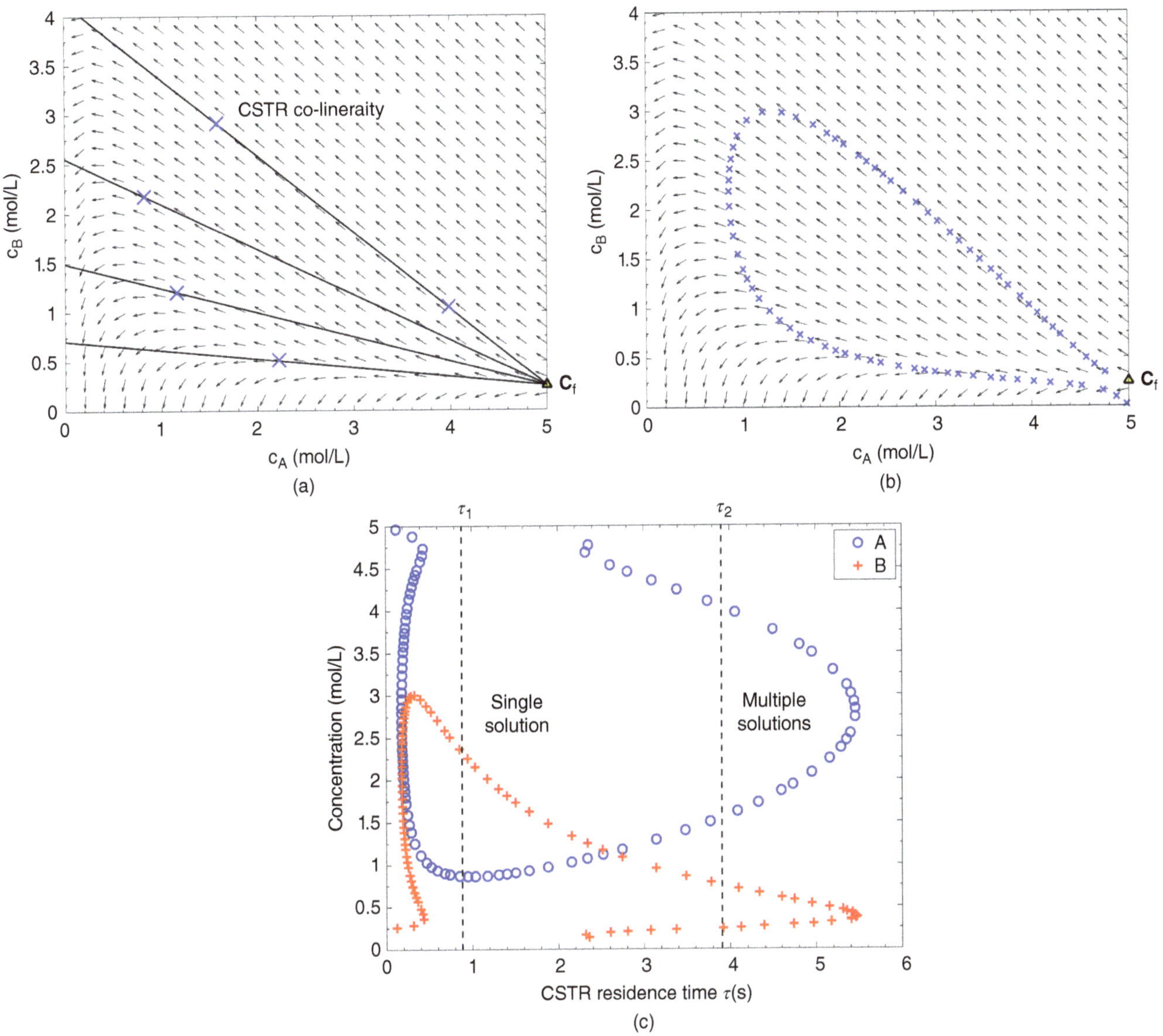

Figure 4.19 (a) Computing CSTR solutions in the autocatalytic system by searching for colinear points, (b) full CSTR locus for the autocatalytic system, and (c) concentration profiles for components A and B achieved in the CSTR for different residence times.

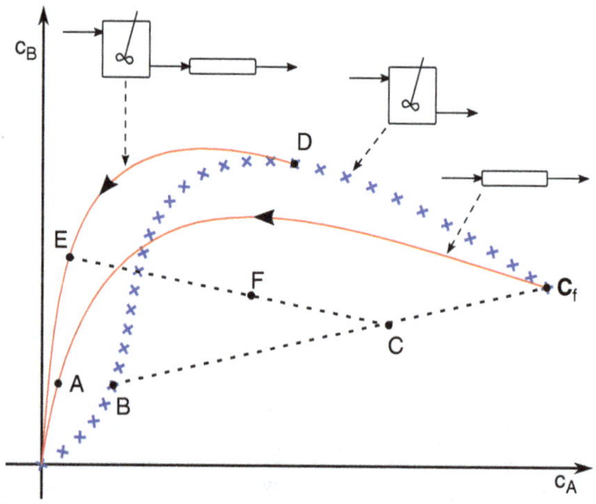

Figure 4.21 Reactor structures for a hypothetical system.

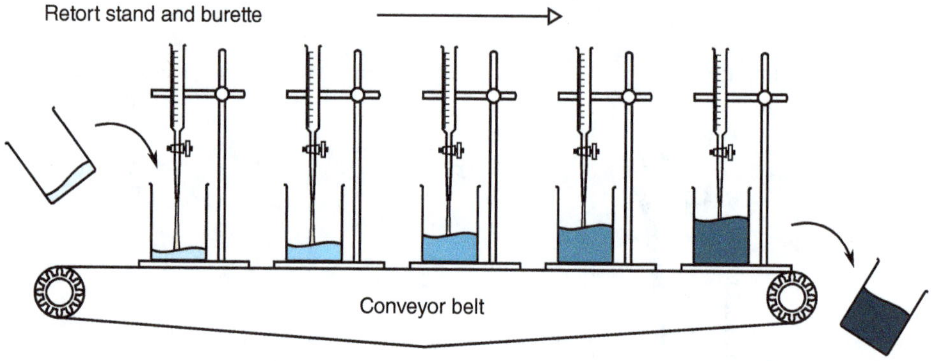

Figure 4.26 A DSR may be viewed as a beaker and retort stand combination on a conveyor belt.

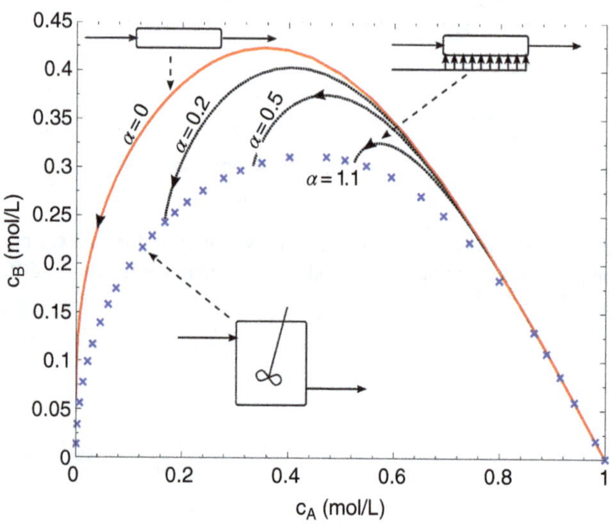

Figure 4.29 DSR trajectories for constant values of α over the integration range.

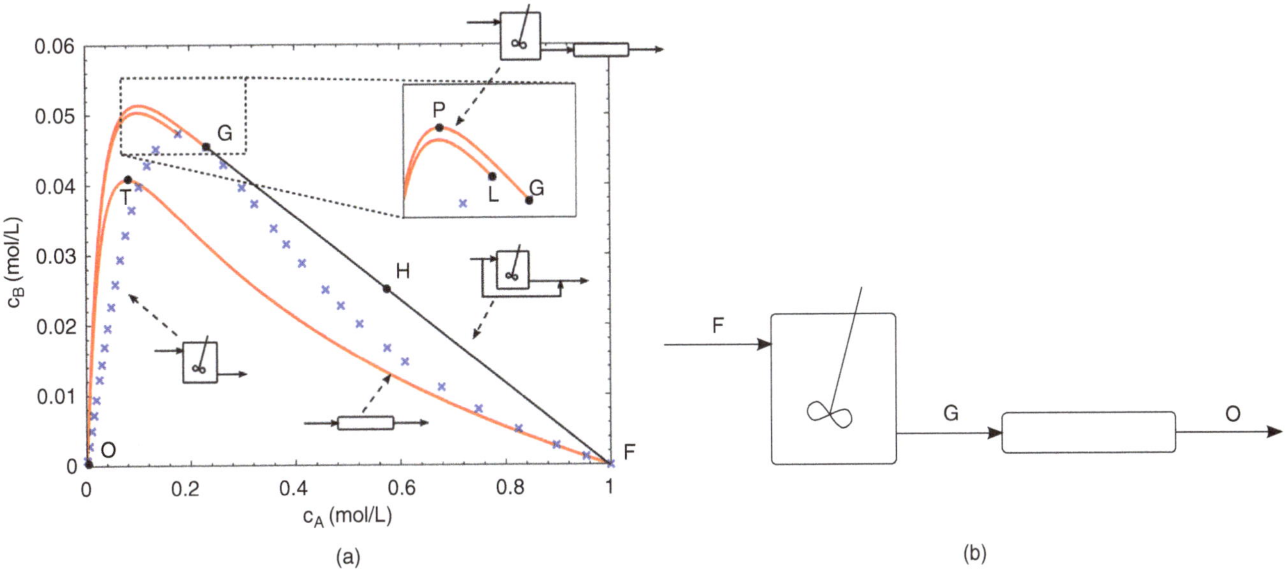

Figure 5.9 (a) Optimal reactor AR boundary structures and (b) optimal reactor network required to generate the AR.

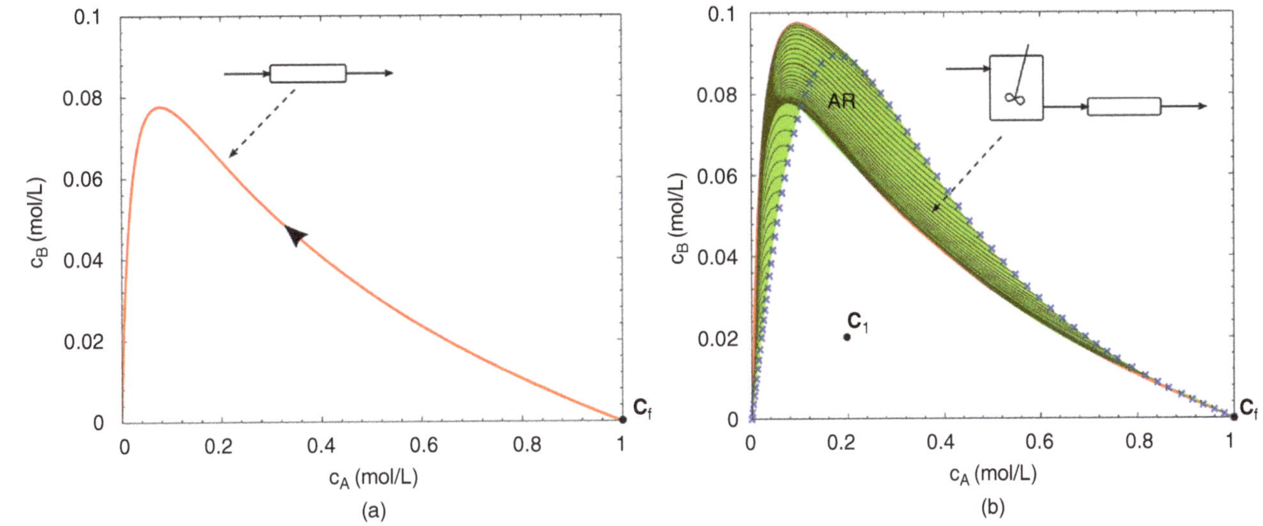

Figure 5.14 (a) The set of achievable concentrations for a PFR from the feed and (b) the AR for a CSTR–PFR structure without mixing.

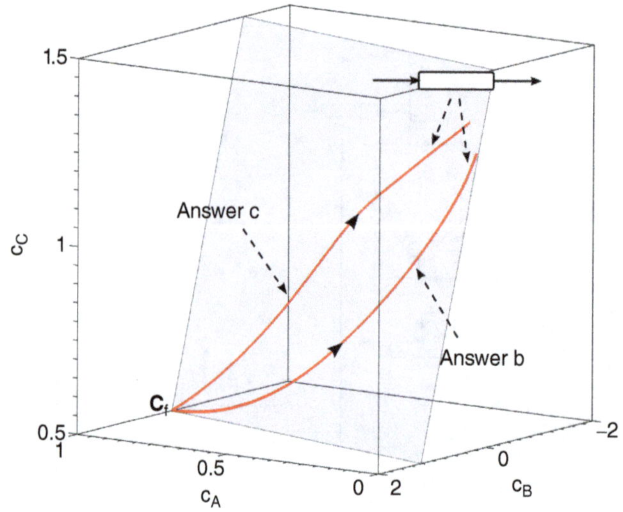

Figure 6.3 PFR trajectory associated with different kinetics, but the same feed point and reaction stoichiometry.

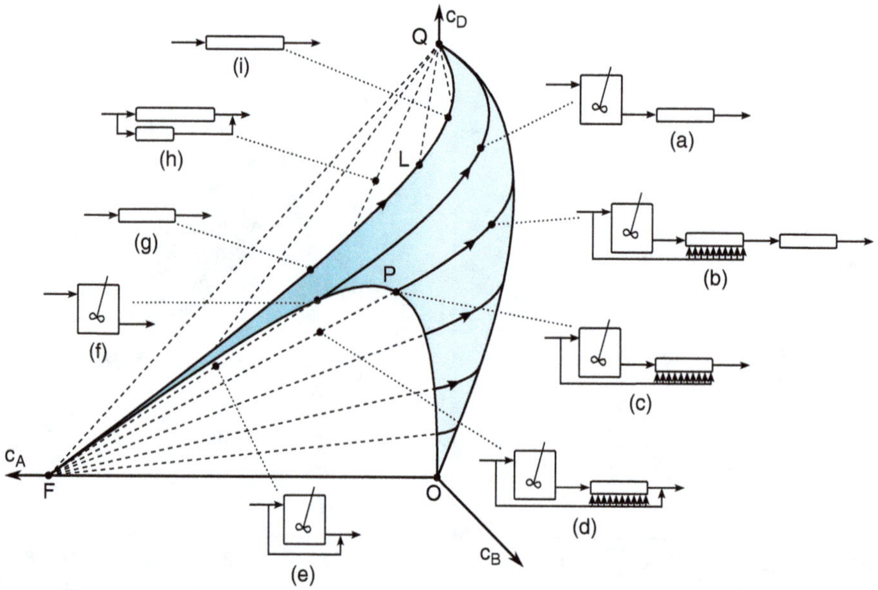

Figure 6.6 The AR boundary for the Van de Vusse system. Feinberg and Hildebrandt (1997). Reproduced with permission of Elsevier.

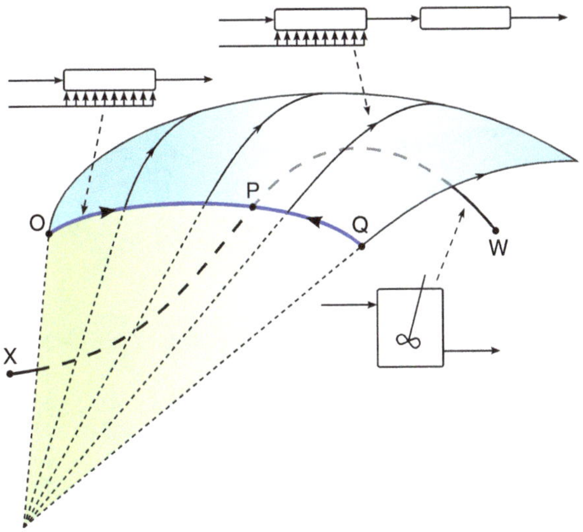

Figure 6.10 Connectors on the AR boundary manifest themselves either as critical CSTR effluent compositions, or critical DSR solution trajectories.

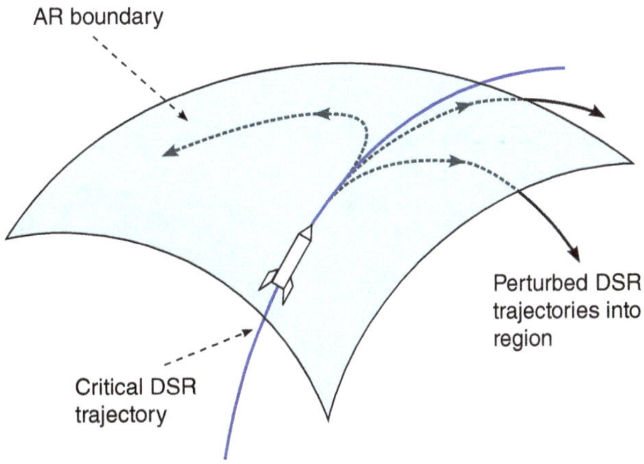

Figure 6.13 DSR trajectories on the AR boundary. All nonoptimal changes in the α policy serve to move the resulting trajectories into the region. The DSR is thus *not* locally controllable on the AR boundary.

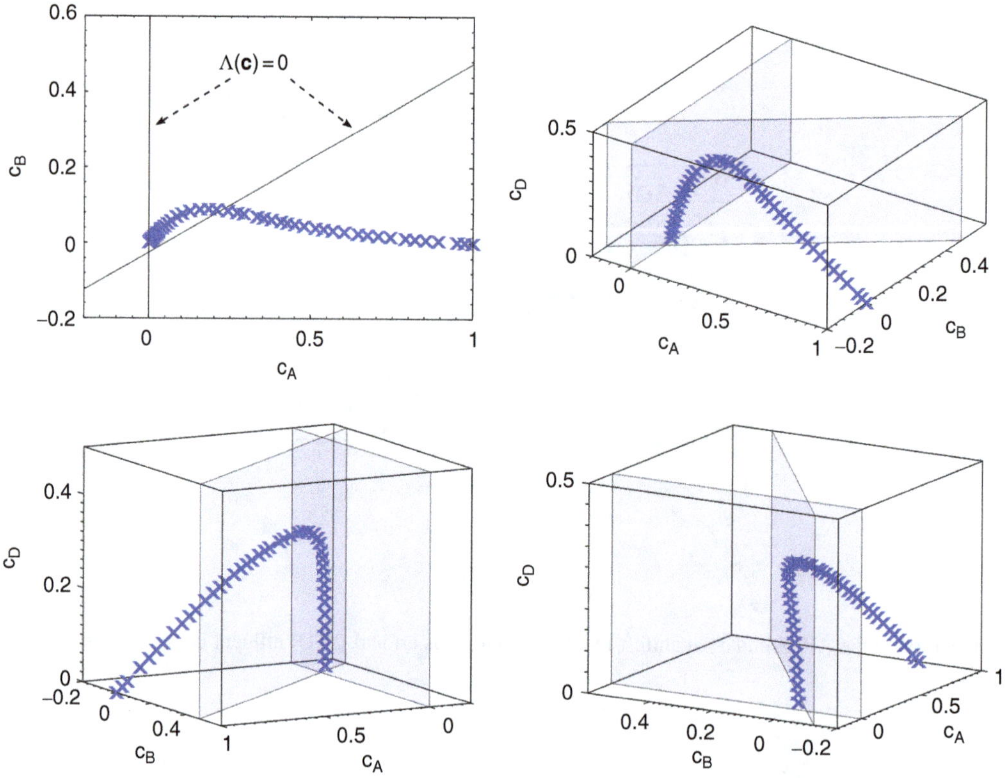

Figure 6.17 Critical CSTR surface ($\Lambda(\mathbf{C}) = 0$) and CSTR locus. The CSTR locus intersects $\Lambda(\mathbf{C}) = 0$ at two points, indicating a CSTR from the feed is critical at two points.

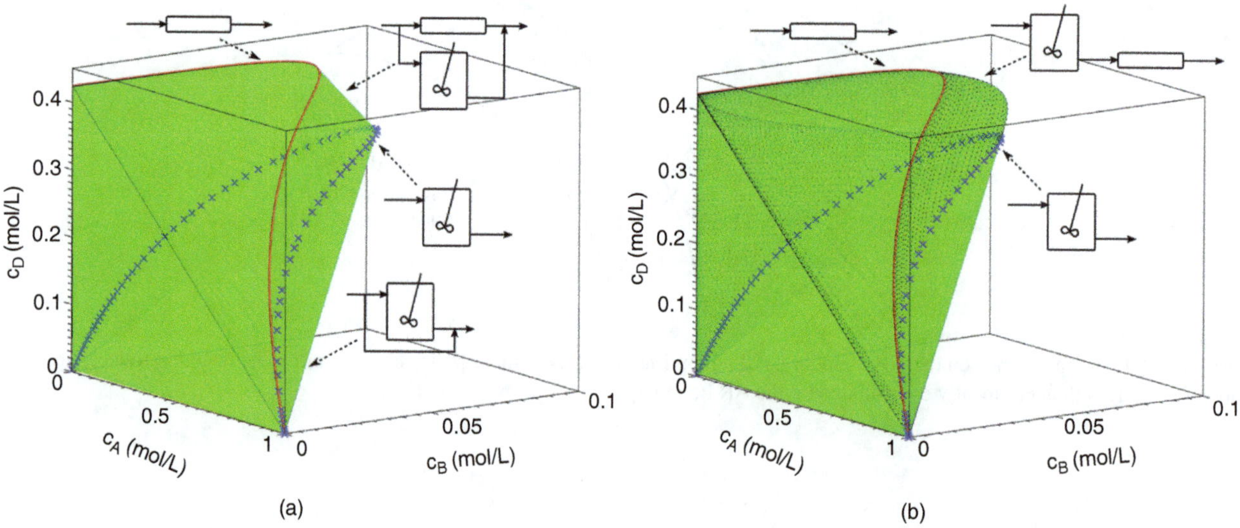

Figure 7.2 (a) Convex hull of a PFR and CSTR from the feed and (b) convex hull of the CSTR–PFR reactor structure from the feed. Note that this is not the full AR for the three-dimensional Van de Vusse system.

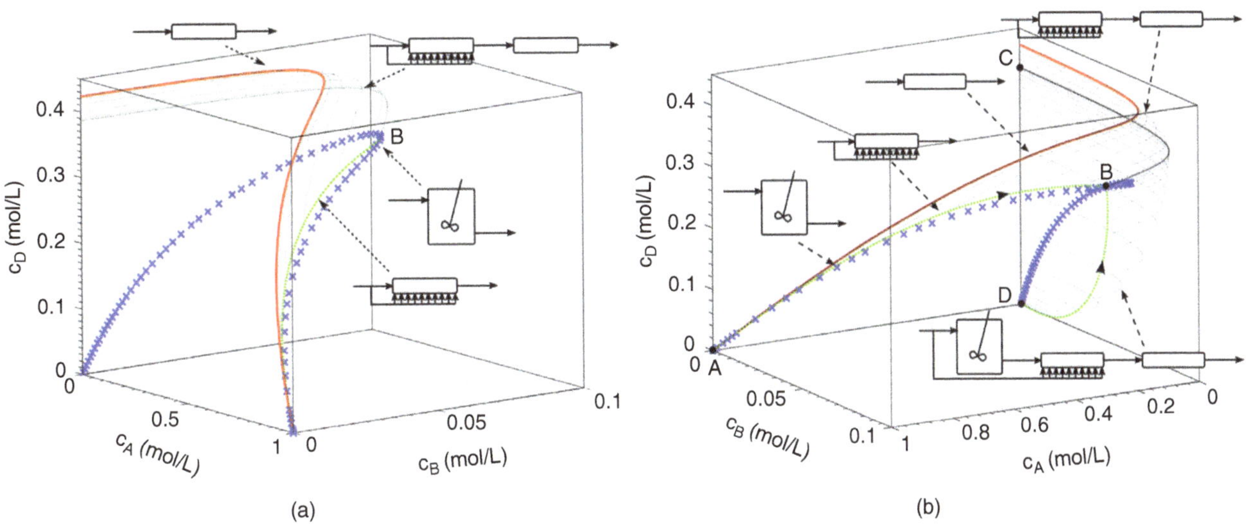

Figure 7.5 (a) Unfilled candidate region for the three-dimensional Van de Vusse kinetics including a critical DSR trajectory from the feed point and (b) the full AR for the three-dimensional Van de Vusse system in $c_A-c_B-c_D$ space. Mixing lines have been removed from the plot to make interpretation of the AR boundary structures easier to identify.

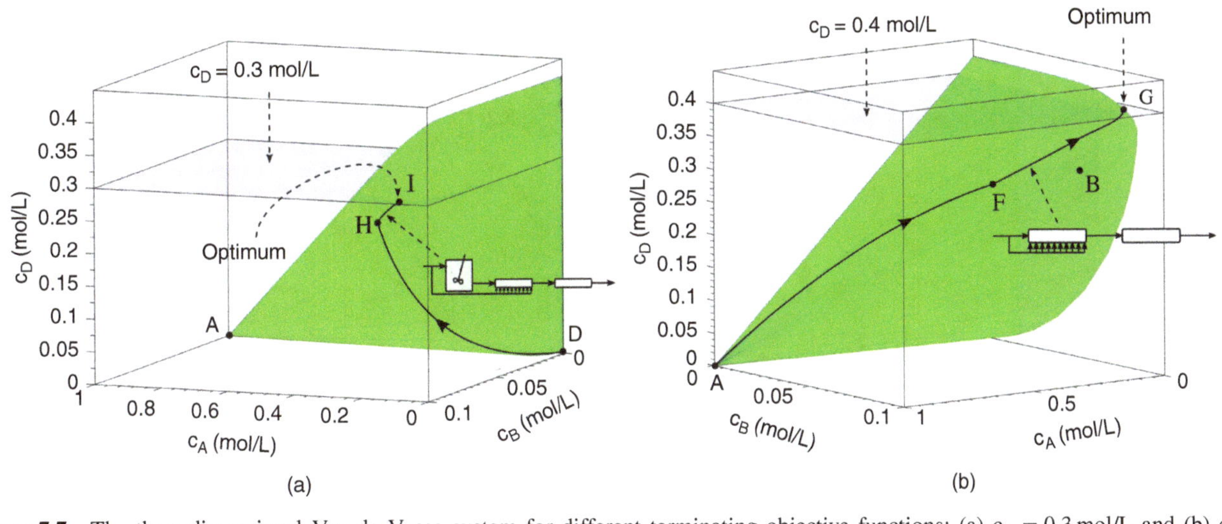

Figure 7.7 The three-dimensional Van de Vusse system for different terminating objective functions: (a) $c_D = 0.3$ mol/L and (b) $c_D = 0.4$ mol/L.

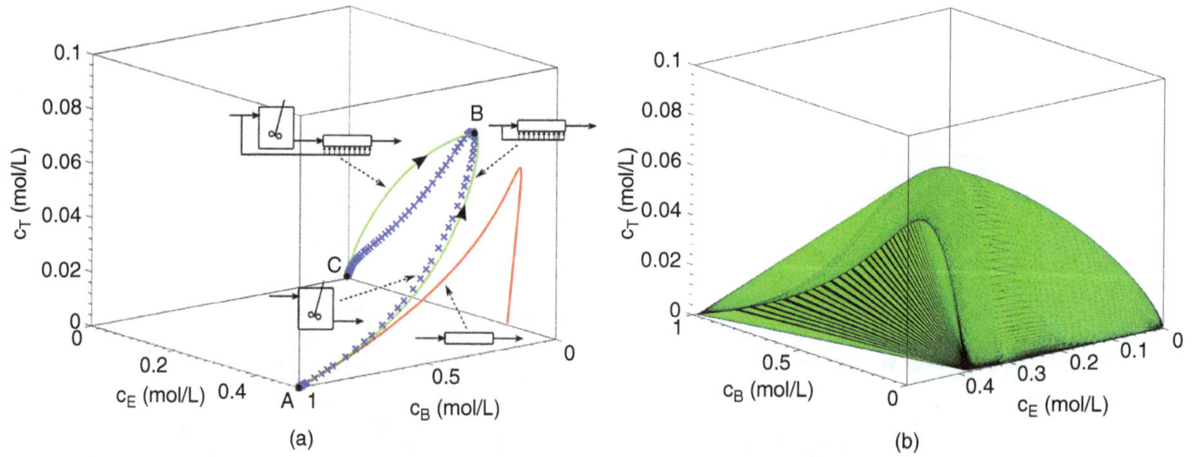

Figure 7.12 (a) Critical DSR trajectories from the feed (curve AB) and from the CSTR equilibrium point (curve CB). (b) Full AR for the BTX system.

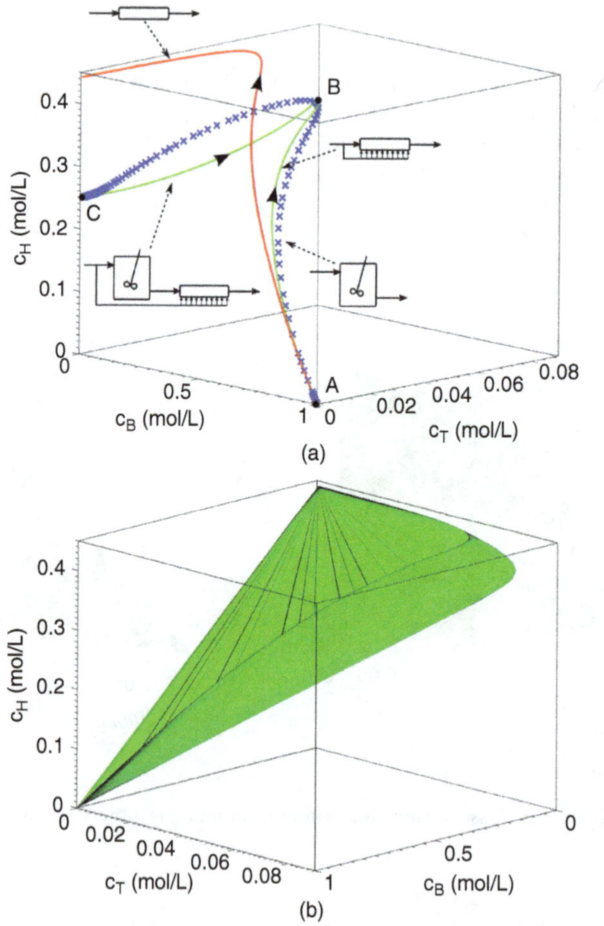

Figure 7.15 (a) Optimal reactor structures in $c_B-c_T-c_H$ space and (b) filled AR for the BTX system in $c_B-c_T-c_H$ space.

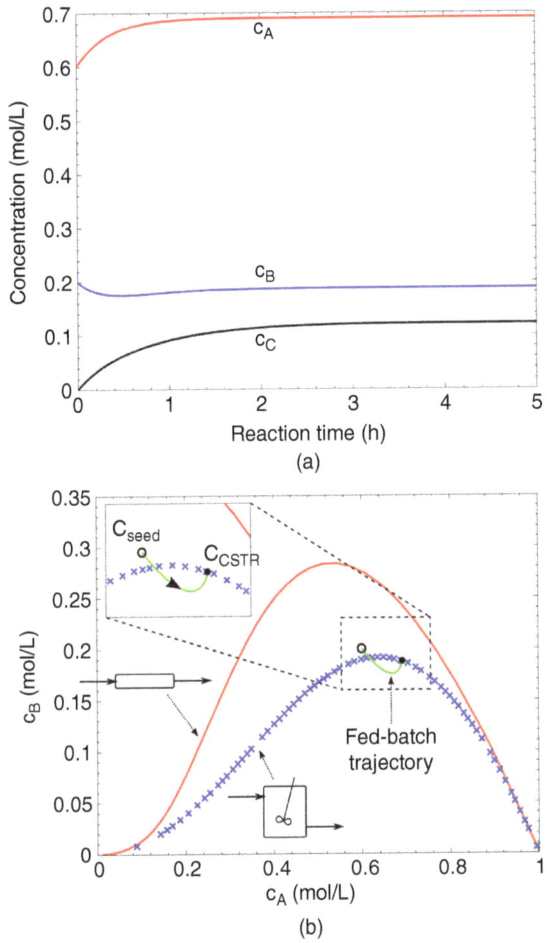

Figure 7.36 (a) Concentration profiles for components A, B, and C, achieved in a constant α fed-batch initiated at the seed concentration. (b) Trajectories and locus achieved in the equivalent fed-batch.

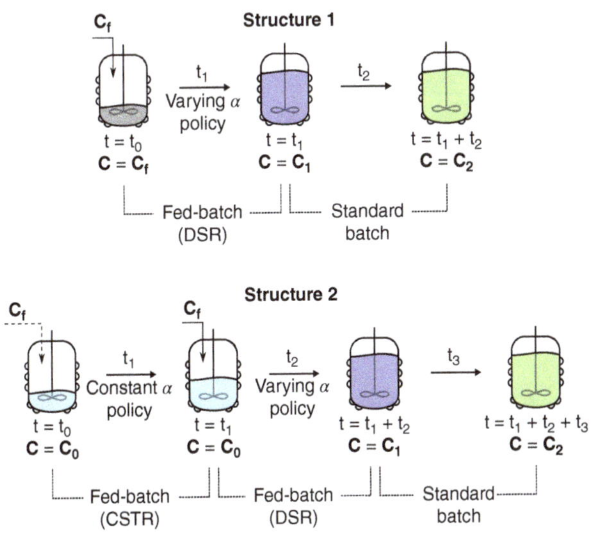

Figure 7.37 Optimal batch reactor structures for the Van de Vusse system. Adapted from Ming et al. (2013).

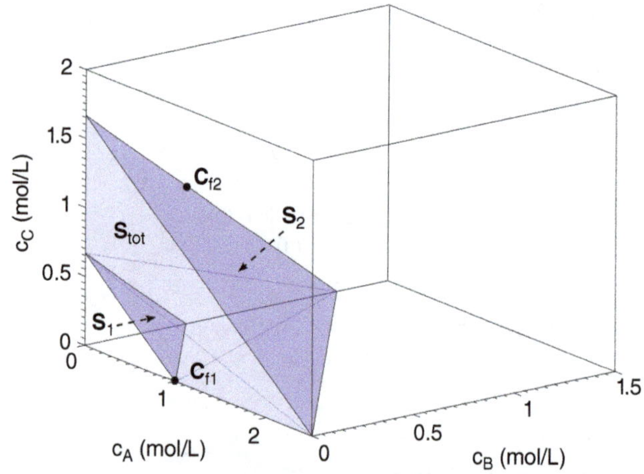

Figure 8.5 Stoichiometric subspace for multiple feeds.

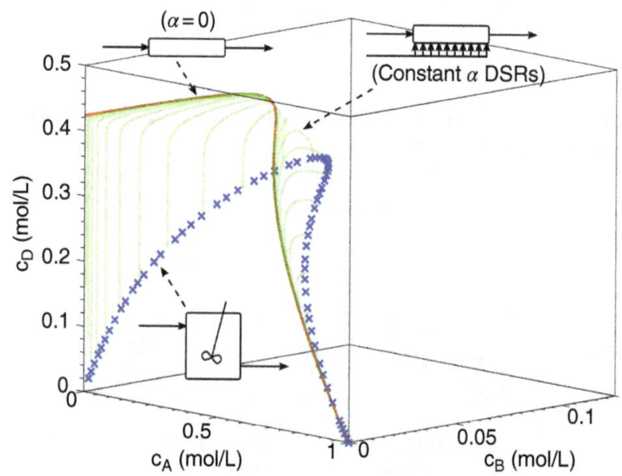

Figure 8.10 DSR trajectories for constant α values using the feed point as the DSR sidestream composition. The CSTR locus from the feed point is also shown for comparison. It is clear that the equilibrium points for the DSR trajectories coincide with the CSTR locus points. Van de Vusse kinetics is used here.

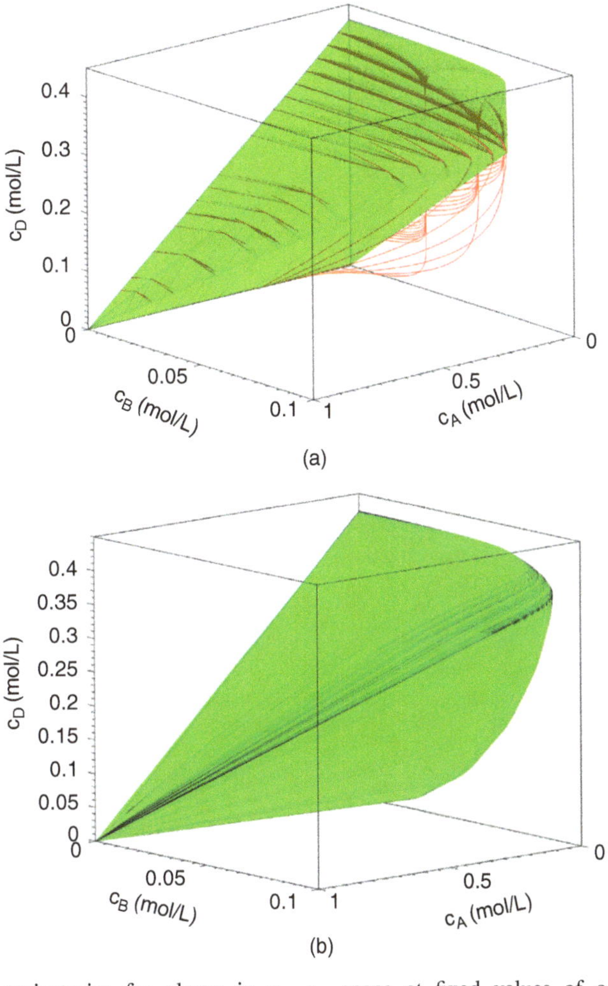

Figure 8.17 (a) Iso-compositional trajectories for planes in c_B–c_D space at fixed values of c_A. The trajectories all lie within a two-dimensional plane. (b) Final candidate region produced for the Van de Vusse system using the iso-state method.

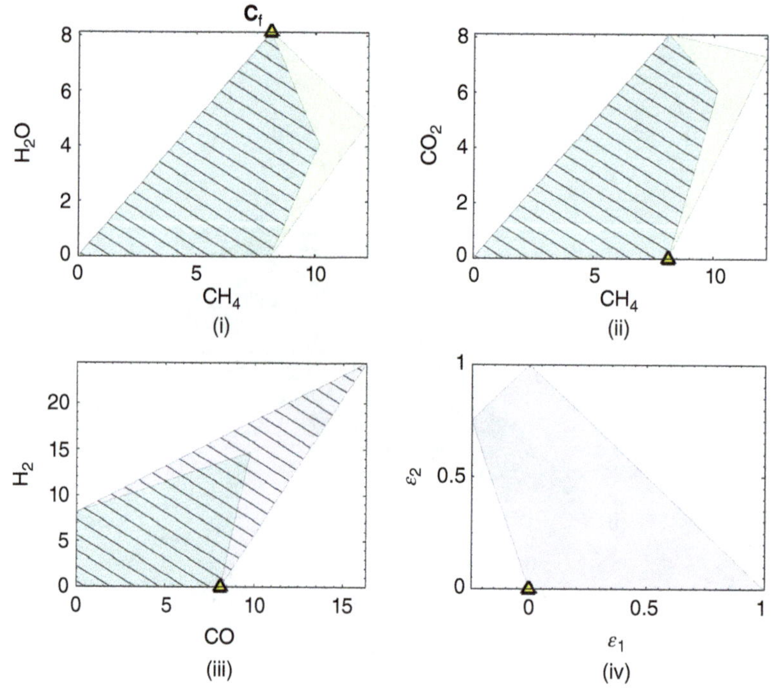

Figure 9.5 Comparison of stoichiometric subspaces in concentration space obtained for the methane steam reforming system for constant density (hatched regions) to the region obtained via mass fractions (clear region).

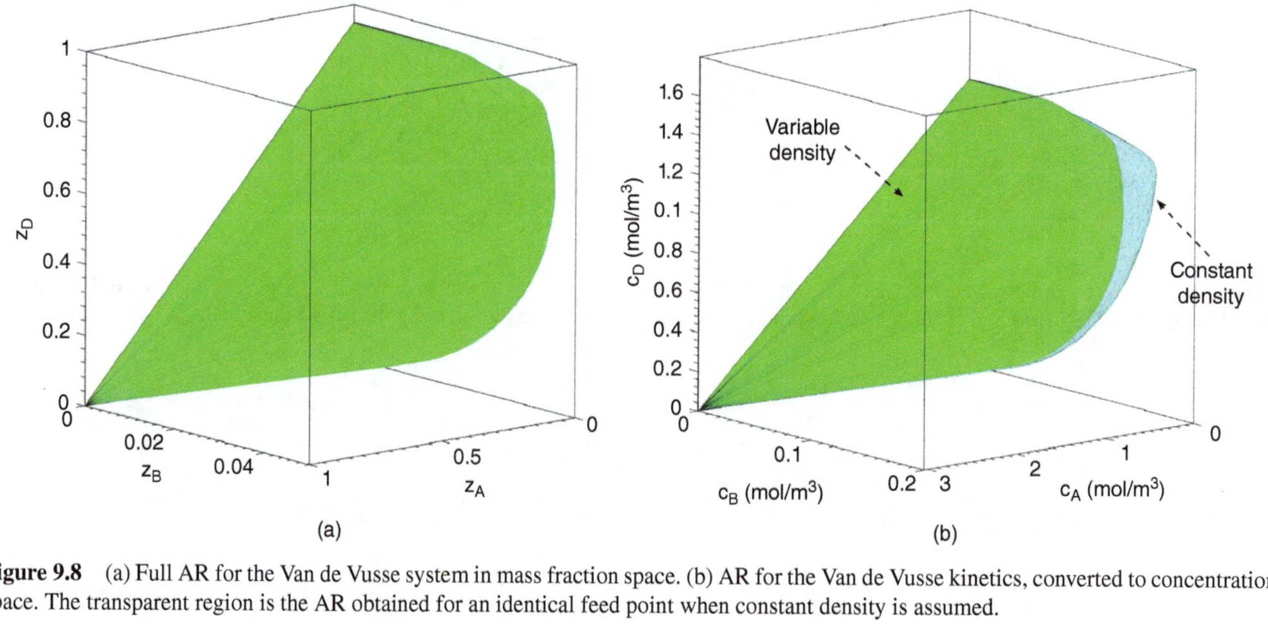

Figure 9.8 (a) Full AR for the Van de Vusse system in mass fraction space. (b) AR for the Van de Vusse kinetics, converted to concentration space. The transparent region is the AR obtained for an identical feed point when constant density is assumed.

8

AR CONSTRUCTION ALGORITHMS

8.1 INTRODUCTION

All of the problems addressed up to now have been carried out, largely, from an analytical standpoint. The problems were simple enough that the associated rate and concentration data could be plotted in, at most, a three-dimensional space. This is convenient as it allows for visualization of the data, making AR construction significantly easier. Visualization of the data is an important part of AR theory when starting out, as it allows for the interpretation of the rate field, where PFRs, CSTRs, and DSRs can be observed how they interact in state space. However, the types of systems that are often encountered in practice are seldom cast in two or three dimensions only. Many problems, particularly those involving biological systems, polymerization, and milling (also known as comminution), contain a large set of interacting species, which are related by complex, coupled reactions. When the dimension of the problem becomes large, then it is no longer practically feasible to rely on visualization and interpretation of data solely by graphical means. And even if visualization of higher dimensional systems is possible, constructing candidate regions by hand is tiresome, particularly when critical reactors must be determined. Automated AR construction methods that are suitable for implementation on a modern computer are then very useful in this regard.

In this chapter, we wish to touch on a number of automated AR construction schemes. At the time of writing, research in AR theory has witnessed a shift toward the development of numerical AR construction algorithms, with less emphasis placed on general AR theory. These developments have arisen primarily out of a practical need to determine candidate regions for complex, higher dimensional problems, which are not easily computed by hand, but which are still important for practical problems of interest. AR construction methods provide a numerical basis wherefrom theoretical predictions may be compared with in the search for a sufficiency condition. Inasmuch as how computer simulation is often employed to support a new theory in many scientific disciplines, numerical AR methods fulfill a similar function in attainable regions research.

> **SIDE NOTE: Overview of AR construction methods**
>
> Our intention in this chapter is not to provide detailed descriptions of current AR construction algorithms. Rather, this chapter serves to provide an *overview* of the different methods available and, importantly, the underlying concepts that make AR construction possible.
>
> Automated AR computation is a popular research field at present, and advances continue to be made. We hope that this chapter will not only make the current AR construction literature more understandable but also promote new ideas and approaches that further advance the field.

8.2 PRELIMINARIES

8.2.1 Hyperplanes

A *hyperplane* H(**n**, b) is defined by the following equation:

$$\mathrm{H}(\mathbf{n},\ \mathrm{b}) = \{\mathrm{C} \in \mathbb{R}^n | \mathbf{n}^{\mathrm{T}}\mathrm{C} = \mathrm{b}\}$$

Two vectors **C** and $\mathbf{C_0}$ lie in the space spanned by the hyperplane H(**n**, b) if the following equation holds:

$$\mathbf{n}^{\mathrm{T}}(\mathbf{C} - \mathbf{C}_0) = 0$$

Here, **n** is a vector, which is often termed the *normal vector*, that is perpendicular to all vectors lying in the hyperplane H(**n**, b), and b is a scalar quantity defined by

$$\mathrm{b} = \mathbf{n}^{\mathrm{T}}\mathbf{C}_0$$

Attainable Region Theory: An Introduction to Choosing an Optimal Reactor, First Edition. David Ming, David Glasser, Diane Hildebrandt, Benjamin Glasser, and Matthew Metzger.

Companion website: http://attainableregions.com

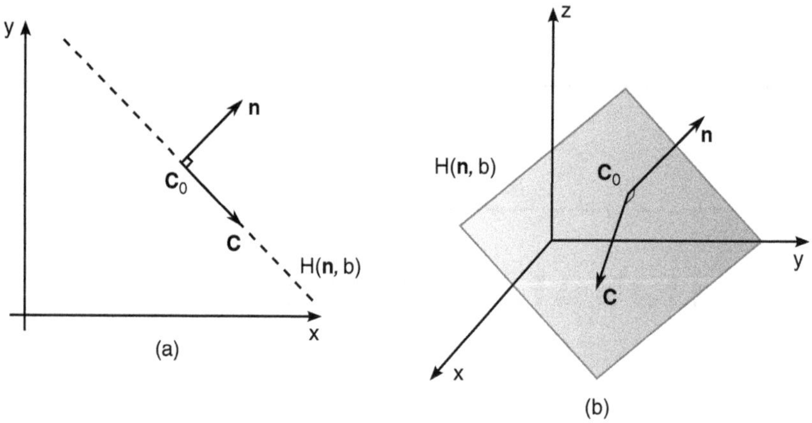

Figure 8.1 Geometric interpretation of the hyperplane equation: (a) Hyperplanes in $\mathbb{R}^2$ are simply straight-line segments (a one-dimensional subspace) and (b) hyperplanes in $\mathbb{R}^3$ are planes (a two-dimensional subspace).

since $\mathbf{C}_0$ lies in H(**n**, b). This criterion is shown graphically in Figure 8.1(a) and (b) for hyperplanes in $\mathbb{R}^2$ and $\mathbb{R}^3$, respectively. A hyperplane is an $(n-1)$-dimensional subspace of $\mathbb{R}^n$.

A hyperplane separates a space $\mathbb{R}^n$ into two *half-spaces*. This property of hyperplanes is useful in AR construction as any convex polytope, P, that is either closed or unbounded may be constructed from a collection of appropriately orientated hyperplanes in $\mathbb{R}^n$. When P is described as a set of hyperplanes——which are mathematically represented as a set of inequality constraints—it is said to be described in the *H-representation*. Similarly, when P is described in terms of its vertices, it is said to be described in the *V-representation*.

CONCEPT: Polytopes

A *polytope* is a geometric object with flat sides. Squares and triangles are examples of polytopes in $\mathbb{R}^2$ (polygons are polytopes in a plane), whereas cubes and pyramids are examples of polytopes in $\mathbb{R}^3$. The convex hull of a set of points residing in $\mathbb{R}^n$ is therefore also an example of an n-dimensional polytope (also known as an n-polytope). Although polytopes have flat sides, we can often approximate curved objects by a polytope with sufficiently many flat sides, and thus we can approximate candidate ARs as polytopes with many hyperplanes.

A candidate AR construction method that utilizes hyperplanes to carve away unachievable space shall be discussed in Section 8.5.2. Linear constraints, such as non-negativity constraints on component concentrations and flow rates, may also be expressed in the form of a hyperplane equation. Hyperplanes therefore also arise in establishing bounds in state space. In Section 8.6, superstructure methods shall be described for the computation of candidate ARs. These methods, at their core, rely on the solution of a large linear programming (LP) problem that is based on higher dimensional hyperplane equations, which are derived from non-negativity constraints.

EXAMPLE 1: Hyperplanes

Consider the hyperplanes H_1, H_2, and H_3, given by the following vectors in $\mathbb{R}^2$:

H(**n**, b)	**n**	$\mathbf{C}_0$
H_1	$[-1, 0]^T$	$[0, 0]^T$
H_2	$[0, -1]^T$	$[0, 0]^T$
H_3	$[1, 1]^T$	$[0.5, 0.5]^T$

Answer the following questions:

(a) Calculate the scalar b_i's associated with each hyperplane.
(b) Express the set of hyperplanes as a system of linear equations.
(c) Plot the resulting hyperplanes in x–y space.

Answer a:

The scalar quantity b_i of a hyperplane H_i is calculated from the hyperplane normal $\mathbf{n}_i$ and $\mathbf{C}_0$. Hence,

$$b_i = \mathbf{n}_i^T \mathbf{C}_0$$

Substituting for each hyperplane gives

$$b_1 = \begin{bmatrix}-1 & 0\end{bmatrix}\begin{bmatrix}0\\0\end{bmatrix} = 0$$

$$b_2 = \begin{bmatrix} 0 & -1 \end{bmatrix} \begin{bmatrix} 0 \\ 0 \end{bmatrix} = 0$$

$$b_3 = \begin{bmatrix} 1 & 1 \end{bmatrix} \begin{bmatrix} 0.5 \\ 0.5 \end{bmatrix} = 1$$

Answer b:

In two dimensions, a hyperplane is represented by a straight line passing through $\mathbf{C}_0$. We can find expressions for each hyperplane in terms of variables x and y by noting that the hyperplane must obey the following relation:

$$\mathbf{n}^{\mathrm{T}}(\mathbf{C} - \mathbf{C}_0) = 0$$
$$\mathbf{n}^{\mathrm{T}}\mathbf{C} = \mathbf{n}^{\mathrm{T}}\mathbf{C}_0$$
$$\mathbf{n}^{\mathrm{T}}\mathbf{C} = \mathrm{b}$$

Here $\mathbf{C}$, is given by the vector $\mathbf{C} = [\mathrm{x}, \mathrm{y}]^{\mathrm{T}}$. Substituting these values into the above expression gives

$$\begin{bmatrix} -1 & 0 \end{bmatrix} \begin{bmatrix} \mathrm{x} \\ \mathrm{y} \end{bmatrix} = 0$$

$$\begin{bmatrix} 0 & -1 \end{bmatrix} \begin{bmatrix} \mathrm{x} \\ \mathrm{y} \end{bmatrix} = 0$$

$$\begin{bmatrix} 1 & 1 \end{bmatrix} \begin{bmatrix} \mathrm{x} \\ \mathrm{y} \end{bmatrix} = 1$$

Multiplying out the matrices gives the desired system of linear equations:

$$-\mathrm{x} = 0$$
$$-\mathrm{y} = 0$$
$$\mathrm{x} + \mathrm{y} = 1$$

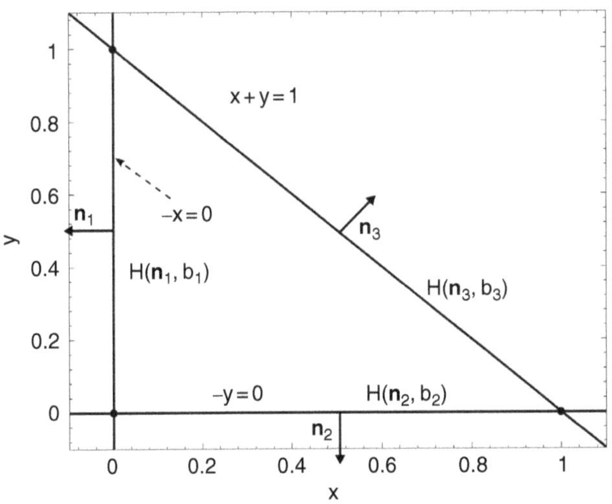

Figure 8.2 Hyperplanes in x–y space.

The coefficients for x and y in each linear equation correspond to the component values in the associated hyperplane normal vector, $\mathbf{n}_i$.

Answer c:

Now that we have expressed each hyperplane as a linear equation, the system may easily be plotted in x–y space. The results are shown in Figure 8.2. Observe that each hyperplane is represented as a straight-line equation in $\mathbb{R}^2$.

8.2.2 Computing the Stoichiometric Subspace S

8.2.2.1 Introduction

RECALL: The stoichiometric subspace S

Hereafter, it is assumed that the reader is comfortable with the idea of the stoichiometric subspace **S**, which is discussed in Chapter 6. In this section, we shall describe how to numerically calculate the bounds of **S** (which is a convex polytope that is expressed as a collection of hyperplane constraints).

Suppose that a system of reactions with associated kinetics is available, obeying a certain reaction stoichiometry. Since expressions for the rate of formation are known and available, it is possible to compute the AR for the system in concentration space. The particular region computed, defined by the kinetics for the system, will exist as a convex polytope residing in $\mathbb{R}^n$.

Suppose now that a different set of kinetic expressions for the same system of reactions and feed point is supplied. Since the size and shape of the AR is defined by the kinetics, the corresponding AR for the new kinetics may differ to the original region computed. Although the particular form of the kinetics may change the shape of the AR, all kinetics irrespective of individual form must still obey mass balance constraints defined by the reaction stoichiometry. Hence, the stoichiometric subspace **S** does not change by the introduction of new kinetics.

Imagine now that a different feed point is specified. Mass balance constraints on the system may be invoked that are consistent with the new feed, which in turn affects that shape of **S**. Thus, although **S** does not change with different kinetics, it is affected by a change in *feed and stoichiometry*.

CONCEPT: Stoichiometric subspace or mass balance triangle?

In AR theory (as well as in distillation theory), the region of concentrations that obey mass balance constraints is often referred to as the *mass balance triangle*—this is because **S** takes the shape of a scalene or right-angled triangle in $\mathbb{R}^2$, and a tetrahedron in $\mathbb{R}^3$. We prefer to use the term "stoichiometric subspace", however, as **S** is not always represented by these shapes, particularly for unusual reaction stoichiometry.

Recall from Chapter 6 that **S** is calculated by expressing each independent reaction in terms of the extent of reaction vector, $\boldsymbol{\varepsilon}$, and the stoichiometric coefficient matrix **A**.

$$\mathbf{C} = \mathbf{C}_\mathrm{f} + \mathbf{A}\boldsymbol{\varepsilon} \tag{8.1}$$

Non-negativity constraints may then be enforced for each component present in the system.

$$\mathbf{C} \geq \mathbf{0} \tag{8.2a}$$

or

$$\mathbf{C}_\mathrm{f} + \mathbf{A}\boldsymbol{\varepsilon} \geq \mathbf{0} \tag{8.2b}$$

Geometrically, Equation 8.2b describes a set of hyperplane constraints (a set of inequality constraints) that define the stoichiometric subspace in $\mathbb{R}^n$.

SIDE NOTE: Computing S from null ($\mathbf{A}^\mathrm{T}$)

The following sections will describe how it is possible to compute the stoichiometric subspace by identifying the bounding constraints in extent space that form the feasible region, which is a function of the reaction stoichiometry and feed point. From this information, it is possible to compute the vertices of the region via *vertex enumeration*, which is described in Section 8.2.2.2.

An alternative method for computing **S** exists that relies on computing a basis for the null space of the stoichiometric matrix **A**. Vectors belonging to null ($\mathbf{A}^\mathrm{T}$) are normal to **S**, and these vectors may be related to bounding hyperplane constraints that define **S**. This method is described in Feinberg (1987, 2000a, 2000b).

8.2.2.2 *Worked Example*

a) Methane Steam Reforming Consider the simple two-dimensional reaction scheme for methane steam reforming:

$$CH_4 + H_2O \rightarrow CO + 3H_2$$

$$CO + H_2O \rightarrow CO_2 + H_2$$

The system involves five components. For simplicity, only forward reactions are considered (although the stoichiometric subspace does not change if reverse reactions are considered as well). The stoichiometric coefficient matrix **A** may be formed in the usual manner giving

$$\mathbf{A} = \begin{bmatrix} -1 & 0 \\ -1 & -1 \\ 1 & -1 \\ 3 & 1 \\ 0 & 1 \end{bmatrix}$$

Row 1 in matrix **A** represents CH_4, row 2 is H_2O, row 3 is CO, row 4 is H_2, and row 5 is CO_2. Two independent reactions are involved in this system, and the entire system may be expressed by specification of two components only.[1] Suppose that a feed vector, containing 1.0 mol/L of CH_4, H_2O, and CO is available. The feed vector $\mathbf{C}_\mathrm{f}$ is given by

$$\mathbf{C}_\mathrm{f} = \begin{bmatrix} 1 & 1 & 1 & 0 & 0 \end{bmatrix}^\mathrm{T} \mathrm{mol/L}$$

In order to compute **S** using the method of extents, one must first express each component concentration in terms of its initial concentration and the extent of reaction. In general, each component may be viewed to participate in both reactions. Species concentrations are expressed in terms of an initial feed concentration $\mathbf{C}_\mathrm{f}$ and two extents of reaction, ε_1 and ε_2, as follows:

$$\begin{aligned} c_{CH_4} &= c^0_{CH_4} - \varepsilon_1 + 0\varepsilon_2 \\ c_{H_2O} &= c^0_{H_2O} - \varepsilon_1 - \varepsilon_2 \\ c_{CO} &= c^0_{CO} + \varepsilon_1 - \varepsilon_2 \\ c_{H_2} &= c^0_{H_2} + 3\varepsilon_1 + \varepsilon_2 \\ c_{CO_2} &= c^0_{CO_2} + 0\varepsilon_1 + \varepsilon_2 \end{aligned}$$

Note that we have assumed constant density in the system in order to arrive at an expression directly in terms of molar concentration. This has been done for simplicity, although it is more correct to begin with moles and then convert to concentration or mass fractions if need be.

Observe that the coefficients of ε_1 and ε_2 belong to the stoichiometric coefficient matrix **A**. It follows that the system may be written in matrix form as follows:

$$\begin{bmatrix} c_{CH_4} \\ c_{H_2O} \\ c_{CO} \\ c_{H_2} \\ c_{CO_2} \end{bmatrix} = \begin{bmatrix} c^0_{CH_4} \\ c^0_{H_2O} \\ c^0_{CO} \\ c^0_{H_2} \\ c^0_{CO_2} \end{bmatrix} + \begin{bmatrix} -1 & 0 \\ -1 & -1 \\ 1 & -1 \\ 3 & 1 \\ 0 & 1 \end{bmatrix} \begin{bmatrix} \varepsilon_1 \\ \varepsilon_2 \end{bmatrix}$$

[1]This may be verified by computing rank(**A**).

or

$$\mathbf{C} = \mathbf{C}_f + \mathbf{A}\boldsymbol{\varepsilon} \qquad (8.1)$$

Here, $\boldsymbol{\varepsilon} = [\varepsilon_1, \varepsilon_2]^T$. In order to compute **S**, the set of equality constraints defined by Equation 8.1 must be converted into a set of inequality constraints, which may be achieved by enforcing non-negativity constraints for each component.

$$\mathbf{C} \geq \mathbf{0} \qquad (8.2a)$$

Expressing each component in terms of $\mathbf{C}_f$ and $\boldsymbol{\varepsilon}$, we obtain

$$\mathbf{C}_f + \mathbf{A}\boldsymbol{\varepsilon} \geq \mathbf{0} \qquad (8.2b)$$

SIDE NOTE: Equations and inequalities

Equations describe distinct points in space, whereas inequalities describe regions in space (half-spaces). Values that satisfy an equality constraint describe a line in $\mathbb{R}^2$ or a surface in $\mathbb{R}^3$. By comparison, values satisfying an inequality constraint describe a region in space (an area in $\mathbb{R}^2$ and volume in $\mathbb{R}^3$).

For example, the equation $x + y = 0$ describes a line in $\mathbb{R}^2$, whereas the equivalent inequality $x + y < 0$ describes a region whereby any combination of $x + y$ less than zero is satisfied. Similarly, the hyperplane equation $H(\mathbf{n}, b) = \gamma$ separates space into two half-spaces given by $H(\mathbf{n}, b) < \gamma$ and $H(\mathbf{n}, b) > \gamma$.

For a reactive system involving n components, n inequality constraints may be formed. The system of inequalities, described by Equation 8.2b, defines the feasible region of all stoichiometrically compatible concentrations in *extent space*. An extent vector $\boldsymbol{\varepsilon}_i$ that satisfies the system of inequalities, given by Equation 8.2b, is one that satisfies mass balance constraints defined by the reaction stoichiometry. The corresponding concentration, $\mathbf{C}_i$, is then said to be feasible—as it lies within the stoichiometric subspace—although it may still not be attainable if it lies outside of the AR.

If the positions of the extreme points of **S** can be identified in extent space, then Equation 8.1 may be invoked to solve for the corresponding points in concentration space. Computing the extreme points of a convex polytope, defined by a set of hyperplane constraints, is termed *vertex enumeration*.

CONCEPT: Vertex and facet enumeration

A convex polytope, P, in $\mathbb{R}^n$ may be described independently, both in terms of its vertices and in terms of its facets. P is said to be given in the vertex representation (the V-representation), when the polytope is described in terms of a collection of vertices. Similarly, P is said to be given in the hyperplane representation (the H-representation) when P is described by a collection of hyperplane constraints of the following form:

$$\mathbf{Ax} \leq \mathbf{b}$$

If there are L hyperplane constraints orientated in $\mathbb{R}^n$, then $\mathbf{A}$ is an $L \times n$ matrix, $\mathbf{x}$ is an $n \times 1$ vector, and $\mathbf{b}$ is an $L \times 1$ vector. Note that imposing non-negativity constraints on the component concentrations is equivalent to expressing the system in the above form. The *vertex enumeration problem* constitutes finding the vertices (corners) of P when P is described by the H-representation. Conversely, the *facet enumeration problem* arises when P is described by the V-representation, and we wish to determine the associated hyperplane constraints in H-representation. We have already encountered the facet enumeration problem on numerous occasions, disguised under the form of the convex hull of a set of points. Vertex and facet enumeration are, in the sense of convex optimization, dual problems. Solution of the one automatically results in solution of the other. Computing the convex hull of a set of points is equivalent to solving the facet enumeration problem for a given convex polytope describing a set of achievable states in state space. Figure 8.3 gives a visual representation of the two problems.

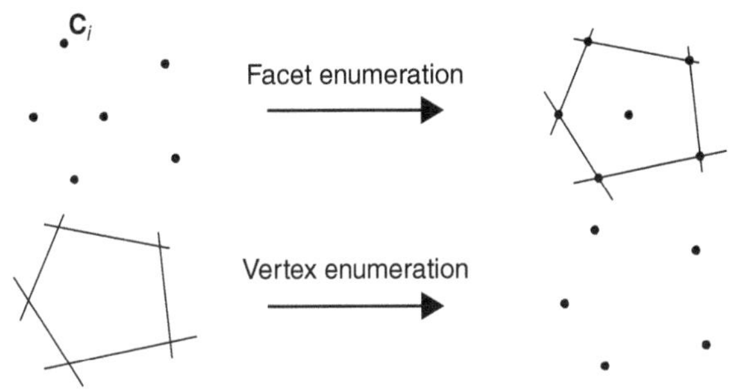

Figure 8.3 Geometric interpretation of vertex and facet enumeration.

Computing the extreme points of **S**, when working with extents, is equivalent to finding the intersection points that produce feasible vertices from a system of inequality constraints. We are, in fact, solving the vertex enumeration problem. Solving vertex and facet enumeration problems is convenient when standard numerical packages are used. In particular, the `con2vert()` and `vert2con()` MATLAB functions are extremely useful for solving vertex and facet enumeration problems, respectively.[2]

[2]Both routines can be found on the MATLAB file exchange (http://www.mathworks.com/matlabcentral/fileexchange/).

EXAMPLE 2: Practice with vertex enumeration

Use the `con2vert()` function to convert the following sets of inequality constraints to vertices. That is, convert H-representation to V-representation.

$$\text{(a)} \quad \mathbf{A} = \begin{bmatrix} 1 & 1 \\ -1 & 0 \\ 0 & -1 \\ 0 & 1 \end{bmatrix} \qquad \mathbf{b} = \begin{bmatrix} 1 \\ 0 \\ 0 \\ 0.5 \end{bmatrix}$$

$$\text{(b)} \quad \mathbf{A} = \begin{bmatrix} 1 & 0 \\ 1 & 1 \\ -1 & 1 \\ -3 & -1 \\ 0 & -1 \end{bmatrix} \qquad \mathbf{b} = \begin{bmatrix} 1 \\ 1 \\ 1 \\ 0 \\ 0 \end{bmatrix}$$

$$\text{(c)} \quad \mathbf{A} = \begin{bmatrix} 1 & 0 & 2 \\ -1 & 1 & 0 \\ 0 & -1 & 0 \\ 0 & 0 & -1 \end{bmatrix} \qquad \mathbf{b} = \begin{bmatrix} 1 \\ 0.5 \\ 0.1 \\ 0 \end{bmatrix}$$

$$\text{(d)} \quad \mathbf{A} = \begin{bmatrix} -1 & 0 & -1 & 0 \\ -1 & 0 & 0 & 1 \\ 1 & -2 & 0 & -1 \\ 0 & 1 & -1 & 0 \\ 0 & 0 & 1 & -1 \\ 0 & 0 & 0 & 1 \end{bmatrix} \qquad \mathbf{b} = \begin{bmatrix} 7 \\ 1 \\ 0 \\ 0 \\ 0 \\ 0 \end{bmatrix}$$

Answer a:

The vertex enumeration problem involves computing the extreme points of the convex polytope, defined by the inequality constraints,

$$\mathbf{Ax} \leq \mathbf{b}$$

Writing out explicitly for the values of **A** and **b** supplied gives

$$\begin{bmatrix} 1 & 1 \\ -1 & 0 \\ 0 & -1 \\ 0 & 1 \end{bmatrix} \begin{bmatrix} x_1 \\ x_2 \end{bmatrix} \leq \begin{bmatrix} 1 \\ 0 \\ 0 \\ 0.5 \end{bmatrix}$$

or

$$x_1 + x_2 \leq 1$$
$$-x_1 \leq 0$$
$$-x_2 \leq 0$$
$$x_2 \leq 0.5$$

which shows that $\mathbf{Ax} \leq \mathbf{b}$ corresponds to a system of linear inequalities in x_1–x_2 space. Inputting the supplied values for **A** and **b** into MATLAB, and solving the vertex enumeration problem, using the `con2vert()` function, gives the following result:

$$\mathbf{X}=\begin{bmatrix}0 & 0\\ 0 & 0.5\\ 1 & 0\\ 0.5 & 0.5\end{bmatrix}$$

Here, **X** is a matrix generated by the solution to the `con2vert()` function in MATLAB. Each row in **X** indicates a unique extreme point of the convex polytope, described by the system of inequalities. Therefore, the convex polytope contains four extreme points: $\{[0, 0]^T, [0, 0.5]^T, [1, 0]^T, \text{and } [0.5, 0.5]^T\}$.

Answer b:

The system of inequalities is now given by

$$\begin{bmatrix}1 & 0\\ 1 & 1\\ -1 & 1\\ -3 & -1\\ 0 & -1\end{bmatrix}\begin{bmatrix}x_1\\ x_2\end{bmatrix}\le\begin{bmatrix}1\\ 1\\ 1\\ 0\\ 0\end{bmatrix}$$

which describes a two-dimensional polytope in x_1–x_2 space. Upon closer inspection of the system, note that this system corresponds to the methane steam reforming system discussed at the beginning of this section. To see this, note that matrix **A** is equivalent to the negative of the stoichiometric coefficient matrix for the methane steam reforming reactions:

$$\begin{aligned}x_1 &\le 1\\ x_1+x_2 &\le 1\\ -x_1+x_2 &\le 1\\ -3x_1-x_2 &\le 0\\ -x_2 &\le 0\end{aligned}$$

where each row in the system corresponds to a component in the system of reactions (row 1 corresponds to CH_4, row 2 to H_2O, and so on), and the right-hand side is the feed concentration. x_1 and x_2 thus correspond to the extents of reaction ($x_1=\varepsilon_1$ and $x_2=\varepsilon_2$). The system described here is equivalent to Equation 8.2b—$\mathbf{A}\varepsilon\le\mathbf{C}_f$. Inputting the values for **A** and **b** into MATLAB produces

$$\mathbf{X}=\begin{bmatrix}1 & 0\\ -0.25 & 0.75\\ 0 & 1\\ 0 & 0\end{bmatrix}$$

There are four extreme points to the convex polytope for this system, which indicates that there are four extreme points in extent space for the methane steam reforming reaction: $\{[1, 0]^T, [-0.25, 0.75]^T, [0, 1]^T, \text{and } [0, 0]^T\}$.

Answer c:

The system of inequalities described in this situation represent the three-dimensional Van de Vusse reaction with a feed point given by $\mathbf{C}_f=[c_{Af}, c_{Bf}, c_{Cf}, c_{Df}]^T=[1, 0.5, 0.1, 0]^T$ mol/L,

$$\begin{gathered}A\rightarrow B\rightarrow C\\ 2A\rightarrow D\end{gathered}$$

Thus, solving the facet enumeration problem with **A** and **b** represents finding the extreme points of the stoichiometric subspace for the three-dimensional Van de Vusse reaction in extent space.

$$\mathbf{A}\varepsilon\le\mathbf{C}_f$$

where **A** is the negative of the stoichiometric coefficient matrix for the Van de Vusse reactions.

$$\begin{bmatrix} 1 & 0 & 2 \\ -1 & 1 & 0 \\ 0 & -1 & 0 \\ 0 & 0 & -1 \end{bmatrix} \begin{bmatrix} \varepsilon_1 \\ \varepsilon_2 \\ \varepsilon_3 \end{bmatrix} \leq \begin{bmatrix} 1 \\ 0.5 \\ 0.1 \\ 0 \end{bmatrix}$$

ε_1, ε_2, and ε_3 represent the extent of reaction for reactions 1, 2, and 3, respectively. Solving the vertex enumeration problem, by inputting the values into `con2vert()` in MATLAB, gives

$$\boldsymbol{\varepsilon} = \begin{bmatrix} 1 & 1.5 & 0 \\ 1 & -0.1 & 0 \\ -0.6 & -0.1 & 0.8 \\ -0.6 & -0.1 & 0 \end{bmatrix}$$

This indicates that for a feed point of $\mathbf{C}_f = [1, 0.5, 0.1, 0]^T$ mol/L, the stoichiometric subspace for the three-dimensional Van de Vusse reaction contains four extreme points in extent (ε_1–ε_2–ε_3) space: $\{[1, 1.5, 0]^T, [1, -0.1, 0]^T, [-0.6, -0.1, 0.8]^T,$ and $[-0.6, -0.1, 0]^T\}$ mol/L.

Answer d:

This system describes six inequalities involving four variables; the region is a four-dimensional convex polytope in x_1–x_2–x_3–x_4 space. Inputting the values into `con2vert()` in MATLAB gives

$$\mathbf{X} = \begin{bmatrix} -4 & -3 & -3 & -3 \\ -4 & -0.5 & -3 & -3 \\ -1.5 & -0.5 & -0.5 & -0.5 \\ -6.5 & -0.5 & -0.5 & -5.5 \\ -5.25 & -1.75 & -1.75 & -1.75 \end{bmatrix}$$

Each row in **X** describes a vertex of the polytope, and therefore the polytope is composed of five vertices in $\mathbb{R}^4$.

EXAMPLE 3: Practice with facet enumeration

Show that vertex enumeration and facet enumeration are duals of each other by converting the extreme points from Example 2(a) to a system of inequality constraints. Use the `vert2con()` function.

Answer:

The vertices in Example 2(a) are given by

$$\mathbf{X} = \begin{bmatrix} 0 & 0 \\ 0 & 0.5 \\ 1 & 0 \\ 0.5 & 0.5 \end{bmatrix}$$

Solving the facet enumeration problem, by using the `vert2con()` function in MATLAB, we obtain the following:

$$\mathbf{A} = \begin{bmatrix} 0 & 4 \\ 0 & -4 \\ 2.667 & 2.667 \\ -2.667 & 0 \end{bmatrix} \qquad \mathbf{b} = \begin{bmatrix} 2 \\ 0 \\ 2.667 \\ 0 \end{bmatrix}$$

When written out explicitly as a system of inequalities, this gives

$$4x_2 \leq 2$$
$$-4x_2 \leq 0$$
$$2.667x_1 + 2.667x_2 \leq 2.667$$
$$-2.667x_1 \leq 0$$

which describes the equivalent set of inequalities as in Example 2(a)—the given inequalities are scalar multiples of that given in Example 2(a).

We can hence calculate the extreme points of the region defined by Equation 8.2b in extent space by vertex enumeration. The set of extents forming the extreme points of the stoichiometric subspace is found to be $\{[1.0, 0.0]^T, [-0.25, 0.75]^T, [0.0, 1.0]^T, [0.0, 0.0]^T\}$. From this set, equivalent extreme points in concentration space may be computed using Equation 8.1. The stoichiometric subspace resides as a two-dimensional subspace in $\mathbb{R}^5$. This subspace may be projected onto different component spaces for visualization. A number of example component pairs are shown in Figure 8.4(a).

The shaded region represents the space wherein all concentrations that are compatible with the reaction stoichiometry must reside. The shaded triangle indicates the feed point, $\mathbf{C}_f$, in space. Note that this space has been computed *without* information related to the kinetics, and therefore it is consistent irrespective of the kinetics specified.

Observe that when $\mathbf{S}$ is projected onto component planes, the shape of the projected region is different. All of the regions displayed belong to the same stoichiometric subspace for the CH_4 reforming reaction and feed point, however. Irrespective of the components used to view $\mathbf{S}$, all concentrations not shown in the figure may be determined by mass balance. For comparison, the region in the bottom right of Figure 8.4(a) shows $\mathbf{S}$ in ε_1–ε_2 space. Note that although species concentrations must remain positive, the extent of reaction is allowed to take on negative values. This is a further reason why it is preferred to construct candidate ARs in concentration space and not in extent space.

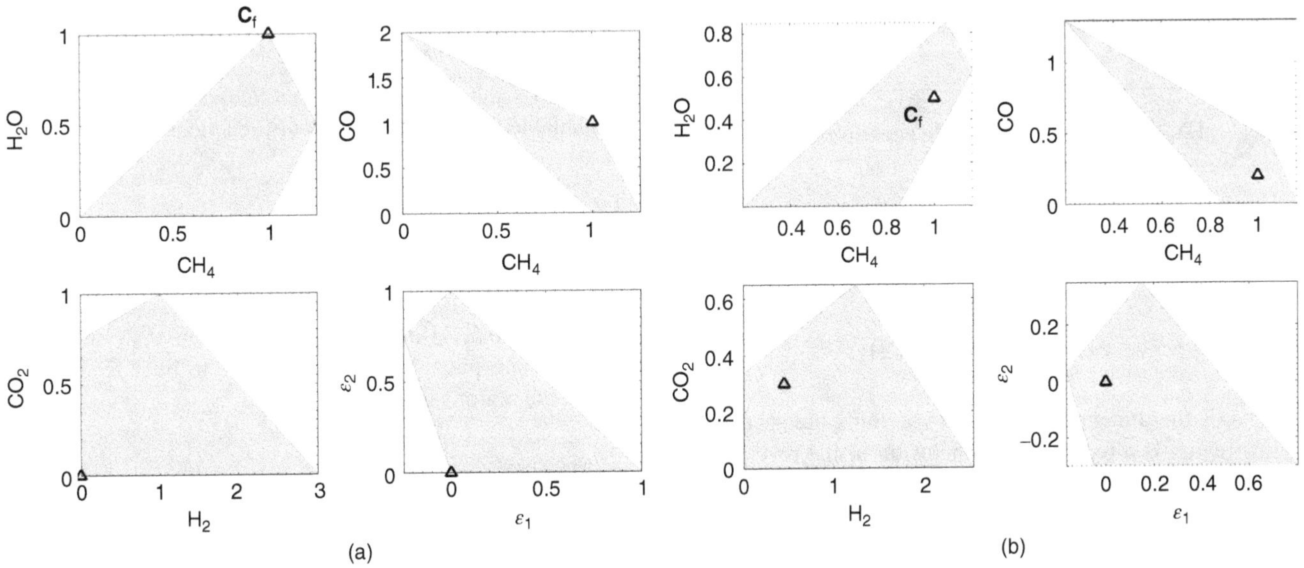

Figure 8.4 (a) Stoichiometric subspace for the steam reforming system projected onto different component spaces: CH_4–H_2O (top left), CH_4–CO (top right), and H_2–CO_2 (bottom left). The stoichiometric subspace, projected onto ε_1–ε_2 space, is also given (bottom right). (b) Stoichiometric subspace for methane steam reforming now given for a different feed point.

b) Different Feed Point Suppose now that a different feed point is chosen for computing **S**. This new subspace must differ from the first seeing as **S** is a function of both the reaction stoichiometry and the feed point specified. To illustrate this, we will carry out the same procedure with the feed point

$$\mathbf{C}_f = [1.0,\ 0.5,\ 0.2,\ 0.45,\ 0.3]^T \text{mol/L}$$

When **S** associated with this feed point is computed, the regions in Figure 8.4(b) are obtained. The set of extents belonging to this region is given by $\{[0.8, -0.3]^T, [0.15, 0.35]^T, [-0.05, -0.3]^T, [-0.1625, 0.0375]^T\}$.

In this instance, **S** is smaller than previously computed. Observe also that whereas in the previous example the feed point is situated on a point on the boundary of the stoichiometric subspace, the feed point lies within the new region.

CONCEPT: Changing the shape of S

S is a function of the reaction stoichiometry and feed point only. Thus, if different kinetics is supplied for the same set of reactions and feed point, the stoichiometric subspace will remain the same (even if the associated ARs may be different for the two systems).

Knowing how to compute **S** is a powerful technique for designers applying AR theory. Understanding the shape of **S** allows one to place an upper bound on the AR, so that the limits of achievability may be determined for *any* kinetics, for the same reaction stoichiometry and feed point.

EXAMPLE 4: Reversible reactions

Consider now a slight modification to the methane steam reforming reactions as follows:

$$CH_4 + H_2O \rightleftharpoons CO + 3H_2$$

$$CO + H_2O \rightarrow CO_2 + H_2$$

What is the dimension of **S**? Do you think the shape of this region is affected by the inclusion of the reversible reaction?

Answer:

In this instance, there is an additional reverse reaction, and thus there are three extents of reaction. We may again compute **S** for this new system, and compare the shape of the region to the original system when no reversible reaction is present. The same feed point shall be used as in the first example.

The stoichiometric coefficient matrix **A** can be formed, given now as follows:

$$\mathbf{A} = \begin{bmatrix} -1 & 1 & 0 \\ -1 & 1 & -1 \\ 1 & -1 & -1 \\ 3 & -3 & 1 \\ 0 & 0 & 1 \end{bmatrix}$$

The size of **A** is now 5×3, indicating that there are three reactions present in the system. Note that column 2 is a negative of column 1, and thus columns 1 and 2 are linearly dependent. The space spanned by the columns of **A** describe a two-dimensional plane in $\mathbb{R}^5$—the vectors belonging to columns 1 and 2 of **A** both describe the same line in $\mathbb{R}^5$. The rank of **A** is therefore still two, and thus the dimension of **S** for the new system, including the reverse reaction, is still two. **S** is therefore unaffected by the presence of the reverse reaction.

It follows that the two systems are identical in terms of reaction stoichiometry, and thus the shape of **S** for the two systems is the same. Equilibrium-limited kinetics are often expressed as reversible reactions. Although the specific reaction kinetics might change, **S** is unchanged for the same feed point.

EXAMPLE 5: Multiple feeds

Suppose that two feed concentrations $\mathbf{C}_{f1}$ and $\mathbf{C}_{f2}$ are available for the following set of reactions:

$$2A \rightarrow B$$

$$A + B \rightarrow 2C$$

where $\mathbf{C}_{f1} = [c_{Af},\ c_{Bf},\ c_{Cf}]^T = [1,\ 0,\ 0]^T$ mol/L and $\mathbf{C}_{f2} = [0,\ 0.5,\ 1]^T$ mol/L. Compute the overall stoichiometric subspace, $\mathbf{S}_{tot}$, corresponding to both feeds, and plot the regions in c_A–c_B–c_C space.

Answer:

The set of concentrations belonging to the overall stoichiometric subspace, $\mathbf{S}_{tot}$, can be found by first computing $\mathbf{S}_1$ and $\mathbf{S}_2$, which are the stoichiometric subspaces for $\mathbf{C}_{f1}$ and $\mathbf{C}_{f2}$, and then calculating the convex hull of $\mathbf{S}_1$ and $\mathbf{S}_2$. The stoichiometric coefficient matrix **A** for the set of reactions is given by

$$\mathbf{A} = \begin{bmatrix} -2 & -1 \\ 1 & -1 \\ 0 & 2 \end{bmatrix}$$

Here, rows 1, 2, and 3 of **A** correspond to components A, B, and C, respectively. Next, vertices of individual stoichiometric subspaces $\mathbf{S}_1$ and $\mathbf{S}_2$ may be found in a standard fashion by solving the vertex enumeration problem for the two separate system of inequalities describing $\mathbf{S}_1$ and $\mathbf{S}_2$, which are given by

$$\mathbf{C}_{f1} + \mathbf{A}\boldsymbol{\varepsilon} \geq 0 \text{ and } \mathbf{C}_{f2} + \mathbf{A}\boldsymbol{\varepsilon} \geq 0$$

When these systems are solved, $\mathbf{S}_1$ and $\mathbf{S}_2$ are given by the following set of vertices:

$$\mathbf{S}_1 = \left\{ \begin{bmatrix} 1 \\ 0 \\ 0 \end{bmatrix}, \begin{bmatrix} 0 \\ 0.5 \\ 0 \end{bmatrix}, \begin{bmatrix} 0 \\ 0 \\ 2/3 \end{bmatrix} \right\} \text{ and}$$

$$\mathbf{S}_2 = \left\{ \begin{bmatrix} 2.5 \\ 0 \\ 0 \end{bmatrix}, \begin{bmatrix} 0 \\ 1.25 \\ 0 \end{bmatrix}, \begin{bmatrix} 0 \\ 0 \\ 5/3 \end{bmatrix} \right\}$$

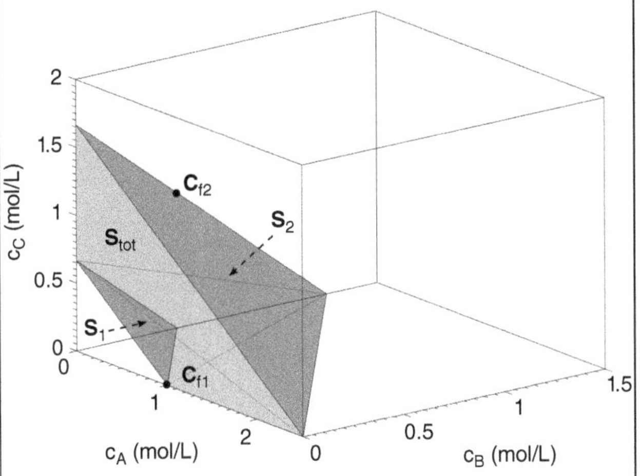

Figure 8.5 Stoichiometric subspace for multiple feeds. (*See color plate section for the color representation of this figure.*)

Determining the overall set of stoichiometrically compatible concentrations is now achieved by computing the convex hull of the set of points belonging to the union of $\mathbf{S}_1$ and $\mathbf{S}_2$:

$$\mathbf{S}_{\text{tot}} = \text{conv}(\{\mathbf{S}_1 \cup \mathbf{S}_2\})$$

$$= \text{conv}\left(\left\{ \begin{bmatrix} 1 \\ 0 \\ 0 \end{bmatrix}, \begin{bmatrix} 0 \\ 0.5 \\ 0 \end{bmatrix}, \begin{bmatrix} 0 \\ 0 \\ 2/3 \end{bmatrix}, \begin{bmatrix} 2.5 \\ 0 \\ 0 \end{bmatrix}, \begin{bmatrix} 0 \\ 1.25 \\ 0 \end{bmatrix}, \begin{bmatrix} 0 \\ 0 \\ 5/3 \end{bmatrix} \right\}\right)$$

We find that all six vertices in $\mathbf{S}_1$ and $\mathbf{S}_2$ belong to $\mathbf{S}_{\text{tot}}$. All three stoichiometric subspaces are shown in Figure 8.5.

Note that $\mathbf{S}_{\text{tot}}$ is a convex polytope that resides in a *three-dimensional space*, whereas $\mathbf{S}_1$ and $\mathbf{S}_2$ alone are both two-dimensional subspaces, indicating that the AR from a single feed $\mathbf{C}_{f1}$ or $\mathbf{C}_{f2}$ is two–dimensional—they both exist as *planes* in c_A–c_B–c_C space. Thus, for multiple feeds, the dimension of the AR could exceed the number of independent reactions, as mixing between different stoichiometric subspaces is possible.

SIDE NOTE: Stoichiometric subspace with residence time

We have shown, in previous chapters, how residence time may be incorporated into AR constructions. Since ARs that involve residence time are technically unbounded, there is no natural upper bound as there is with species concentrations. In practice, however, we institute a sufficiently large residence time in these constructions to adequately capture all the necessary features of the AR. The AR thus resides in the range $0 \leq \tau \leq \tau_{\text{max}}$.

For example, a triangle, representing **S** in two-dimensional concentration space, might be viewed as a triangular prism in **C**–τ space as in Figure 8.6. A similar argument applies to temperature, when temperature can be assumed to obey a linear mixing law.

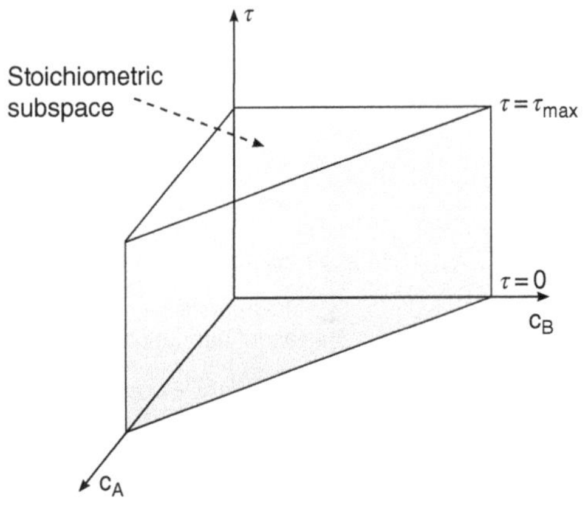

Figure 8.6 Stoichiometric subspace with residence time.

8.3 OVERVIEW OF AR CONSTRUCTION METHODS

8.3.1 Introduction

AR construction algorithms are similar to many numerical methods in engineering in that there are often different methods for solving the same problem. There is no definitive AR construction method that is suitable for all problem types, and it is the duty of the designer to pick the best method for the problem at hand. In general, AR construction methods may be classified into two categories:

1. **Geometric methods**. These methods rely on constructing the AR from a geometric viewpoint, viewing the AR as a convex polytope in state space, and applying the familiar geometric properties of reaction and mixing, as discussed in previous chapters.
2. **Superstructure methods**. The superstructure approach was briefly mentioned in Chapter 1. A reactor superstructure is designed to approximate the performance characteristics for various scenarios. Combining the superstructure in different arrangements produces a range of outputs that is, in turn, an approximation to the AR. Superstructure methods result in a system of algebraic equations and must be solved using a non-LP method.

Geometric methods map closer to the geometric nature of the AR itself, and therefore these methods are also often easier to understand. Within the category of geometric construction methods, we can again categorize methods into two broad subclasses:

1. Inside-out methods
2. Outside-in methods

The difference between inside-out and outside-in methods will be described in Section 8.3.2. In Figure 8.7, a schematic classifying the general types of AR construction algorithms available is given for reference.

The algorithms described in this chapter deal mostly with candidate regions in concentration space alone. At present, some of the methods discussed are able to handle nonisothermal problems, such as those given in Chapter 7. Other methods possess the ability to compute candidate regions in residence time space (unbounded problems). As further research into AR construction methods continues, a broader range of techniques for the determination of the AR may be developed.

> **SIDE NOTE: Necessary assumptions about the system**
>
> Hereafter, we shall assume that constructions occur in concentration space. It will also be assumed that kinetics in a functional form—for use in the reactor equations—is available, and that a known feed point from which to generate the AR from has also been specified.

> **SIDE NOTE: Simplest AR construction method**
>
> Perhaps the earliest account of an automated AR construction method appears in the PhD thesis of McGregor (1998). In this work, AR construction is achieved not by a specific technique, but rather by a constructive, trial and error approach, given reaction kinetics and a feed point. In Figure 8.7, McGregor's recommendations are listed as one potential geometric method, although the recommendations are not discussed here. Interested readers are referred to McGregor (1998) for further details.

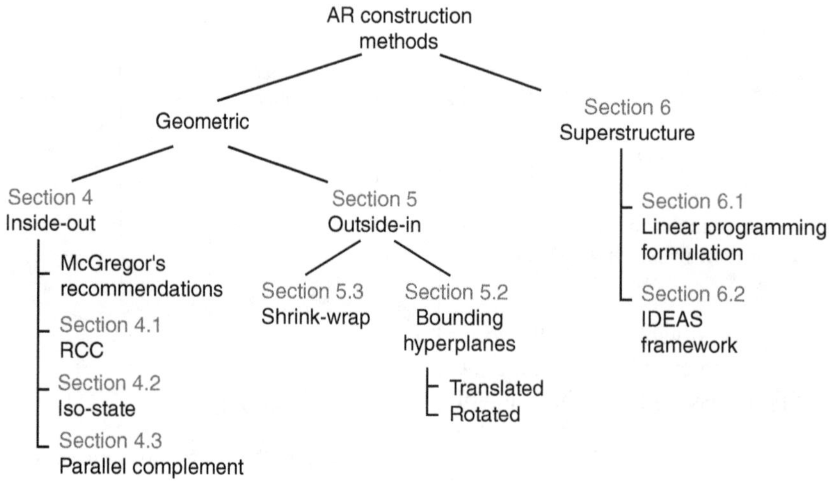

Figure 8.7 Summary of AR construction algorithms.

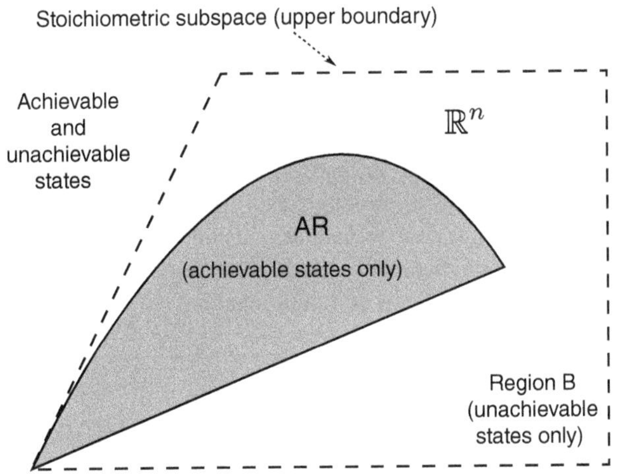

Figure 8.8 The AR may be enclosed by a larger region, containing both attainable and unattainable points.

8.3.2 Inside-Out versus Outside-In Methods

8.3.2.1 Achievable and Unachievable Concentrations It will be helpful to keep the following picture in mind when geometric AR construction methods are discussed later on. Recall that the AR may be viewed as a convex polytope P residing in $\mathbb{R}^n$. If the AR can be enclosed by a larger region that also resides in $\mathbb{R}^n$, then the larger region will contain both achievable and unachievable points. Inside-out and outside-in methods differ in the manner whereby this space is treated. Figure 8.8 demonstrates this pictorially for a hypothetical AR residing in $\mathbb{R}^n$.

The AR is the convex, shaded, region in the center of Figure 8.8. The surrounding space, given in the figure as region B, is the remaining set of (unachievable) concentrations in $\mathbb{R}^n$. If an upper bound for the AR can be found—containing both attainable and unattainable points—then the AR may be enclosed within the larger region. This is the stoichiometric subspace **S** discussed in Section 8.2.2.

8.3.2.2 Inside-Out Methods AR construction using inside-out methods is carried out by successively building the AR from the feed point, in an *additive* manner. These methods use information regarding the known attainable points of the current region to grow the region outward—the AR does not decrease in size with each iteration. Figure 8.9(a) gives a simplified schematic of the construction process for a typical inside-out construction method. For any particular iteration k,

- The current candidate region contains only achievable points. In the limit of infinitely many iterations, the true AR is obtained.
- The region is not smaller than the region generated at a previous iteration $k-1$

The set of points given by the candidate region in an inside-out method is thus guaranteed to be attainable. Termination using an inside-out method typically occurs when no further expansion of the region is possible. Inside-out methods are often easier to understand than outside-in methods. Construction of candidate regions using an inside-out approach is typically carried out using known process vectors and reactor types.

8.3.2.3 Outside-In Methods AR construction via an outside-in approach is best viewed as the inverse problem of inside-out methods. Whereas inside-out methods are concerned with growing a candidate region at each construction step, outside-in methods seek to reduce the size of the region at each construction step. These methods are *subtractive* in nature. Outside-in methods compute candidate ARs

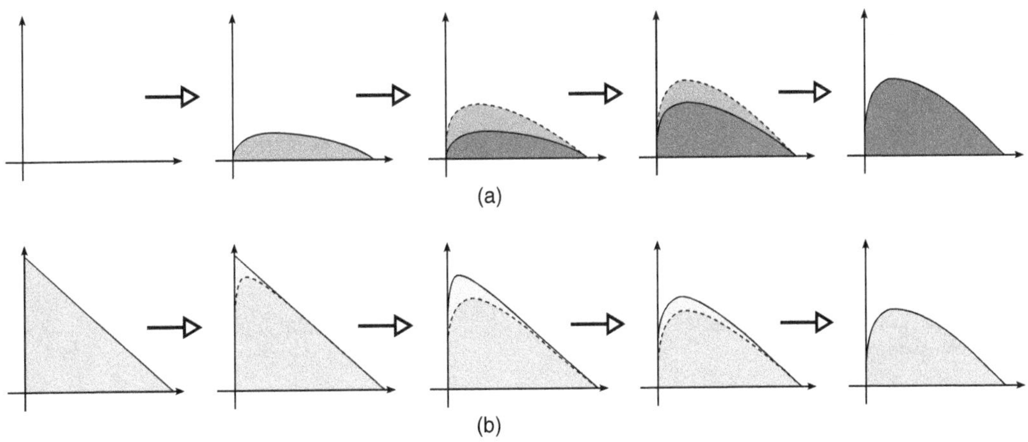

Figure 8.9 (a) The inside-out construction process and (b) the outside-in construction process.

by beginning with the largest possible region—generally **S**—and then progressively removing unachievable states. The goal of a typical outside-in method is to therefore eliminate unachievable points from the region. At the end of each iteration step k,

- The current candidate region contains both achievable and unachievable states. In the limit of infinitely many elimination stages, the true AR is obtained.
- The remaining region is not larger than that obtained in the previous iteration $k-1$.

Figure 8.9(b) shows a simplified representation of the construction process for a typical outside-in method. This approach provides a close approximation to the true AR, provided that many eliminations are performed. Termination of an outside-in algorithm is established when no further reduction in the region may be achieved.

Since outside-in methods rely on the removal of unattainable points from a region, there is no guarantee that the resulting candidate region will only contain attainable points.[3] Additionally, since the remaining region has been found from removal of space, there is often little known about how the remaining region is achieved, because we have not used any information that describes how to construct (or build) the remaining region. This uncertainty makes it difficult to determine the corresponding reactor structure for the computed region when outside-in methods are used.

Outside-in methods are generally more robust than inside-out methods. It is easier to address complex kinetics involving multiple steady states with outside-in methods. Since these methods begin with a region that encompasses the AR, the resulting constructions (even the crudest approximations to the AR) will always contain the full set of achievable points. There are many times, particularly during performance targeting, when one is more interested in determining the size of the AR rather than understanding the physical reactor structure needed to achieve the associated states. In these situations, it is not essential to know the optimal reactor structure and outside-in methods may be better suited for the task.

[3]A *perfect* outside-in method would be able to achieve this. However, this result cannot be validated, in practice, unless it is compared with the results of an alternate construction method.

SIDE NOTE: Uncertainty in construction and validation

At the time of writing, a sufficiency condition for the AR has still not yet been established. Both inside-out and outside-in methods suffer from uncertainty in construction as a result. A stronger result might be achieved if a construction can be validated using two independent construction methods. Validation could be achieved either theoretically or computationally.

8.4 INSIDE-OUT CONSTRUCTION METHODS

8.4.1 The Recursive Constant Control Policy Algorithm

8.4.1.1 Introduction In Chapter 7, under the context of batch reaction, it is demonstrated how the fed-batch reactor may be used to approximate the behavior of both the PFR and CSTR, and how the fed-batch reactor is the batch analogue of a DSR. It is therefore possible to construct a candidate AR, composed of all three fundamental reactor types, using *only* DSR trajectories. This is the basic premise behind the recursive constant control (RCC) policy algorithm (Seodigeng et al., 2009).

8.4.1.2 Basic Idea

a) Approximating PFRs We begin with the DSR equation as follows:

$$\frac{d\mathbf{C}}{d\tau} = \mathbf{r}(\mathbf{C}) + \alpha(\mathbf{C}^0 - \mathbf{C}) \tag{8.3}$$

From Chapter 7, we note that if $\alpha = 0$ for the entire DSR length, then the resulting trajectory is that of a PFR with the same feed point. The DSR equation is identical to the PFR equation when $\alpha = 0$.

$$\frac{d\mathbf{C}}{d\tau} = \mathbf{r}(\mathbf{C}) \tag{8.4}$$

b) Approximating CSTRs For $\alpha \geq 0$, the DSR equation may be integrated for sufficiently large integration times to approximate the equilibrium state relative the initial condition specified. The particular equilibrium point achieved in the DSR is hence strongly linked to the value of α specified over the integration range. This is expressed mathematically as follows:

$$\frac{d\mathbf{C}}{d\tau} = \mathbf{0} = \mathbf{r}(\mathbf{C}) + \alpha(\mathbf{C}^0 - \mathbf{C})$$

Isolating the DSR effluent concentration vector **C** then gives

$$\mathbf{C} = \mathbf{C}^0 + \frac{1}{\alpha}\mathbf{r}(\mathbf{C})$$

This equation is almost identical to the CSTR equation when the sidestream feed composition is set to the feed concentration of the CSTR ($\mathbf{C}^0 = \mathbf{C}_f$) and the reciprocal of the equilibrium value of α is set to the residence time of the CSTR.

$$\frac{1}{\alpha} = \tau_{CSTR} \text{ (at equilibrium)} \tag{8.5}$$

From this observation, the following important result is obtained:

A CSTR concentration is obtained in a DSR if the DSR is operated at the equilibrium point with a constant α policy. The CSTR solution appears as the DSR equilibrium concentration with the feed point to the CSTR given by the sidestream composition $\mathbf{C}^0$, and the CSTR residence time given by the reciprocal value of α.

Both PFR and CSTR behavior may be approximated by correctly controlling the value of α and $\mathbf{C}^0$ in the DSR. Thus, all three fundamental reactor structures may be approximated with the RCC method. This, in turn, allows one to generate (in theory) the entire set of boundary structures in concentration space.

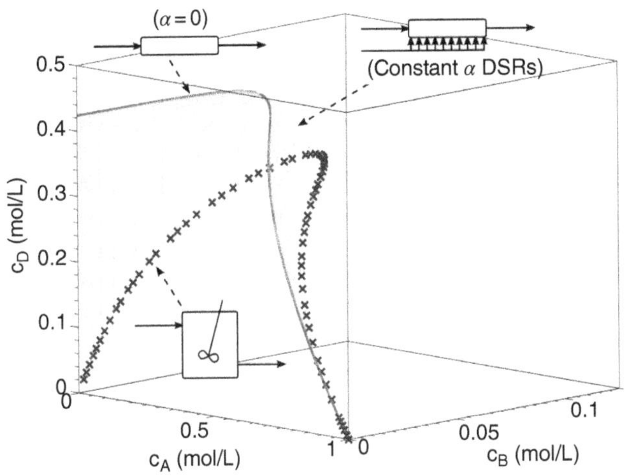

Figure 8.10 DSR trajectories for constant α values using the feed point as the DSR sidestream composition. The CSTR locus from the feed point is also shown for comparison. It is clear that the equilibrium points for the DSR trajectories coincide with the CSTR locus points. Van de Vusse kinetics is used here. (*See color plate section for the color representation of this figure.*)

c) Candidate AR Construction In Chapter 6, we described how critical DSR trajectories follow a highly specific α policy. The nonlinear nature of the equations is difficult to compute analytically, which makes implementation of the critical DSR conditions difficult within an AR construction scheme. The RCC method attempts to avoid this difficulty by using DSR trajectories where α is fixed at a constant value for the entire integration period of the DSR. The α policies used in the DSR integrations are therefore *constant control policies*, which is where the RCC method derives its name from.

Assume that a valid DSR sidestream mixing concentration $\mathbf{C}^0$ is available. It is possible to generate a range of positive α values, each corresponding to a unique DSR trajectory. If the DSR integration time is chosen long enough so that the associated equilibrium concentration can be approached for each trajectory, then the CSTR locus for the concentration $\mathbf{C}^0$ may be approximated. Figure 8.10 shows DSR trajectories associated for a range of constant α values for the Van de Vusse system. All DSR trajectories shown in Figure 8.10 are those that belong to a DSR with $\mathbf{C}_f = [1, 0, 0]^T$ mol/L as the feed and mixing points—the different trajectories each correspond to a different constant α value, and hence each trajectory terminates at a different DSR equilibrium point along the CSTR locus. The feed point $\mathbf{C}_f$ is used as the sidestream concentration to the DSR, that is, $\mathbf{C}^0 = \mathbf{C}_f$.

Although the shape of the DSR trajectories are different to the PFR (expect for when $\alpha = 0$), the equilibrium concentration of the DSR lies on the CSTR locus for the same system starting at the feed point $\mathbf{C}_f$. The DSR trajectories are therefore able to reach a larger set of concentrations and expand the region as a result. The RCC method aims to exploit this nature of constant α DSR trajectories at many points in space to approximate the AR.

> **CONCEPT: Flexibility of the DSR equation**
>
> The flexible nature of the DSR expression means that we can use it to approximate both CSTR and PFR solutions. By integrating many constant α DSR trajectories, a range of achievable reactor outputs can be obtained.

8.4.1.3 The Algorithm

a) Description It is straightforward to formulate an AR construction algorithm utilizing constant α DSR trajectories as a basic building block. The RCC method is broken down into four main phases, which are summarized as follows:

1. Initialization phase (optional): Given a feed point $\mathbf{C}_f$ and kinetics, solve for the PFR trajectory and CSTR locus from the feed. Points generated from this step may then be used for both additional DSR starting points and sidestream mixing points. This phase is optional because the growth phase incorporates constant α DSR trajectories that approximate both PFR ($\alpha = 0$) and CSTR (equilibrium DSR) points.
2. Initial growth phase: In this stage, the initial region is expanded from the feed point, $\mathbf{C}_f$, with DSR trajectories. This is achieved by first generating N *constant* values for α, and then integrating the DSR expression for very large DSR residence times (between $0 \leq \tau_{DSR} \leq \infty$), for each constant value of α, to approximate the DSR equilibrium point.
   - Recall that $\alpha = 1/\tau_{CSTR}$ at the DSR equilibrium point (when the value of τ_{DSR} is very large).
   - For $\alpha = 0$, the DSR equation simplifies to the PFR expression.
   - For $\alpha = \infty$, the DSR equation is dominated by the mixing term $(\mathbf{C}^0 - \mathbf{C})$.
   - For any positive value of α between 0 and ∞, the DSR solution trajectory approaches a CSTR effluent concentration.
   - Each of these α values is associated with a constant α DSR solution trajectory. The union of points generated from steps 1 and 2 is denoted here by the set **X**.
3. Iteration phase: Points in conv(**X**) serve as initial conditions for both DSR trajectories and sidestream mixing points. Select a new DSR feed $\mathbf{C}_f^{[k]}$ and size-stream concentration $\mathbf{C}^{0[k]}$ using points in conv(**X**).
   - For each $\mathbf{C}_f^{[k]}$ and $\mathbf{C}^{0[k]}$ specified before, loop through the generated range of N α values and integrate the DSR equation to produce N DSR solution trajectories corresponding to $\mathbf{C}_f^{[k]}$ and $\mathbf{C}^{0[k]}$.

     i. This step is achieved in practice by programming three nested `for` loops on a computer—two `for` loops are required for selecting $\mathbf{C}_f^{[k]}$ and $\mathbf{C}^{0[k]}$ in conv(**X**), and one `for` loop is needed for each value of α.

     ii. Store all DSR trajectories generated in this phase by the set **Y**.

   - Update the total set of achievable points in **X** by combining with points from set **Y** (i.e., let $\mathbf{X} = \{\mathbf{X} \cup \mathbf{Y}\}$). Compute conv(**X**) and repeat stage 3 until no further increase in the size of the candidate region can be achieved.
   - This is the most computationally demanding step of the algorithm.
4. Polish phase: Generate PFR trajectories from each point in conv(**X**) to produce a final (smooth) boundary.

The RCC algorithm successively generates a large number of constant α DSR trajectories from all extreme points, using all other extreme points as sidestream mixing points. Since only extreme points of the candidate region are needed, this reduces the number of points considered. The CSTR locus is computed indirectly through the DSR equation, and there is no need to employ a specialized nonlinear solver to compute CSTR solutions. This is a desirable property of the method, since only a convex hull algorithm and ODE integrator are needed to implement the RCC algorithm in practice. The algorithm is computationally demanding, however, owing to the fact that a large number of potential points for each DSR trajectory are generated during the iteration phase.

b) Complexity We can show that for M extreme points of the current polytope and N values in the α range, the RCC algorithm requires NM^2 ODE integrations of the DSR equation.[4] This suggests that a large number of integrations must be carried out for adequate approximation of the AR to be achieved.

In order to reduce the computational workload, it is often necessary to only select a sample of all extreme points in conv(**X**) as feed points $\mathbf{C}_f^{[k]}$ for use in the DSR integrations. We often also designate specific points to act mixing points $\mathbf{C}^{0[k]}$ (i.e., feed and equilibrium points), since, from Chapter 6, it is known that critical DSR trajectories must be fed with sidestream concentrations that reside on the true AR boundary, which also cannot originate from protrusions (concentrations from PFR trajectories). A trade-off between construction time and computational accuracy must often be established when implementing the RCC algorithm in practice.

[4] If there are M extreme points in conv(**X**), each extreme point can act as both a DSR feed point $\mathbf{C}_f^{[k]}$ and a mixing point $\mathbf{C}^{0[k]}$ (M×M combinations). For each combination of $\mathbf{C}_f^{[k]}$ and $\mathbf{C}^{0[k]}$, there are N constant α values that must be integrated, which results in NM^2 total combinations.

ILLUSTRATION: RCC method applied to the Van de Vusse system

Figure 8.11 shows construction results for the three-dimensional Van de Vusse system at the four construction phases within the RCC algorithm. The initial region is generated with a PFR from the feed. The growth and iteration stages serve to significantly enlarge the region. To mitigate some of the computation burden, only a subset of the total extreme points are used to generate further constant α DSR trajectories. This results in the crude approximations given in the growth and iteration stages. A closer approximation of the region may be achieved if more extreme points (and more constant α trajectories) are used. Considering that only a subset of extreme points have been used in the growth phase, the final polishing phase serves well to generate a smoothed, convex shape as the final candidate region, which is in agreement with theory.

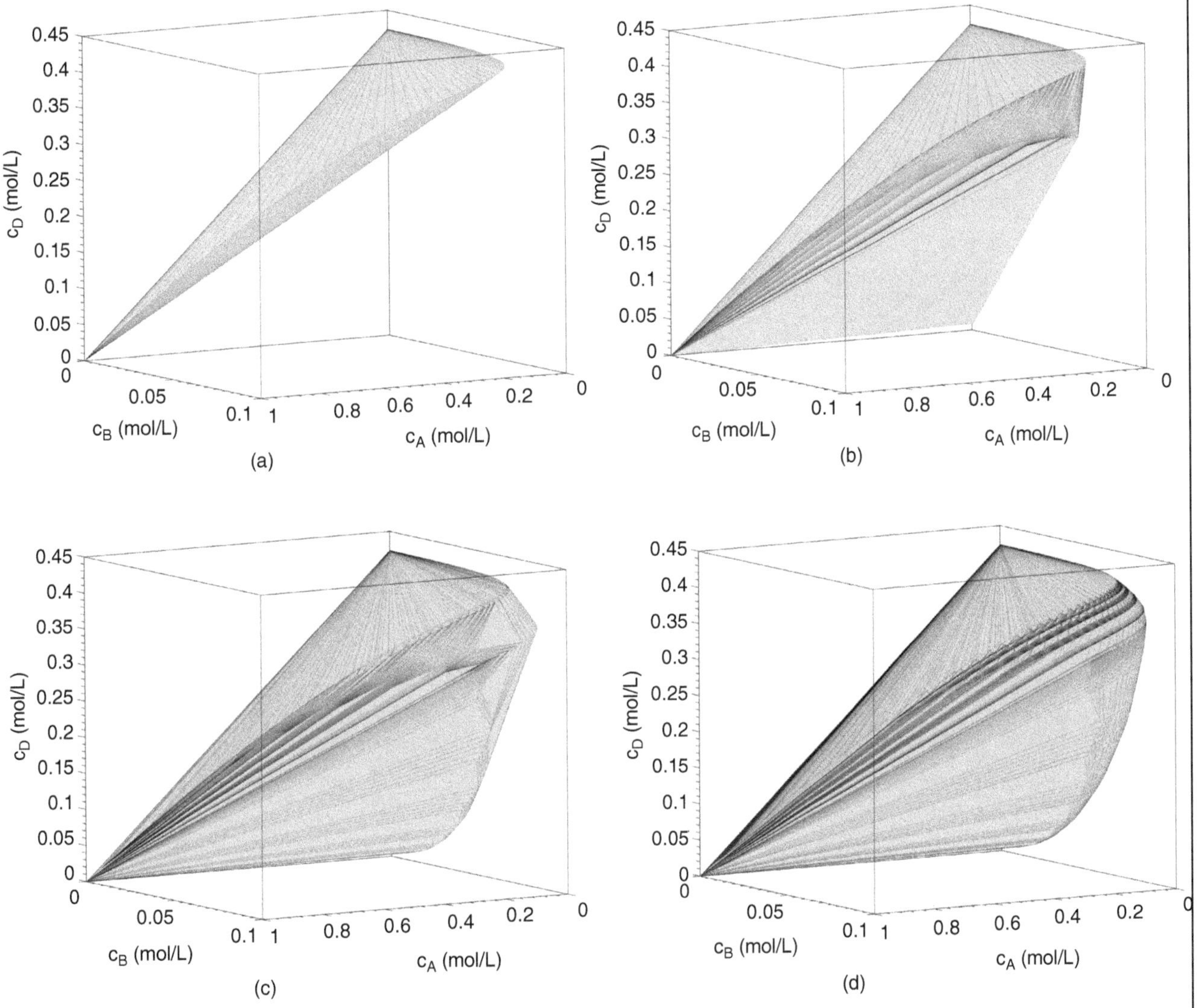

Figure 8.11 The RCC method applied to the three-dimensional Van de Vusse. The method may be summarized into four broad construction phases: (a) Initialization (PFR from the feed point), (b) growth (initial DSR trajectories from the feed), (c) iteration (constant α DSR trajectories from the extreme points), and (d) polish (PFR trajectories from the convex hull points).

8.4.1.4 Multiple CSTR Solutions In Chapter 5, we discuss the isola example. The isola example is difficult to solve because it is possible to generate a candidate region that appears to be the AR, but which is not the true AR unless multiple CSTR steady states are considered. Since the RCC method relies solely on the integration of the DSR expression to achieve growth, there is no possibility for jumps in space to occur, for the integration itself relies on local continuity in the DSR function. For this reason, the RCC method does not allow for multiple CSTR solutions to be discovered unless different starting points for the DSR expression are utilized for the same mixing concentration, which is often difficult to identify in practice.

Figure 8.12(a) and (b) shows construction results for the isola example, discussed in Chapter 5, using the RCC method.

The RCC method fails to find the concentrations associated with the isola. If one is content with only generating a subset of achievable points, then the RCC method is a simple method that is easily implemented on a computer. The method is suitable for systems where multiple CSTR steady states are not present, although, even if this is not satisfied, the method still allows for the determination of candidate regions that are different to those obtained by simple repeated application of CSTR and PFR reactor structures.

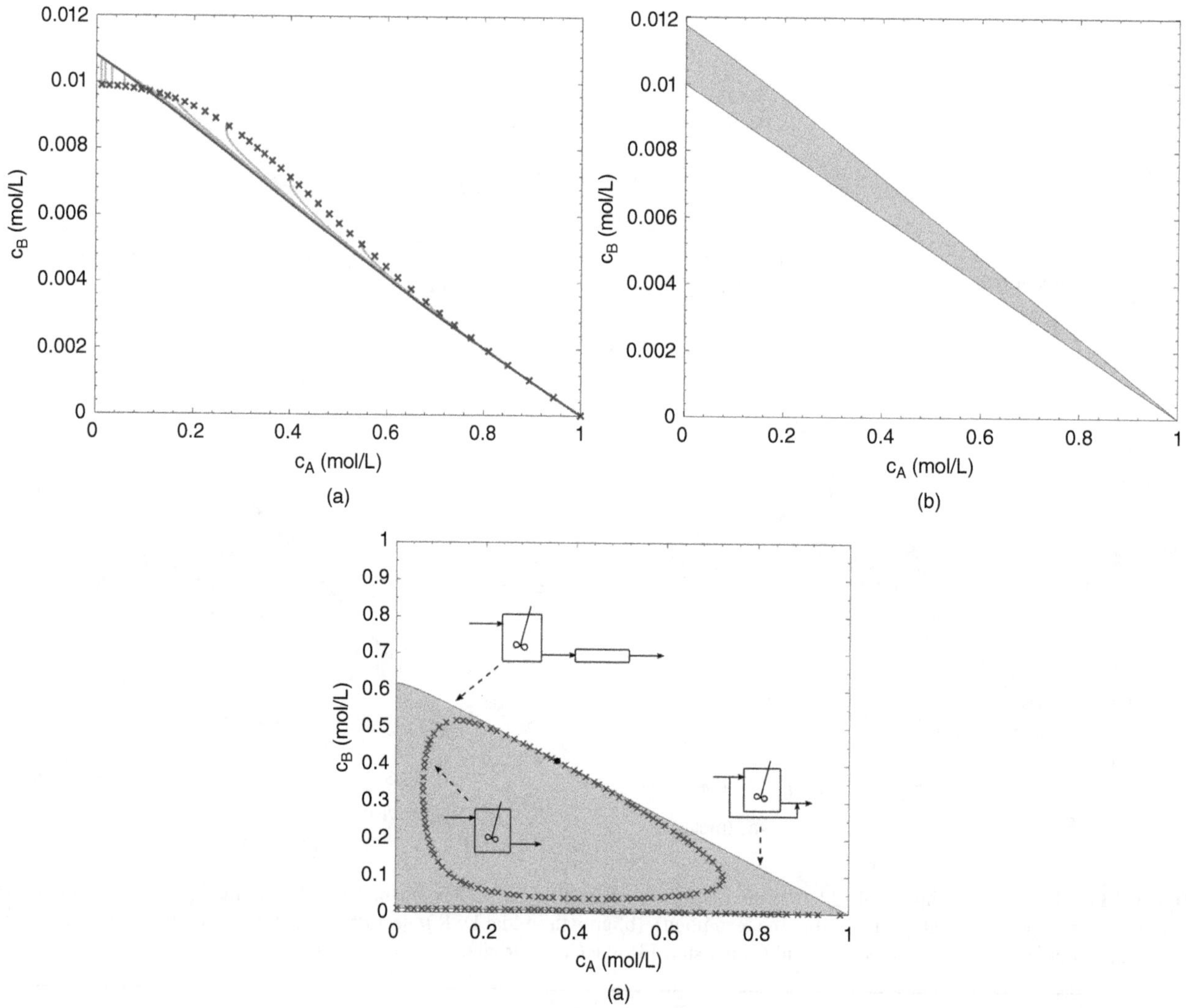

Figure 8.12 The RCC method for the isola example from Chapter 5: (a) Constant α DSR trajectories from the feed point, (b) the candidate region obtained by the RCC method, and (c) the true AR for the isola example. Note that the RCC method fails to identify the isola in this instance.

8.4.2 The Iso-State Method

8.4.2.1 ***Introduction*** When a system involves only two independent reactions, then the corresponding AR is two-dimensional, and it is always possible to construct an AR in concentration space using combinations of CSTRs and PFRs only. This is a desirable property. Working with DSRs is always more challenging, since the need to specify an α policy and mixing point is often a nontrivial task. If a method of utilizing only CSTRs and PFRs for more than two independent reactions can be found, then determining critical DSR trajectories is no longer required. This is the central idea behind the iso-state method (Rooney et al., 2000).

If it is possible to construct a candidate region given a feed point and kinetics, then it is also possible to *project* the AR onto a lower dimensional subspace. In Figure 8.13, the AR for the Van de Vusse system, associated with the feed point $[1, 0, 0]^T$ mol/L, is given. The region is convex and resides in $\mathbb{R}^3$; however, the shape of the region is different depending on which plane that the polytope is projected onto. This procedure resembles shining a torch on an object in a direction perpendicular to the plane of interest. The resulting shape of the AR boundary in the c_A–c_B plane corresponds to the two-dimensional AR for the Van de Vusse system, which is studied in Chapter 5.

A relevant question to ask is then: if it is possible to construct a candidate AR and then project the same construction onto lower dimensional subspaces, then is it possible to perform the reverse process and assemble a candidate AR from multiple lower dimensional constructions?

The iso-state method attempts to construct a candidate region in $\mathbb{R}^n$ (where $n \geq 2$), by a successive assembly process involving smaller two-dimensional subproblems, where each subproblem forms a "slice" of the larger region. The reasons for using two-dimensional slices, as opposed to higher dimensional slices, are as follows:

1. Flexibility. Higher dimensional systems can be broken down into a number of smaller and simpler two-dimensional problems. For systems involving exactly two independent reactions, the resulting AR can still be computed with the method. Two-dimensional constructions provide a minimum working dimension from which candidate regions may be determined.
2. Simplicity. When sub-constructions are performed in two dimensions specifically, there is no reliance on the standard DSR equation, and the difficulties associated with their use might be avoided.

8.4.2.2 ***Basic Idea*** Construction via two-dimensional slices is intuitive in principle. Candidate ARs are convex polytopes that always reside in the full space defined by the dimension of the stoichiometric subspace **S**. When projections are performed onto a lower dimensional subspace, information associated with the axis parallel to the direction of projection is lost in the process. Figure 8.14 demonstrates this with two simple shapes that both have a circular base.

This process is similar to shining a torch on the object onto different planes—different shapes (shadows) are obtained when the direction of projection is altered. It follows that if we are to use lower dimensional projections to construct a higher dimensional shape, then various projections, from different directions, must be performed.

For instance, reconstruction of the sphere in Figure 8.14(a) might occur by taking various slices of the sphere at different points along each axis, producing a number of concentric circles of varying radius. For an object in $\mathbb{R}^3$, two-dimensional projections along the x, y, and z axes may be employed. A number of slicing planes are required at discrete points along the axis length. For example, three slices along each axis in $\mathbb{R}^3$ results in nine two-dimensional sub-constructions in total. This illustration is shown more clearly in Figure 8.15.

Building higher dimensional candidate ARs using information from two-dimensional planes follows a similar methodology. In general, if we wish to construct a region in $\mathbb{R}^n$ using two-dimensional slices, then a binomial combination of $\binom{n}{2}$ sub-planes must be considered. If each axis is split into L slices, then $\mathrm{L} \times n!/(2!(n-2)!)$ sub-constructions must be performed at each assembly step. This is computationally intensive but necessary to ensure that the full shape is resolved.

8.4.2.3 Iso-Compositional Reactors

a) Overview It is not possible to use the standard CSTR and PFR equations to perform the constructions in the iso-state method. This is because the fundamental reactor equations are necessarily of full dimension—both the PFR solution trajectory and the CSTR locus result in curves that reside in the full space. The rate field is described in $\mathbb{R}^n$ and is not, in general, only two-dimensional (unless there are only two independent reactions present in the system).

In order to construct candidate regions using two-dimensional sub-constructions, variations of the familiar reactor equations are needed to construct PFR trajectories and CSTR loci that are maintained within a two-dimensional plane. Rooney et al. (2000) refer to these reactors as *iso-compositional reactors*. Solutions resulting from iso-compositional reactors produce concentrations that lie entirely in a two-dimensional plane in $\mathbb{R}^n$. There is, in fact, no difference in form between the iso-compositional reactor equation belonging to an iso-compositional CSTR (iCSTR)

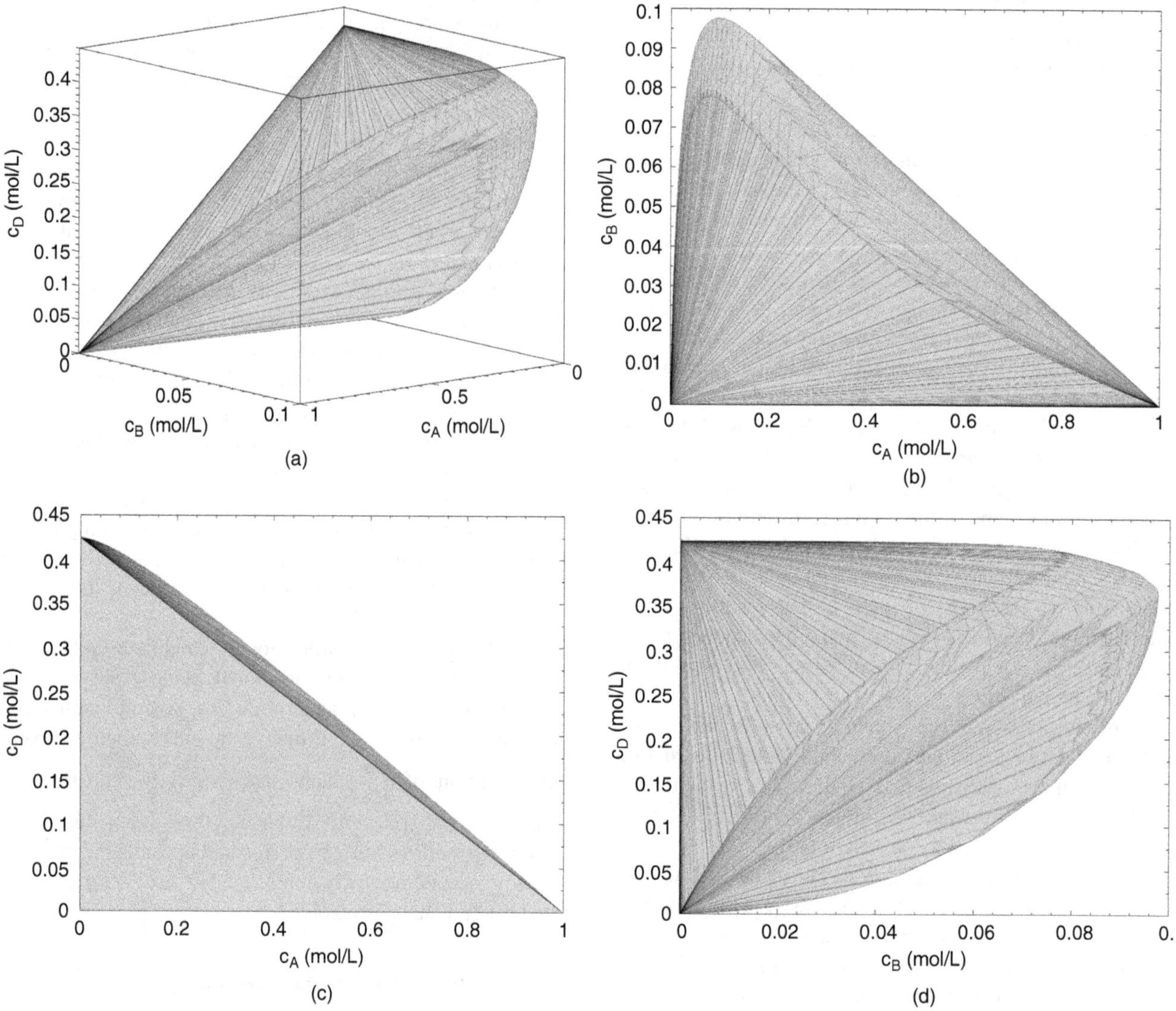

Figure 8.13 (a) Three-dimensional Van de Vusse AR, projected onto different component planes: (b) c_A–c_B space, (c) c_A–c_D space, and (d) c_B–c_D space.

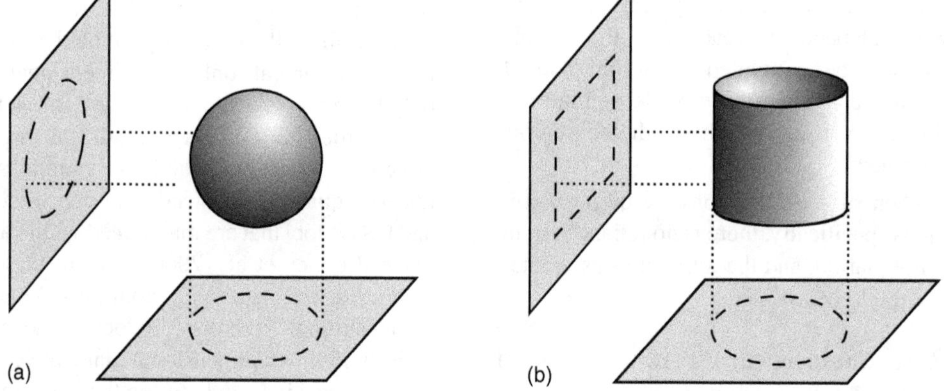

Figure 8.14 (a, b) Two shapes, projected onto different planes. The shape of the resulting projection differs between objects and the plane that it is projected on.

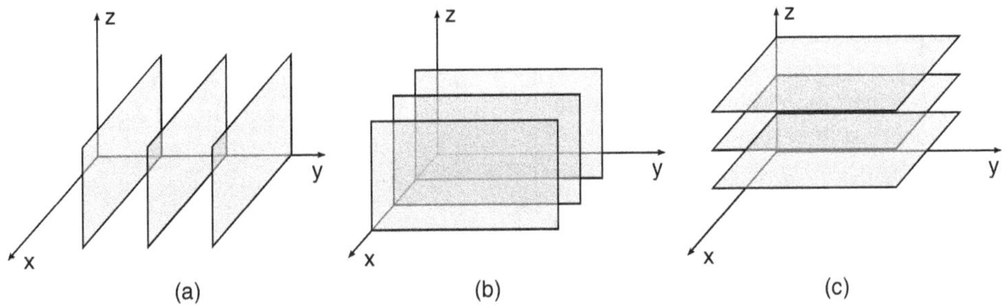

Figure 8.15 Two-dimensional slices in $\mathbb{R}^3$ taken in the (a) x–z plane, (b) y–z plane, and (c) x–y plane.

and the equation for an iso-compositional PFR (iPFR), as both equations are simply variants of the DSR equation with a unique sidestream feeding rate policy. This is different to the α policy for a conventional critical DSR operating on the AR boundary.

In order to maintain a solution in a two-dimensional plane in $\mathbb{R}^n$, the following relation must hold[5]

$$\left(\frac{\partial c_i}{\partial \tau}\right)_{c_j} = r_i(\mathbf{C}) + \beta(c_i^0 - c_i) \tag{8.6}$$

Here, β is a non-negative scalar quantity that controls how much mixing occurs relative to reaction, and $\mathbf{C}^0$ is an *attainable* mixing concentration from a point *outside* of the plane under consideration—points from outside the current plane must be chosen to ensure that we do not need infinite amounts of $\mathbf{C}^0$ in order to maintain a trajectory in the plane. Note that the form of the iso-compositional reactor equation closely resembles the conventional DSR equation. However, the former is a partial differential equation, whereas the latter is an ordinary differential equation. β—which is similar to the α parameter used in the DSR expression—may be used to help steer solutions, obtained from the iso-compositional reactor, so that they are constrained to lie in a two-dimensional plane.

ILLUSTRATION: The origins of Equation 8.6

It will be useful to explain the construction process of the iso-state method with a clear picture of the reactor behavior in mind. Figure 8.16 shows a hypothetical construction of a two-dimensional plane in $\mathbb{R}^3$. (The same idea also applies to higher dimensional systems.)

Assume that a DSR is available, and that a mixing point for the sidestream $\mathbf{C}^0$ and initial concentration $\mathbf{C}_f$ are specified. The standard DSR expression may then be given as follows:

$$\frac{d\mathbf{C}}{d\tau} = \mathbf{r}(\mathbf{C}) + \beta(\mathbf{C}^0 - \mathbf{C}) \tag{8.7a}$$

Note that we have replaced the traditional symbol for α with β to signify that this DSR is a somewhat specialized case of the conventional DSR expression. Our goal is to control the DSR in a manner that allows us to produce effluent concentrations that all lie within a two-dimensional subspace (a plane) of $\mathbb{R}^n$. One might envision the trajectory of such a DSR to be something like the dashed curve drawn in Figure 8.16. This effect can be achieved by the appropriate use of the β parameter and the sidestream mixing concentration $\mathbf{C}^0$.

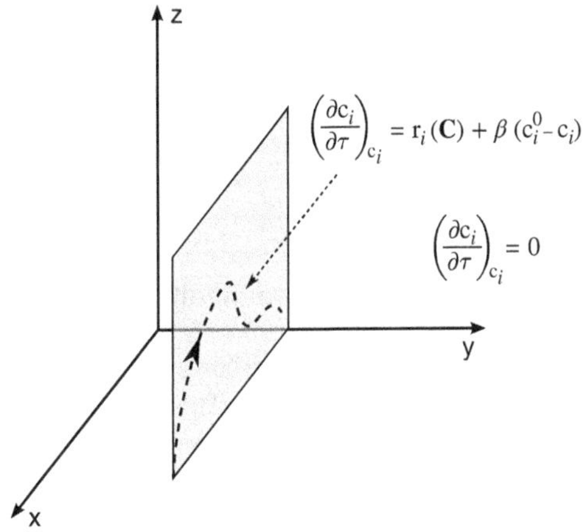

Figure 8.16 Reaction in a two-dimensional plane in $\mathbb{R}^3$. Reactors that produce effluent concentrations that lie within a plane of a fixed composition are termed iso-compositional reactors. These reactors may be viewed as DSR with specialized sidestream policies.

To maintain a trajectory in a specific plane—in other words, a certain slice along one of the component axes—the concentration of a specified component must remain *constant*. (In Figure 8.16, component y is held constant.) If we wish to produce trajectories that lie

[5] In the original paper Rooney et al. (2000), a slightly different notation is used. Discussion is made with reference to the vector version of Equation 8.6 instead, given by $\partial\mathbf{C}/\partial\tau = \mathbf{r}(\mathbf{C}) + \beta(\mathbf{C}^0 - \mathbf{C})$.

only on a plane perpendicular to component j, the concentration of component j must remain at a fixed value.[6] The change in concentration of component j with respect to residence time must therefore be zero:

$$\frac{dc_j}{d\tau} = r_j(\mathbf{C}) + \beta(c_j^0 - c_j) = 0 \tag{8.7b}$$

Keep in mind that this is the differential equation belonging to component j only. The resulting expression for β that maintains component j at a fixed value is then found by solving Equation 8.7b directly for β,

$$\beta = \frac{-r_j(\mathbf{C})}{(c_j^0 - c_j)} \tag{8.7c}$$

with the requirement that $\beta \geq 0$. The remaining components in the system are then given by

$$\left(\frac{\partial c_i}{\partial \tau}\right)_{c_j} = r_i(\mathbf{C}) + \beta(c_i^0 - c_i) \tag{8.6}$$

$$\left(\frac{\partial c_i}{\partial \tau}\right)_{c_j} = r_i(\mathbf{C}) - \left(\frac{r_j(\mathbf{C})}{c_j^0 - c_j}\right)(c_i^0 - c_i) \tag{8.7}$$

where $i \neq j$. Components belonging to c_i are the remaining components in the system that are free to change in the plane of interest.[6] The value for β in Equation 8.6 is not arbitrary, but rather it is the expression given specifically by Equation 8.7c that ensures component j is constant in the plane.

SIDE NOTE: Slight ambiguity in notation

Note that we have taken slight liberty with the use of the partial derivative in Equation 8.6. That is, Equation 8.6 is simply used to designate that components i in the system are those that are free to change by the DSR expression when component j is held fixed. It does *not* mean that we have taken the partial derivative of the DSR expression with respect to component i keeping component j constant. We have merely orchestrated the value of β in a specific way—given by Equation 8.7c—that maintains component j at a fixed, user-nominated, value.

b) Iso-Compositional CSTRs The RCC method demonstrates how the DSR expression may be used to achieve effluent concentrations belonging to both PFRs and CSTRs. Equation 8.6 may be used in a similar manner to achieve the same effect in a two-dimensional sub-plane for CSTRs as well. Thus, integration of Equation 8.6, for a sufficiently long residence time, results in

$$\left(\frac{\partial c_i}{\partial \tau}\right)_{c_j} = 0$$

If vector $\mathbf{C}_s$ is denoted as the set of components contained in the sub-plane, and $\mathbf{C}_j$ as the remaining set of all other components in the system, then we can define an iCSTR as one that obeys

$$\left(\frac{\partial \mathbf{C}_s}{\partial \tau}\right)_{\mathbf{C}_j} = \mathbf{0} \tag{8.8}$$

Only components belonging to $\mathbf{C}_s$ must satisfy the CSTR equation in order to produce a CSTR locus that lies in the sub-plane under consideration. The remaining components that do not participate in the subspace are free to change value, given that they are not restricted to lie in the plane. Solutions to Equation 8.8 are either those that belong to traditional CSTR solutions—in other words, all components in the system that satisfy the conventional CSTR condition—or those that are obtained as the equilibrium points to Equation 8.6, which are obtained by integration for a sufficiently long residence time. Similar to traditional CSTRs, it is also possible to solve for the roots of Equation 8.8.

Any available mixing point $\mathbf{C}^0$ may be chosen to act as the feed point in the iCSTR, provided that β is non-negative:

$$\beta = \frac{-r_s}{(C^0 - C)} \geq 0$$

c) Iso-Compositional PFRs Solution trajectories belonging to iPFRs are defined as those found from the integration of Equation 8.6. That is, there is no difference between an iPFR solution trajectory and the solution trajectory obtained by the DSR satisfying Equation 8.6. Indeed, the notion of an iPFR is merely a convenient way to express a solution trajectory that lies in a sub-plane in $\mathbb{R}^n$. Hence, even though these trajectories belong to the DSR expression, they are referred to as iPFR trajectories.[7]

The solution approach differs to solving for the roots of Equation 8.6 when computing iCSTR solutions. The use of the β parameter in this instance is to maintain an iso-compositional trajectory in a defined plane.

Finally, note that mixing is unaffected by the use of sub-planes. Mixing between achievable points—whether these points are contained in the same plane or not—is physically and mathematically identical to that found in previous chapters.

[6] In Figure 8.16, this would correspond to the x–z plane, perpendicular to the y-axis.

[7] This is out of historical convention.

> **CONCEPT: Higher dimensional regions from two-dimensional planes**
>
> We can assemble higher dimensional polytopes by taking the convex hull of points generated from many two-dimensional constructions. This is done using iso-compositional reactors that operate in a specific two-dimensional "slice" within $\mathbb{R}^n$. Many planes, orientated in various positions in space, are required in order to fully capture the form of the higher dimensional polytope.

8.4.2.4 ***The Algorithm*** The iso-state method attempts to produce a candidate AR by successive application of a number of two-dimensional constructions. That is, the method of construction involves using iso-compositional reactors, generated in two-dimensional planes, and then combining individual constructions together to form a higher dimensional polytope representing a candidate AR. The method may be described by the following four main construction phases:

1. Initialization: In this phase, a starting polytope is generated that is known to be feasible. This polytope is then utilized in the growth phase of the construction process. The choice of starting region is arbitrary in practice, although it is common that the convex hull belonging to a PFR and CSTR from the feed be used as the initial region.
2. Growth: After the initial candidate AR has been determined, the region is grown by successive application of two-dimensional constructions as follows:
   a. A number of two-dimensional "slices" for each component axis are generated. These planes are used to cut the region generated in step 1 at specific points. These points are used as feeds to iPFRs and iCSTRs.
   b. For each intersection point with the region in the plane, iPFRs and iCSTRs may be generated. The solutions all must lie in the plane and serve to expand the region defined by the plane. This process is done for all planes that intersect the region in all component axes.
   c. For each plane, the new points generated in step 2b may be used to compute a new convex region for the plane. Do this for all planes.
3. Collation: The two-dimensional convex regions generated in step 2c are combined. This allows for a new convex region in $\mathbb{R}^n$ to be computed using the results obtained from step 2. Once achieved, the growth phase may be repeated using the larger convex region generated by collation. This process is repeated until no further expansion of the region is possible.
4. Polishing: PFR trajectories are generated from the extreme points of the polytope given by step 3. This ensures that the resulting region terminates in a PFR, suggested by AR theory, and is similar to the RCC method described in Section 8.4.1. PFRs generated in this step occur in the full space and not in the sub-planes.

> **ILLUSTRATION: Results of the iso-state method, applied to three-dimensional Van de Vusse kinetics**
>
> Figure 8.17(a) and (b) shows construction results obtained by the iso-state method for three-dimensional Van de Vusse kinetics. These results are those obtained using 25 two-dimensional planes for each component

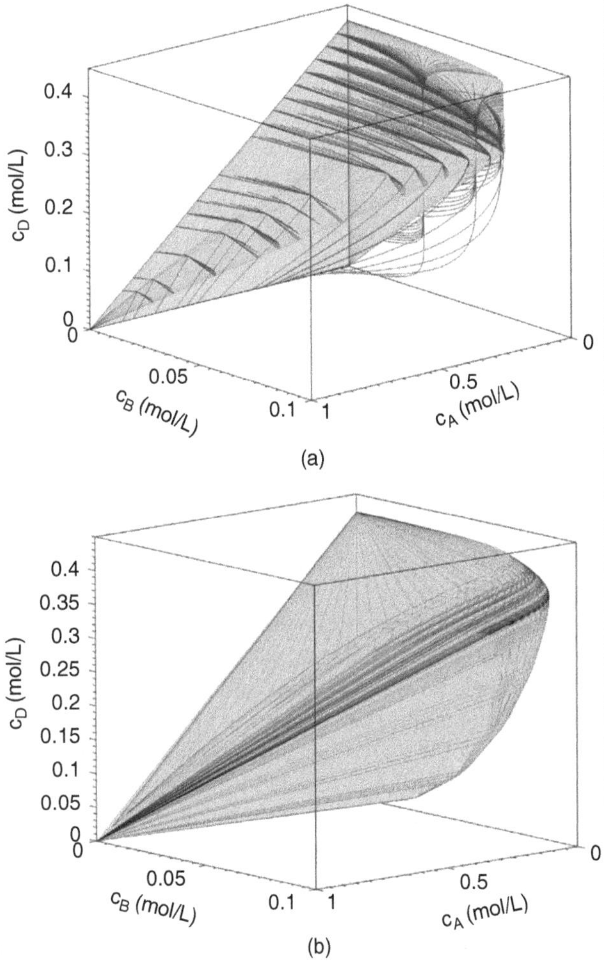

Figure 8.17 (a) Iso-compositional trajectories for planes in c_B–c_D space at fixed values of c_A. The trajectories all lie within a two-dimensional plane. (*See color plate section for the color representation of this figure.*) (b) Final candidate region produced for the Van de Vusse system using the iso-state method.

axis. The initial region used in the initialization phase is a PFR and CSTR from the feed point. The mixing point used in Equation 8.6 was the feed point ($\mathbf{C}^0 = \mathbf{C}_f = [1, 0, 0]^T$ mol/L).

In Figure 8.17(a), the result of integrating Equation 8.6 in the c_B–c_D plane is shown for different values of c_A. Note that no further expansion of the original region is obtained for c_A values greater than approximately 0.5 mol/L. However, a large set of iso-compositional concentrations are obtained for $c_A < 0.3$ mol/L.

The final candidate region is displayed in Figure 8.17 (b), after polishing PFRs have been added to the region. Observe that in this instance, the iso-state method shows good agreement with the theoretically constructed region given in Figure 8.13.

The use of iso-compositional reactors is easy to program in practice, although, the method requires the solution of many two-dimensional subproblems in order to generate a reasonable approximation of the candidate region. Since Equation 8.6 must be solved in order to obtain iCSTR solutions, the same difficulties experienced by the RCC method are also shared with this method. If iCSTR solutions are obtained via an integration of Equation 8.6, then it may be difficult to identify CSTR solutions involving multiple steady states. Nevertheless, the iso-state method is a geometrically straightforward method to understand.

8.4.3 The Complement Method

8.4.3.1 ***Introduction*** Multiple steady states and higher dimensional constructions make it difficult to know when the true AR has been found. In Chapter 5, a problem involving multiple CSTR steady states is described (the isola problem) to demonstrate the difficulty with solving the CSTR equation for nonlinear systems. A problem that is often encountered with CSTR solutions is that even when a candidate region has been constructed that appears to obey all properties of the AR, there may still be concentrations in the *complement space* that could further enlarge the current region—unless all points in the complement region are checked, AR property 7 from Chapter 4 cannot be validated.

In the following sections, we wish to describe an AR construction method that specifically employs information from the complement region. This method is robust in that it is able to handle a wide variety of problem types. The method is also parallel in nature, which allows computation to be split over multiple computing nodes for addressing large problems that involve many independent reactions.

8.4.3.2 ***Basic Idea*** This method is similar to the LP formulations of Kauchali et al. (2002), which is discussed in Section 8.6.1. The method of construction is additive in nature, and therefore it may be classified as an inside-out method. Construction is best understood in terms of a generalized iteration in the following way.

Assume that iteration k of the algorithm has been established. Up to this point, a set of points, which are known to be achievable, has already been generated. The current set constitutes the extent of the region of achievability (it is a candidate AR for iteration k), and has been found from previous iterations $k-1$, $k-2$, etc. By solving the facet enumeration problem, we can describe the current set of points as a set of hyperplane constraints in the H-representation. This is designated as region P_k, indicating that the candidate region is at iteration k of the construction. Let $\mathbf{X}_k$ represent the set of achievable points belonging to the polytope P_k. Assume also that the stoichiometric subspace, $\mathbf{S}$, has been calculated previously.

Assume now that a set of N states $\{\mathbf{C}_1, \mathbf{C}_2, \ldots, \mathbf{C}_N\}$ is generated in the complement region $\mathbf{S}\backslash\mathbf{X}_k$. In other words, N concentration vectors that lie in $\mathbf{S}$ (but not in P_k) are generated so that $\mathbf{C}_1, \mathbf{C}_2, \ldots, \mathbf{C}_N \in \mathbf{S}\backslash\mathbf{X}_k$. Suppose that there is a point $\mathbf{C}$ in this set that when the rate evaluated at $\mathbf{C}$, $\mathbf{r}(\mathbf{C})$, is extended backward, it intersects the boundary of region P_k. We shall denote the point of intersection as $\mathbf{C}^*$. Since $\mathbf{C}$ and $\mathbf{C}^*$ lie on a straight line in the direction of $\mathbf{r}(\mathbf{C})$, it follows that the vector $(\mathbf{C}-\mathbf{C}^*)$ is collinear with $\mathbf{r}(\mathbf{C})$. Furthermore, since the ray has been extended *backward* from $\mathbf{C}$, there exists a positive scalar τ such that

$$\mathbf{C} - \mathbf{C}^* = \tau\mathbf{r}(\mathbf{C})$$

This equation is clearly the CSTR equation operating at an effluent concentration $\mathbf{C}$ for a feed composition $\mathbf{C}^*$ and positive residence time τ. $\mathbf{C}$ must also be achievable by the CSTR equation, and therefore it may be included in the set of achievable points. Conversely, if $\mathbf{C}$ does not have a rate vector that intersects P_k, then $\mathbf{C}$ is not achievable by the CSTR condition at the current iteration. This is the basic procedure of the method.

SIDE NOTE: Achievability using the CSTR condition

If a point $\mathbf{C}$ in the complement region does not satisfy the CSTR condition, then it simply means that it is not achievable, by the CSTR condition, *at the current iteration*. However, $\mathbf{C}$ might still be attainable at a later point:

1. When a different region P_i is generated that does intersect the boundary;
2. Or if *combinations* of different $\mathbf{C}$s can be combined to intersect P_k.

The second point shall be explained briefly later.

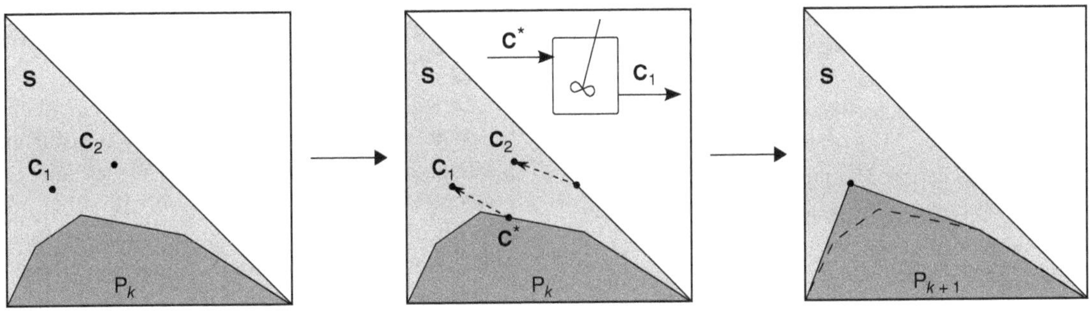

Figure 8.18 The AR construction process for one iteration of the complement method. Rate vectors that intersect the region are included in the list of achievable points and a newer, larger region is then computed. Ming (2014). Reproduced with permission of Elsevier.

An iterative method of computing a set of achievable points can now be constructed: at each iteration, a set of achievable points $\mathbf{X}_k$ is available that represents a convex region P_k. The points may be converted into a set of hyperplane constraints. This list is used to check for intersections with the boundary later on in the construction procedure. A list of N points is then generated in the complement region $\mathbf{S} \backslash \mathbf{X}_k$. For each point $\mathbf{C}_i$ generated in $\mathbf{S} \backslash \mathbf{X}_k$, the rate vector $\mathbf{r}(\mathbf{C}_i)$ is evaluated and checked to see if it satisfies the CSTR condition. If a point is found that satisfies the achievability condition, then it is recorded and appended to the list of achievable states $\mathbf{X}_k$. An updated convex polytope P_{k+1} incorporating the latest set of achievable points $\mathbf{X}_{k+1}$ may then be computed. Figure 8.18 provides a schematic of the process for the evaluation of two points, $\mathbf{C}_1$ and $\mathbf{C}_2$, in the complement region. $\mathbf{C}_1$ is achievable, whereas $\mathbf{C}_2$ is not at the current iteration.

The list of potentially feasible points in $\mathbf{S} \backslash \mathbf{X}_k$ is generated randomly, and as a result, the method is non-deterministic in nature. This allows for scaling of the number of points, generated at each iteration, depending on the computational resources available. Moreover, the generation of points in a random fashion means that generating distinct points in space beforehand (i.e., on a well-defined grid), and the associated "bookkeeping" of achievable and unachievable points is not required. This potentially allows for higher dimensional problems to be addressed in an easier fashion—the number of potential points generated on an evenly spaced grid in $\mathbb{R}^n$ scales by an exponent of base n. The storage requirement for this is large for high dimensions using an evenly spaced grid, whereas less storage is required if points are generated, non-deterministically, as in the current description.

a) Computing Feasible Intersections Checking for feasible concentrations is performed by the matrix inequality.

$$\mathbf{A}^{[k]}\mathbf{C}_i \leq \mathbf{b}^{[k]} \tag{8.9}$$

Here $\mathbf{A}^{[k]}$ and $\mathbf{b}^{[k]}$ are matrices representing region P_k as a list of hyperplane constraints in state space. $\mathbf{C}_i$ is hence a feasible point if it satisfies all hyperplane constraints given by Equation 8.9. Intersections with the current region are performed with the same system of inequality constraints. For each potential concentration $\mathbf{C}_i$ generated in the complement region $\mathbf{S} \backslash \mathbf{X}_k$, it is possible to express the intersection point $\mathbf{C}_i^*$ as a linear combination of the rate vector at $\mathbf{C}_i$ and a scalar variable τ.

$$\mathbf{C}_i^* = \mathbf{C}_i - \tau\mathbf{r}(\mathbf{C}_i)$$

$\mathbf{C}_i^*$ is a feasible intersection point if the following two criteria are met:

1. τ takes on a positive value.
2. $\mathbf{C}_i^*$ is in the feasible region of the current polytope—it satisfies 8.9.

If such a point $\mathbf{C}_i^*$ can be found, the corresponding point $\mathbf{C}_i$ is therefore also achievable, and it may be included into the existing set of achievable points $\mathbf{X}_k$.

> **CONCEPT: Geometric form of the CSTR equation for AR construction**
>
> The CSTR equation can be used to find points in the complement region that expand the region. Many points can be checked at once, which means that the AR can be computed from many CSTR evaluations in parallel.

b) Initial Region The initial candidate region P_0 may be generated easily by computing a PFR trajectory from the feed point. The convex hull of the resulting PFR solution trajectory can therefore be used as a first approximation for use in subsequent iterations. In addition to the PFR solution trajectory from the feed, it is also beneficial to include any known equilibrium points in the initial list of achievable points. This assists the algorithm in finding achievable states when the rate vector is near or equal to the zero vector.

c) Terminating Criteria Since the size of the candidate AR is unknown during construction, it is difficult to determine when the true AR has been obtained. Nevertheless, it is possible to compute the volume of the current polytope at each iteration. The volume of the candidate region is therefore tracked and monitored for any changes in growth. The percentage AR volume change between an iteration interval may then be used as a measure of construction accuracy. Since the construction method is nondeterministic, it is preferable to track the average volume change over a number of iterations to account for any fluctuations in volume growth generated by the random discovery of feasible points.

d) Temperature-Dependent Kinetics and Unbounded Regions Since constructions occur based on a CSTR condition using points generated in the full space of the complement region, the determination of regions involving complex kinetics, potentially containing multiple steady states, is easily accommodated for.

Kinetics with a parameter such as temperature may also be handled by the method. For these cases, a random temperature value is generated within a user specified range $T_{min} \leq T \leq T_{max}$ for every random point generated in the complement region. The particular temperature value produced guides the direction of the rate vector associated with the generated point in space. Regardless of this, the attainability condition must still be satisfied if the point is achievable and thus little modification to the underlying method is required when temperature-dependent kinetics are used as a result. We have already noted in Chapter 7 that it is generally not possible to construct candidate ARs in temperature space—unless temperature obeys a linear mixing law—and thus temperature cannot be treated as a component in the state vector **C** in general.

It is also possible to include residence time in the state vector **C** and build ARs in residence time space. This is included in the usual manner, with the corresponding τ component in **r**(**C**) given as unity ($d\tau/d\tau = 1$).

SIDE NOTE: Parallel computation

Note that checking for feasibility occurs independently of other points generated in the complement region. It is therefore possible to compute multiple intersections simultaneously to speed up computations if necessary. This is in fact the preferred method of construction. The particular method has been implemented on multi-core CPUs and graphics processing units (GPUs) for addressing large systems.

SIDE NOTE: Combinations of rate vectors (talking CSTRs)

Although points found in the complement that satisfy Equation 8.9 are ones that are feasible, points that do not satisfy Equation 8.9 on their own may still be feasible when considered in *combination* with other points. That is, linear combination of multiple rate vectors that each does not satisfy Equation 8.9 may be combined to produce a resulting rate vector that does satisfy Equation 8.9. Systems that exhibit this behavior are rare, for one might already imagine the highly specific behavior required by the rate function to achieve this behavior. We should nonetheless be mindful of it in the discussions.

A common name for the system is the "talking CSTRs" system, which derives its name from the reactor structure required to produce this effect. This is shown in Figure 8.19.

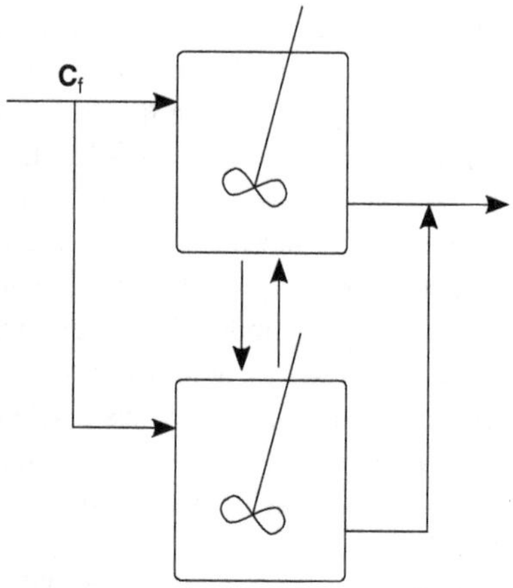

Figure 8.19 Talking CSTRs. Ming (2014). Reproduced with permission of Elsevier.

This behavior may be accounted for after all single points have been found in a "second phase" type of construction, but this second step is not considered here.

8.4.3.3 ***The Algorithm*** An algorithmic description of the method is easily devised once a geometric understanding of the method has been established. It is assumed that a feed point $\mathbf{C}_f$ and stoichiometric coefficient matrix **A** have been specified. The method is broken down into three main phases as follows:

1. Initialization: In this phase, an initial region P_0 representing an initial feasible set of points $\mathbf{X}_0$, used for feasibility checking, is generated. The set $\mathbf{X}_0$ is generated by computing the convex hull of a PFR trajectory from the feed point $\mathbf{C}_f$. This phase also computes the stoichiometric subspace **S** for the reaction stoichiometry and feed point, used to generate potential states in the complement region $\mathbf{S}\backslash\mathbf{X}_k$.

2. Growth: This phase constitutes the bulk of the construction processes. The following steps are carried out iteratively:
   a. Generate N random points in the complement region $\mathbf{S}\backslash\mathbf{X}_k$. If temperature-dependent kinetics is specified, generate a set of temperatures between the temperature range $T_{min} \leq T \leq T_{max}$. The stoichiometric subspace $\mathbf{S}$ and current region $\mathbf{X}_k$ is used here to ensure that all concentration vectors generated in the list reside in the complement region $\mathbf{S}\backslash\mathbf{X}_k$.
   b. For each point $\mathbf{C}_i \in \mathbf{S}\backslash\mathbf{X}_k$ generated, evaluate the associated rate vector $\mathbf{r}(\mathbf{C}_i)$ and determine feasibility. This step may be performed in parallel to speed up the construction process. Determine the subset of the N points in $\mathbf{S}\backslash\mathbf{X}_k$ that are feasible and let $\mathbf{Y}$ represent this set. Append $\mathbf{Y}$ to the current set of extreme points $\mathbf{X}_k$ describing the current polytope so that $\mathbf{X}_{k+1} = \mathbf{X}_k \cup \mathbf{Y}$.
   c. Compute the convex hull of the updated list of points $\text{conv}(\mathbf{X}_{k+1})$ and determine a new list of hyperplane constraints.
   d. Compute the volume of the new convex region P_{k+1} and determine whether the volume change from polytopes P_k to P_{k+1} meets minimum growth stopping criteria. Repeat steps a–c until this occurs. Terminate when the growth criteria are met.
3. Polish (optional): PFR trajectories may be initiated from the extreme points of the latest polytope P_k, producing a smoothed region composed of only PFR trajectories and mixing lines. This stage is optional as curvature can often be approximated sufficiently well by use of many sampling points. Although the use of PFR trajectories produces a more complete, smoothed region, this stage is also computationally intensive if there are many extreme points—many points in $\text{conv}(\mathbf{X}_k)$—where many PFR trajectories are to be integrated.

ILLUSTRATION: Summary of results for three-dimensional Van de Vusse kinetics

A summary of the construction results with the complement method is shown in Figure 8.20. This shows the current polytope P_k, representing the candidate AR, at different snapshots during construction. The circles indicate points generated in the complement region $\mathbf{S}\backslash\mathbf{X}_k$ that satisfy the CSTR attainability condition. Note that a large set of points are initially generated that satisfy feasibility. This serves to rapidly grow the region initially. The final result after 100 iterations is in good agreement with the AR generated from theory.

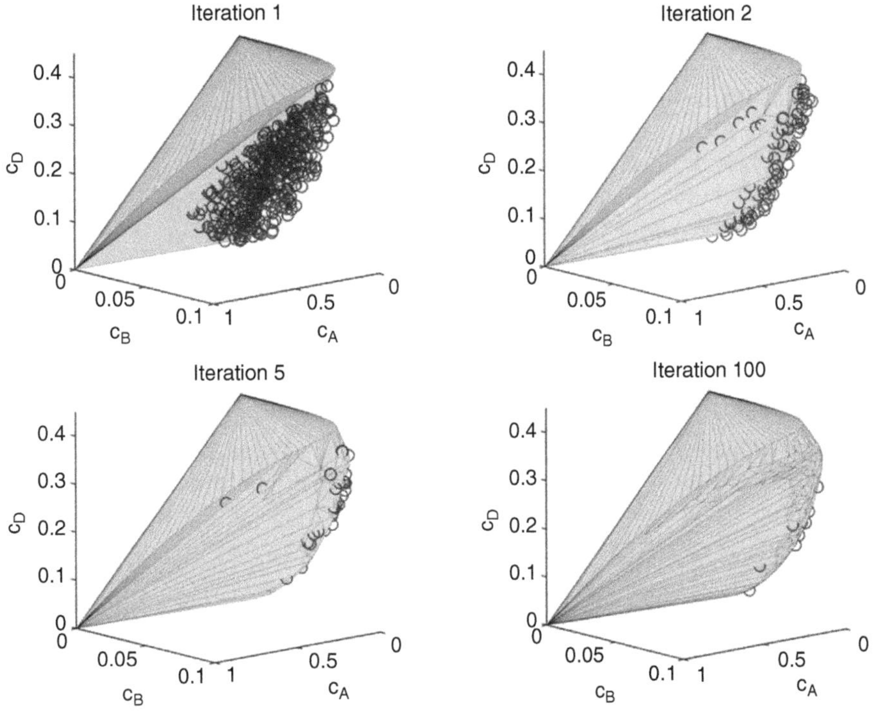

Figure 8.20 Results for three-dimensional Van de Vusse kinetics, taken at snapshots during construction. Ming (2014). Reproduced with permission of Elsevier.

The complement method for computing candidate ARs is simple to understand, and relatively easy to implement in practice; however, it requires the use of a vertex and facet enumeration routine to switch between H-representation and V-representation of the candidate region. The method is able to handle a large set of problems including temperature-dependent kinetics and unbounded regions in higher dimensional space, although it is not able to reconstruct an associated optimal reactor structure, nor handle the talking reactors case. Nevertheless, the method is a fairly robust AR construction technique, and a preferred choice for many AR constructions.

8.5 OUTSIDE-IN CONSTRUCTION METHODS

8.5.1 Overview

The inside-out algorithms discussed in Section 8.4 use information about the current AR to *increase* the set of achievable states at each construction step. Outside-in methods, by comparison, carry out constructions by initiating construction with a large region that encloses the true AR. The set is then *decreased* until no further reduction may be carried out. Whereas inside-out methods achieve constructions via an additive procedure, outside-in methods achieve construction via a sequence of eliminations instead.

8.5.2 The Method of Bounding Hyperplanes

8.5.2.1 Basic Idea

a) Overview The method of bounding hyperplanes (Abraham and Feinberg, 2004) is considered as one of the first contributions to outside-in AR construction algorithms. Description of the method is best facilitated by considering a hypothetical reaction network in $\mathbb{R}^2$. Assume that a feed point $\mathbf{C}_f$ and kinetics is available. If the stoichiometric subspace **S** can be computed, then the set of extreme points describing **S** may be converted to the H-representation. The principle idea behind the method of bounding hyperplanes is to introduce additional hyperplanes that successively refine the region of achievable states. The method of bounding hyperplanes is an iterative process. Let $\mathbf{X}_k$ represent the set of points that are contained within the current polytope P_k at iteration k of the construction process.

Hyperplanes are moved into the current polytope P_k with the purpose of "cutting away" unattainable space from the region. The resulting polytope P_{k+1} is defined by the collection of all hyperplanes and smaller than the original polytope P_k, and thus it provides a closer approximation to the true AR. In the limit of infinitely many elimination steps, the remaining set of points is an approximation to the AR. Figure 8.21 shows a schematic of the construction sequence for the method.

The stoichiometric subspace **S** is initially described by polytope P_0 and comprises of three bounding hyperplanes in Figure 8.21. Eliminations may be carried out by introducing a hyperplane H at the corners (vertices) of the current polytope. The hyperplane is moved into the region until it is no longer possible to do so without violating the achievability condition, which is described later. When this occurs, hyperplane movement is terminated and H is added to the current list of bounding hyperplanes, which now represents a smaller region P_1. Since hyperplanes are assumed to be added in such a way as to only eliminate unachievable states, the remaining space given by P_1 is a closer approximation to the true AR. Each addition of a new hyperplane results in further removal of unattainable space.

The central approach to construction is to iterate over each *corner* of the current polytope P_k and introduce new hyperplanes that eliminate unattainable space. Sharp corners of the polytope are slowly smoothed out by the introduction of additional bounding planes, and the level of accuracy obtained is hence a strong function of the number of unique hyperplanes that are introduced. Curvature of a region, such as that generated by a PFR manifold on the AR boundary, may be approximated by the use of many bounding hyperplanes.

b) Eliminating Unachievable Points Eliminations in the bounding hyperplanes method are carried out using careful orientation of the hyperplane H so as to partition sections of the current polytope into two half-spaces:

1. A space that contains only unachievable states.
2. The remaining space, containing the true AR and other unachievable states that have yet to be eliminated.

Point 1 is feasible due to a condition, developed by Abraham and Feinberg (2004), which guarantees the *denial* of

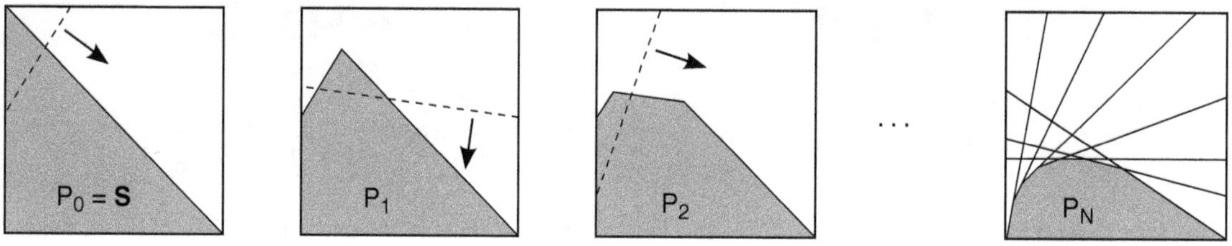

Figure 8.21 Graphical representation of the method of bounding hyperplanes. Hyperplanes are introduced at the corners of the polytope.

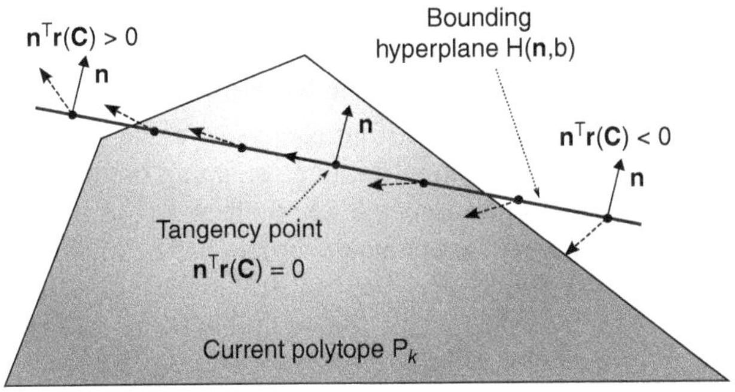

Figure 8.22 Rate vectors evaluated relative to a bounding hyperplane.

achievability. This condition follows a proof that is similar in principle to the complement principle of Chapter 6. Hence, points on the separated space must be checked for denial of achievability relative to H.

In particular, this elimination condition states that if a rate vector $\mathbf{r}(\mathbf{C}^*)$, evaluated at a point $\mathbf{C}^*$, on a hyperplane points *into* the hyperplane, then $\mathbf{C}^*$ is not achievable.[8] By moving hyperplanes inward starting from the extreme points of the polytope, we ensure that only unachievable states are removed from the space.

> **SIDE NOTE: Rate vector relative to the hyperplane**
>
> A numerical condition for the rate vector $\mathbf{r}(\mathbf{C})$ orientation relative to the current hyperplane H($\mathbf{n}$, b) may be established by computing the angle, θ, between $\mathbf{r}(\mathbf{C})$ and the hyperplane normal vector $\mathbf{n}$. Hence,
>
> 1. $\mathbf{n}^T\mathbf{r}(\mathbf{C}) < 0$ indicates that $\mathbf{r}(\mathbf{C})$ points *into* the hyperplane ($\theta > 90°$).
> 2. $\mathbf{n}^T\mathbf{r}(\mathbf{C}) = 0$ indicates that $\mathbf{r}(\mathbf{C})$ is tangent to the hyperplane ($\theta = 90°$).
> 3. $\mathbf{n}^T\mathbf{r}(\mathbf{C}) > 0$ indicates that $\mathbf{r}(\mathbf{C})$ points *out of* the hyperplane ($\theta < 90°$).
>
> This condition is true for all dimensions.

Figure 8.22 demonstrates how rate vectors pointing into a region are unachievable, and may therefore be excluded from the current region P_i.

Three distinct classes of rate vectors, relative to the bounding hyperplane H($\mathbf{n}$, b), can be identified in Figure 8.22:

1. Rate vectors that point into H($\mathbf{n}$, b), satisfying the condition $\mathbf{n}^T\mathbf{r}(\mathbf{C}) < 0$. These points are unachievable according to the achievability condition of Abraham and Feinberg (2004), which can be eliminated.
2. Rate vectors that point out of H($\mathbf{n}$, b), satisfying the condition $\mathbf{n}^T\mathbf{r}(\mathbf{C}) > 0$. No assumption can be made about the attainability of these points (provided they lie within the current polytope P_k), and thus these points must be retained.
3. Rate vectors tangent to H($\mathbf{n}$, b), satisfying the condition $\mathbf{n}^T\mathbf{r}(\mathbf{C}) = 0$. Points that satisfy this condition may or may not be achievable.

[8]The full proof may be found in Abraham and Feinberg (2004).

c) Stopping Criteria Stopping criteria are important in ensuring that only unattainable concentrations are removed from the space at each elimination step. In practice, a hyperplane H($\mathbf{n}$, b) is moved into the current polytope P_k at discrete steps in the opposite direction of the hyperplane normal. Movement is achieved by varying the values of b used in H($\mathbf{n}$, b). Further movement into the region is possible provided the following stopping criteria are not encountered:

1. H($\mathbf{n}$, b) excludes any feed points. Feed points are attainable by definition and must always be included in the current polytope P_k.
2. H($\mathbf{n}$, b) excludes any equilibrium points. Similar to feed points, equilibrium points are also attainable (in the limit of infinite residence time) and thus must also be included in P_k. Note that at an equilibrium concentration $\mathbf{C}^\#$, $\mathbf{r}(\mathbf{C}^\#) = \mathbf{0}$ and thus $\mathbf{n}^T\mathbf{r}(\mathbf{C}^\#) = 0$, which implies that $\mathbf{C}^\#$ will produce the same result as if $\mathbf{r}(\mathbf{C}^\#)$ is tangent to $\mathbf{n}$.
3. H($\mathbf{n}$, b) does not exclude any rate vectors that are tangent to the hyperplane normal. That is, $\mathbf{n}^T\mathbf{r}(\mathbf{C}) \geq 0$.

Note that no assumption is made regarding the attainability of points where rate vectors point out of or are tangent to the hyperplane. It is not possible to discern from the achievability condition, and so points satisfying $\mathbf{n}^T\mathbf{r}(\mathbf{C}) \geq 0$ cannot be excluded from the region on this basis alone.

> **CONCEPT: Removing unachievable points with hyperplanes**
>
> Hyperplanes can be used to divide a region into two half-spaces. If the hyperplanes can be orientated in such a way as to exclude unachievable states from a larger region (containing both achievable and unachievable states), then this can be used as a method of AR construction.

8.5.2.2 ***The Algorithm*** The actual method of bounding hyperplanes may now be described. The following algorithm describes a general overview of the method:

1. Initial polytope computation: Given a feed point $\mathbf{C}_f$ and reaction stoichiometry, compute the stoichiometric subspace **S** in the H-representation. Let P_k represent the convex polytope at iteration k of the construction process. Initially, P_0 is equal to the stoichiometric subspace **S**.
2. Vertex enumeration and vertex selection:
   a. Using the set of hyperplane constraints generated from step 1, compute the set vertices **V** of P_k by solving the vertex enumeration problem. Points in **V** are extreme points (corners) of the convex polytope, which act as starting positions in space from which new hyperplanes may be introduced.
   b. From the set of vertices **V** generated from step 2, pick a vertex $\mathbf{v}_i \in \mathbf{V}$ from which to introduce a new hyperplane. In practice, the vertex $\mathbf{v}_i$ is associated with the sharpest corner of the current polytope P_k.
   c. Introduce a new hyperplane H with normal vector equal to the average value of the hyperplanes that form the vertex. The new hyperplane position must be such that the vertex $\mathbf{v}_i$ lies in H. Let **H** represent the set of points lying in the hyperplane H.
3. Elimination stage:
   a. Begin moving the newly introduced hyperplane into the region P_k.
   b. Discretization stage: For each movement into the region, check for points on the hyperplane H to determine if rate vectors point out of the region. This is done by discretizing the hyperplane H into N grid points. Let $\mathbf{C}_i \in \text{H} \cap \mathbf{X}_k$ be one of the N grid points lying in H within the current set of points $\mathbf{X}_k$. For each $\mathbf{C}_i$, evaluate the rate vector $\mathbf{r}(\mathbf{C}_i)$

      i. If no $\mathbf{r}(\mathbf{C}_i)$ in the N list of grid points can be found where $\mathbf{n}^T\mathbf{r}(\mathbf{C}_i) \geq 0$, then all points $\mathbf{C}_i$ on the hyperplane are *unachievable*. Return to step 3a and repeat until one of the three stopping criteria is met.

      ii. Conversely, if one of the three stopping criteria is met, this signifies the end of movement for the current hyperplane. Proceed to step 4.

4. Update polytope:
   a. Record the position and orientation in space of the current hyperplane. This hyperplane forms a new inequality constraint that will be added to the current set of constraints. For each additional hyperplane added to the list, a tighter bound on the region of achievable points P_{k+1} is produced, and a closer approximation to the true AR is achieved.
   b. Return to step 2. If no corners in **V** can be picked without violating the three stopping criteria, then construction is complete. Store the list of hyperplanes and terminate the program. The final list of hyperplanes may be converted back into a list of vertices that form a convex hull of points bounding the region of achievable points.

In practice, a large amount of initialization and secondary computation, necessary in setting up eliminations, is needed. Once implemented, however, the method is fairly robust particularly when dealing with complicated kinetics.

It is also not possible to determine the optimal reactor structure using this method. Only an upper limit of all achievable points may be computed.

SIDE NOTE: Computing the Cartesian product and the curse of dimensionality

When a list of points generated on an n-dimensional grid must be determined, this is equivalent to computing the *Cartesian product* of the space. To see how the number of points scales with larger dimensions, consider the following: For an equal spacing of N points in each axis of a D-dimensional space, the number of points that must be evaluated is N^D. The number of grid points generated is thus proportional to the exponent of the system dimension, and directly proportional to the number of grid points in each axis.

The task of hyperplane discretization is therefore computationally demanding when a large number of grid points must be generated and evaluated. This requirement scales exponentially for larger dimensions, and is often termed "the curse of dimensionality" (Bellman, 2010). Constructing the AR in higher dimensions is hindered significantly because of this task.

AR constructions of systems involving five independent reactions have been investigated using the method of bounding hyperplanes (Abraham, 2005). Systems involving a larger number of independent reactions are difficult to construct due to the increased computational complexity at present, which is primarily due to the evaluation of points obtained from higher dimensional Cartesian products.

ILLUSTRATION: Summary of results for the three-dimensional Van de Vusse kinetics

A summary of construction results using the bounding hyperplanes for three-dimensional Van de Vusse kinetics is given in Figure 8.23. We show the progression of construction at different stages of refinement. The starting convex polytope is given by the stoichiometric subspace **S**, with vertices at $\{[0, 0, 0]^T, [0, 1, 0]^T, [1, 0, 0]^T, \text{and } [0, 0, 0.5]^T\}$. A new hyperplane is introduced at the corner of the current polytope and moved into the region. In this example, a large portion of the polytope is unattainable and removed within the first 10 iterations of the method. In the bottom-right pane of Figure 8.23, the region after 180 iterations is shown in detail. The overall shape of the resulting polytope matches that given by theory, although the computed region is slightly larger than the true AR. A closer approximation may be achieved if smaller elimination steps are employed, although increased accuracy significantly increases construction time.

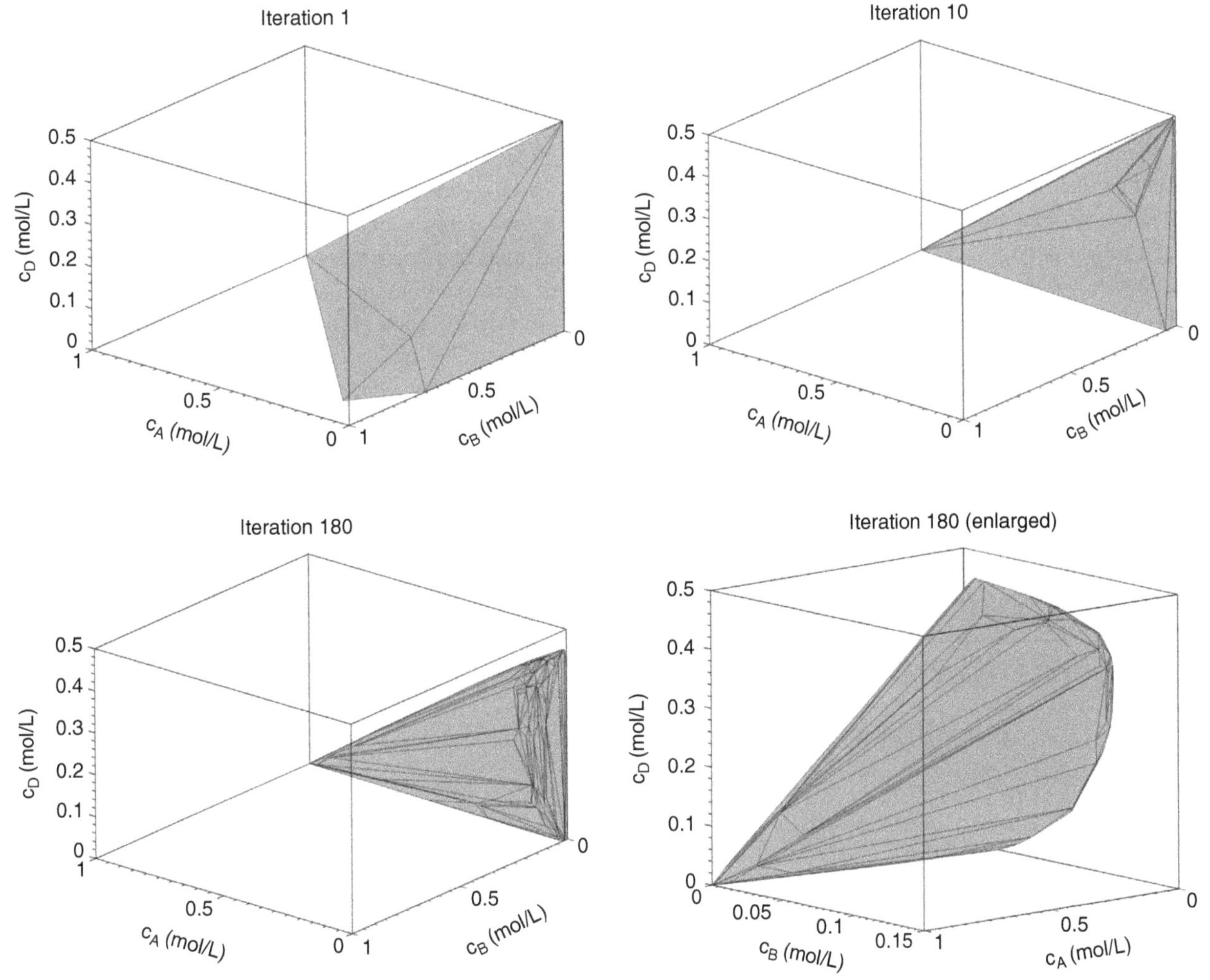

Figure 8.23 Summary of construction results for the three-dimensional Van de Vusse kinetics, using the method of bounding hyperplanes.

8.5.2.3 ***The Method of Rotated Hyperplanes*** In Section 8.5.2.2, we described how the method of bounding hyperplanes introduces new hyperplanes at the sharpest corner of the current polytope P_k. In this way, eliminations are brought about by a hyperplane translation through space with fixed orientation. An alternate method for removing unattainable concentrations is to view eliminations via a plane rotation (Ming et al., 2010). In this approach, the position of the hyperplane in space is fixed, but is allowed to rotate about existing vertices of the current convex polytope. Hyperplanes are introduced at a vertex of the polytope and then constantly *rotated into* the region. Rotations are achieved by continual modification to the hyperplane normal at a fixed position in space, as opposed to a stepwise movement into the region as in the method of bounding hyperplanes. This process is graphically shown in Figure 8.24.

A positive consequence of moving hyperplanes via rotations is that there is no need to perform vertex enumeration. Elimination via rotations does not require knowledge of the position of the polytope corners in order to introduce new hyperplanes. When employing rotations, new hyperplanes are introduced as copies of a current hyperplane at existing corners of the polytope. The positions of these corners are already known and vertex enumeration may be avoided as a result. The existing vertices are then employed as pivot points wherefrom new hyperplanes may be rotated about.

For problems that are bound by the vertex enumeration step, construction times may be improved via rotations. This approach allows for either faster computation times for the same level of accuracy, or more detail to be added for the same computation time. In Figure 8.25, a comparison of construction results for the two-dimensional Van de Vusse kinetics is shown.

An additional benefit of computing candidate ARs in this manner is that there is no need to assume that the feasible region be bounded. Candidate ARs in residence time space may therefore be determined with this method, such as that shown in Figure 8.26(a).

Candidate AR construction via bounding hyperplanes (either via translation or rotation) may be extended to allow for constructions involving a parameter that may affect the direction of the rate vector. A system involving temperature-dependent kinetics is a well-known example of this. In Figure 8.26(b), the rotated hyperplanes method is used to compute an AR with temperature-dependent kinetics. At each point of evaluation, a temperature range between 300 and 1000 K is generated and rate vectors are checked for tangency with the hyperplane. This allows for

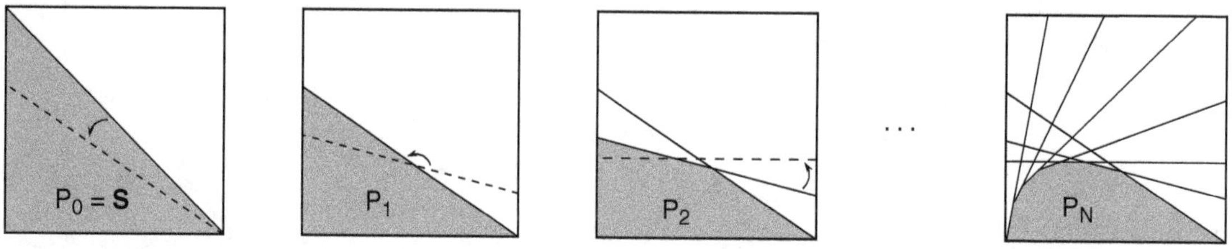

Figure 8.24 Computing candidate ARs using bounding hyperplanes, rotated about an edge.

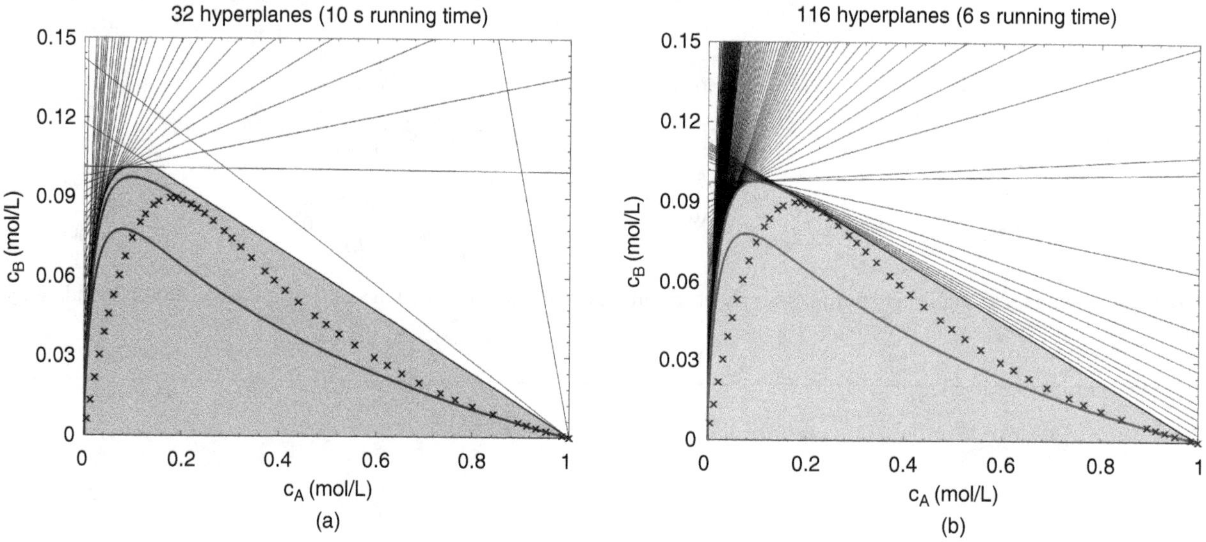

Figure 8.25 Comparison of constructions between (a) bounding hyperplanes and (b) rotated hyperplanes.

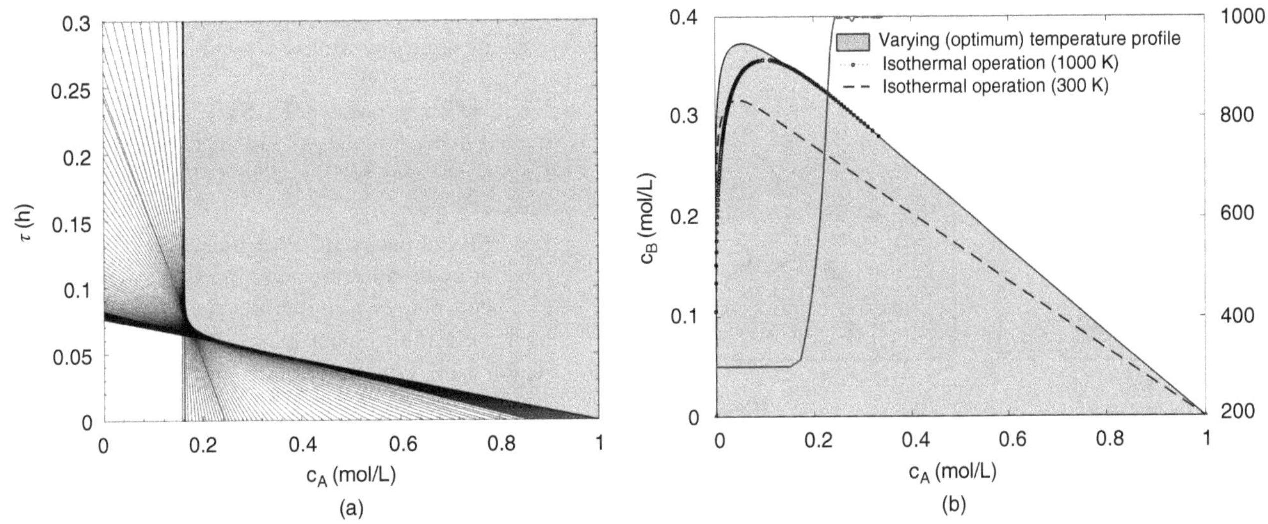

Figure 8.26 Alternate construction scenarios using rotated hyperplanes: (a) Unbounded constructions in τ space and (b) nonisothermal constructions with temperature-dependent kinetics.

the generation of the optimal temperature profile associated with the candidate region.

In two dimensions, rotations are fixed. However, in higher dimensions, there are more planes of rotation that a given hyperplane can be rotated about in space, and thus it is more challenging to develop a systematic method for rotation in higher dimensions.

8.5.3 The Shrink-Wrap Algorithm

8.5.3.1 Introduction The shrink-wrap algorithm (Manousiouthakis et al., 2004) is an outside-in AR construction technique that is geometrically similar to the complement method, described in Section 8.4.3, and the LP formulation, to be discussed in Section 8.6.1. The method is based on results obtained from the Infinite DimEnsionAl State-space (IDEAS) framework, which will also be discussed in Section 8.6.2. Whereas the IDEAS framework is a superstructure method for constructing the AR boundary in a point-by-point fashion, the shrink-wrap algorithm carries out construction using the geometric properties of CSTRs and PFRs.

The shrink-wrap method derives its name from the manner in which AR construction is carried out, and the geometric resemblance of this process to that of wrapping shrink-wrap over an object.[9] Candidate ARs are constructed by the successive removal of unattainable points from the stoichiometric subspace. This is in contrast to the IDEAS approach, which grows regions outward.

8.5.3.2 Basic Idea Construction is carried out in a manner that might be best viewed as stretching plastic film (shrink-wrap) over an object, thus forming a convex polytope in space. The first iteration of the algorithm begins with computing the stoichiometric subspace **S**, or an equivalent region, that encloses the true AR. The boundary of the enclosing region is then progressively reduced in size by an elimination procedure. Since the method is outside-in in nature, the basic idea of the method is to search for unachievable concentrations that may be discarded from an initial bounding region. Repeated removal of unattainable space leads to an approximation of the true AR. The shrink-wrap method uses the extreme points of the current convex polytope as starting points for elimination. Consider then Figure 8.27.

> **SIDE NOTE: Similarities between shrink-wrap method and other superstructure methods**
>
> In later sections, we discuss superstructure methods (LP formulations and the IDEAS framework) that are similar in nature to the shrink-wrap method. A major difference between the methods is that the shrink-wrap method is geometric in nature.
>
> The total connectivity model, used in the LP formulation, only considers combinations of CSTRs. The shrink-wrap method, in contrast, allows for combinations of both PFRs and CSTRs.

A candidate AR, represented by the shaded region, is given at iteration k of a typical construction process in the shrink-wrap method. The convex polytope, P_k, represents an approximation to the AR at iteration k in the construction, and $\mathbf{X}_k$ is the set of extreme points contained within P_k. The

[9] Recall from Chapter 2 that the convex hull of an object resembles that of when plastic film is pulled tightly over an object.

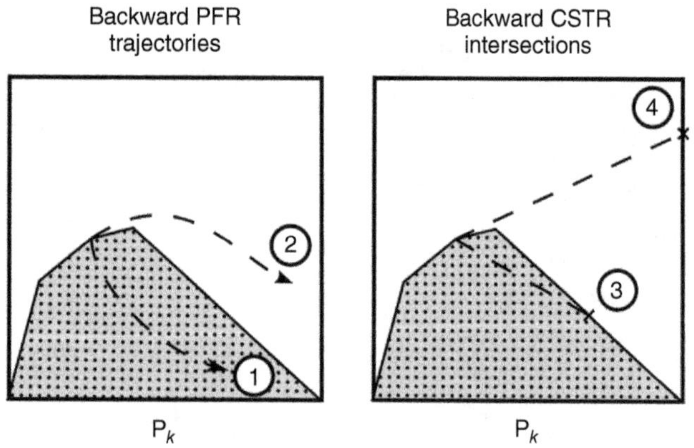

Figure 8.27 The shrink-wrap method. Invalid extreme points are eliminated via boundary intersections.

goal of the shrink-wrap method is to progressively contract the bounding region of extreme points by evaluating each extreme point of P_k and determining whether it is achievable or not. This is accomplished by solving the CSTR and PFR equations *backward* with respect to residence time, and checking if the resulting backward solutions intersect the current region at a *different* point. If no such points can be identified at the extreme point considered, then it may be proved—via the IDEAS framework—that the extreme point is not achievable (Manousiouthakis et al., 2004). This extreme point may then be removed from the set belonging to P_k. These conditions are represented geometrically by points 2 and 4 in Figure 8.27 for a backward PFR and CSTR, respectively. Alternatively, if a CSTR or PFR from the extreme point does intersect the region at another point, then nothing further can be claimed about the achievability of the point, and the point must be retained in the set—that is, the current extreme point is not eliminated. This is shown in Figure 8.27 by points 1 and 3 for a PFR and CSTR, respectively.

This reasoning outlines a basic elimination procedure for the construction of candidate ARs. Each extreme point $\mathbf{C}_i \in \mathbf{X}_k$, on the current convex polytope P_k, is checked for feasibility via backward CSTR solutions and PFR trajectories. If either the backward CSTR or PFR trajectory from $\mathbf{C}_i$ intersects the current polytope boundary, then it is retained. Otherwise, $\mathbf{C}_i$ is removed from the current set of extreme points $\mathbf{X}_k$ and an updated convex polytope P_{k+1} is generated. This process is repeated for all extreme points until no further eliminations are possible.

Checking points obtained from backward CSTR and PFR points is identical to checking if points lie within the convex region defined by a set of inequality constraints. The feasibility condition given by Equation 8.9 may be used to check if points intersect the current region P_k,

$$\mathbf{A}^{[k]}\mathbf{C}_i \leq \mathbf{b}^{[k]}$$

where Equation 8.9 is the same feasibility condition used in the complement method from Section 8.4.3.2. Thus

- For CSTRs, a backward CSTR solution is parameterized using a scalar, τ, and checked using the same conditions described in section "Computing Feasible Intersections".
- For PFRs, a backward PFR trajectory is obtained by integrating the PFR equation for a negative value of τ. Each concentration in the backward PFR trajectory is then checked for intersection with the current feasible region using Equation 8.9.

> **CONCEPT: Eliminating unachievable states from backward solution of CSTRs and PFRs**
>
> We can use the geometric nature of both CSTRs and PFR to determine whether a point is achievable or not. This is accomplished by checking whether the resulting backward PFR or CSTR solutions intersect the current polytope boundary or not, which works in an opposite way to the complement method described in Section 8.4.3.

8.5.3.3 ***The Algorithm*** An overview of the shrink-wrap method for candidate AR construction is given in the following:

1. Initialization: Compute a convex region that encloses the AR. Typically, this will be the stoichiometric subspace **S** defined by the reaction stoichiometry, although an n-dimensional bounding box may also be employed. This convex set constitutes the first approximation of the AR. Discretize the space in **S** by a grid of points (lattice points representing distinct concentrations in space). A finer grid results in a closer approximation of the AR at the expense of an increased computational workload. Let $\mathbf{X}_k$ denote this set of grid points in **S** at iteration k of the procedure.
2. Compute extreme points: Grid points in **X** constitute the set of evaluation points for the elimination stage considered in step 3. The extreme points of **X** are determined by computing the convex hull of **X**, conv(**X**).
3. Elimination: For each extreme point $\mathbf{C}_i \in \mathbf{X}$ generated in step 2, compute a backward PFR trajectory and CSTR intersection point. Backward PFR trajectories are generated by the integration of the PFR expression using a negative value of τ, whereas CSTR feed points are determined by computing the intersection points with the list of constraints defining the boundary. This procedure is similar to that described in Section 8.4.3.2 using the complement method.

a. If either the PFR backward trajectory or CSTR points intersect the current boundary P_k at a point different to $\mathbf{C}_i$, then retain the extreme point $\mathbf{C}_i$ in the current list $\mathbf{X}_k$. A point of intersection indicates that $\mathbf{C}_i$ may be feasible and thus it cannot be removed. Also retain all feed points $\mathbf{C}_f$, for all feeds must be attainable by definition of the AR.
b. If the backward PFR trajectory and CSTR points yield no intersection with the current boundary P_k, then this implies that $\mathbf{C}_i$ is not feasible. Thus, remove $\mathbf{C}_i$ from the current set $\mathbf{X}_k$.
c. Compute an updated set of extreme points $\mathbf{X}_{k+1}$ that remain from the elimination phase in step 3. Repeat step 3 until no further extreme points $\mathbf{C}_i$ can be removed from the set.

An approximation to the AR is then generated by the remaining extreme points after no further eliminations are possible. The accuracy of construction is determined by the number of points defined on the grid.

The bulk of the computation arises from step 3 of the algorithm. A large portion of the computational overhead results from checking for intersections on the current polytope boundary P_k. A useful feature of the method is that, unlike the IDEAS algorithm of Burri et al. (2002) and LP formulations of Kauchali et al. (2002), there is no need to solve a large LP problem. Accordingly, this assists in a straightforward implementation of the method on a computer without the need for specialized optimization software. Necessary components of the algorithm involve, simply, the use of a convex hull algorithm and an ODE integration procedure. The remaining pieces of code may be implemented using basic matrix algebra.

The extent to which a candidate AR may be constructed is directly related to the number of grid points used to discretize the space. This complexity scales by exponent of the dimension of the system under consideration. For instance, if a constant value of N points is used for each component axis, and construction is performed in $\mathbb{R}^D$, then at most D^N points must be evaluated in the worst case. For each point considered, both CSTR and PFR backward solutions must be generated and checked for intersection with the boundary.

The shrink-wrap method is generally more successful at handling complicated kinetics (where multiple steady states may exist) when compared to competing methods. Since CSTR solutions are found geometrically as opposed to through the solution of a potentially difficult system of nonlinear equations, the method is attractive for highly nonideal systems.

ILLUSTRATION: Two-dimensional Van de Vusse kinetics

Construction results for the two-dimensional Van de Vusse system, using the shrink-wrap method, are supplied in Figure 8.28. These results are generated using a simplified version of the algorithm, which only uses CSTR points for simplicity and demonstration. (In practice, it may be preferable to incorporate backward PFR trajectories into the construction, although this adds to the computational complexity.)

In Figure 8.28, the AR is generated in c_A–c_B space for three different grid sizes (25×25, 50×50, and 100×100) to demonstrate the effect of the number of points considered and the corresponding accuracy of construction.

As the number of grid points is made larger, a closer approximation to the AR is obtained. Note that the shrink-wrap method always underestimates the true size of the AR. The nature of the elimination procedure, detailed before in the algorithm description, means that extreme points that do not intersect the current polytope must be eliminated, yet we also understand from Chapter 6 that the true AR boundary must be composed of PFRs and mixing surfaces (all points on the AR boundary must be *exposed points* that do not intersect the boundary anywhere else). The criterion for termination in the shrink-wrap method is when all extreme points on the current polytope have been visited, and none of them can be removed because they intersect the boundary somewhere else. Thus, this suggests that the shrink-wrap method will always remove the outermost boundary point of the true AR, even for high-resolution grids (using a small grid spacing).

Curvature on the AR boundary may be approximated with a sufficiently large number of points, as is observed by a comparison between the candidate regions obtained corresponding to a 25×25 grid and the region resulting from a 100×100 grid. Note that similar region sizes result from the 25×25 and 50×50 grids; however, the 50×50 grid provides slightly more curvature; greater detail could be obtained using backward PFR trajectories also.

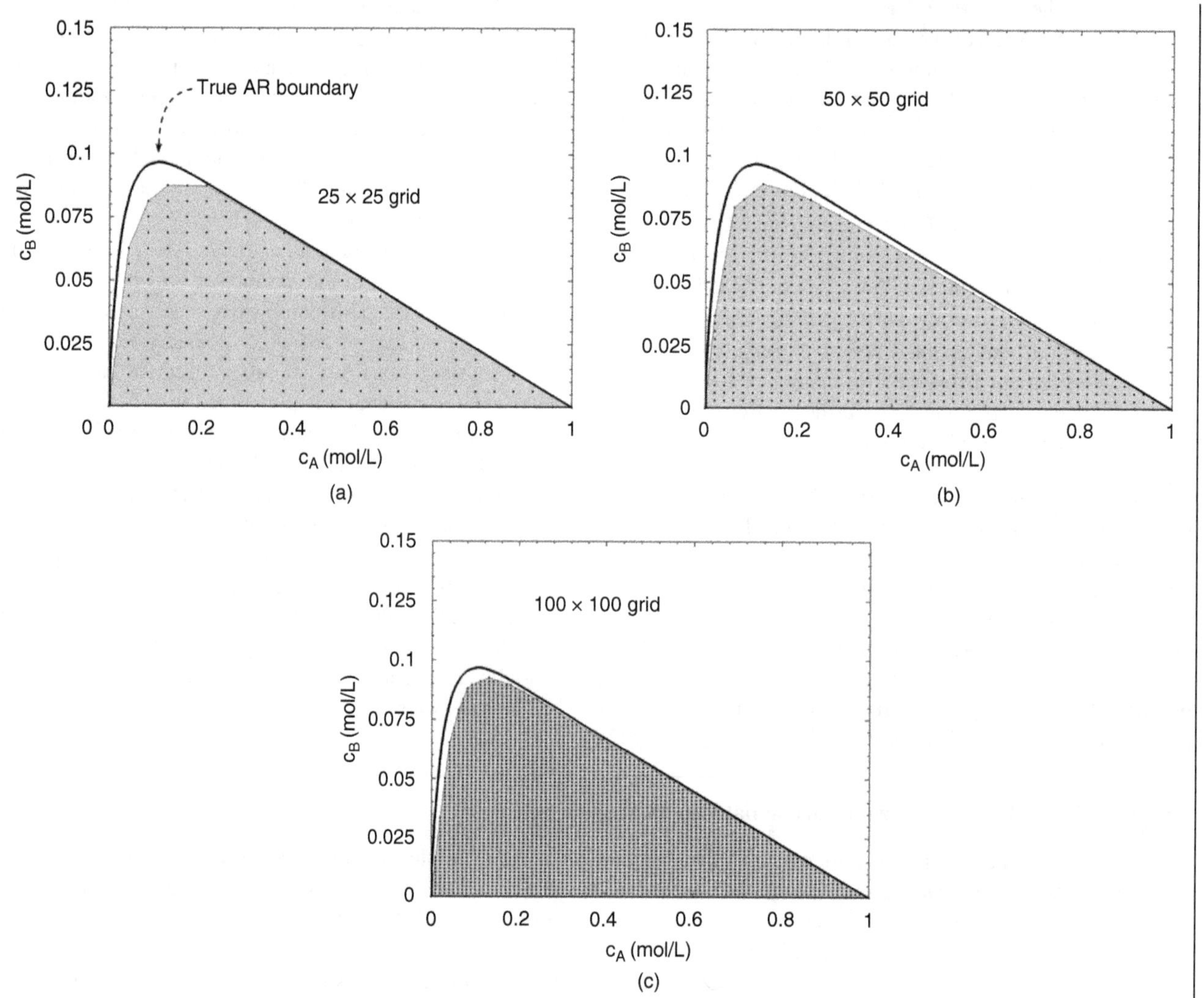

Figure 8.28 Candidate AR constructed using the shrink-wrap method. Results are generated for (a) 25 × 25, (b) 50 × 50, and (c) 100 × 100 grid sizes.

8.6 SUPERSTRUCTURE METHODS

> **SIDE NOTE: Additional reading**
>
> AR construction using superstructure formulations are generally targeted toward mathematically sophisticated readers, which require the use of mathematical optimization techniques. Brief overviews of two superstructure methods are described later, although readers interested in this field of work should consult Burri et al. (2002), Davis et al. (2008), Kokossis and Floudas (1990), Manousiouthakis et al. (2004), Posada and Manousiouthakis (2008), Rooney and Biegler (2000), and Zhou and Manousiouthakis (2006, 2007, 2009) for further details.

8.6.1 LP Formulations

8.6.1.1 Basic Idea Candidate AR construction using the LP method is geometrically identical to the complement method from Section 8.4.3, which is simply an application of AR property 7 from Chapter 4. Consultation of Figure 8.18, from Section 8.4.3.2, is therefore useful in understanding the construction approach of the LP method.

A key difference between the complement method and the LP method is that the latter requires the solution of a linear program, whereas the former is a direct application of the CSTR attainability condition. In the LP approach, all points on the AR boundary are computed *simultaneously*—via the solution of a large linear program—in a single calculation step. In order for this result to be achieved, the candidate region boundary points must be expressed in terms of all other boundary points in space using a superstructure formulation, which is termed the *total connectivity model*.

8.6.1.2 The Total Connectivity Model The connectivity model is a reactor superstructure formulation that attempts to approximate different reactor types using a network of small CSTRs. Combination of CSTRs in series and parallel allows for the approximation of different fundamental reactor types, specifically:

1. **PFR approximation**: A PFR may be approximated by connecting many smaller CSTRs in series. In doing so, the PFR equation is discretized by via Euler-style approximations in residence time.
2. **DSR approximation**: A DSR may be approximated by a number of CSTRs in series, with intermediate mixing between CSTR effluent streams and a specified sidestream composition.

Figure 8.29 demonstrates physically how the appropriate arrangement of many small CSTRs might be used to approximate different reactor types. Furthermore, elaborate reactor structures may also be modeled by combining these two base reactor types (together with the conventional CSTR).

Each CSTR used in the total connectivity model must be arranged in a flexible manner to allow for a wide range of configurations to be achieved. Inlet and outlet streams around a generalized CSTR building block used in the total connectivity model are shown in Figure 8.30(a). Several allowances are made in the total connectivity model including

- The feeding of each CSTR from all other CSTRs in the network.
- Recycle and bypass within individual CSTRs.
- Recycle and bypass from one CSTR with all other CSTRs.
- Connection of the overall plant feed to each CSTR.
- An overall bypass stream over the entire network, given by $Q_{0,\text{out}}$, which is shown in Figure 8.30(b).

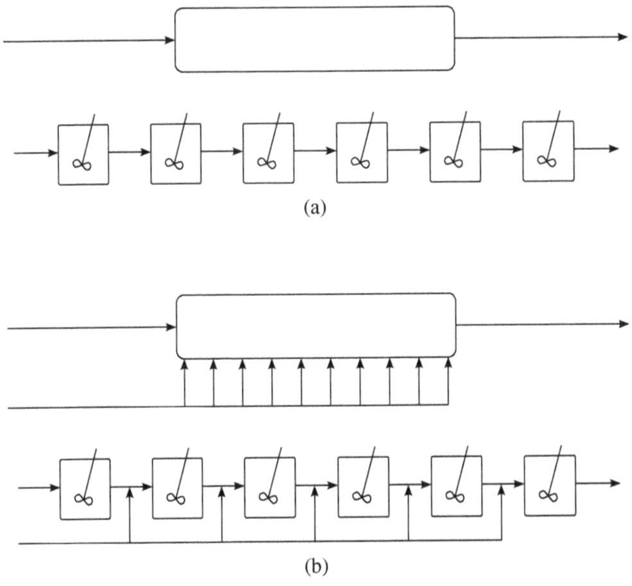

Figure 8.29 The total connectivity model. Approximation of PFRs (a) and DSRs (b) using a number of CSTRs. Kauchali et al. (2002). Reproduced with permission of Elsevier.

Concentration space is then discretized into N grid points $\mathbf{C}_i$ where $1 \leq i \leq N$. The corresponding rate vector evaluated at $\mathbf{C}_i$, $\mathbf{r}_i = \mathbf{r}(\mathbf{C}_i)$ may then also be determined given the system kinetics. The total connectivity model assumes that a CSTR operates at each grid point. When a mass balance around the ith CSTR is performed, the following system of equations, considered in conjunction with Figure 8.30(a), are generated:

$$F_i c_i = F_i c_i^0 + V_i r_i \tag{8.10a}$$

where

$$F_i = Q_{0i} + \sum_{j=1}^{N} Q_{ji} \tag{8.10b}$$

$$F_i = Q_{i,\text{out}} + \sum_{j=1}^{N} Q_{ij} \tag{8.10c}$$

$$F_i c_i^0 = Q_{0i} c_0 + \sum_{j=1}^{N} Q_{ji} c_j \tag{8.10d}$$

Equation 8.10b and 8.10c refer to the inlet and outlet volumetric flow rates around the ith CSTR, respectively. Equation 8.10d expresses to the total molar flow rate feeding the ith CSTR from all other CSTRs in the network. In general, the feed concentration to each CSTR is unknown because of mixing from all other CSTRs. We therefore express $\mathbf{C}_i^0$ in terms of the effluent concentrations $\mathbf{C}_j$ from all other CSTRs. Observe from the form of the equations that bypassing and recycling within an individual CSTR occurs when $i = j$ for Q_{ij}. From Figure 8.30(b), volumetric flow rates over the entire system may also be expressed in terms of individual CSTR flow rates as follows:

$$Q_{\text{feed}} = Q_{0,\text{out}} + \sum_{i=1}^{N} Q_{0i} = 1 \tag{8.10e}$$

$$Q_{\text{out}} = Q_{0,\text{out}} + \sum_{i=1}^{N} Q_{i,\text{out}} = 1 \tag{8.10f}$$

Equation 8.10e and 8.10f express the total volumetric flow rate into and out of the entire reactor network, respectively. The total plant flow rates, Q_{feed} and Q_{out}, are normalized to unity. Finally, the exit concentration from the network $\mathbf{C}_{\text{out}}$ may be expressed as follows:

$$Q_{\text{out}} c_{\text{out}} = Q_{0,\text{out}} c_f + \sum_{i=1}^{N} Q_{i,\text{out}} c_i \tag{8.10g}$$

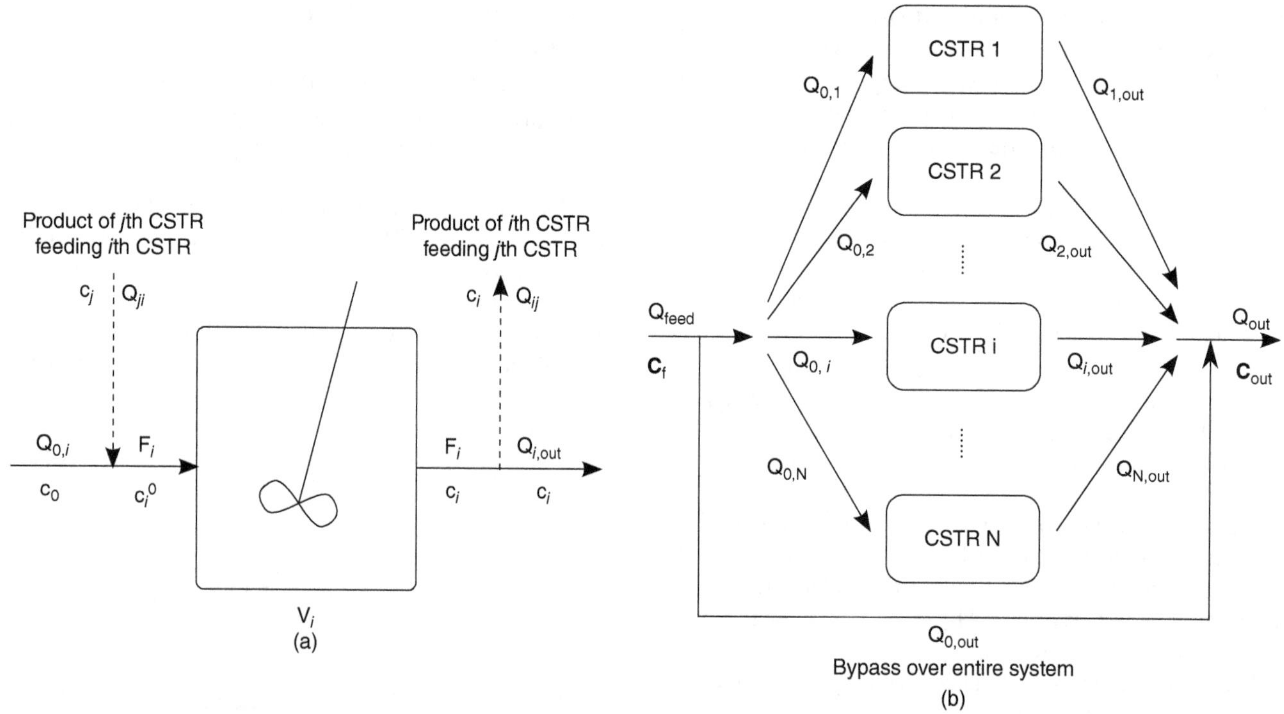

Figure 8.30 The total connectivity model: (a) Streams around the *i*th CSTR in the total connectivity model and (b) overview of the total connectivity model.

Here, $Q_{out} = 1$. Each concentration in space may be discretized into a grid of N points wherein each point is assumed to be a possible CSTR effluent concentration. Hence, the concentrations expressed in Equation 8.10a–8.10g and corresponding rate vectors are all *known*. This results in a system of *linear* equations wherein F, Q, and V are the only unknowns, which may be found using an LP solver. The number of CSTRs used in the model can be made arbitrarily large by generating a finer grid of points in space, which allows for the approximation of curvature resulting from DSR trajectories and PFR manifolds.

CONCEPT: A system of linear equations

When a feed point $\mathbf{C}_f$ and residence time τ for a system of nonlinear kinetics is supplied, it is generally difficult to solve for the corresponding effluent concentration $\mathbf{C}$ in a CSTR. However, the converse problem is much easier to solve—given an effluent concentration $\mathbf{C}$ and a residence time τ, it is much easier to find the corresponding $\mathbf{C}_f$ in a CSTR. Specification of $\mathbf{C}$ allows for the computation of $\mathbf{r}(\mathbf{C})$, which is the source of nonlinearity in the CSTR expression.

EXAMPLE 6: Solving for a network of CSTRs

Consider the following reactor structure involving three CSTRs in series.

The reaction for the system is assumed to follow the Van de Vusse reaction scheme. For simplicity,[10] assume that we are only interested in components A and B in the system, so that the component rate expressions are given by

$$r_A = -k_1 c_A - k_3 c_A^2$$
$$r_B = k_1 c_A - k_2 c_B$$

where $k_1 = 1.0\,s^{-1}$, $k_2 = 1.0\,s^{-1}$, and $k_3 = 10.0$ L/(mol.s). It is assumed that the feed concentration to the system is pure in component A, given by $\mathbf{C}_f = [c_{Af}, c_{Bf}]^T = [1, 0]^T$ mol/L. The volumetric flow rate of the feed to the first CSTR is $Q_1 = 1$ L/s. Constant density may be assumed. Answer the following questions.

[10]If component C were to be considered, an additional set of equations and unknowns would need to be incorporated into the problem.

Question 1: Solving for exit concentrations

Assuming that the following volumetric flow rates and CSTR reactor volumes are given, compute the concentration vectors $\mathbf{C}_2$ to $\mathbf{C}_4$ in Figure 8.31. It is understood that the CSTRs have the following volumes: $V_1 = 0.1$ L, $V_2 = 0.2$ L, $V_3 = 0.3$ L.

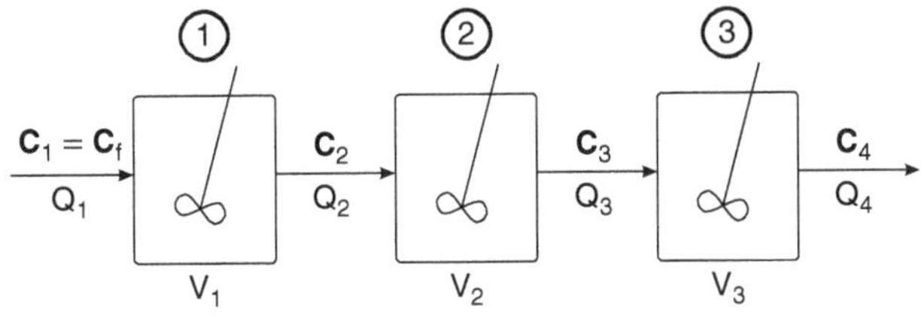

Figure 8.31 Three CSTRs in series.

Answer 1:

In order to compute the concentration vectors in streams two to four, a mass balance must be performed around each CSTR. The resulting mass balance expressions for each CSTR, generalized for component *i* in the system, are given as follows:

$$Q_2c_{2,i} = Q_1c_{1,i} + V_1r_i(\mathbf{C}_2) \quad \text{(CSTR 1 balance)}$$
$$Q_3c_{3,i} = Q_2c_{2,i} + V_2r_i(\mathbf{C}_3) \quad \text{(CSTR 2 balance)}$$
$$Q_4c_{4,i} = Q_3c_{3,i} + V_3r_i(\mathbf{C}_4) \quad \text{(CSTR 3 balance)}$$

Here, component *i* may represent components A, B, C, or D in the Van de Vusse reaction. Since we have assumed that density is constant and that all three CSTRs are connected in series with no additional bypass or recycles, the feed volumetric flow rates to each CSTR are also constant. Hence, $Q_1 = Q_2 = Q_3 = Q_4 = 1$ L/s.

The rate expressions are generalized to show that each component may be a function of all components in the system. Exit concentrations are unknown, and hence they must be solved for simultaneously. This may be achieved with a *nonlinear* solver.

For instance, the system of equations corresponding to components A and B for the reactor network are as follows:

$$Q_2c_{2,A} = Q_1c_{1,A} + V_1(-k_1c_{2,A} - 2k_3c_{2,A}^2)$$
$$Q_3c_{3,A} = Q_2c_{2,A} + V_2(-k_1c_{3,A} - 2k_3c_{3,A}^2)$$
$$Q_4c_{4,A} = Q_3c_{3,A} + V_3(-k_1c_{4,A} - 2k_3c_{4,A}^2)$$

For component B,

$$Q_2c_{2,B} = Q_1c_{1,B} + V_1(k_1c_{2,A} - k_2c_{2,B})$$
$$Q_3c_{3,B} = Q_2c_{2,B} + V_2(k_1c_{3,A} - k_2c_{3,B})$$
$$Q_4c_{4,B} = Q_3c_{3,B} + V_3(k_1c_{4,A} - k_2c_{4,B})$$

Note that the system is a coupled set of *nonlinear equations*—a very difficult system to solve in general. Solution of the system using Newton's method, for example, may be employed if an appropriate initial guess is supplied. The following concentrations are obtained after performing the solution:

$$\mathbf{C}_2 = [0.4837, \ \ 0.0440]^{\mathrm{T}}\,\mathrm{mol/L}$$
$$\mathbf{C}_3 = [0.2287, \ \ 0.0748]^{\mathrm{T}}\,\mathrm{mol/L}$$
$$\mathbf{C}_4 = [0.1149, \ \ 0.0840]^{\mathrm{T}}\,\mathrm{mol/L}$$

Question 2: Solving for flow rates and volumes

Consider now the case where we are given the concentrations for $\mathbf{C}_2$ to $\mathbf{C}_4$. Assume that the concentrations are now given by the following set:

$$\mathbf{C}_2 = [0.3394,\ \ 0.0679]^{\mathrm{T}}\mathrm{mol/L}$$

$$\mathbf{C}_3 = [0.2203,\ \ 0.0817]^{\mathrm{T}}\mathrm{mol/L}$$

$$\mathbf{C}_4 = [0.0929,\ \ 0.0853]^{\mathrm{T}}\mathrm{mol/L}$$

The feed concentration to the network is assumed to remain unchanged so that $\mathbf{C}_f = [1, 0]^{\mathrm{T}}$ mol/L.

Answer 2:

The system of equations for component i is again given by

$$Q_2 c_{2,i} = Q_1 c_{1,i} + V_1 r_i(\mathbf{C}_2)$$

$$Q_3 c_{3,i} = Q_2 c_{2,i} + V_2 r_i(\mathbf{C}_3)$$

$$Q_4 c_{4,i} = Q_3 c_{3,i} + V_3 r_i(\mathbf{C}_4)$$

Since the component concentrations for all streams have now been specified, the unknowns are now the volumetric flow rates and CSTR volumes. There are hence six unknowns (Q_2 to Q_4 and V_1 to V_3). In addition, the following three relations are available from the constant density assumption around each CSTR:

$$Q_2 = Q_1$$

$$Q_3 = Q_2$$

$$Q_4 = Q_3$$

This system contains a total of six unknowns in six equations. A beneficial property of this system, however, is that the volumetric flow rates and CSTR volumes appear *linearly* in the mass balance expressions. We therefore have a system of linear equations that must be solved. This may be done by performing elementary row operations on the appropriate matrices.

$$\mathbf{A} = \begin{bmatrix} c_{2,i} & 0 & 0 & -r_i(\mathbf{C}_2) & 0 & 0 \\ -c_{2,i} & c_{3,i} & 0 & 0 & -r_i(\mathbf{C}_3) & 0 \\ 0 & -c_{3,i} & c_{4,i} & 0 & 0 & -r_i(\mathbf{C}_4) \\ 1 & 0 & 0 & 0 & 0 & 0 \\ -1 & 1 & 0 & 0 & 0 & 0 \\ 0 & -1 & 1 & 0 & 0 & 0 \end{bmatrix}$$

and

$$\mathbf{b} = \left[Q_1 c_{1,i}\ 0\ 0\ Q_1\ 0\ 0\right]^{\mathrm{T}}$$

The first three columns in A represent coefficients belonging to Q_2, Q_3, and Q_4, whereas columns 4–6 represent coefficients belonging to V_1, V_2, and V_3, respectively. This results in the following system of linear equations for each component present in the system

$$\mathbf{Ax} = \mathbf{b}$$

with

$$\mathbf{x} = [Q_2,\ Q_3,\ Q_4,\ V_1,\ V_2,\ V_3]^{\mathrm{T}}$$

Solution of these equations, which may be achieved using a method such as Gaussian elimination, results in the following values for the corresponding volumetric flow rates and volumes:

$$\mathbf{x} = \mathbf{A}^{-1}\mathbf{b} = [1,\ 1,\ 1,\ 0.25,\ 0.10,\ 0.48]^{\mathrm{T}}$$

Once solution vector **x** is "unpacked," $Q_2 = 1$ L/s, $Q_3 = 1$ L/s, $Q_4 = 1$ L/s, $V_1 = 0.25$ L, $V_2 = 0.10$ L, and $V_3 = 0.48$ L. Observe that it is significantly easier to solve for Q and V given values for **C** (the CSTR effluent concentration), than it is to solve for **C** given values for Q and V. This nature of the CSTR equation is how a system of linear equations may arise (and solved using an LP solver) in AR construction. Nevertheless, a very large number of linear equations and unknowns must often be generated to adequately model a system in detail, which increases rapidly with an increase in the dimension of the system. This complexity often limits the size of the problems that can be feasibly considered.

8.6.1.3 Conversion to an LP Problem Once the connectivity model equations have been established, the system of linear equations may be transformed into an LP problem. This is achieved by introducing non-negativity constraints on all flow rates and reactor volumes for each CSTR. An objective function must be supplied in the linear program. Two vectors, $\mathbf{a}_1$ and $\mathbf{a}_2$, are introduced to act as objective functions for the LP problem. The following LP may then be formulated:

$$\text{Maximize} \quad \mathbf{a}_1^T\mathbf{C}_{\text{out}}$$

$$\text{Subject to}$$

$$\mathbf{C}_{\text{min}} \le \mathbf{a}_2^T\mathbf{C}_{\text{out}} \le \mathbf{C}_{\text{max}}$$

$$F_i,\ F_i c_i^0,\ Q_{ij},\ Q_{oi},\ Q_{i,\text{out}},\ Q_{o,\text{out}},\ V_i \ge 0 \qquad (8.11)$$

with equality constraints given by Equation 8.10a–8.10g

By appropriate specification of the component entries in vectors $\mathbf{a}_1$ and $\mathbf{a}_2$, the LP defined by Equation 8.11 can be solved for a fixed grid size, by specifying the outlet concentrations for all but one component. Solution of the LP generates a single point on the AR boundary, and the AR boundary may be constructed in a point-wise manner by resolving the LP for a different set of outlet concentrations. (This is similar to the iso-state method in that the AR boundary is constructed by taking many slices at fixed points along individual axes combinations, and solving for the relevant boundary point.) A major drawback to solving this LP in practice is that a large system must be solved, even for lower dimensional problems. Kauchali et al. (2002) show that for a two-dimensional system with 100 grid points, an LP on the order of 1×10^4 equations and 1×10^8 variables must be solved. This number increases significantly as the number of components and independent reactions increases. Thus, even for small problems, direct solution of the LP given in Equation 8.11 is often too large for practical use. In Section 8.6.1.4, we describe how the computational burden may be reduced by reformulating the LP equations.

8.6.1.4 An Improved LP Formulation The method given in Section 8.6.1.3 does not make use of information regarding the complement region, and the full solution must be obtained simultaneously in a single step. Assume then that a candidate AR, represented by the convex polytope P and the corresponding set of point **X**, is available. It is assumed that P has been generated previously. The set of concentrations that lie in the complement region are then given by **S\X**, where **S** is the stoichiometric subspace. We assume that **S\X** is discretized by a grid of M points. Our interest will be not so much in the interior of the candidate region P but rather on the boundary points of P (conv(**X**)).

Next, suppose that conv(**X**) is composed of K distinct points, and these distinct points may be used as feeds to the M points in **S\X**. The mass balance equations, given by Equation 8.10a–8.10g, consider the network of CSTRs when only a *single* plant feed is available. Since points on the boundary of P are also now available, the original LP formulation may be extended to include *multiple feed points,* using the K boundary points in conv(**X**). Each point on the boundary of P acts in a similar situation to that given in Figure 8.30. Hence, the LP formulation may be extended by performing a mass balance around the *i*th CSTR operating in the complement region.

$$F_i c_i = F_i c_i^0 + V_i r_i \qquad (8.12a)$$

Here, $c_i \in \mathbf{S\backslash X}$ where $i = 1, 2, \ldots, M$. The feed and molar flow rates to the *i*th CSTR may also be generalized to now include feeding from all K boundary points in conv(**X**).

$$F_i = \sum_{k=1}^{K} Q_{k,i} + \sum_{j=1}^{M} Q_{ji} \qquad (8.12b)$$

$$F_i = Q_{i,\text{out}} + \sum_{j=1}^{M} Q_{ij} \qquad (8.12c)$$

$$F_i c_i^0 = \sum_{k=1}^{K} Q_{k,i} c_k + \sum_{j=1}^{M} Q_{ji} c_j \qquad (8.12d)$$

Note that these equations are similar to the original mass balance equations for a single feed. The summation over *k* indicates that there are K feed points from the candidate region boundary and M other CSTRs from the complement region. The total plant feed flow rates may then also be updated to include K additional feeds:

$$Q_{\text{feed}} = \sum_{i=1}^{M}\sum_{k=1}^{K} Q_{k,i} + \sum_{k=1}^{K} Q_{k,\text{out}} = 1 \tag{8.12e}$$

$$\sum_{i=1}^{M}\sum_{k=1}^{K} Q_{k,i} - \sum_{i=1}^{M} Q_{i,\text{out}} = 0 \tag{8.12f}$$

Note that $Q_{k,\text{out}}$ indicates that each potential feed point on the boundary of P is also allowed to bypass to the M CSTR points in the complement space. Finally, the overall product concentration from the network c_{out} is then

$$Q_{\text{out}}c_{\text{out}} = Q_{0,\text{out}}c_f + \sum_{i=1}^{N} Q_{i,\text{out}}c_i \tag{8.12g}$$

where $Q_{\text{out}} = 1$ because it has been normalized to unity. From this set of equations, a similar LP problem may be formulated, which is slightly different to that given in Section 8.6.1.3.

Since a candidate AR has already been generated previously, CSTRs operating in $\mathbf{S \backslash X}$ serve to validate whether the candidate region is the true AR or not. If it is possible to find a positive CSTR volume V_i operating in the complement region, then the corresponding CSTR is feasible and this may be used to expand the region. Conversely, if all CSTRs in the complement have negative volume, then the candidate region is the true AR. An alternate LP can be solved, using the sum of CSTR volumes in the complement region, as an objective function:

$$\text{Maximize} \quad \sum_{i=1}^{M} V_i$$

$$\text{Subject to}$$

$$\mathbf{c}_{\min} \le \mathbf{a}_2^{T}\mathbf{c}_{\text{out}} \le \mathbf{c}_{\max}$$

$$F_i,\ F_i c_i^0,\ Q_{ij},\ Q_{ki},\ Q_{i,\ \text{out}},\ Q_{k,\text{out}},\ V_i^0 \ge 0 \tag{8.13}$$

with equality constraints given by Equation 8.12a–8.12g

V_i's in this instance refer to the CSTRs operating in the complement region. This is similar in form to the LP given in Section 8.6.1.3, however, now extended to K feed points and with a different objective function.

Implementation of the LP algorithm can be done in a straightforward manner on a computer using standard LP solvers. A large system of grid points and corresponding constraint and connectivity relations must be enforced in order to compute candidate regions to within reasonable accuracy.

Computing candidate ARs using LP methods is computationally feasible for lower dimensional problems. For systems involving many independent reactions, potentially *billions* of variables may be needed in an LP formulation. Nevertheless, a benefit of solving LP problems is that a globally optimal solution may be found (as the solution of an LP problem is always globally optimal).

Since computation of the AR is carried out via many small CSTRs in the total connectivity model, interpretation of the associated optimal reactor structure is not clear using the method. This is a trait observed for many superstructure methods. Nevertheless, candidate ARs for systems with highly nonlinear kinetics may be found with this approach since the method does not solve for exit concentrations, but rather volumes and flow rates.

8.6.2 IDEAS Approach

8.6.2.1 Introduction In contrast to the total connectivity model, where a superstructure representation is formulated specifically for the determination of the AR, the IDEAS framework (Burri et al., 2002) caters to the solution of more generalized reactor network synthesis problems. Candidate AR construction is one of many outputs that the IDEAS framework is capable of performing.

The IDEAS approach is a reactor superstructure method that represents all reactor networks as the combination of two generalized blocks. When the system is viewed in this manner, the resulting equations describing the problem can be made *linear*. An advantage of this is that traditionally nonlinear reactor network problems may then be solved via an LP technique, such as that described by the LP formulations in Section 8.6.1. And as a result, the solution to the linear system is guaranteed to be globally optimal.

A key difference between the IDEAS framework and the LP formulations in Section 8.6.1 is that the former allows for both CSTRs and PFRs in the superstructure model, whereas the latter considers only CSTRs. DSRs are not considered in the IDEAS framework, although, these may be approximated by using many PFRs in series with sidestream addition.

8.6.2.2 Basic Idea The IDEAS framework describes a generalized reactor superstructure that may be used for addressing many reactor network synthesis problems. Construction of a candidate AR occurs by successive solution of a number of LP subproblems for different objective functions. The solution corresponding to each LP problem results in a different point on the AR boundary. Therefore, the computation of candidate ARs is resolved in a point-wise manner. The accuracy of the construction is determined by the number of unknown variables used in each LP problem, whereas the number of points computed for the AR boundary depends on the number of individual LPs solved.

The IDEAS framework seeks to represent a generalized reactor network as the combination of two blocks (Burri et al., 2002):

1. **A reactor operator (ROP) block**: The ROP block manages all operations related to reaction occurring in the system, as well as the particular reactors associated with these reactions. The ROP uses inlet stream information—in the form of flow rates, concentration, enthalpy (temperature), and residence time—and transforms it to outlet stream flow rates—in the form of concentrations, temperatures, and so on—associated with the reactor model.
2. **A distribution network (DN) block**: The DN block coordinates the distribution of inlet and effluent streams throughout the network via mixing, splitting, recycling, and bypassing operations. The purpose of the DN blocks acts as "switchboard" for all streams in the system. This leads to a system of linear constraints when the concentrations in the ROP block are specified beforehand.

Figure 8.32 provides a schematic representation of the ROP and DN blocks, applied to the IDEAS framework.

A large amount of generalization may be incorporated into the system via the IDEAS framework. In general, the following four characteristic connections exist:

1. Exist streams that connect the inlet and outlet of the DN block (Z_{yu});
2. The DN inlet to the ROP inlet (Z_{wu});
3. The ROP outlet to the DN outlet (Z_{yx});
4. The ROP outlet to the ROP inlet (Z_{wx}).

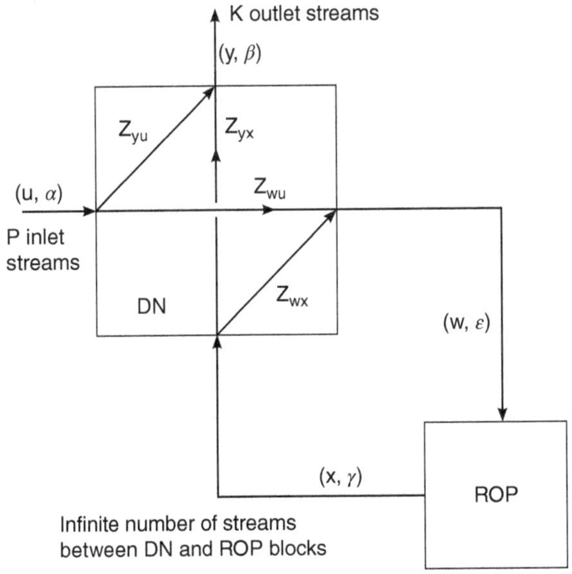

Figure 8.32 ROP and DN blocks for use in the IDEAS framework. Burri et al. (2002). Reproduced with permission of Elsevier.

Mass balance relations may then be derived for all the streams shown in Figure 8.32, connecting all intensive properties of the system to one another. Non-negativity constraints are also imposed, resulting in the following system of equations:

$$u(k) = \sum_{i=1}^{K} Z_{yu}(i,k) + \sum_{i=1}^{\infty} Z_{wu}(i,k) \quad k = 1, \dots, P$$

$$y(i) = \sum_{k=1}^{P} Z_{yu}(i,k) + \sum_{k=1}^{\infty} Z_{yx}(i,k) \quad i = 1, \dots, K$$

$$w(i) = \sum_{k=1}^{P} Z_{wu}(i,k) + \sum_{k=1}^{\infty} Z_{wx}(i,k) \quad i = 1, \dots, \infty$$

$$x(k) = \sum_{i=1}^{K} Z_{yx}(i,k) + \sum_{k=1}^{\infty} Z_{wx}(i,k) \quad k = 1, \dots, \infty$$

This system of equations describes the distribution of streams between the DN and ROP blocks. Variables $u(k)$, $y(i)$, $w(i)$, and $x(k)$ indicate flow rates of the streams present in the system. $u(k)$ and $y(i)$ represent inlet and outlet streams of the DN block, respectively, whereas $w(i)$ and $x(k)$ represent inlet and outlet streams of the ROP block, respectively.

It is assumed that there are P inlet streams into the DN block and K streams exiting the DN block, and that there are an *infinite* number of streams available that connect the DN and ROP blocks together. These streams are given by $w(i)$ and $x(k)$ from above, and are also represented mathematically by the infinite sums in the equations for $u(k)$, $y(i)$, $w(i)$, and $x(k)$. Species molar flow rates exiting the DN block may then be written giving

$$\delta(i)w(i) = \sum_{k=1}^{P} \alpha(k) Z_{wu}(i,k) + \sum_{k=1}^{\infty} \gamma(k) Z_{wx}(i,k) \; i = 1, \dots, \infty$$

$$\beta(i)y(i) = \sum_{k=1}^{P} \alpha(k) Z_{yu}(i,k) + \sum_{k=1}^{\infty} \gamma(k) Z_{yx}(i,k) \quad k = 1, \dots, K$$

$$y_{min}(i) \leq y(i) \leq y_{max}(i) \qquad i = 1, \dots, K$$

$$\beta_{min}(i)y(i) \leq \sum_{k=1}^{P} \alpha(k) Z_{yu}(i,k)$$

$$+ \sum_{k=1}^{\infty} \gamma(k) Z_{yx}(i,k) \leq \beta_{max}(i)y(i) \qquad k = 1, \dots, K$$

$$x(i) = w(i) \qquad i = 1, \dots, \infty$$

$$u, y, w, x, Z_{yu}, Z_{wu}, Z_{yx}, Z_{wx} \geq 0$$

where δ represents the mixed stream leaving the DN feeding the ROP block and β represents the mixed stream exiting the system. α and γ represent the inlet concentrations into the DN block from the process feed and ROP block, respectively.

These equations only relate connections between the ROP and DN blocks. The particular reactor model employed in the ROP block must also be specified. PFR and CSTR equations are used in the IDEAS formulation, but not DSRs. The IDEAS framework utilizes a "technology flag"—specified by the variable ε in Figure 8.32—that is the inlet stream to the ROP block to denote whether a CSTR or PFR is used (Burri et al., 2002). The ε variable also contains values for the residence time of the associated reactor.

SIDE NOTE: An alternative view of the IDEAS framework

An alternate formulation of the IDEAS framework, which gives the same results, exists that we do not discuss here. Details of the alternate formulation can be found in the paper by Burri et al. (2002).

Next, the associated objective function, J, is given by

$$J = \sum_{i=1}^{\infty} \left[\begin{array}{c} \upsilon(i)\,\mathrm{u}(i) + \psi(i)\mathrm{y}(i) + \omega(i)\mathrm{w}(i) + \chi(i)\mathrm{x}(i) \\ + \sum\limits_{\mathrm{l=u,x}} \sum\limits_{\mathrm{m=y,w}} \sum\limits_{k=1}^{\infty} \xi_{\mathrm{ml}}(i,k)\mathrm{z}_{\mathrm{ml}}(i,k) \end{array} \right] \tag{8.14}$$

where υ, ψ, ω, χ, and ξ are cost coefficients used to form different objective functions. Again, the IDEAS framework is a generalized reactor superstructure framework. Despite the complexity, Equation 8.14 simply attempts to describe a generalized objective function for the optimization of a number of different objective functions. For the purposes of AR construction, Equation 8.14 is simplified.

The generalized nature of the IDEAS equations allows for a wide range of different optimization criteria to be investigated. The construction of candidate ARs is thus a specialized subproblem of the objective function formulation provided before. For example, if it is desired that the AR for a two component system A and B be found, then one might start by specifying the outlet concentration of component A and then solving the LP problem that maximizes the corresponding concentration in component B. Graphically, this approach is equivalent to starting from the feed point in $\mathbb{R}^2$, selecting a concentration c_A, and then solving for the corresponding maximum concentration c_B that satisfies the IDEAS constraints. Upon solution of the problem, a concentration on the AR boundary is achieved. This procedure may then be repeated for a different c_A value until the entire boundary is traced out.

8.6.2.3 ***Finite Dimensional Approximations*** The complexity of the IDEAS formulation resides in the number of possible junctions (the Z_{ij}'s) that may interact with other junctions throughout the network. In theory, infinitely many possible outputs may be formulated from the network, leading to an infinitely large number of potential variables to be solved.

This implies that an infinite-size LP must be solved, which represents the true system. In practice, the IDEAS framework must be approximated. Finite approximation of the IDEAS equations are as follows:

$$\nu_{\mathrm{L}} = \inf\left(-\mathrm{c}_2^{\alpha}(1)\,\mathrm{Z}_{\mathrm{yu}} - \sum \mathrm{c}_2^{\gamma}(k)\mathrm{Z}_{\mathrm{yx}}(1,k)\right)$$

$$\mathrm{u}(1) = \sum_{i=1}^{\mathrm{K}} \mathrm{Z}_{\mathrm{yu}}(1,1) + \sum_{i=1}^{\mathrm{L}} \mathrm{Z}_{\mathrm{wu}}(i,1)$$

$$\mathrm{y}(1) = \sum_{i=1}^{\mathrm{P}} \mathrm{Z}_{\mathrm{yu}}(1,1) + \sum_{k=1}^{\mathrm{L}} \mathrm{Z}_{\mathrm{yx}}(1,k)$$

$$\mathrm{w}(i) = \mathrm{Z}_{\mathrm{wu}}(i,1) + \sum_{k=1}^{\mathrm{L}} \mathrm{Z}_{\mathrm{wx}}(1,k) \quad k = 1, \dots, \mathrm{L}$$

$$\mathrm{x}(k) = \mathrm{Z}_{\mathrm{yx}}(1,k) + \sum_{k=1}^{\mathrm{L}} \mathrm{Z}_{\mathrm{wx}}(1,k) \quad k = 1, \dots, \mathrm{L}$$

$$\delta(i)\mathrm{w}(i) = \alpha(1)\mathrm{Z}_{\mathrm{wu}}(i,1) + \sum_{k=1}^{\mathrm{L}} \gamma(k)\mathrm{Z}_{\mathrm{wx}}(i,k) \quad i = 1, \dots, \mathrm{L}$$

$$\mathrm{y}_{\min}(1) \le \mathrm{y}(1) \le \mathrm{y}_{\max}(1)$$

$$[\mathrm{C}_1^{\beta}(1)]_{\min}\mathrm{y}(1) \le \mathrm{C}_1^{\alpha}(1)\mathrm{Z}_{\mathrm{yu}}(1,1) + \sum_{k=1}^{\mathrm{L}} \mathrm{C}_1^{\gamma}(k)\mathrm{Z}_{\mathrm{wx}}(i,k) \le [\mathrm{C}_1^{\beta}(1)]_{\max} y(1)$$

$$\mathrm{x}(i) = \mathrm{w}(i) \quad i = 1, \dots, \mathrm{L}$$

$$\mathrm{u}(1), \mathrm{y}(1), \mathrm{w}(i), \mathrm{x}(i), \mathrm{Z}_{\mathrm{yu}}(1,1), \mathrm{Z}_{\mathrm{wu}}(1,1), \mathrm{Z}_{\mathrm{yx}}(1,i), \mathrm{Z}_{\mathrm{wx}}(i,k) \ge 0 \quad i = 1, \dots, \mathrm{L}$$

$$\mathrm{C}_1^{\beta}(1) = [\mathrm{C}_1^{\beta}(1)]_{\min} = [\mathrm{C}_1^{\beta}(1)]_{\max}$$

Here, L is a large integer used to approximate infinity. Burri et al. (2002) prove that if $\mathrm{L} \to \infty$, the infinite dimensional formulation of the IDEAS problem is obtained. Variable L may therefore be used as a parameter to adjust the accuracy of the constructions. However, this also results in a larger LP that must be solved.

ILLUSTRATION: IDEAS results for two-dimensional Van de Vusse kinetics

Figure 8.33 provides an example of construction resulting from the IDEAS formulation for the well-known two-dimensional Van de Vusse reaction.

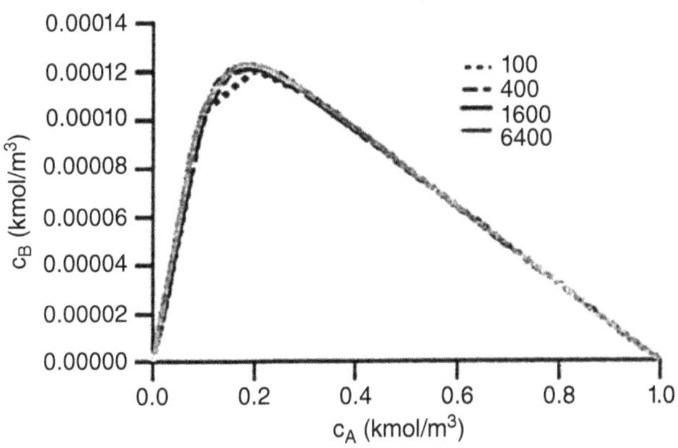

Figure 8.33 Two-dimensional AR for Van de Vusse kinetics, constructed using the IDEAS framework, as a function of grid size. Burri et al. (2002). Reproduced with permission of Elsevier.

The AR is constructed in this instance by arranging the IDEAS solution as an iterative LP problem involving maximization of component B, for varying concentrations of component A. Burri et al. (2002) solved four separate finite LP problems, each associated with a different value for the infinite approximation parameter L. These are given by L = 100, L = 400, L = 1600, and L = 6400. Further information regarding the specific details of the solution may be found in the actual paper by Burri et al. (2002).

Observe that as the value of L is increases, the resulting construction approaches the true shape of the AR boundary. The IDEAS framework is able to approximate curvature on the AR boundary given a large enough value for L, although the actual problem that must be solved is considerably larger in size due to the increased number of potential cross-flow junctions that must be considered in the optimization.

8.7 CHAPTER REVIEW

Due to the geometric nature of AR construction, computing candidate ARs on a computer can be challenging, particularly for higher dimensions. This challenge stems primarily from the following three factors:

1. It is often not easy to visualize higher dimensional data in an intuitive manner.
2. There is a large data requirement for higher dimensional constructions, and computing higher dimensional convex polytopes places a large computational burden on the designer—there are often too many points to consider in higher dimensions for AR computation to be feasible without using some manner of automation.
3. AR theory currently does not have a sufficiency condition for when the true AR has been found.

We must therefore resort to automated (numerical) AR construction techniques to carry out our investigations. Yet, automated AR construction itself is not perfect, and all construction methods described in this chapter possess both strengths and weaknesses. Furthermore, all construction methods generally cannot decode the associated reactor structure at present.

The theoretical nature of the AR boundary has been well understood (Feinberg, 2000a, 2000b; Feinberg and Hildebrandt, 1997). Yet, these concepts are not easily computed numerically, nor can they be easily applied in a practical design environment. There exists a clear need to continue developing better (i.e., faster, more accurate, and more robust) AR construction schemes.

Automated AR construction has witnessed meaningful growth over the past two decades, although a *bulletproof* AR construction method—suitable for all design scenarios and complexities—may still be a long way away, and which may require significant advances both in fundamental theory and computational power before it is available.

With such a method, greater insights into practical and theoretical aspects of AR theory could be undertaken, such as the search for a numerical sufficiency condition (obtained from a numerical calculation) as opposed to a theoretical sufficiency condition (obtained from an analytic proof). The development of a hybrid AR construction method—a method that utilizes both inside-out and outside-in steps to construct the AR—might allow for both more accurate AR constructions and also help in handling degenerate kinetics, which would find use as a validation tool in AR theory. Access to a reliable AR construction technique may also open up easier teaching opportunities for chemical reactor design and analysis, as a performance targeting tool.

The steady advance of computational power will allow for more complicated systems to be studied in the future. Existing AR construction methods may need to be adapted to fully utilize new computing architectures, and so there is also a practical need to continue adapting methods for various computing environments.

REFERENCES

Abraham, T.K., 2005. Kinetic Bounds on Attainability in the Reactor Synthesis Problem. Ohio State University, Ohio.

Abraham, T.K., Feinberg, M., 2004. Kinetic bounds on attainability in the reactor synthesis problem. Ind. Eng. Chem. Res. 43, 449–457.

Bellman, R., 2010. Dynamic Programming, Princeton Landmarks in Mathematics. Princeton University Press, Princeton.

Burri, J.F., Wilson, S.D., Manousiouthakis, V.I., 2002. Infinite DimEnsionAl State-space approach to reactor network synthesis: Application to attainable region construction. Comput. Chem. Eng. 26, 849–862.

Davis, B.J., Taylor, L.A., Manousiouthakis, V.I., 2008. Identification of the attainable region for batch reactor networks. Ind. Eng. Chem. Res. 47, 3388–3400.

Feinberg, M., 1987. Chemical reaction network structure and the stability of complex isothermal reactors—I. The deficiency zero and deficiency one theorems. Chem. Eng. Sci. 42, 2229–2268.

Feinberg, M., 2000a. Optimal reactor design from a geometric viewpoint—III. Critical CFSTRs. Chem. Eng. Sci. 55, 3553–3565.

Feinberg, M., 2000b. Optimal reactor design from a geometric viewpoint. Part II. Critical sidestream reactors. Chem. Eng. Sci. 55, 2455–2479.

Feinberg, M., Hildebrandt, D., 1997. Optimal reactor design from a geometric viewpoint—I. Universal properties of the attainable region. Chem. Eng. Sci. 52, 1637–1665.

Kauchali, S., Rooney, W.C., Biegler, L.T., Glasser, D., Hildebrandt, D., 2002. Linear programming formulations for attainable region analysis. Chem. Eng. Sci. 57, 2015–2028.

Kokossis, A.C., Floudas, C.A., 1990. Optimization of complex reactor networks—I. Isothermal operation. Chem. Eng. Sci. 45, 595–614.

Manousiouthakis, V.I., Justanieah, A.M., Taylor, L.A., 2004. The shrink-wrap algorithm for the construction of the attainable region: An application of the IDEAS framework. Comput. Chem. Eng. 28, 1563–1575.

McGregor, C., 1998. Choosing the Optimal System Structure Using the Attainable Region Approach for Systems Involving Reaction and Separation. University of the Witwatersrand, Johannesburg.

Ming, D., 2014. A parallel attainable region construction method suitable for implementation on a graphics processing unit (GPU). Comput. Chem. Eng. 67, 103–120.

Ming, D., Hildebrandt, D., Glasser, D., 2010. A revised method of attainable region construction utilizing rotated bounding hyperplanes. Ind. Eng. Chem. Res. 49, 10549–10557.

Posada, A., Manousiouthakis, V., 2008. Multi-feed attainable region construction using the shrink-wrap algorithm. Chem. Eng. Sci. 63, 5571–5592.

Rooney, W.C., Biegler, L.T., 2000. Multiperiod reactor network synthesis. Comput. Chem. Eng. 24, 2055–2068.

Rooney, W.C., Hausberger, B.P., Biegler, L.T., Glasser, D., 2000. Convex attainable region projections for reactor network synthesis. Comput. Chem. Eng. 24, 225–229.

Seodigeng, T., Hausberger, B., Hildebrandt, D., Glasser, D., 2009. Recursive constant control policy algorithm for attainable regions analysis. Comput. Chem. Eng. 33, 309–320.

Zhou, W., Manousiouthakis, V.I., 2006. Non-ideal reactor network synthesis through IDEAS: Attainable region construction. Chem. Eng. Sci. 61, 6936–6945.

Zhou, W., Manousiouthakis, V.I., 2007. Variable density fluid reactor network synthesis—construction of the attainable region through the IDEAS approach. Chem. Eng. J. 129, 91–103.

Zhou, W., Manousiouthakis, V.I., 2009. Corrigendum to "On dimensionality of attainable region construction for isothermal reactor networks" [Comput. Chem. Eng. 32 (3) (2008) 439–450] (DOI:10.1016/j.compchemeng.2007.02.013). Comput. Chem. Eng. 33, 927.

9

ATTAINABLE REGIONS FOR VARIABLE DENSITY SYSTEMS

9.1 INTRODUCTION

All of the systems and associated AR constructions provided thus far have been computed, primarily, in concentration space. Understanding concepts related to concentration is more intuitive, we believe. Thus, it is simpler to base discussions in concentration space, as opposed to other process variables that also obey linear mixing laws. The AR has also been developed historically with isothermal constant density systems in mind, although this constraint is relaxed in Chapter 7. Yet, concepts such as those from Chapters 6 to 8 demonstrate that even when these constraints are relaxed, the resulting theory might still be quite complex.

Many systems of interest typically occur under nonconstant density conditions, and heterogeneous reactions are possibly more common in industry than constant density systems. It is hence beneficial to understand how such systems might be improved via AR analysis.

In this chapter, we wish to demonstrate how the existing set of AR theory, developed in previous chapters, may be adapted to include variable density systems as well. Certainly, many of the founding ideas, involving reaction concentration and mixing, may be adapted for use with mass fractions instead. The use of mass fractions is an important concept in nonconstant density systems, and thus an adequate understanding of the idea is required.

We will begin by discussing a number of important formulae for converting common process variables involving moles to equivalent quantities involving mass fraction. These concepts are not difficult to understand, however, they are fundamental to how the computation of ARs in mass fraction space must be organized. Discussion of how the stoichiometric subspace may be computed and how residence time may be incorporated in mass fraction space is also provided. From this, a number of examples are provided that demonstrate the theory. In particular, isothermal and nonisothermal unbounded gas phase systems shall be investigated.

In the following discussions, it is assumed that an appropriate equation of state is available. This is necessary for describing how the system temperature, pressure, and volume are related to each other. It is also assumed that molar masses for all components participating in the system are available, which will be used to convert between species mass and moles.

SIDE NOTE: Analogy between Q in concentration space and G in mass fraction space

In the following sections, a number of conversions for well-known molar quantities will be given. The conversions often depend on the component mass fractions z_i, and by association, the total mass flow rate G. Although these conversions are not difficult to understand, it is usually easy to determine the analogous mass fraction version of the quantity by substituting the total volumetric flow rate Q_{tot} for G_{tot} instead. The resulting expressions are usually similar to their concentration equivalents when this substitution is made.

9.2 COMMON CONVERSIONS TO MASS FRACTION SPACE

9.2.1 Preliminary Notation

9.2.1.1 Overview In the following sections, a number of common conversions will be discussed. It is essential to keep in mind that the end goal with all of these conversions is to relate the process variable of interest to an expression involving the mass fraction of the system and, frequently, the system temperature T and total pressure P. Such a system might look like something given in Figure 9.1.

Attainable Region Theory: An Introduction to Choosing an Optimal Reactor, First Edition. David Ming, David Glasser, Diane Hildebrandt, Benjamin Glasser, and Matthew Metzger.

Companion website: http://attainableregions.com

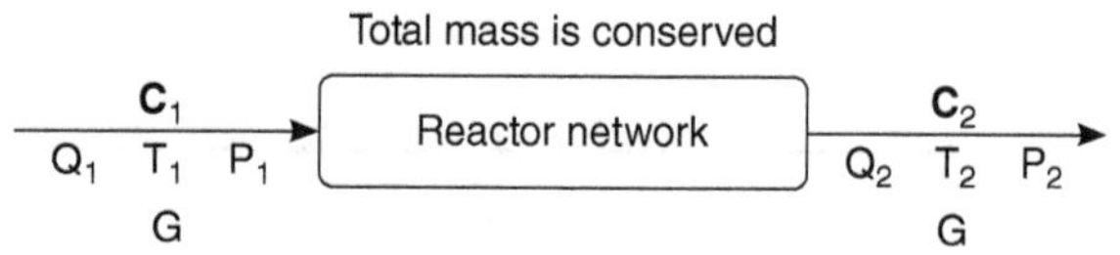

Figure 9.1 Reactor network involving a variable density system.

A stream containing n components is fed into a reactor network at a total volumetric flow rate of Q_1. The effluent stream leaving the network is at a total volumetric flow rate Q_2. Since the system is no longer assumed to obey the constant density assumption, $Q_1 \neq Q_2$. We use the symbol G to signify the *total* mass flow rate of a stream, whereas g_i signifies the component mass flow rate of i within a mixture. The total mass flow rate is then found to be the sum of all component mass flows in the stream:

$$G = \sum_{i=1}^{n} g_i$$

Since mass is always conserved, the total mass flow of material in stream 1 in Figure 9.1 is equal to the total mass flow in stream 2, or $G_1 = G_2 = G$.

9.2.1.2 Mass Fraction and the Mass Fraction Vector The mass fraction of component i, z_i, in a stream with a total mass flow G is defined as

$$z_i = \frac{g_i}{G} \tag{9.1a}$$

z_i may be similarly defined for batch systems in terms of the mass of component i, m_i, and the total mass of the system, m_{tot}:

$$z_i = \frac{m_i}{m_{tot}} \tag{9.1b}$$

If n components are present in the system, then n mass fractions exist. Only $n-1$ mass fractions are independent, since all mass fractions in the system must sum to unity

$$\sum_{i=1}^{n} z_i = 1$$

The *mass fraction vector*, **z**, is then defined in a similar manner to the concentration vector **C** as follows:

$$\mathbf{z} = [z_1, z_2, \ldots, z_n]^T$$

That is, for any mixture containing n species, we may associate the mixture with a mass fraction vector **z**. From a geometric viewpoint, **z** is the vector in $\mathbb{R}^n$ mass fraction space associated with a unique magnitude and direction.

9.2.1.3 Molecular and Average Molecular Weight The molecular weight of component i shall be denoted by W_i. It follows that the molar flow rate of component i, n_i, in the system may be converted into an equivalent mass using the molecular weight belonging to component i

$$n_i = \frac{g_i}{W_i}$$

This may also be expressed in vector form by

$$\mathbf{n} = \mathbf{W}^{-1}\mathbf{g}$$

where $\mathbf{n} = [n_1, n_2, \ldots, n_n]^T$, $\mathbf{g} = [g_1, g_2, \ldots, g_n]^T$, and **W** is an $n \times n$ matrix containing the species molecular weights W_i contained on the main diagonal:

$$\mathbf{W} = \begin{bmatrix} W_1 & 0 & \cdots & 0 \\ 0 & W_2 & \cdots & 0 \\ \vdots & \vdots & \ddots & \vdots \\ 0 & 0 & \cdots & W_n \end{bmatrix} \tag{9.2}$$

W is called the *molecular weight matrix*. We denote the *average molecular weight* of a mixture or a process stream by W_{avg}. An expression for the average molecular weight of a mixture is easily given in terms of the species mole fractions x_i as follows:

$$W_{avg} = x_1 W_1 + x_2 W_2 + \cdots + x_n W_n$$

It is easily shown that W_{avg} may also be expressed in terms of the component mass fractions:

$$W_{avg}(\mathbf{z}) = \frac{1}{z_1/W_1 + z_2/W_2 + \cdots + z_n/W_n} \tag{9.3}$$

$W_{avg}(\mathbf{z})$ is written to show explicitly how the average molecular weight of a mixture is a function of all species mass fractions participating in the system.

EXAMPLE 1: Deriving the Average Molecular Weight of a Mixture from Component Mass Fractions

The previous discussion stated that the average molecular weight of a mixture W_{avg} may be expressed in terms of both component mole fractions x_i and component mass fractions z_i. Since mass fractions are used extensively in this chapter, it is useful to understand how W_{avg} may be expressed in terms of z_i. Hence from the mole fraction definition of W_{avg}, show that $W_{avg}(\mathbf{z}) = (z_1/W_1 + z_2/W_2 + \cdots + z_n/W_n)^{-1}$

Answer:

From the definition of average molecular weight in terms of x_i,

$$W_{avg} = x_1W_1 + x_2W_2 + \cdots + x_nW_n$$

Mole fractions may be expressed in terms of component molar quantities n_i:

$$W_{avg} = \frac{n_1W_1 + n_2W_2 + \cdots + n_nW_n}{n_1 + n_2 + \cdots + n_n}$$

n_i can be expressed in terms of the component mass m_i and component molecular weight W_i by $n_i = g_i/W_i$. Additionally, m_i may be expressed in terms of the total mass flow rate G and component mass fraction z_i by $g_i = Gz_i$. Substitution into the above expression gives

$$W_{avg} = \frac{G(z_1 + z_2 + \cdots + z_n)}{G(z_1/W_1 + z_2/W_2 + \cdots + z_n/W_n)}$$

Cancelling G and recognizing the fact that all mass fractions must sum to unity, we obtain

$$W_{avg}(\mathbf{z}) = \frac{1}{z_1/W_1 + z_2/W_2 + \cdots + z_n/W_n} \tag{9.3}$$

SIDE NOTE: Why not use Mole Fractions?

It might not be apparent why mass fractions are used over mol fractions in variable density systems. Mole fractions behave in a manner similar to mass fractions (for example, both sum to unity), and mole fractions are commonly used in rate expressions for gas phase reactions, such as when species partial pressures are used. Although it is possible to construct the AR in mole fraction space (see Zhou and Manousiouthakis, 2007), mass fractions are often easier to handle from a computational perspective. We therefore prefer to use mass fractions when the AR for a nonconstant density system is to be determined.

To briefly motivate why mass fractions are preferred, consider the synthesis of ammonia by the following reaction:

$$N_2 + 3H_2 \rightarrow 2NH_3$$

A reactor feed containing N_2 and H_2 in stoichiometric proportion would result in a total of 4 mol in the feed. If 100% conversion is assumed, then the total number of moles in the product stream would only be 2. The total number of moles is hence halved from the feed to the product in this instance. Certainly, total moles are only conserved under the appropriate reaction stoichiometry.

Defining a species mole fraction x_i in the feed may contain different relative proportions of component i when compared to the product stream. The mole fraction x_i in the feed is hence not directly comparable to x_i in the product stream. If we were to mix product and feed streams, the resulting mixture may not obey a linear mixing law as a consequence.

9.2.2 Conversions Involving Molar Quantities

9.2.2.1 Moles The rate of reaction is commonly expressed in terms of molar concentration, and thus it is important to have a method for expressing molar quantities in terms of component mass fractions. It was previously shown that the molar flow rate of a component i in a mixture may be expressed in terms of z_i, simply in terms of the molecular weight and mass flow rate of i

$$n_i = \frac{g_i}{W_i}$$

The mass flow rate may be written in terms of z_i to give

$$n_i = \frac{Gz_i}{W_i} \tag{9.4}$$

Similar expressions may be obtained for batch systems as well. The total molar flow rate of all components in a mixture is given by

$$n_{tot} = n_1 + n_2 + \cdots + n_n$$

Substitution of Equation 9.4 into this expression gives

$$n_{tot} = G\left(\frac{z_1}{W_1} + \frac{z_2}{W_2} + \cdots + \frac{z_n}{W_n}\right)$$

This may be expressed in terms of $W_{avg}(\mathbf{z})$ to give

$$n_{tot} = \frac{G}{W_{avg}(\mathbf{z})} \tag{9.5}$$

9.2.2.2 Concentration The molar concentration of component i, c_i, is expressed in terms of the molar flow rate, n_i, and the total volumetric flow rate of the mixture Q_{tot}

$$c_i = \frac{n_i}{Q_{tot}}$$

which, from Equation 9.4, gives

$$c_i = \frac{Gz_i}{W_i}\frac{1}{Q_{tot}} \tag{9.6}$$

For molar concentrations, an expression for the volume change in terms of the system process variables (P, T) is required. For gases, an appropriate equation of state of the following form may be available:

$$Q_{tot} = \frac{Zn_{tot}RT}{P} \tag{9.7}$$

Here, Z is the compressibility factor for the mixture; R, T, and P are the universal gas constant, reactor temperature, and pressure, respectively. Substituting Equations 9.5 and 9.7 into Equation 9.6 and simplifying gives

$$c_i = z_i \frac{W_{avg}(\mathbf{z})}{W_i} \frac{P}{ZRT} \tag{9.8}$$

For mixtures occurring in the liquid phase, similar expressions may be formulated with nonlinear mixing rules in terms of the component activity coefficients and fugacities, for these are often expressed in terms of species mole fractions x_i (Sandler, 2006).

EXAMPLE 2: Ideal Gas

Two gas streams, containing ideal gases A, B, and C, are mixed continuously at 5 bar and 450 K. Stream 1 has a composition given by the molar flow rate vector $\mathbf{n}_1 = [n_{1A}, n_{1B}, n_{1C}]^T = [0.5, 5.5, 3.1]^T$ mol/h. Stream 2 has a molar flow rate vector given by $\mathbf{n}_2 = [1.2, 1.0, 0.12]^T$ mol/h. The molecular weights of A, B, and C are 143, 63, and 100 g/mol, respectively. Calculate the mass fraction and concentration vectors for all three streams.

Answer:

Since the molar flow rates are given for streams 1 and 2, the mixture stream (stream 3) is found by mass balance as follows:

$$\begin{aligned} \mathbf{n}_3 &= \mathbf{n}_1 + \mathbf{n}_2 \\ &= [1.7, 6.5, 3.22]^T \text{ mol/h} \end{aligned}$$

Mass fraction vectors may be found by computing the equivalent species mass flow rates using molecular weights and then dividing by the total mass flow rates in each stream. This gives $\mathbf{m}_1 = [71.5, 346.5, 310.0]^T$ g/h, $\mathbf{m}_2 = [171.6, 63.0, 12.0]^T$ g/h, and $\mathbf{m}_3 = [243.1, 409.5, 322.0]^T$ g/h, corresponding to $G_1 = 728.0$ g/h, $G_2 = 246.6$ g/h, and $G_3 = 974.6$ g/h, respectively. The corresponding mass fraction vectors are then as follows:

$$\mathbf{z}_1 = [0.0982, \ 0.4760, \ 0.4258]^T$$
$$\mathbf{z}_2 = [0.6959, \ 0.2555, \ 0.0487]^T$$
$$\mathbf{z}_3 = [0.2494, \ 0.4202, \ 0.3304]^T$$

Computing the equivalent concentration vectors can be found using Equation 9.8 when the compressibility factor is equal to unity (Z = 1).

$$c_i = z_i \frac{W_{avg}(\mathbf{z})}{W_i} \frac{P}{RT}$$

Conversion to concentration using Equation 9.8 requires the average molecular weight of the mixture, which depends on the mass fraction vector $\mathbf{z}_i$. Using Equation 9.3, we may compute the average molecular weight for streams 1, 2, and 3, which are 80.0, 106.29, and 85.34 g/mol, respectively. Substituting these values along with the system temperature and pressure into Equation 9.8 results in the following concentration vectors:

$$\mathbf{C}_1 = [7.34, \ 80.77, \ 45.52]^T \text{mol/m}^3$$
$$\mathbf{C}_2 = [69.12, \ 57.60, \ 6.91]^T \text{mol/m}^3$$
$$\mathbf{C}_3 = [19.89, \ 76.06, \ 37.68]^T \text{mol/m}^3$$

9.2.2.3 Mole Fraction Now that expressions for species molar flow rates have been established, it is simple to express species mole fractions x_i in terms of species mass fractions z_i. The mole fraction x_i of component i in a mixture is given by

$$x_i = \frac{n_i}{n_{tot}}$$

Substitution of Equations 9.4 and 9.5 for n_i and n_{tot} thus gives

$$x_i = z_i \frac{W_{avg}(\mathbf{z})}{W_i} \tag{9.9}$$

The expression for converting mass fractions to mole fractions given in Equation 9.9 is for a single component. In practice, calculations often involve the use of vectors over individual species. From Equation 9.9, it is simple to express the conversion from mass fractions to mole fractions in terms of vectors. Hence, the conversion using $\mathbf{z}$ and $\mathbf{W}$ is

$$\mathbf{x} = \mathbf{W}^{-1}\mathbf{z}W_{avg}(\mathbf{z})$$

whereas the conversion from **x** to **z** is

$$\mathbf{z} = \frac{1}{W_{avg}(\mathbf{z})}\mathbf{W}\mathbf{x}$$

EXAMPLE 3: Expressing Mole Fractions in Terms of z_i

A particular reaction involving components A, B, and C is given as follows:

$$A + 2B \rightarrow 2C$$

The rate of reaction for component A is given by

$$r_A = -k_1 \frac{P_c^2}{1 - KP_A P_B^2}$$

where k_1 and K are rate and equilibrium constants, respectively. P_A, P_B, and P_C are the partial pressures of components A, B, and C, respectively. Develop on expression for the rate of reaction for component A in terms of mass fractions, suitable for AR use. The reactor is assumed to be operated under isobaric conditions.

Answer:

The rate expression is written in terms of species partial pressures P_i. These quantities may be expressed in terms of the species mole fractions and total pressure P as follows:

$$P_i = x_i P$$

From Equation 9.9, we may write x_i in terms of z_i to give

$$P_i = z_i \frac{W_{avg}(\mathbf{z})}{W_i} P$$

Substituting this for P_A, P_B, and P_C then gives

$$r_A = -k_1 \frac{(z_C/W_c)^2 (W_{avg}(z)P)^2}{1 - K(z_A/W_A)(z_B/W_B)^2 (W_{avg}(z)P)^3}$$

This is the expression desired. Note that since the reactor is operated at constant pressure, P is left in the expression as an input parameter. Had the reactor been operated under different conditions (e.g., isothermal operation), an equation of state such as in Equation 9.7 may be employed.

9.2.3 Average Density

Although density does not directly involve the use of moles, an expression for the average density of a mixture ρ_{avg} may be expressed in terms of moles and an equation of state. The density of a mixture may be defined in terms of z_i and the pure component densities ρ_i as follows:

$$\rho_{avg} = \frac{1}{z_1/\rho_1 + z_2/\rho_2 + \cdots + z_n/\rho_n}$$

This is suitable when pure component densities are available, which are a function of the system temperature and pressure. The average density of a mixture may also be expressed in terms of the total mass and volumetric flow rates

$$\rho_{avg} = \frac{G}{Q_{tot}}$$

If a suitable equation of state is available, then Q_{tot} may be expressed in terms of G and $W_{avg}(\mathbf{z})$ as in Section 9.2.2.2. For example, assuming the mixture is in the gas phase, this simplifies to the following:

$$\rho_{avg} = W_{avg}(\mathbf{z})\frac{P}{ZRT} \tag{9.10}$$

This is a useful expression for conversions from total volumetric flow rates to total mass flow rates.

9.2.4 Mixing and Reaction

9.2.4.1 Mixing Computing the AR for a reactive system relies on the fact that mixing follows a linear mixing law. When variable density systems are of interest, it is no longer possible to use concentration, and hence mass fractions must be employed instead. Mixing two streams in mass fraction space always obeys a linear mixing law, even when the density of the system under investigation is not constant. To see this, consider Figure 9.2.

Two chambers of equal volume, each containing gas species A and B, are compressed into a single chamber. As a result, the average density of chamber 2 increases due to the compression. The mixture concentration in chamber 1 differs to that in chamber 2. That is, the concentration vector in chamber 1, $\mathbf{C}_1$ is different from $\mathbf{C}_2$. Since density is longer constant, it is not possible to express the mixture concentration in chamber 2 as a linear combination of concentration vectors $\mathbf{C}_1$ and $\mathbf{C}_2$. Instead, the mass of components A and B are conserved, and hence a mass balance may be performed.

Similar to the discussion in Chapter 2, a component mass balance for A gives

$$m_{tot,A} = m_{1A} + m_{2A}$$

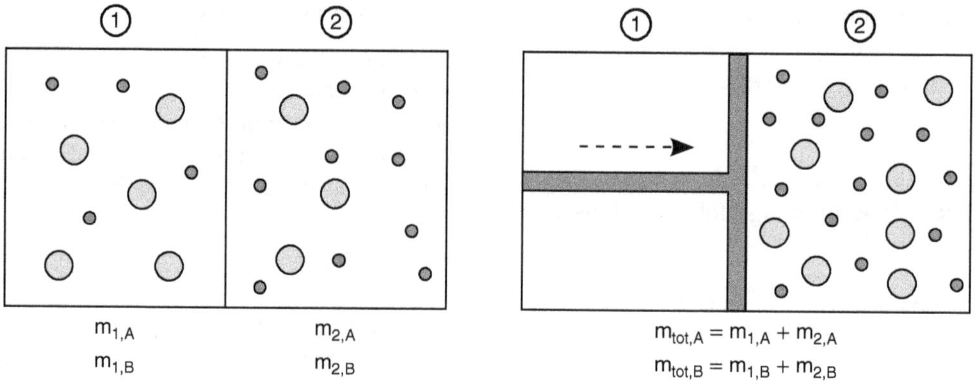

Figure 9.2 Gases compressed into a single chamber.

where subscripts 1 and 2 indicate chambers 1 and 2. Expressing the mass in terms of the total mass and $z_{A,}$

$$z_A^* m_{tot} = z_{1,A} m_{1,tot} + z_{2,A} m_{2,tot}$$

z_A^* represents the mass fraction of A in the combined mixture. Dividing by m_{tot} and making the substitution $\lambda = m_{1,tot}/m_{tot}$, we obtain

$$z_A^* = \lambda z_{1,A} + (1-\lambda) z_{2,A}$$

A similar procedure may be carried out with component B, giving

$$z_B^* = \lambda z_{1,B} + (1-\lambda) z_{2,B}$$

The system of equations may be combined into mass fraction vectors

$$\begin{bmatrix} z_A^* \\ z_B^* \end{bmatrix} = \lambda \begin{bmatrix} z_{1,A} \\ z_{1,B} \end{bmatrix} + (1-\lambda) \begin{bmatrix} z_{2,A} \\ z_{2,B} \end{bmatrix}$$

These can be written more compactly in vector notation as follows:

$$\mathbf{z}^* = \lambda \mathbf{z}_1 + (1-\lambda)\mathbf{z}_2 \tag{9.11}$$

Hence, the geometric interpretation of this is that a mixture resulting from combinations of mixtures $\mathbf{z}_1$ and $\mathbf{z}_2$ produces a mass fraction vector $\mathbf{z}^*$ that lies on a straight line between $\mathbf{z}_1$ and $\mathbf{z}_2$ in $\mathbf{z}_A$–$\mathbf{z}_B$ space. Linear mixing is therefore maintained in mass fraction space.

9.2.4.2 ***Reaction*** An important function in determining the final shape and size of the AR is dependent upon the kinetics of the system under investigation. Certainly, the nature of the rate field for a system will ultimately dictate the recommended optimal reactor structure for a given duty.

It is important to be able to express the rate of reaction for a component $r_i(\mathbf{C})$, as well as the rate vector $\mathbf{r}(\mathbf{C})$, in terms of species mass fractions. The previous sections have demonstrated how common molar quantities may be expressed in terms of mass fractions. It is hence sufficient to state that there exists an equivalent rate field in mass fraction space, which may be computed from the original rate field in concentration space. Given a rate vector involving the concentrations of n species, $\mathbf{r}(c_1, c_2, \ldots, c_n)$, it is possible to express each component concentration in terms of an equivalent mass fraction vector $\mathbf{z} = [z_1, z_2, \ldots, z_n]^T$, so that $\mathbf{r}(\mathbf{z}) = \mathbf{r}(c_1(\mathbf{z}), c_2(\mathbf{z}), \ldots, c_n(\mathbf{z}))$.

Note that the underlying nature of the problem has not changed. The original rate expressions are simply expressed in terms of more appropriate mass fraction quantities. The dimension of $\mathbf{r}(\mathbf{z})$ is equal to the dimension of $\mathbf{r}(\mathbf{C})$, since $\mathbf{r}(\mathbf{z})$ is the same as $\mathbf{r}(\mathbf{C})$, simply expressed in terms of $\mathbf{z}$ instead of $\mathbf{C}$.

EXAMPLE 4: Expressing Reaction Rate in Terms of Mass Fraction

Consider the equation

$$A + B \rightarrow 2C$$

The molar rate of reaction for component A is given by

$$r_A = -k_1 c_A c_B$$

What is the rate equation for component A in terms of mass fractions?

Answer:

Since the rate expression is given in terms of concentration, we can use Equation 9.6 to express r_A in terms of mass fractions. Thus,

$$c_A = \frac{G z_A}{W_A} \frac{1}{Q_{tot}}$$

$$c_B = \frac{G z_B}{W_B} \frac{1}{Q_{tot}}$$

Substituting this into the rate expression gives the following:

$$r_A = -k_1\left(\frac{Gz_A}{W_A}\frac{1}{Q_{tot}}\right)\left(\frac{Gz_B}{W_B}\frac{1}{Q_{tot}}\right)$$

$$= -k_1\left(\frac{z_A z_B}{W_A W_B}\right)\left(\frac{G}{Q_{tot}}\right)^2$$

Had an equation of state been available, we could express Q_{tot} in terms of system temperature and pressure. For example, the following expression for r_A is obtained if the system obeys the modified ideal gas equation incorporating the compressibility factor

$$r_A = -k_1\left(\frac{z_A z_B}{W_A W_B}\right)\left[W_{avg}(\mathbf{z})\frac{P}{ZRT}\right]^2$$

Observe that the units of $\mathbf{r}(\mathbf{z})$ remain the same. The component rate expressions are in [concentration/time], but are simply written in terms of $\mathbf{z}$ instead of $\mathbf{C}$.

9.2.5 Residence Time in Mass Fraction Space

Constructions involving residence time are also possible using mass fractions. To see this, consider the reactor structure (Figure 9.3) involving two parallel reactors R_1 and R_2.

It is assumed that the density of the streams in Figure 9.3 is different. Product streams from R_1 and R_2 (streams 1 and 2, respectively) are combined to form a total product stream (stream 3). The total mass flow to the system in stream 3 is given by $G_{tot} = G_1 + G_2$.

To show that residence time in mass fraction space obeys a linear mixing law, the traditional definition for τ written in terms of volumetric flow rate is no longer suitable. Instead, we define, in an analogous fashion, the equivalent *mass fraction residence time* for a reactor *i* as follows:

$$\sigma = \frac{V_i}{G_i} \tag{9.12}$$

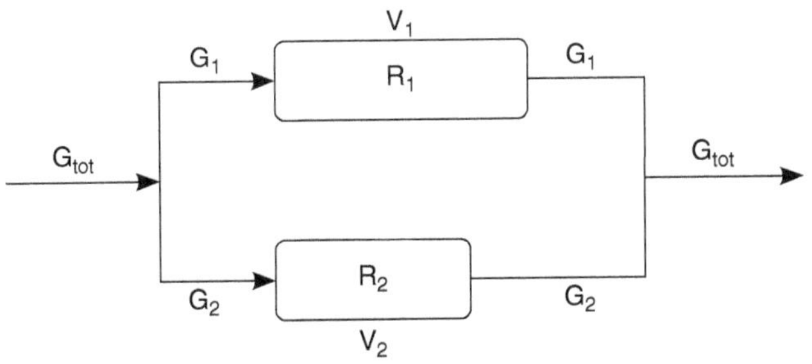

Figure 9.3 Parallel reactors in a variable density system.

Here, V_i and G_i are the reactor volume and total mass flow in the exit (or the feed), respectively. The total reactor volume for the reactor structure in Figure 9.3 is hence computed as the sum of individual reactor volumes,

$$V_{tot} = V_1 + V_2$$

Next, the average mass fraction residence time σ^* is defined as the total reactor volume divided by the total mass flow,

$$\sigma^* = \frac{V_{tot}}{G_{tot}}$$

Expressing all reactor volumes in a similar manner thus gives

$$\sigma^* G_{tot} = \sigma_1 G_1 + \sigma_2 G_2$$

Dividing by G_{tot} and making the substitution $\lambda = G_1/G_{tot}$ gives

$$\sigma^* = \lambda\sigma_1 + (1-\lambda)\sigma_2 \tag{9.13}$$

Therefore, σ obeys a linear mixing law.

EXAMPLE 5: Converting τ to σ

Develop an expression for the mass fraction residence time σ in terms of standard volumetric residence time τ.

Answer:
Converting between τ and σ is achieved in a straightforward manner when the mixture density is known. Hence, the two residence time expressions are

$$\tau_i = \frac{V_i}{Q_i}$$

and

$$\sigma_i = \frac{V_i}{G_i}$$

Combining both expressions and eliminating V_i thus gives

$$\tau_i Q_i = \sigma_i G_i$$

Expressing σ_i in terms of τ_i and identifying that $\rho_i = G_i/Q_i$, the following result is obtained:

$$\sigma_i = \frac{\tau_i}{\rho_{avg}} \tag{9.14}$$

If an equation of state is available, ρ_{avg} may be expressed in terms of P, T, and Z, as given in Section 9.2.3.

SIDE NOTE: Units of σ

Observe that the units of σ are no longer dimensionally consistent with the units of time. Instead, the units of σ are given by [volume of reactor × time/total mass]. Even though the units of σ are different to τ, the underlying mathematical and geometric behavior of σ is equivalent to residence time. Hence, σ is still a useful measure of the volume of a reactor network, and thus it is an important variable when determining reactor structures with minimum total volume in variable density systems.

9.2.6 Fundamental Reactor Types

Next we consider how the three fundamental reactor types found in AR theory may be expressed in terms of the mass fraction vector $\mathbf{z}$. These expressions are derived directly from the molar versions of the same expression, and converted using the mass fraction formulae given in Sections 9.2.1–9.2.4.

9.2.6.1 PFR and Batch Reactor

a) Continuous The equation describing the behavior of component i undergoing reaction in a PFR is given by

$$\frac{dn_i}{dV} = r_i(\mathbf{C})$$

From Equation 9.4, this becomes

$$\frac{G}{W_i}\frac{dz_i}{dV} = r_i(\mathbf{C})$$

Invoking the definition of mass fraction residence time σ and recognizing that $r_i(\mathbf{C}) = r_i(\mathbf{C}(\mathbf{z}))$ results in

$$\frac{dz_i}{d\sigma} = W_i r_i(\mathbf{z})$$

which is the PFR equation in terms of the mass fraction vector $\mathbf{z}$. For n components occurring in the system, a similar procedure may be formed. The vector version of this expression is thus given by

$$\frac{d\mathbf{z}}{d\sigma} = \mathbf{W}\mathbf{r}(\mathbf{z}) \tag{9.15}$$

b) Batch A similar expression for a batch reactor may be found. The vector form of the batch reactor is

$$\frac{d\mathbf{n}}{dt} = V\mathbf{r}(\mathbf{C})$$

where V is the volume of the batch vessel. Substituting the appropriate conversion expressions gives

$$\frac{d\mathbf{z}}{dt} = \frac{V}{m_{tot}}\mathbf{W}\mathbf{r}(\mathbf{C})$$

where m_{tot} is the total mass of the mixture. m_{tot} and V can also be combined to give

$$\frac{d\mathbf{z}}{dt} = \frac{1}{\rho_{avg}}\mathbf{W}\mathbf{r}(\mathbf{C}) \tag{9.16}$$

where ρ_{avg} is the average density of the mixture, as defined in Section 9.2.3.

9.2.6.2 CSTR

The CSTR expression operating in mass fraction space is derived in a manner similar to the traditional CSTR expression in molar concentration space. A species molar balance around the CSTR operating at steady state gives

$$n_i = n_{i,f} + Vr_i(\mathbf{C})$$

Here, V is the volume of the CSTR. The volumetric flow rate of the feed and effluent streams are no longer equal, and thus mass fractions must be employed instead. Using Equation 9.4, the molar CSTR expression may be written in terms of G, W_i, and z_i to give

$$G\frac{z_i}{W_i} = G\frac{z_{i,f}}{W_i} + Vr_i(\mathbf{C})$$

Rearranging for the species mass fraction z_i in the CSTR product stream results in

$$z_i = z_{i,f} + \frac{V}{G}W_i r_i(\mathbf{C})$$

Similar to that given in the development for the PFR expression, $r_i(\mathbf{z}) = r_i(\mathbf{C}(\mathbf{z}))$ which is still the molar reaction rate now expressed in terms of the mass fraction vector $\mathbf{z}$ instead of the concentration vector $\mathbf{C}$. Also, the mass fraction residence time σ may be substituted in to give

$$z_i = z_{i,f} + \sigma W_i r_i(\mathbf{z})$$

This is the mass fraction equivalent expression for a CSTR, for a single component i. A similar procedure may be carried out on all species components participating in the system, and combined compactly into a single vector expression, resulting in

$$\mathbf{z} = \mathbf{z}_f + \sigma\mathbf{W}\mathbf{r}(\mathbf{z}) \tag{9.17}$$

Observe that the geometric behavior of the mass fraction version of the CSTR expression shares identical traits to the constant density molar CSTR expression. In the case of mass fractions, the vector $(\mathbf{z} - \mathbf{z}_f)$ must be collinear with the vector $\mathbf{W}\mathbf{r}(\mathbf{z})$ for $\mathbf{z}$ to be a feasible CSTR state in mass fraction space (provided that $\sigma \geq 0$). Hence, the same geometric interpretations for CSTRs in mass fraction space apply.

9.2.6.3 DSR A DSR operating in mass fraction space is best derived in a manner similar to that given in Chapter 4. However unlike the PFR and CSTR, it is preferable to begin with an expression in terms of species mass flow rates, g_i, instead of species molar flow rates n_i. Hence, consider a component mass balance over a small differential slice of the DSR, operated at steady state. The sidestream concentration is written in terms of a sidestream mass fraction z_i^0 and an addition rate policy β. The units of β are [(mass of sidestream)/(volume of reactor × time)]. A mass balance over a volume slice dV hence gives

$$\frac{dg_i}{dV} = W_i r_i(\mathbf{C}) + \beta z_i^0$$

Note that the rate of reaction for component i is written with respect to the molar expression out of convenience—rate expressions are often given in terms of moles and not in terms of mass. Next, the species molar flow rate may be written in terms of G and z_i by Equation 9.1a

$$\frac{d[Gz_i]}{dV} = W_i r_i(\mathbf{C}) + \beta z_i^0$$

Since a sidestream is present in the DSR, the total mass flow G is no longer constant. Expanding the left-hand side derivative gives

$$G\frac{dz_i}{dV} + z_i\frac{dG}{dV} = W_i r_i(\mathbf{C}) + \beta z_i^0$$

An expression for the change in G as a function of DSR reactor volume V may be derived giving

$$\frac{dG}{dV} = \beta$$

Substituting this result into the DSR expression gives

$$G\frac{dz_i}{dV} = W_i r_i(\mathbf{C}) + \beta(z_i^0 - z_i)$$

This is the mass fraction equivalent version of the DSR expression given by Equation 4.8. Similar to the discussion in Chapter 4, a coordinate transformation may be performed on the above expression relating the DSR residence time σ in terms of V and G.[1] This gives

$$\frac{dz_i}{d\sigma} = W_i r_i(\mathbf{C}) + \beta(z_i^0 - z_i)$$

In a similar fashion to the PFR and CSTR, a vector version of the DSR expression is available

$$\frac{d\mathbf{z}}{d\sigma} = \mathbf{Wr(z)} + \beta(\mathbf{z}^0 - \mathbf{z}) \qquad (9.18)$$

[1] Recall from Chapter 4, a DSR can be envisioned as a number of fed-batch reactors moving at constant velocity down a conveyor belt.

Note that this is analogous to the DSR expression in concentration space. The variable β used in Equation 9.18 is the mass fraction equivalent of α. Note the geometric interpretation of a DSR operating in mass fraction space is also equivalent—a DSR solution trajectory in mass fraction space is resultant from the linear combination of a reactive component $\mathbf{Wr(C)}$ and a mixing vector $(\mathbf{z}^0 - \mathbf{z})$.

9.2.7 Computing the Stoichiometric Subspace

Determining the stoichiometric subspace of a system of reactions in mass fraction space is carried out in a manner similar to that performed in concentration space. In fact, the stoichiometric subspace represented in mass fraction space is slightly more convenient, for all species mass fractions are necessarily bounded by $0 \leq z_i \leq 1$.

Similar to the procedure carried out in Chapter 8, computation of the stoichiometric subspace **S** begins with the stoichiometric coefficient matrix **A**. The dimension of **S** in mass fraction space is equivalent to that in concentration space, and it is found by computing the rank of **A**. This is determined by the number of independent reactions present in the system. For n components in d reactions, the size of **A** is $(n \times d)$.

To compute the bounds of **S**, the vertices of the polytope described by the non-negativity molar constraints must be found. For each component i in the system, we may express the molar flow rate n_i in terms of an initial feed value and extent of reaction. Expressing each component in this manner and writing in vector notation results in

$$\mathbf{n} = \mathbf{n}_f + \mathbf{A}\boldsymbol{\varepsilon}$$

Here, $\boldsymbol{\varepsilon}$ is the molar extent of reaction, **A** is the stoichiometric coefficient matrix, and $\mathbf{n}_f$ is the feed molar flow rate vector $\mathbf{n}_f = [n_{1f}, n_{2f}, \ldots, n_{nf}]^T$. This can be converted to a system of mass fractions giving

$$\mathbf{z} = \mathbf{z}_f + \frac{\mathbf{WA}}{G}\boldsymbol{\varepsilon} \qquad (9.19)$$

Non-negativity constraints may be imposed on all species molar flow rates to give

$$\mathbf{n} \geq \mathbf{0}$$

$$\mathbf{n}_f + \mathbf{A}\boldsymbol{\varepsilon} \geq \mathbf{0}$$

We may express **n** and $\mathbf{n}_f$ in terms of species mass fractions and the total mass flow rate to give

$$\mathbf{z}_f + \frac{\mathbf{WA}}{G}\boldsymbol{\varepsilon} \geq 0 \qquad (9.20)$$

The set of inequality constraints given by Equation 9.20 describes a feasible region in *molar* extent space in which all species mass fractions must reside. The region described by Equation 9.20 is therefore the stoichiometric subspace **S**. Vertices of this region may be computed in the usual manner, such as via a vertex enumeration algorithm.

Understanding the bounds imposed by the reaction stoichiometry in mass fraction space plays an important role in helping to deploy standard AR construction schemes (in concentration space) for use in variable density systems.

SIDE NOTE: Residence time component in the rate vector

In order to compute ARs in mass fraction residence time space, we must also be able to express σ as a component in the rate vector. From the definition of reaction rate, we know

$$\mathbf{Wr}(\mathbf{C}) = \frac{d\mathbf{z}}{d\sigma}$$

Here, $\mathbf{Wr}(\mathbf{C})$ is the equivalent rate vector in mass fraction space. Assuming that we are working in residence time space, the state vector $\mathbf{z}$ will contain one σ as one of its entries

$$\mathbf{z} = \begin{bmatrix} z_A & z_B & \cdots & \sigma \end{bmatrix}^T$$

Substitution of the state vector into the definition of the rate vector gives

$$\mathbf{Wr}(\mathbf{C}) = \begin{bmatrix} dz_A/d\sigma \\ dz_B/d\sigma \\ \vdots \\ d\sigma/d\sigma \end{bmatrix} = \begin{bmatrix} W_A r_A(\mathbf{z}) \\ W_B r_B(\mathbf{z}) \\ \vdots \\ 1 \end{bmatrix}$$

From this expression, we observe that when constructions involving σ are carried out, the component corresponding to mass fraction residence time is equal to one. That is, $d\sigma/d\sigma = 1$. This is analogous to residence time constructions involving concentration, where the analogous component in the molar rate vector is also one ($d\tau/d\tau = 1$)

EXAMPLE 6: Stoichiometric Subspace for CH_4 Steam Reforming in Mass Fraction Space

In Chapter 8, we showed how the stoichiometric subspace for the methane steam reforming reaction can be computed in concentration space. Since the reaction occurs in the gas phase, it is more appropriate to determine the stoichiometric bounds in mass fraction space. This approach is preferable as the density of the mixture is no longer required to be constant.

Compute the stoichiometric subspace for the CH_4 steam reforming reaction and compare it to the answer obtained in Chapter 8. Assume that a feed molar vector of $\mathbf{n}_f = [1, 1, 1, 0, 0]^T$ kmol/s is available, and that the gas mixture obeys the ideal gas assumption to simplify calculations. Assume a constant pressure and temperature of P = 101 325 Pa and T = 500 K, respectively.

Answer:

The methane steam reforming system involves two independent reactions in five components as follows:

$$CH_4 + H_2O \rightarrow CO + 3H_2$$
$$CO + H_2O \rightarrow CO_2 + H_2$$

The stoichiometric coefficient matrix **A** for this set of reactions is given by

$$\mathbf{A} = \begin{bmatrix} -1 & 0 \\ -1 & -1 \\ 1 & -1 \\ 3 & 1 \\ 0 & 1 \end{bmatrix}$$

The rows of **A** each correspond to a different component: row 1 for CH_4, row 2 for H_2O, row 3 for CO, row 4 for H_2, and row 5 for CO_2. To determine the stoichiometrically feasible region in mass fraction space, Equation 9.20 is used.

$$\mathbf{z}_f + \left(\frac{\mathbf{WA}}{G}\right)\boldsymbol{\varepsilon} \geq \mathbf{0}$$

W is the molecular weight matrix as defined by Equation 9.2:

$$\mathbf{W} = \begin{bmatrix} W_{CH_4} & 0 & 0 & 0 & 0 \\ 0 & W_{H_20} & 0 & 0 & 0 \\ 0 & 0 & W_{CO} & 0 & 0 \\ 0 & 0 & 0 & W_{H_2} & 0 \\ 0 & 0 & 0 & 0 & W_{CO_2} \end{bmatrix} = \begin{bmatrix} 16 & 0 & 0 & 0 & 0 \\ 0 & 18 & 0 & 0 & 0 \\ 0 & 0 & 28 & 0 & 0 \\ 0 & 0 & 0 & 2 & 0 \\ 0 & 0 & 0 & 0 & 44 \end{bmatrix}$$

The feed mass fraction vector $\mathbf{z}_f$ may be computed using **W** and $\mathbf{n}_f$. The feed vector written in terms of mass flow rates is given by

$$\begin{aligned}\mathbf{m}_f &= \mathbf{W}\mathbf{n}_f \\ &= \begin{bmatrix}16 & 18 & 28 & 0 & 0\end{bmatrix}^T\end{aligned}$$

The total mass flow is then determined as the sum of the elements in $\mathbf{m}_f$, $G = 16 + 18 + 28 = 62$ kg/s. The feed mass fraction vector is then

$$\begin{aligned}\mathbf{z}_f &= \mathbf{m}_f/G \\ &= \begin{bmatrix}0.2581 & 0.2903 & 0.4516 & 0 & 0\end{bmatrix}^T\end{aligned}$$

Substituting $\mathbf{z}_f$ together with **W** and **A** into the inequality expression results in the following system of inequality constraints:

$$\begin{bmatrix}0.2581 \\ 0.2903 \\ 0.4516 \\ 0 \\ 0\end{bmatrix} + \frac{1}{62}\begin{bmatrix} 16 & 0 & 0 & 0 & 0 \\ 0 & 18 & 0 & 0 & 0 \\ 0 & 0 & 28 & 0 & 0 \\ 0 & 0 & 0 & 2 & 0 \\ 0 & 0 & 0 & 0 & 44 \end{bmatrix}\begin{bmatrix} -1 & 0 \\ -1 & -1 \\ 1 & -1 \\ 3 & 1 \\ 0 & 1 \end{bmatrix}\begin{bmatrix}\varepsilon_1 \\ \varepsilon_2\end{bmatrix} \geq \begin{bmatrix}0 \\ 0 \\ 0 \\ 0 \\ 0\end{bmatrix} \quad (9.21)$$

This system of inequalities describes a feasible region, in molar extent space, that all species must obey if they are to be compatible with the feed point and reaction stoichiometry. The system of inequalities may then be solved to determine the extreme points $\boldsymbol{\varepsilon} = [\varepsilon_1, \varepsilon_2]^T$ of the polytope described by Equation 9.21 using a standard vertex enumeration algorithm.

Calculating the extreme points results in the following set of **molar** extents defining the stoichiometric subspace: $\{[0.0161, 0.0]^T, [0.0, 0.0]^T, [-0.004, 0.0121]^T, [0.0, 0.0161]^T\}$. This list may then be converted back to mass fraction space using Equation 9.19.

Since rank(**A**) = 2, the stoichiometric subspace in this instance is two-dimensional residing in $\mathbb{R}^5$ mass fraction space. Due to the dimension of the component space, it is not possible to view the entire region in a single plot; however, two-dimensional projections onto different component spaces may be performed. This is shown in Figure 9.4.

Observe that the shape of the regions is identical to those given in Chapter 8 in concentration space. The values of the extreme points are now given in terms of species mass fractions and molar extents of reaction.

To compare the regions obtained in Chapter 8 to those given here, mass fraction vectors belonging to the stoichiometric subspace must be converted to equivalent concentration vectors. Since the components are assumed to behave as ideal gases, the conversion from species mass fractions to species molar fractions is straightforward. Hence, species concentration c_i for an ideal gas mixture is given by

$$c_i = z_i \frac{W_{avg}}{W_i} \frac{P}{RT}$$

or when written in vector form

$$\mathbf{C} = \mathbf{W}^{-1}\mathbf{z}\frac{W_{avg}(\mathbf{z})P}{RT}$$

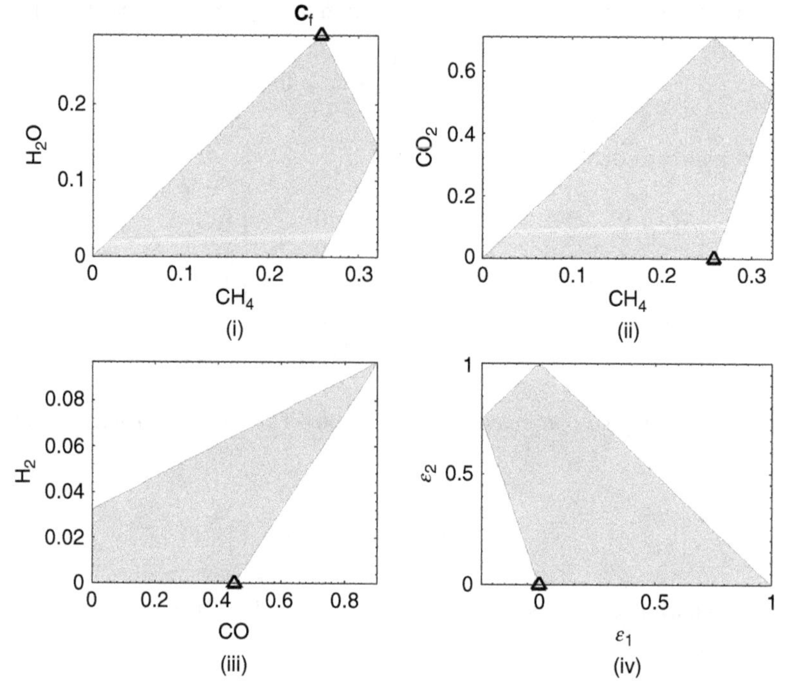

Figure 9.4 Stoichiometric subspace for methane steam reforming for different component pairs in mass fraction space: (i) CH_4–H_2O, (ii) CH_4–CO_2, (iii) CO–H_2, and (iv) ε_1–ε_2.

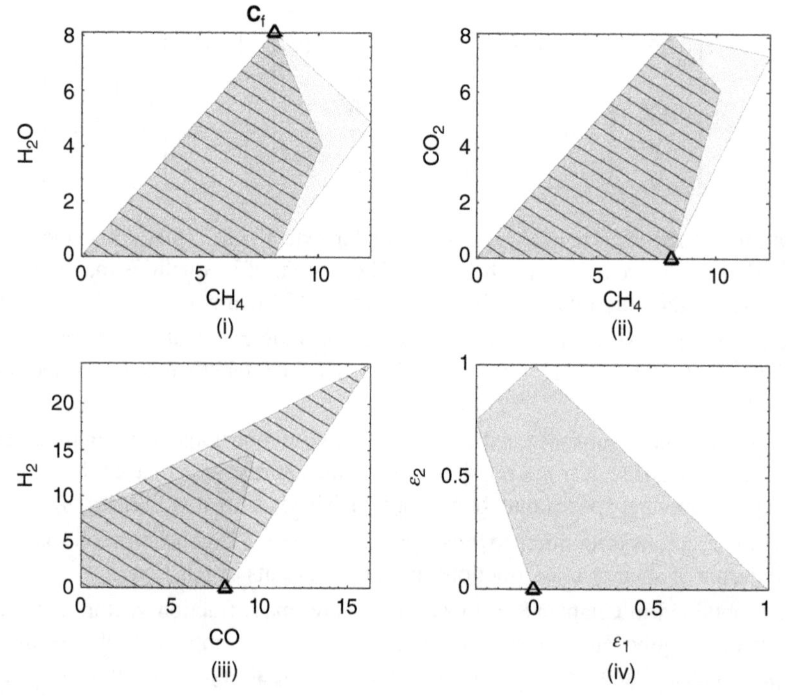

Figure 9.5 Comparison of stoichiometric subspaces in concentration space obtained for the methane steam reforming system for constant density (hatched regions) to the region obtained via mass fractions (clear region). (*See color plate section for the color representation of this figure.*)

where $\mathbf{W}^{-1}$ is the inverse matrix of the molecular weight matrix $\mathbf{W}$. From this, the bounds computed in mass fraction space can be converted to concentrations.

To compute the equivalent stoichiometric subspace in concentration space, the feed molar flow rates must be converted to a feed concentration vector $\mathbf{C}_f$. The methods described in Chapter 8 for constant density systems may then be applied. The results of the constant density system and the region obtained via mass fractions are shown together in Figure 9.5.

The clear region is the stoichiometric subspace obtained by mass fractions, whilst the hatched region is obtained assuming constant density. It is clear that the projections are different for the two methods. Had kinetics been available for the construction of the AR for this system, the resulting feasible set may result in different achievability constraints in the constant density case.

9.3 EXAMPLES

9.3.1 Three-Dimensional Van de Vusse Kinetics

9.3.1.1 ***Introduction*** Throughout the discussions in this book, a common example used to demonstrate the applicability of different concepts used in AR has been the Van de Vusse system. In this example, we wish to return to the Van de Vusse system and briefly investigate what the AR for this reaction scheme might be if the reactions were to occur under a nonconstant density situation. Specifically, it will be assumed that the components are all ideal gases.

9.3.1.2 ***Problem Specification*** For convenience, the Van de Vusse reaction scheme is given in the following:

$$A \rightarrow B \rightarrow C$$

$$2A \rightarrow D$$

The system involves three independent reactions with four components. It follows that the AR is a three-dimensional subspace in $\mathbb{R}^4$. Later on, it will be useful to provide a comparison of the AR generated in this chapter in mass fraction space, to that produced in Chapter 7 originally in concentration space. For this reason, the AR shall be generated in z_A–z_B–z_D space. Components C may be found by mass balance. The mass fraction and rate vectors are then defined as $\mathbf{z} = [z_A, z_B, z_D]^T$ and $\mathbf{r}(\mathbf{z}) = [r_A(\mathbf{z}), r_B(\mathbf{z}), r_D(\mathbf{z})]^T$. It is assumed that the feed available is pure in component A. The feed molar flow rate vector is hence given as $\mathbf{n}_f = [1, 0, 0]^T$. Since the feed is pure in A, it follows that the mass fraction feed vector be given as $\mathbf{z}_f = [1, 0, 0]^T$.

Species rate expressions are assumed to obey mass action kinetics, expressed in terms of concentration as follows:

$$r_A = -k_1c_A - 2k_3c_A^2$$
$$r_B = k_1c_A - k_2c_B$$
$$r_D = k_3c_A^2$$

Using the conversion equations given in Section 9.2, these rate expressions may be transformed into equivalent rate expressions in terms of mass fraction.

9.3.1.3 ***Conversion to Mass Fraction Space*** Since the goal of this example is to produce an AR in mass fraction space, the species rate expressions must be converted to equivalent expressions involving the new mass fraction vector. In Section 9.2.2.2, it is discussed that in order to convert a species concentration, c_i, to a species mass fraction, z_i, a relation is required that relates how the system volume changes with changes in the state vector. This is easily realized for the current problem, since it is assumed that components A through D behave as ideal gases. The ideal gas equation may therefore be used to relate c_i to z_i. From Equation 9.8, using a compressibility factor of $Z = 1$, the species concentration may be expressed in terms of species mass fraction as follows:

$$c_i = z_i\frac{W_{avg}(\mathbf{z})}{W_i}\frac{P}{ZRT}$$

When written in vector form, this gives

$$\mathbf{C} = \mathbf{W}^{-1}\mathbf{z}\frac{W_{avg}(\mathbf{z})P}{RT}$$

Using this equation, the rate expressions given in the previous section may be utilized to describe the kinetics in mass fraction space.

Actual values for the species concentrations are dependent on the particular values of system temperature T and pressure P, and the species molecular weights. For demonstration purposes, it shall be assumed that the system temperature and pressure are held constant at T = 400 K and P = 0.1 atm. For the current example, the molecular weight matrix is specified as

$$\mathbf{W} = \begin{bmatrix} 10 & 0 & 0 & 0 \\ 0 & 10 & 0 & 0 \\ 0 & 0 & 10 & 0 \\ 0 & 0 & 0 & 20 \end{bmatrix}$$

Note that $\mathbf{W}$ is the molecular weight matrix containing all species occurring in the system, and simply the component molecular weights for which the AR is generated in. It follows that rows 1–4 of $\mathbf{W}$ correspond to components A through D accordingly.

From these values, Equations 9.7, 9.17, and 9.18 may be employed to generate PFR, CSTR, and DSR reactor structures with corresponding output states in mass fraction space. The AR is expected to exist as a convex polytope in $\mathbb{R}^3$.

9.3.1.4 Results

a) Stoichiometric Subspace Before starting construction, it may be helpful to understand the absolute bounds for the Van de Vusse system. The stoichiometric subspace provides an indication of the region in which the AR must be contained in for the feed point specified. From Section 9.2.7, the stoichiometric subspace may be computed from the stoichiometric coefficient matrix $\mathbf{A}$ and the feed point $\mathbf{z}_f$. Matrix $\mathbf{A}$ is given for this system as follows:

$$\mathbf{A} = \begin{bmatrix} -1 & 0 & -2 \\ 1 & -1 & 0 \\ 0 & 1 & 0 \\ 0 & 0 & 1 \end{bmatrix}$$

The feasible region is then given by the system of linear inequalities defined by Equation 9.20. In a procedure similar to that given in the steam reforming example, the vertices of the stoichiometric subspace may be computed using a vertex enumeration program. The results of the computation are shown in Figure 9.6.

Observe that the region is represented by a tetrahedron in $\mathbb{R}^3$ mass fraction space. The benefit of generating regions in mass fraction space is that the results are always scaled between 0 and 1.

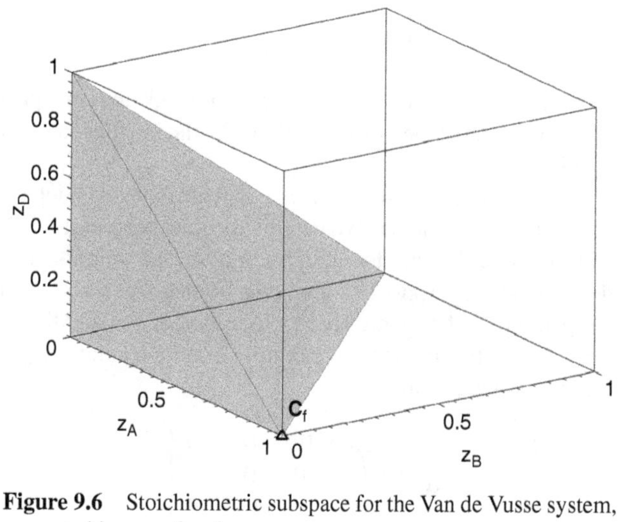

Figure 9.6 Stoichiometric subspace for the Van de Vusse system, generated in mass fraction space.

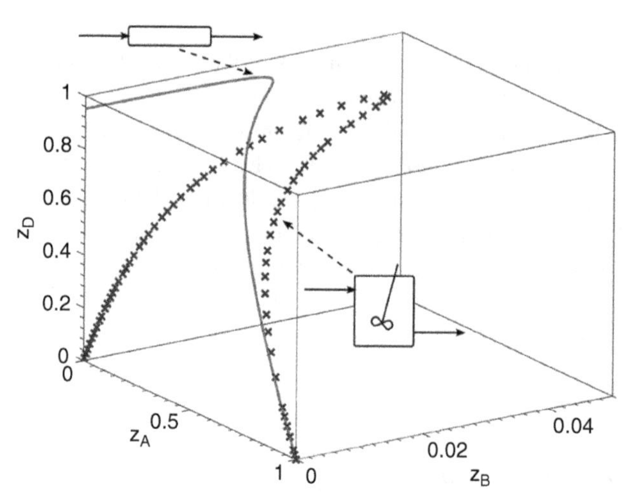

Figure 9.7 PFR trajectory and CSTR locus from the feed point.

b) PFR and CSTR Locus The PFR trajectory and CSTR locus from the feed may be computed from Equations 9.15 and 9.17, respectively. Hence, the PFR solution trajectory is obtained by integrating the differential equation (Eq. 9.15) using $\mathbf{z}_f$ as the initial condition and σ as the integration parameter. The CSTR locus is obtained by successive solution of a system of nonlinear equations (Eq. 9.17). The results of both solutions are displayed in Figure 9.7.

The PFR trajectory follows a similar shape to that given in Chapter 7 in concentration space. The CSTR locus from the feed also exhibits a similar behavior.

Since the system is three-dimensional in nature, the "VdelR" condition may be used to compute the critical α policy for the system. Since the system is no longer expressed in terms of concentrations, however, the specific critical α policy given by Equation 7.3 cannot be used. Rather, the Jacobian matrix expressed in terms of mass fractions $\mathbf{dr}(\mathbf{z})$ must be computed and then used to compute $\varphi(\mathbf{z})$, as in Chapter 7. The computations are somewhat lengthy, and hence this approach will not be adopted here. Instead, the parallel complement automated AR construction method discussed in Chapter 8 shall be employed, providing a quick means to compute the AR for use in comparisons.

c) The AR in Mass Fraction Space The construction results obtained by the parallel complement method in mass fraction space are provided in Figure 9.8(a). It is clear that the shape of the candidate region obtained bears a close resemblance to the shape of the AR obtained for the same system in concentration space. Straight-line sections corresponding to mixing lines can be seen, similar to that found in the equivalent concentration construction.

Linear mixing lines are still visible, even though the system is assumed to occur in the gas phase. Had an objective function been available, the function could be expressed in terms of the state variables and overlaid onto the space to

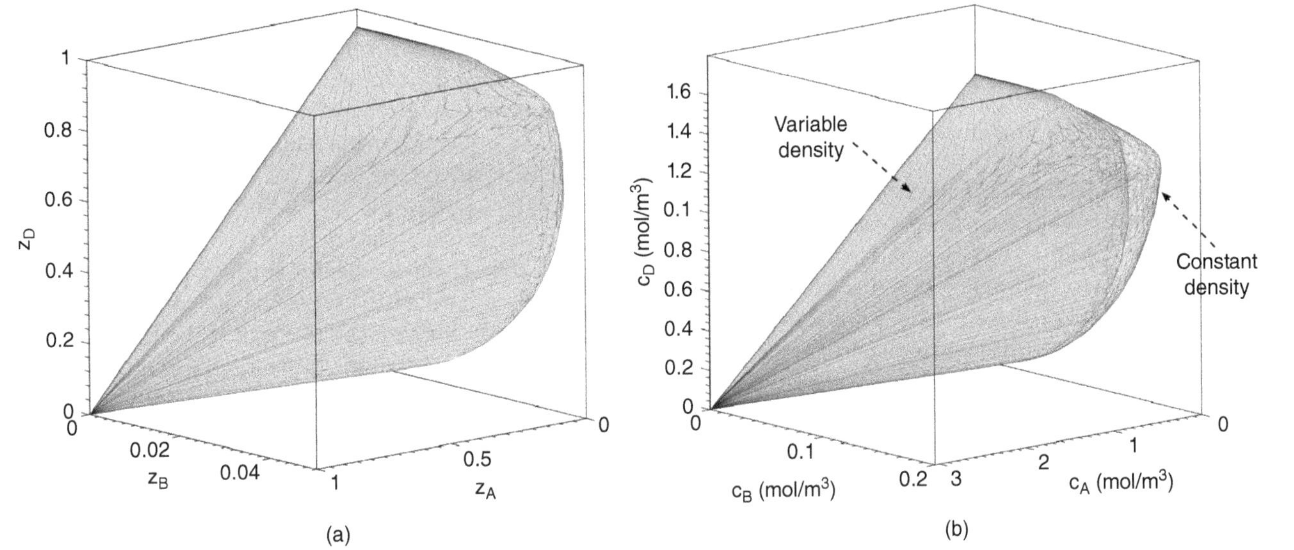

Figure 9.8 (a) Full AR for the Van de Vusse system in mass fraction space. (b) AR for the Van de Vusse kinetics, converted to concentration space. The transparent region is the AR obtained for an identical feed point when constant density is assumed. (*See color plate section for the color representation of this figure.*)

determine its intersection with the AR boundary. Hence, the approach to optimization remains unchanged, even though the system is inherently different in this instance.

d) Comparison to Constant Density System The points generated for the AR from Figure 9.8(a) in mass fraction space may be converted back to concentration space via Equation 9.8. Hence, the AR can be recast in a more familiar space once the AR in mass fraction space has first been identified. At the same time, the AR for the Van de Vusse system might also be generated directly in concentration space, if constant density is assumed. Using the molar feed vector $\mathbf{n}_f$ and the ideal gas equation of state, an equivalent feed concentration may be found from Equation 9.8. Once the feed point is specified, construction of the AR in concentration space follows the procedure outlined in Chapter 7. Comparison can then be made between the mass fraction approach and the constant density approach. This comparison is show in Figure 9.8(b).

The solid region represents the AR obtained via mass fractions, whereas the opaque region is that given by the constant density assumption for the same feed point and kinetics. Although the shapes of the two regions are similar, they are not the same. The region obtained via mass fractions is slightly smaller. It achieves a lower concentration in component B (~0.15 mol/m^3 compared to ~0.20 mol/m^3 for the constant density case), although a slightly larger concentration is achieved in component D (~1.45 mol/m^3 compared to ~1.42 mol/m^3). Had the constant density assumption been used directly, recommendations based on the AR may be different, as the region predicts a slightly different set of bounds for the set of achievable concentrations.

9.3.2 Steam Reforming and Water–gas Shift Reaction

9.3.2.1 ***Introduction*** All of the examples up to this point have approached construction of the AR for various systems under different conditions. In the initial chapters of the book, we investigated lower dimensional systems under constant density, isothermal operation in concentration space. We have slowly relaxed many of these assumptions throughout the course of the book. In this final example, we wish to show how the construction of the AR for a more realistic system might be addressed.

In Section 9.2.7, the stoichiometric subspace for the CH_4 steam reforming reaction was computed. In reality, the system of equations given involves the CH_4 reforming reaction, as well as the water–gas shift reaction. Both of these reactions are important, for instance, in Fischer–Tropsch synthesis reactors (Anderson et al., 1984; Dry, 2002). It follows that it would be useful to understand the limits of achievability for this system.

In this section, the AR for the CH_4 steam reforming and water–gas shift reaction will be investigated. This system involves two independent reactions involving five components. Ordinarily, generation of the AR for this system would involve the construction of a two-dimensional AR. For this example, the interest will also be in understanding the minimum reactor volume achievable. Reactions of all components occur in the gas phase under nonisothermal conditions. The ideal gas equation of state is hence not a suitable one for this system. Instead, the Peng–Robinson equation of state shall be employed for this purpose.

Hence, in this example, the goal is to determine the AR for a nonisothermal, variable density system, involving a non-ideal gas mixture in combined mass fraction space and residence time space.

SIDE NOTE: Model Accuracy

The following discussion is merely shown to demonstrate how one might go about approaching a non-ideal system. The specific kinetics and equations of state used are shown for illustrative purposes alone, and are not indicative of how these reactions are carried out in reality. We ultimately want to demonstrate that even when the system is fairly complex, given realistic data and operating assumptions, the ideas discussed earlier may still be employed to compute the AR.

9.3.2.2 Problem Specification The system of reactions to be investigated is the same as that given in Section 9.2.7. For convenience, the reactions are repeated here as follows:

$$CH_4 + H_2O \rightarrow CO + 3H_2 \tag{9.22a}$$

$$CO + H_2O \rightarrow CO_2 + H_2 \tag{9.22b}$$

The stoichiometric subspace for this system has already been established, both in concentration space (Chapter 6) and mass fraction space (Section 9.2.7). The same numbering convention shall be adopted as used for the stoichiometric subspace calculation. Hence, rows 1–5 correspond to CH_4 to CO_2, respectively.

It is already known that the AR for this system must reside in a two-dimensional subspace of $\mathbb{R}^5$. Specification of any two components allows for computation of all other remaining components in the system.

Synthesis gas, commonly known as syngas, is a gas mixture containing CO and H_2 and is an important intermediate product for use in Fischer–Tropsch synthesis reactors. The AR generated in z_{CO}–z_{H_2} space may be useful as a first step to understanding the operating limits of Fischer–Tropsch reactors.

Since the minimum reactor volume is also to be investigated, the associated AR should include residence time. This increases the dimension of the problem. Hence, the AR to be computed will reside in z_{CO}–z_{H_2}–σ space. The mass fraction and rate vectors are therefore defined as $\mathbf{z} = [z_{CO}, z_{H_2}, \sigma]^T$ and $\mathbf{r}(\mathbf{z}) = [r_{CO}(\mathbf{z}), r_{H_2}(\mathbf{z}), 1]^T$, respectively.

The molecular weight matrix for this system is given by

$$\mathbf{W} = \begin{bmatrix} 16 & 0 & 0 & 0 & 0 \\ 0 & 18 & 0 & 0 & 0 \\ 0 & 0 & 28 & 0 & 0 \\ 0 & 0 & 0 & 2 & 0 \\ 0 & 0 & 0 & 0 & 44 \end{bmatrix}$$

Here, the diagonal entries correspond to CH_4, H_2O, CO, H_2, and CO_2 and are in units of g/mol. It shall be assumed that a feed stream containing equimolar quantities of CH_4 and H_2O is available, given by the molar flow rate vector $\mathbf{n}_f = [1, 1, 0, 0, 0]^T$ mol/s. From this, and the molecular weight matrix, the mass fraction feed vector in CO–H_2 space is hence given by $\mathbf{z}_f = [z_{CO,f}, z_{H_2,f}, 0]^T = [0, 0, 0]^T$.[2]

Finally, it is assumed that the reactor system is operated under isobaric conditions, set at 5 atm. The water–gas shift and CH_4 reforming reactions are assumed to be operated in the temperature range $700\,K \leq T \leq 1000\,K$. The AR will therefore be generated for the same temperature range.

9.3.2.3 Kinetics The reaction given by Equation 9.22a describes methane reforming with steam, whereas Equation 9.22b describes the water-gas shift reaction. The kinetic expression used for the CH_4 reforming reaction shall be given by the following simple relation:

$$r_1 = k_1(T)c_{CH_4}c_{H_2O} \tag{9.23}$$

Here, $k_1(T) = k_0\exp(-E/(RT))$ and c_{CH_4} and c_{H_2O} are the molar concentrations of CH_4 and H_2O, respectively. Similarly, the rate expression for the water–gas shift reaction is given by

$$r_2 = k_2(T)c_{CO}c_{H_2O} \tag{9.24}$$

with $k_2(T) = k_0\exp(-E/(RT))$. Note that more realistic expressions for both reactions may be supplied if required. Rate expressions for CO and H_2 are thus given by the following reaction stoichiometry:

$$r_{CO} = r_1 - r_2$$

$$r_{H_2} = 3r_1 + r_2$$

Table 9.1 summarizes values for the rate constants used for this example.

9.3.2.4 The Peng–Robinson Equation of State To demonstrate the use of a non-ideal gas mixture, the Peng–Robinson equation of state will be employed to describe the

TABLE 9.1 Rate Constant for Methane Steam Reforming and Water-Gas Shift Reactions

	CH_4 Reforming	Water–Gas Shift
k_o (mol/L/h)	1.10×10^{-3}	2.35×10^{-3}
E (J/mol)	500	800

[2]Note that although the mass fraction feed vector in CO–H_2 space is the zero vector $[0, 0, 0]^T$, the full mass fraction feed vector $\mathbf{z}_f = [0.4706, 0.5294, 0, 0, 0]^T$ containing all components is needed in calculating the AR.

variation of density with respect to pressure and temperature of a species concentration (Peng and Robinson, 1976; Sandler, 2006). In particular, the Peng–Robinson equation for a pure component is given by

$$P = \frac{RT}{\underline{V} - b} - \frac{a(T)}{\underline{V}(\underline{V} + b) + b(\underline{V} - b)} \tag{9.25}$$

Note that $\underline{V}$ is the specific molar volume, defined as follows:

$$\underline{V} = V/n_{tot}$$

Variables a(T) and b are specific for each pure component in the system, which are functions of component critical temperatures and pressures are as follows:

$$a(T) = \frac{0.45724R^2T_c^{\ 2}}{P_c}\alpha(T) \tag{9.26a}$$

$$b = \frac{0.07780RT_c}{P_c} \tag{9.26b}$$

$$\alpha(T) = \left[1 + k\left(1 - \sqrt{\frac{T}{T_c}}\right)\right]^2 \tag{9.26c}$$

$$k = 0.37464 + 1.54226\omega - 0.26992\omega^2 \tag{9.26d}$$

T_c, P_c, and ω, respectively, are the critical temperature, critical pressure, and acentric factor specific to the component of interest. These may be found in any suitable handbook of thermodynamic data, such as Perry et al. (2008).

Seeing as we are interested in modeling a system containing gases, the Peng–Robinson equation may be used. However, the original form of the Peng–Robinson equation deals with pure components and not mixtures containing multiple components. Thus, the equation of state must first be adapted to allow for the description of gas mixtures. That is,

$$P = \frac{RT}{\underline{V} - b_{mix}} - \frac{a_{mix}(T)}{\underline{V}(\underline{V} + b_{mix}) + b_{mix}(\underline{V} - b_{mix})} \tag{9.27}$$

Here, a_{mix} and b_{mix} are those belonging to the mixture. These values are found via the pure component values and are related using the following mixing rules:

$$a_{mix} = \sum_{i=1}^{n}\sum_{j=1}^{n} x_i x_j a_{ij} \tag{9.28a}$$

$$b_{mix} = \sum_{i=1}^{n} x_i b_i \tag{9.28b}$$

$$a_{ij} = \sqrt{a_{ii}a_{jj}}(1 - k_{ij}) \tag{9.28c}$$

Here, a_{ii} and b_i are the pure component parameters, defined by Equation 9.26a and b, and k_{ij} is the binary interaction parameter between component *i* and *j*.

TABLE 9.2 Pure Component Properties for the Steam Reforming and Water–Gas Shift Reaction

Component	T_c (K)	$P_c \times 10^{-6}$ (Pa)	Acentric Factor ω
CH_4	190.56	4.59	0.011
H_2O	647.13	21.94	0.334
CO	132.92	3.49	0.048
H_2	33.19	1.32	−0.215
CO_2	304.21	7.39	0.224

Values are taken from Perry et al. (2008).

TABLE 9.3 Binary Interaction Parameters for Use in the Peng–Robinson Equation of State

	CH_4	H_2O	CO	H_2	CO_2
CH_4	0	0.5[a]	0.03[b]	0.0156[b]	0.0919[b]
H_2O		0	0[c]	0[c]	0.1896[d]
CO			0	0.0919[b]	0.03[a]
H_2				0	−0.1622[b]
CO_2					0

Note that $k_{ij} = k_{ji}$.
[a]Obtained from Dhima et al. (1999).
[b]Obtained from Aspen Properties Database (2013).
[c]The binary interaction parameter for this pair was unknown at the time of writing.
[d]Obtained from Ávila-Méndez et al. (n.d.).

Values for critical temperature, pressure, and acentric factor for all five components participating in the system are given in Table 9.2. Values for the binary interaction parameters used in Equation 9.28c are given in Table 9.3. Note that the values found in Table 9.3 were obtained from different sources. Where the interaction parameters are unknown, the k_{ij} values are zero for the purposes of demonstration. More accurate predictions may be obtained by fitting the relevant vapor–liquid equilibrium data, if available.

From these values, more accurate P–V–T behavior for the system may be determined, which may be used in constructing the AR.

9.3.2.5 Conversion to Mass Fraction Space Since the rate expressions for CH_4 reforming and the water–gas shift reaction are given in terms of species concentration, the Peng–Robinson equation of state must be employed to express the rate expressions in terms of species mass fractions instead.

Given a system temperature and pressure, the mixture volume may be solved from the equation of state. This is achieved by rearranging Equation 9.27 for the specific volume. Since the particular equation of state is of a cubic form with respect to volume, multiple volume solutions exist. In particular, the following function is solved:

$$P\underline{V}^3 + (Pb_{mix} - 2RT)\underline{V}^2 + (a_{mix} - 3Pb^3_{mix})\underline{V} + (Pb^3_{mix} - a_{mix}b_{mix}) = 0 \quad (9.29)$$

The largest (real) volume solution is retained and used in the calculation of species concentration. Concentration may then be expressed in terms of specific volume as follows:

$$c_i = \frac{n_i}{V(T,P)} = \frac{x_i}{\underline{V}(T,P)}$$

Using Equation 9.9, species concentrations are then expressed in terms of mass fractions giving

$$c_i = \frac{1}{\underline{V}(T,P)} \frac{z_i}{W_i} W_{avg}(\mathbf{z}) \quad (9.30)$$

Substitution of Equation 9.30 into the rate expressions given in Section 9.3.2.3 results in rate expressions in terms of mass fractions. These may be used to compute the AR in z_{CO}–z_{H_2}–σ space.

9.3.2.6 Results The results of the computation, using the parallel complement AR construction method, are shown in Figure 9.9(a). Note that the AR computed for this system is smaller than the stoichiometric subspace computed in Section 9.2.7. In particular, although the stoichiometric subspace predicts a maximum CO mass fraction of approximately 0.9, the maximum achievable CO mass fraction predicted by the AR is approximately 0.21. This value is achievable for a mass fraction residence time of approximately 0.5 m^3.h/g as predicted by the AR construction. Figure 9.9(b) compares the size of the AR in relation to the stoichiometric subspace.

Moreover, although the result may not appear significant, the fact that the AR can be computed at all bears special meaning: the AR for a non-ideal system involving temperature dependence and minimum reactor volume has been found. From the results of the construction, the AR in mass fraction space can be converted to an equivalent AR in concentration space. An appropriate objective function may then be overlaid to determine its intersection with the boundary. Optimization of the system may then follow. From this, deeper insights into the limits of the system can be obtained.

9.4 CHAPTER REVIEW

A number of simple transformations were explained in this chapter that help to express common quantities used in AR theory in terms of species mass fractions. Mass fractions in AR computations are useful for describing industrial systems, as the conservation of mass guarantees that mixing in mass fraction space always obeys a linear mixing law. This result, in turn, allows for the use of AR theory in variable density systems (when molar concentration does not mix linearly).

Conversion of systems to mass fraction space often requires an appropriate equation of state to relate the temperature, pressure, and composition of a mixture. Use of

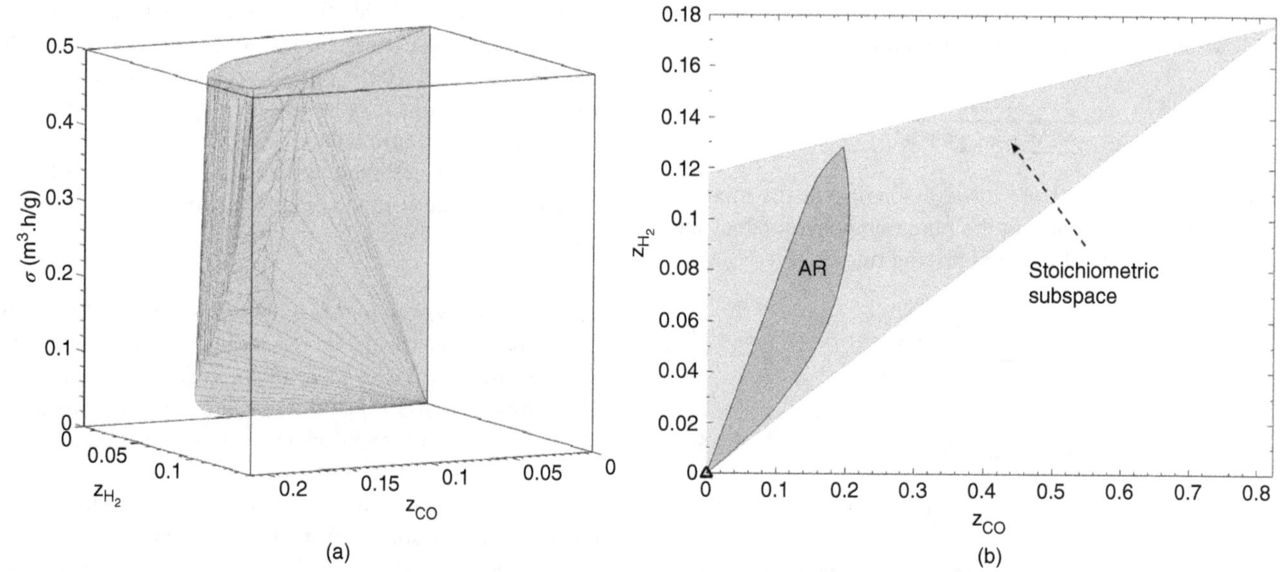

Figure 9.9 (a) AR for the methane steam reforming reaction in mass fraction space. (b) Comparison of the stoichiometric subspace and two-dimensional projection of the AR onto z_{CO}–z_{H_2} space.

this equation of state, along with species molecular weights, allows for straightforward conversion between constant density and variable density systems. In fact, many of the topics discussed in previous chapters may be generalized to variable density systems as well. Reaction and mixing find equivalent geometric interpretations in terms of the mass fraction vector **z**, and hence fundamental reactor types also exhibit mass fraction analogues. Furthermore, it is possible to express the residence time of the reactor network in terms of mass fraction residence time, σ, so that minimum reactor volume problems may be determined in variable density systems.

The use of mass fractions in AR theory allows designers to investigate a larger set of systems. Developing an understanding for how to transform molar quantities into mass fraction equivalent space is valuable in tackling systems of industrial and economic relevance using AR theory.

REFERENCES

Anderson, R.B., Kölbel, H., Ralek, M., 1984. The Fischer-Tropsch Synthesis. Academic Press, New York.

Aspen Properties Database, 2013. Aspen Technology, Inc., Burlington, MA.

Ávila-Méndez, G.A., Justo-García, D.N., García-Sánchez, F., García-Flores, B.E., n.d. Prediction of phase behavior for the system methane-carbon dioxide-hydrogen sulfide-water with the PR and PC-SAFT equations of state.

Dhima, A., de Hemptinne, J.-C., Jose, J., 1999. Solubility of hydrocarbons and CO2 mixtures in water under high pressure. Ind. Eng. Chem. Res. 38, 3144–3161.

Dry, M.E., 2002. The Fischer–Tropsch process: 1950–2000. Catal. Today 71, 227–241.

Peng, D.-Y., Robinson, D.B., 1976. A new two-constant equation of state. Ind. Eng. Chem. Fundam. 15, 59–64.

Perry, R.H., Green, D.W., Maloney, J.O., 2008. Perry's Chemical Engineers' Handbook. McGraw-Hill, New York.

Sandler, S.I., 2006. Chemical, Biochemical, and Engineering Thermodynamics. John Wiley & Sons, Hoboken.

Zhou, W., Manousiouthakis, V.I., 2007. Variable density fluid reactor network synthesis-Construction of the attainable region through the IDEAS approach. Chem. Eng. J. 129, 91–103.

10

FINAL REMARKS, FURTHER READING, AND FUTURE DIRECTIONS

10.1 INTRODUCTION

This chapter aims to summarize the concepts that were developed in previous chapters. We also wish to reflect on a small number of self-study and original research topics that interested readers could pursue. The introductory nature of this book has only allowed us to cover a subset of the total work currently available, but we hope that it has been sufficient to pique your interest in the field. Furthermore, AR theory is still a relatively new field of study, and many new breakthroughs await discovery.

10.2 CHAPTER SUMMARIES AND FINAL REMARKS

An important goal of this book has been to understand the fundamental concepts that constitute the foundations of AR theory. These concepts are by no means exhaustive, but we feel they provide a suitable introduction to the fundamentals of AR theory. Our approach has been instructional, rather than purely academic, to emphasize clarity at the expense of mathematical correctness. The foundational concepts of AR theory are often easy to grasp initially; however, these concepts quickly build in complexity, and hasty treatment of the introductory material often results in confusion later on. We urge you to attempt solving as many AR problems as possible provided in this book, as well as apply these concepts to your own problems.

In Chapter 1, we introduced the story of Sam, Alex, and Donald in order to describe a fundamental problem that is often encountered in design—*how do we know when we are the best*? Problems in chemical reactor design (as well as many problems unrelated to chemical reactor design) may often be arranged according to a hierarchy of approaches, which we may not always be mindful of from the beginning. But commitment to an approach, in turn, influences the manner in which solutions are implemented later in the design process. These decisions may greatly shape our understanding of the problem, and also the quality of the recommendations that are generated.

We believe that if designers begin to think about reactor problems in terms of reactor networks—as opposed to single reactors—then more effective designs may be achieved overall. Furthermore, reactive systems involving multiple reactors, termed reactor structures, are necessary when systems involving multiple reactions are involved, and so familiarity with this approach is valuable in achieving the best performance from complex reactive systems.

Yet, even if this information is known and reactor networks are employed in a problem, how do we know if there are not better solutions in existence? We must often have an understanding of a performance target with which designs may be benchmarked against, and AR theory is one such approach that can assist in finding these operating targets.

***Takeaway from Chapter 1*:**

To understand how good our current designs and methods are, we must have an understanding of how good our designs and methods could ever be under all circumstances. Performance targets that we can work toward are therefore very useful, but we must know how to find these targets. Designing with multiple reactors (reactor structures) in mind often provides opportunities to significantly improve the performance of reactive systems. Begin to think from a reactor structure viewpoint rather than strictly from an optimization viewpoint of a set design. Defer optimization until the appropriate reactor structure is first identified. This strategy provides the best opportunity to gain the most performance from a system.

In Chapter 2, we investigated how data from the benzene–toluene–xylene (BTX) system could be visualized

Attainable Region Theory: An Introduction to Choosing an Optimal Reactor, First Edition. David Ming, David Glasser, Diane Hildebrandt, Benjamin Glasser, and Matthew Metzger.

Companion website: http://attainableregions.com

effectively. Viewing data in the phase plane is convenient for understanding which concentrations should be targeted to obtain the highest concentration in toluene. The act of mixing, in particular, inherits a favorable property in concentration space. When two distinct concentrations are mixed together, be that from a set of beaker experiments, or a set of continuous streams in a process plant, the resulting concentration lies on a straight line joining the two concentrations together. Mixing is therefore a linear process.

Mixing ultimately allows for the attainment of new concentrations. Concentrations that might not have been considered, or thought to be unachievable prior, may be obtained via the mixing of two appropriate (achievable) concentrations. Given a set of achievable concentrations, it is possible to fill in the space via mixing. Computing the convex hull of a set of points produces a set of "unique" concentrations—unique in that these points are the only ones required to generate all other points in the set. In this chapter, we also introduced the idea of describing concepts from three perspectives: physical, mathematical, and geometric. We believe the geometric perspective offers the greatest insights into reactor network synthesis.

***Takeaway from Chapter 2*:**

Interpreting concentration and mixing in terms of geometric objects (specifically vectors) is a key principle in AR theory. Mixing is fundamental to achieving new concentrations, as a finite set of achievable points may be used to synthesize an infinite number of mixtures, which is represented as a filled convex region in the phase plane. Mixing is closely linked to the idea of achievability, and therefore it is an important process in AR theory. ARs are essentially a continuum of points defined by a finite set of unique, achievable points. This result is made possible by mixing. Always look for opportunities to mix, especially when the data contain concavities, as this will allow for new concentrations to be achieved.

In Chapter 3, we returned to the BTX beaker experiment and used the ideas of mixing and convex hulls from Chapter 2 to improve the maximum concentration of toluene. This chapter also introduced the idea of a candidate AR specific to the system of interest. The AR for the BTX system was revealed as the limiting case of infinitely many batch mixing and reaction experiments, which were conducted in a serial fashion. Chapter 3 also introduced a number of necessary conditions of the AR. Since mixing is a linear process, the AR must be a closed, compact, and convex set of points in concentration space that is formed by the convex hull of all achievable points.

***Takeaway from Chapter 3*:**

The AR represents the entire set of all achievable concentrations, obtained by the processes of reaction and mixing, which acts as a boundary between all that is achievable to all that is not. The AR is a convex, finite, region in n-dimensional state space. The AR is simply connected and is composed of sections that either belong to mixing (straight) or reaction (curved) surfaces. Although the AR is often thought to exist as a vast, possibly infinite, region in state space, it is in fact finite and well defined, and we can generate the AR using simple reaction and mixing steps. Understanding the shape and size of the AR helps us to understand how to set appropriate performance targets, which generates insight into the operating limits of our system. Any set of achievable points may be interpreted as a candidate AR if mixing is available.

Chapter 4 extended our understanding of the geometric viewpoint to include reaction. Systems of reactions, together with reaction stoichiometry, may be interpreted geometrically in the form of rate vectors, which allows us to view fundamental reactor types as geometric processes.

It is thus also possible to view reaction in the phase plane in a similar manner to concentration and mixing. A system of reactions and associated rate functions describe vector fields (called "rate fields" in AR theory) in the phase plane. All reactors (both novel and classic) serve the geometric purpose of transitioning the state of a system from one point in state space to another along a path defined by the intrinsic performance of the reactor—that is, the specific path taken through the rate field is governed by the particular reactor employed. The feed point and kinetics (rate field) define the size and shape of the AR. The three fundamental reactor types used in AR theory were also described in this chapter, which are characterized by the degree to which each reactor type incorporates mixing. Whereas mixing operations serve to form achievable states through convex combination of achievable points, reaction may be viewed to enlarge the set of achievable points.

***Takeaway from Chapter 4*:**

Reaction may also be interpreted geometrically. Rate vectors are used to describe how reactions can be used to move through concentration space. Three fundamental reactor types are employed in AR theory: the PFR, the CSTR, and the DSR. No other reactor type is required to generate the AR. Each reactor type exhibits its own characteristic geometric interpretation and each performs a crucial function in building different parts of the AR boundary.

A number of simple, isothermal, two-dimensional constructions were described in Chapter 5. In particular, the well-known Van de Vusse and isola systems were studied. The isola problem allowed us to demonstrate the influence that degenerate kinetics have on the construction of the AR. Calculating the CSTR locus is not always easy, particularly when complicated kinetics is present. We also investigated how an AR involving residence time, τ, may be constructed, which are represented as unbounded regions in state space. Residence time constructions are feasible since τ

is geometrically similar to concentration in that τ also obeys a linear mixing law, and therefore it may be incorporated as a component in the concentration vector **C** and rate vector **r**(**C**). Although two-dimensional constructions may at first appear restrictive, many realistic systems of interest can be formulated as two-dimensional problems and solved in this way. Two-dimensional subspaces of higher dimensional problems may be used to obtain a rudimentary idea of attainability, if we do not wish to generate higher dimensional ARs (i.e., with the BTX problem, which was originally introduced as a two-dimensional region in Chapter 3).

***Takeaway from Chapter 5*:**

When there are only two independent reactions that occur in the system, the resulting AR may be represented as an area in $\mathbb{R}^2$. Two-dimensional constructions are easily performed graphically. Only combinations of CSTRs and PFRs are needed to construct the AR in $\mathbb{R}^2$. Once the AR has been computed, optimization with a specified objective function is easily carried out as all answers to all possible objective functions are known. Generating the AR first and optimizing later allows us to find many potential answers to many potential optimization problems, which provides insight into how our recommendations might change when there is a change in objective function. Always attempt to generate the AR first and defer optimization until the end, once the appropriate structures are known.

Chapter 6 summarized a number of important results in higher dimensional AR theory, which are due to significant contributions by Feinberg (2000a, 2000b) and Feinberg and Hildebrandt (1997). The theorems discussed in Chapter 6 are commonly employed in everyday AR applications, but often these results are used without full appreciation of their origins. We discussed how the AR is always composed of mixing and reaction surfaces. Reaction surfaces result from the union of PFR trajectories, often termed a PFR manifold, and all optimal reactor structures on the boundary of the AR terminate with a PFR. Chapter 6 also demonstrates how only combinations of PFRs, CSTRs, and DSRs are needed to construct the AR for any dimension.

We also described how concrete equations for critical DSR and CSTRs may be computed. These expressions are complicated to compute analytically, which are derived from geometric controllability arguments developed by Feinberg (2000a, 2000b). These conditions are intricate, and thus it is often not possible to compute analytic solutions to the equations that describe critical reactors. For three-dimensional systems, a shortcut method involving the "vDelR" condition may be used to find critical α policies. Irrespective of the method used, the conditions for critical reactors are well defined, irrespective of the legitimacy of the kinetics studied, and thus these conditions must be enforced if we wish to attain points on the true AR boundary.

***Takeaway from Chapter 6*:**

The AR is composed of mixing lines and manifolds of PFR trajectories. The final approach to the extreme points of the AR boundary is achieved using PFR solution trajectories—if a desired operating point resides on the AR boundary, a PFR must be incorporated into the reactor structure in order to reach it, and thus PFRs are often the best terminating reactor to use in practice (for any kinetics and feed point). Only combinations of PFRs, CSTRs, and DSRs are required to form the AR. This result is true for all dimensions. Distinct expressions may be derived to compute critical α policies for the DSR profile and critical CSTR residence times. These expressions are intricate and complex in nature, which are ultimately based on the lack of controllability in a critical reactor. This idea is important in understanding the nature of the AR and how to achieve points on the true AR boundary.

In Chapter 7, a number of more complex examples were provided for computing candidate ARs. These examples are by no means exhaustive, but they provide a realistic subset of problem types that may be addressed via AR theory, which include adiabatic reaction, interstage cooling and higher dimensional problems involving minimum residence time. The AR for the well-studied three-dimensional Van de Vusse system was computed using concepts established in Chapter 6. We also generated the true AR for the BTX system, which was initially introduced in Chapter 1. Chapter 7 also described how AR theory may be used in batch systems. This extension of the theory allows us to understand the influence of different batch reactor types on system performance, as well as how these batch reactors should be arranged in time to produce a desired output. Simple transformations to all three fundamental reactor types may be applied that convert continuous reactor structure to an equivalent batch reactor structure.

***Takeaway from Chapter 7*:**

Complex systems involving multiple reactions, temperature dependence, and minimum residence time may be handled using AR theory. The three-dimensional Van de Vusse system is an ideal system to study for those interested in investigating AR theory in greater detail. ARs generated from a reactor-type constraint are useful in understanding operating strategies such as cold-shot and interstage cooling. It is also possible to apply AR theory to batch reactors using simple transformations.

Chapter 8 provided a general overview of automated AR construction techniques. These methods are necessary when very complex, higher dimensional, systems are investigated. AR construction techniques have historically fallen into one of two categories: inside-out and outside-in methods. Inside-out methods are additive in nature. These methods begin with a small achievable region (usually the feed point) and then enlarge the set of achievable concentrations. Outside-in methods work in the opposite manner to inside-out methods. These methods serve to enclose the AR in a larger space and then progressively remove unattainable concentrations from the space. This larger space is typically achieved via computation of the stoichiometric subspace **S**. Outside-in methods are typically better suited for complex kinetics that may contain multiple steady states, but they are also generally more difficult to program, and current outside-in methods do not generally provide the reactor structure associated with the constructed region. Inside-out methods, by comparison, are easier to program and interpret, although current methods often rely on continuity arguments that may result in incomplete answers for systems involving multiple steady states.

Alternate methods for finding the AR, via superstructure methods, were also briefly discussed. The IDEAS framework, in particular, is a generalized AR construction scheme that may also be utilized to solve both AR and non-AR-related problems. Superstructure methods rely on mathematical programming techniques and are hence not always easy to program and implement without prior knowledge and training.

***Takeaway from Chapter 8*:**

Computer algorithms may be developed for the construction of candidate ARs. AR construction methods generally arise in two forms: inside-out (additive) and outside-in (subtractive) methods. Each construction scheme exhibits its own strengths and weaknesses. These methods assist in answering and validating many unanswered problems in AR theory, such as the development of an AR sufficiency condition. These techniques also help us to determine the limits of achievability for higher dimensional problems for which an analytic answer does not exist. Even though the AR serves to help determine an optimal reactor structure, the nature of AR construction is usually achieved via geometric interpretation of the fundamental processes and not through an actual simulation of proposed optimal structures.

Chapter 9 discussed the construction of candidate ARs in variable density systems. Mass fractions are hence useful in broadening the use of AR theory to a wider class of systems (i.e., gas phase reactions). Mass fractions always obey a linear mixing law, and thus these constructions are valid even when the system does not obey constant density. We described a number of useful relations for converting common process variables used in AR theory, such as concentration and mole fraction, in terms of species mass fractions. These conversions are often assisted by use of an appropriate equation of state that relates the system volume to the process variables of interest. Rate expressions involving concentration are then easily viewed in mass fraction space, and the AR may be computed appropriately.

***Takeaway from Chapter 9*:**

AR theory is not restricted to constant density systems. The use of mass fractions allows one to compute the AR, even for variable density problems. This feat is possible because mass fractions always obey a linear mixing law. With an appropriate equation of state, it is possible to express common process variables (including residence time) in terms of species mass fractions, and thus the AR for systems involving molar concentration and mole fraction may still be computed. This theory allows for the investigation of a wider range of realistic reactor network synthesis problems, such as reactions occurring in the gas phase or heterogeneous reactions.

10.3 FURTHER READING

Further applications of AR theory to a number of different fields, as well as the extensions of core ideas (the same ideas discussed in this work), are described in published literature. We provide a number of these publications later for interested readers. These publications are generally mathematical in nature, and assume that the reader is already knowledgeable in AR theory.

10.3.1 AR-related Papers

10.3.1.1 Feinberg's Three Papers In Chapter 6, we highlighted many core ideas related to critical reactors and the shape of the AR boundary in n-dimensional concentration space. These concepts are derived from a series of publications by Feinberg (2000a, 2000b, 1999) and Feinberg and Hildebrandt (1997). These works are mathematically rigorous, yet they are the basis for many of the discussions and results described in this book.

10.3.1.2 Chemical Reactor Theory and Multiple Steady States For readers interested in broader topics relating to the theory of chemical reactor networks, reactor network synthesis, and process synthesis, we recommend a number of papers by Feinberg (2002, 1987) and Schlosser and Feinberg (1994). These papers provide a deeper understanding of how a large system of chemical reactions might be characterized and interpreted. For example, the deficiency one and deficiency zero theorems (Feinberg, 1987) can be used to identify when multiple steady states may arise in isothermal CSTRs from reaction stoichiometry alone, even when the kinetics is not defined.

10.3.1.3 Complementary Concepts and Advanced Topics in AR Theory Godorr et al. (1999) expand on the idea of AR construction for temperature-dependent reaction kinetics. This work describes how it is possible to compute an optimal temperature profile for an AR, which was briefly described in Chapter 7. This process is achieved by an extension of the familiar "vDelR" condition to include an optimal control parameter such as temperature. The method is currently applicable only to systems in $\mathbb{R}^3$, although it may be extended to higher dimensions.

A number of three-dimensional adiabatic reactor network synthesis problems were described in Chapter 7. A paper by Nicol et al. (1997) looks at extending these principles to include heating and cooling utilities. In effect, heat transfer equipment may be incorporated into the optimal reactor structure on the AR boundary. The results are dependent on a number of ideas developed by Godorr et al.(1994). Aspects of this work also involve finding conditions for optimality in four-dimensional space. Further details may be found in the PhD theses of Love (1995) and Nicol (1998).

ARs for systems involving *multiple* rate vectors have also been investigated (Godorr et al., 1994), which might occur when there are multiple catalysts present in the reactor, for example. The form of the resulting critical reactor expressions is similar in form to the iterated Lie bracket expressions obtained in Chapter 6 for critical DSRs with a single rate vector. The present theory is centered on three-dimensional problems, although there is again possibility for extension to higher dimensions.

The relation between the AR boundary and the Pontryagin's maximum principle (a classic optimization procedure) is described in a paper by McGregor et al. (1999). Readers interested in the calculus of variations and optimal control theory are encouraged to investigate these ideas further.

Similar comparisons between the AR and the popular maximum mixedness and segregated reactor models of Zwietering (1959) have also been undertaken. Readers interested in understanding how these reactor models may be interpreted geometrically and compared to the AR should consult Glasser et al. (1994). Wen et al. (2002) describe how selectivity may be incorporated and maximized for the Van de Vusse system. This analysis is similar to the worked examples discussed in Chapter 5.

10.3.1.4 Industrial Examples AR theory has been investigated for a number of industrial examples in the literature. For example, there exists a growing set of literature related to the application of AR theory to milling (size reduction) circuits. The work of Khumalo et al. (2006) and related papers describe these concepts in detail. Candidate regions for minimal residence time in the water–gas shift reaction have been investigated by Kauchali et al. (2004) where the linear programming formulations, which were originally described in Chapter 8, are used by Kauchali et al.(2004) in the analysis. A number of papers (Milne, 2008; Milne et al., 2004, 2009, 2010) provide detailed application of AR to the oxidative dehydrogenation of *N*-butanes (an important reaction used in the production of butenes and butadiene). Scott et al. (2013) apply AR theory to the optimization of residence times to saccharification and fermentation processes for the production of bioethanol.

10.3.2 Non-reactor-Related Papers

The improvement of processes using concepts based on AR theory, but which do not directly involve reactor networks, is also found in the scientific literature. Sung and Maravelias (2007) detail how elements of the AR approach may be used in optimization of production planning. Nisoli et al. (1997) describe a mathematical approach to generating ARs when both reaction and separation processes are involved. This theory could be used, for example, in the design of reactor–separator systems. Agarwal et al. (2008), Gadewar et al. (2004), Hauan and Lien (1998), and Thery et al. (2005) also provide theories for how regions of feasibility could be obtained and used in areas of reactive distillation. Alvarado-Morales et al. (2010) apply AR theory in conjunction with classical optimization techniques to improve the performance of bioethanol processes. Similarly, Muvhiiwa et al. (2015) apply the concept of feasible regions with respect to Gibbs free energy to study thermodynamic ARs (as opposed to kinetic ARs), in order to determine work requirements in an anaerobic biodigester. The papers given in the text are shown simply to reflect how AR theory has been used in current research, although many other applications exist.

10.4 FUTURE DIRECTIONS

Significant advances in the field of AR theory await discovery. In the following sections, we discuss a number of current research problems in the field, and highlight potential research avenues.

10.4.1 The Search for a Sufficiency Condition

The absence of a sufficiency condition—which will signify when the true AR has been determined—presents a large theoretical challenge for both AR theorists and practitioners who employ AR theory to solve reactor synthesis problems. Without a sufficiency condition, there is no certainty that the regions produced are the true AR. This is true even if the region computed has been generated from an automated AR construction method, such as those described in Chapter 8. Only for systems of a simplified or unique nature (i.e., when a rate field contains completely convex PFR trajectories), or for systems that have been well studied (i.e., Van de Vusse kinetics), is one confident that the true AR has been found.

Although Chapter 6 discusses how precise conditions exist for computing critical DSRs and CSTRs, the complex nature of the resulting equations not only provides a rigid set of constraints but also brings about uncertainty related to the particular mathematical form of the kinetics that are meant to approximate the true behavior of the system in question. In this way, a suitable sufficiency condition, which is based on our current understanding of AR theory, may already suffice, but these conditions make it impractical for application to realistic systems of even simple-to-moderate complexity. Hence, if an analytical sufficiency condition is available, yet infeasible, to compute, then perhaps *numerical* validation is required. The need for an alternative sufficiency check, based on the prior computation of a candidate region, may perhaps present a more practical discovery method, which would be applicable in practical design situations.

10.4.2 Applying AR Theory to Real Systems

AR construction and interpretation present a significant challenge for first-time practitioners of AR theory who wish to understand more complex systems in higher dimensions. Improvements in both the general understanding of higher dimensional ARs and the development of robust AR construction tools will allow for the use of AR theory to many more systems.

In addition to applying AR theory to systems where explicit mathematical expressions for kinetics are known, generating candidate regions from experimental data—in the absence of a mathematical model—is also feasible. At present, the adoption of AR theory to experimental work has been limited, although this situation is slowly improving. With advances in general AR theory and construction schemes, using AR theory for experimental investigations will become easier. We have already touched on how AR theory may be used to improve batch reactor structures in Chapter 8. Validation of basic AR theories via experiment would also assist in demonstrating how these concepts may be applied in practice, and also help to explain theoretical concepts.

10.4.3 Automated AR Construction

10.4.3.1 Improvements to Current Methods Over approximately the past decade, AR research has noted a definite shift toward understanding how the AR may be constructed via numerical computation. Chapter 8 describes a number of popular AR construction methods that are currently in use, although a *complete* AR construction algorithm, which is able to fulfill many design scenarios and systems, still does not exist. AR construction is often computationally strenuous with current methods and hardware, and investigations are thus restricted to systems with a small number of independent reactions and construction scenarios. The need for faster, more accurate, and more robust AR construction techniques is therefore still required, and will be important not only for computing more complex candidate regions but also for validating theoretical concepts.

Many AR construction schemes still do not allow for the decoding of the optimal reactor structure associated with the region. Hence, it is still challenging to know what reactor structure is required to reach a particular state on the AR boundary once AR construction has been completed.

Topics in AR construction related to both the determination of regions under a wide variety of construction scenarios (i.e., unbounded and nonisothermal systems involving multiple feeds) and improvements to construction times and accuracy are actively being investigated. Understanding how to construct more accurate candidate regions in shorter computing times, as well as formulating new methods suitable

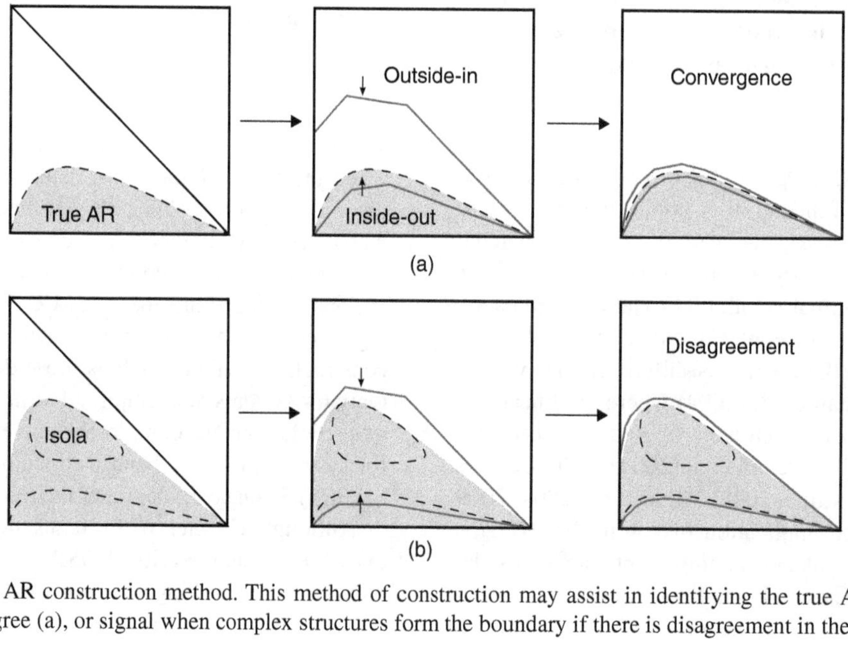

Figure 10.1 A hybrid AR construction method. This method of construction may assist in identifying the true AR boundary when both construction methods agree (a), or signal when complex structures form the boundary if there is disagreement in the two methods (b).

for new computer hardware, will allow for the application of AR theory to broader class of industrially relevant systems. Characterizing conditions for when traditional, computationally demanding, AR constructions may be avoided (i.e., when critical CSTRs and DSRs do not participate) that can be replaced with shortcut heuristic techniques is also being investigated currently.

10.4.3.2 Hybrid AR Construction and Numerical Sufficiency Improvement of AR construction methods may eventually lead to improved understanding of an AR sufficiency condition. Recently, the development of "hybrid" AR construction methods has begun to show promise. These methods incorporate both inside-out and outside-in construction steps in order to determine the AR. An AR construction method utilizing both approaches may serve to solve this long-standing problem. If it is possible to compute the same region via two independent approaches (inside-out and outside-in) to within an acceptable tolerance, such as in Figure 10.1(a), then this approach provides a basis to confirm that the computed region is the true AR. Similarly, even if there is disagreement between the two regions, as in Figure 10.1(b), then this result also indicates that additional investigation is required. This approach of incorporating two different construction methods is similar to spatial branch and bound approaches used in mathematical optimization techniques. The benefit of AR construction via this approach may be highly valued, and development of a satisfactory hybrid method is currently underway.

Thus, the use of a hybrid construction scheme may provide an alternative method for a *numerical* sufficiency condition. Validation of the AR might then only be applicable to the specific system computed, but this is nevertheless valuable, for at least a theoretically reliable estimate of the AR may be established.

REFERENCES

Agarwal, V., Thotla, S., Mahajani, S.M., 2008. Attainable regions of reactive distillation-Part I. Single reactant non-azeotropic systems. Chem. Eng. Sci. 63, 2946–2965.

Alvarado-Morales, M., Hamid, M.K.A., Sin, G., Gernaey, K.V., Woodley, J.M., Gani, R., 2010. A model-based methodology for simultaneous design and control of a bioethanol production process. Comput. Chem. Eng. 34, 2043–2061.

Feinberg, M., 1987. Chemical reaction network structure and the stability of complex isothermal reactors—I. The deficiency zero and deficiency one theorems. Chem. Eng. Sci. 42, 2229–2268.

Feinberg, M., 1999. Recent results in optimal reactor synthesis via attainable region theory. Chem. Eng. Sci. 54, 2535–2543.

Feinberg, M., 2000a. Optimal reactor design from a geometric viewpoint. Part II. Critical sidestream reactors. Chem. Eng. Sci. 55, 2455–2479.

Feinberg, M., 2000b. Optimal reactor design from a geometric viewpoint—III. Critical CFSTRs. Chem. Eng. Sci. 55, 3553–3565.

Feinberg, M., 2002. Toward a theory of process synthesis. Ind. Eng. Chem. Res. 41, 3751–3761.

Feinberg, M., Hildebrandt, D., 1997. Optimal reactor design from a geometric viewpoint—I. Universal properties of the attainable region. Chem. Eng. Sci. 52, 1637–1665.

Gadewar, S., Tao, L., Malone, M., Doherty, M., 2004. Process alternatives for coupling reaction and distillation. Chem. Eng. Res. Des. 82, 140–147.

Glasser, D., Hildebrandt, D., Godorr, S., 1994. The attainable region for segregated, maximum mixed, and other reactor models. Ind. Eng. Chem. Res. 33, 1136–1144.

Godorr, S.A., Hildebrandt, D., Glasser, D., 1994. The attainable region for systems with mixing and multiple-rate processes: finding optimal reactor structures. Chem. Eng. J. Biochem. Eng. J. 54, 175–186.

Godorr, S., Hildebrandt, D., Glasser, D., McGregor, C., 1999. Choosing optimal control policies using the attainable region approach. Ind. Eng. Chem. Res. 38, 639–651.

Hauan, S., Lien, K., 1998. A phenomena based design approach to reactive distillation. Chem. Eng. Res. Des. 76, 396–407.

Kauchali, S., Hausberger, B., Hildebrandt, D., Glasser, D., Biegler, L.T., 2004. Automating reactor network synthesis: Finding a candidate attainable region for the water-gas shift (WGS) reaction. Comput. Chem. Eng. 28, 149–160.

Khumalo, N., Glasser, D., Hildebrandt, D., Hausberger, B., Kauchali, S., 2006. The application of the attainable region analysis to comminution. Chem. Eng. Sci. 61, 5969–5980.

Love, A.D., 1995. Mathematics of the Attainable Region. University of the Witwatersrand, Johannesburg.

McGregor, C., Glasser, D., Hildebrandt, D., 1999. The attainable region and pontryagin's maximum principle. Ind. Eng. Chem. Res. 38, 652–659.

Milne, A.D., 2008. The Application of the Attainable Region Concept to the Oxidative Dehydrogenation of N-butanes in Inert Porous Membrane Reactors. University of the Witwatersrand, Johannesburg.

Milne, D., Glasser, D., Hildebrandt, D., Hausberger, B., 2004. Application of the attainable region concept to the oxidative dehydrogenation of 1-butane in inert porous membrane reactors. Ind. Eng. Chem. Res. 43, 7208.

Milne, D., Seodigeng, T., Glasser, D., Hildebrandt, D., Hausberger, B., 2009. Candidate attainable regions for the oxidative dehydrogenation of n-butane using the recursive constant control (RCC) policy algorithm. Ind. Eng. Chem. Res. 48, 5211–5222.

Milne, D., Seodigeng, T., Glasser, D., Hildebrandt, D., Hausberger, B., 2010. The oxidative dehydrogenation of n-butane in a differential side-stream catalytic membrane reactor. Catal. Today 156, 237–245.

Muvhiiwa, R.F., Hildebrandt, D., Glasser, D., Matambo, T., Sheridan, C., 2015. A thermodynamic approach toward defining the limits of biogas production. AICHE J. 61, 4270–4276.

Nicol, W., 1998. Extending the attainable region technique by including heat exchange and addressing four dimensional problems [PhD thesis]. University of the Witwatersrand, Johannesburg.

Nicol, W., Hildebrandt, D., Glasser, D., 1997. Process synthesis for reaction systems with cooling via finding the attainable region. Comput. Chem. Eng. 21, S35–S40.

Nisoli, A., Malone, M.F., Doherty, M.F., 1997. Attainable regions for reaction with separation. AICHE J. 43, 374–386.

Schlosser, P.M., Feinberg, M., 1994. A theory of multiple steady states in isothermal homogeneous CFSTRs with many reactions. Chem. Eng. Sci. 49, 1749–1767.

Scott, F., Conejeros, R., Aroca, G., 2013. Attainable region analysis for continuous production of second generation bioethanol. Biotechnol. Biofuels 6, 171.

Sung, C., Maravelias, C.T., 2007. An attainable region approach for production planning of multiproduct processes. AICHE J. 53, 1298–1315.

Thery, R., Meyer, X.M., Joulia, X., Meyer, M., 2005. Preliminary design of reactive distillation columns. Chem. Eng. Res. Des. 83, 379–400.

Wen, Z., Chuanguang, Z., Zhishan, Z., Fangyu, H., Chengyue, L., 2002. Strategy of an attainable region partition for reactor network synthesis. Ind. Eng. Chem. Res. 41, 190–195.

Zwietering, T.N., 1959. The degree of mixing in continuous flow systems. Chem. Eng. Sci. 11, 1–15.

APPENDIX A

FUNDAMENTAL REACTOR TYPES

In Chapter 4, we discuss the role of three fundamental reactor types in attainable region (AR) theory. Many readers may already be familiar with these reactors, for they are common in chemical reaction engineering. A small, qualitative, summary of these reactors is provided in the following text.

A.1 THE PLUG FLOW REACTOR

A plug flow reactor (PFR) is a type of continuous reactor in which material inside the reactor is assumed to flow in a "plug-like" manner. The underlying assumption of PFR behavior is that *no axial mixing* occurs between different "plugs" of material in the PFR. Physically, a PFR is often constructed out of a tube. Material is passed in one end of the tube and reacts as it traverses along the tube length. The diameter and length of the tube, as well as the flow rate of material into the tube, determine the effluent concentrations of components at the exit of the PFR. PFRs are represented by the symbol in Figure A.1(a).

When density may be assumed to be constant, we can mathematically describe the evolution of species concentrations inside a PFR by a system of differential equations of the following form:

$$Q\frac{dc_i}{dV_{PFR}} = r_i\left(c_1, c_2, \ldots, c_n\right)$$

Here, c_i is the concentration of component i in the system, Q is the volumetric flow rate through the PFR, V_{PFR} is the volume of the reactive portion of the PFR tube, and r_i is the rate of reaction for component i, which is assumed to be dependent on all components participating in the reaction.

An interesting observation of this equation is that the mathematical behavior of a PFR is identical to that of a batch reactor, and hence batch reactors may be used to understand the performance of a similarly sized PFR. The equation describing the change in concentration for component i in a batch reactor is given by

$$\frac{dc_i}{dt} = r_i\left(c_1, c_2, \ldots, c_n\right)$$

which is mathematically similar in form to the PFR equation.

A.2 THE CONTINUOUS-FLOW STIRRED TANK REACTOR

The continuous-flow stirred tank reactor (CSTR), which is sometimes also more accurately abbreviated as a CFSTR,[1] is a form of a continuous reactor that is often viewed as the opposite of a PFR in terms of mixing. That is, *perfect mixing* is assumed to occur in a CSTR so that the contents inside the CSTR vessel are evenly distributed and that the reactor operates at a *single* concentration. Under this assumption, the CSTR effluent concentration is identical to the concentration inside the reactor itself. (This behavior is very different from

[1] In this book, we prefer to refer to continuous-flow stirred tank reactors as CSTRs as opposed to CFSTRs.

Attainable Region Theory: An Introduction to Choosing an Optimal Reactor, First Edition. David Ming, David Glasser, Diane Hildebrandt, Benjamin Glasser, and Matthew Metzger.

Companion website: http://attainableregions.com

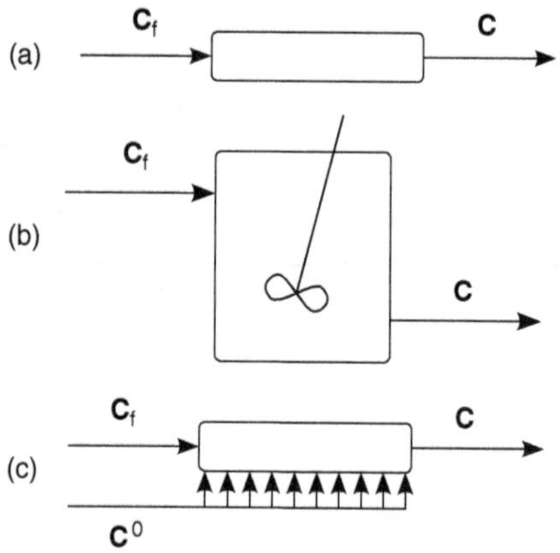

Figure A.1 (a) Reactor symbol for a plug flow reactor (PFR), (b) reactor symbol for a continuous-flow stirred tank reactor (CSTR), and (c) reactor symbol for a differential sidestream reactor (DSR).

a PFR where species concentrations continuously change as they move down the length of the PFR.) Physically, a CSTR is constructed as a tank and fitted with an agitator to promote even mixing throughout the entire tank volume. The reactor symbol for a CSTR is shown in Figure A.1(b).

When density can be assumed constant, a steady-state molar balance for component i over a CSTR gives

$$Q\left(c_i - c_{i,f}\right) = V_{CSTR} r_i\left(c_1, c_2, \dots, c_n\right)$$

where Q and V_{CSTR} are the volumetric flow rate through the CSTR and the CSTR tank volume, respectively, and c_{if} is the feed concentration to the reactor. Observe that whereas the equation describing a PFR is a differential equation, CSTRs are described by a conventional algebraic equation, which do not need to be integrated. Solving for the effluent concentrations in a CSTR requires the solution of a system of nonlinear algebraic equations.

A.3 THE DIFFERENTIAL SIDESTREAM REACTOR

The differential sidestream reactor (DSR) may be viewed as a combination of a PFR (in which no mixing is assumed to occur) and a CSTR (in which perfect mixing perfect mixing is assumed to occur). A DSR is hence best represented as an extension of a PFR with sidestreams along the reactor's length that allow for additional material to be introduced into the reactor during reaction. In practice, this behavior may be achieved by constructing a PFR with additional inlet ports along the reactor tube. The symbol for a DSR used in this book is shown in Figure A.1(c).

Similar to the PFR, species concentrations in a DSR can be modeled by a differential equation of the following form:

$$Q\frac{dc_i}{dV_{DSR}} = r_i(C) + \alpha\left(c_i^0 - c_i\right)$$

Here, Q, V_{DSR}, and c_i are the DSR volumetric flow rate, reactor volume, and species concentration, respectively. Furthermore, c_i^0 represents the concentration of the sidestream—which is assumed to be held at a constant value throughout the reaction—and α is a scalar parameter that used to control the amount of additional sidestream material introduced along the reactor's length.

Similar to batch reactors, fed-batch reactors are mathematically equivalent to DSRs. The differential equation describing the change in concentration in a fed-batch reactor is given by

$$\frac{dc_i}{dt} = r_i(C) + \left[\frac{F(t)}{V(t)}\right]\left(c_i^0 - c_i\right)$$

where F(t) and V(t) are the volumetric feed rate and reactor volume of the fed-batch reactor, respectively. DSRs play an important role in AR theory in higher dimensions, which are described in Chapter 6.

APPENDIX B

MATHEMATICAL TOPICS

B.1 SET NOTATION

It will be helpful to represent groups of concentrations, **C**, symbolically using set notation. Table B.1 provides a list of common notations used throughout this book.

B.2 ASPECTS OF LINEAR ALGEBRA

Although liner algebra is not used extensively in AR theory, having an understanding of a number of basic principles allows us to describe higher dimensional ideas compactly, which allows for convenient mathematical manipulation. Many textbooks on linear algebra are available (Lay, 2012; Strang, 2003). The topics described in the following sections will be particularly useful when reading Chapters 6–9.

B.2.1 General Definitions

B.2.1.1 Vector Norm The p-norm of a vector, $\mathbf{x} = [\mathrm{x}_1, \mathrm{x}_2, \ldots, \mathrm{x}_n]^\mathrm{T} \in \mathbb{R}^n$, is defined as follows:

$$\|\mathbf{x}\|_p = \left(\sum_{i=1}^{n} |\mathrm{x}_i|^p\right)^{1/p}$$

Here, p is a positive integer. The most commonly encountered vector norm is when $p = 2$, which calculates the magnitude (length) of **x**

$$\|\mathbf{x}\|_2 = \|\mathbf{x}\| = \sqrt{x_1^2 + x_2^2 + \cdots + x_n^2}$$

Thus, it is always possible to use $\|\mathbf{x}\|$ to scale a nonzero vector to unit magnitude by diving by its norm.

$$\hat{\mathbf{x}} = \frac{\mathbf{x}}{\|\mathbf{x}\|}$$

B.2.1.2 Inner Product The dot product, or *inner product*, of two vectors $\mathbf{x} = [\mathrm{x}_1, \mathrm{x}_2, \ldots, \mathrm{x}_n]^\mathrm{T} \in \mathbb{R}^n$ and $\mathbf{y} = [\mathrm{y}_1, \mathrm{y}_2, \ldots, \mathrm{y}_n]^\mathrm{T} \in \mathbb{R}^n$, is given by

$$\mathbf{x} \cdot \mathbf{y} = \sum_{i=1}^{n} \mathrm{x}_i \mathrm{y}_i$$

Note that the vector products

$$\mathbf{x}^\mathrm{T}\mathbf{y} = \mathbf{y}^\mathrm{T}\mathbf{x} = \sum_{i=1}^{n} \mathrm{x}_i \mathrm{y}_i$$

are equivalent to $\mathbf{x} \cdot \mathbf{y}$. We prefer to represent the inner product between **x** and **y** by $\mathbf{x}^\mathrm{T}\mathbf{y}$. The angle θ between **x** and **y** may be found from the inner product using the following identity:

$$\mathbf{x}^\mathrm{T}\mathbf{y} = \|\mathbf{x}\|\|\mathbf{y}\|\cos(\theta)$$

B.2.1.3 Tangency and Orthogonality Inner products may be used to determine when **x** and **y** are tangent or orthogonal to each other. Hence, vectors **x** and **y** are *tangent* to each other when

$$\mathbf{x}^\mathrm{T}\mathbf{y} = \|\mathbf{x}\|\|\mathbf{y}\|$$

Similarly, **x** and **y** are *orthogonal* (perpendicular) to each other when

$$\mathbf{x}^\mathrm{T}\mathbf{y} = 0$$

Attainable Region Theory: An Introduction to Choosing an Optimal Reactor, First Edition. David Ming, David Glasser, Diane Hildebrandt, Benjamin Glasser, and Matthew Metzger.

Companion website: http://attainableregions.com

TABLE B.1 Set-Builder Notation

Notation	Meaning
$\mathbf{X} = \{x, \gamma, 3\}$	**X** is a set containing three elements: x, γ, and 3.
$\mathbf{A} \cup \mathbf{B}$	The union of **A** and **B**
$\mathbf{A} \cap \mathbf{B}$	The intersection of **A** and **B**
$\mathbf{C} \in \mathbf{X}$	**C** is an element of **X**
$\mathbf{C} \notin \mathbf{X}$	**C** is not an element of **X**
$\mathbf{A} \subset \mathbf{B}$	**A** is a subset of **B**
$\mathbf{A} \subseteq \mathbf{B}$	**A** is a subset of or equal to **B**
$\mathbf{A} \not\subset \mathbf{B}$	**A** is not a subset of **B**
$\mathbf{A} \nsubseteq \mathbf{B}$	**A** is not a subset nor equal to **B**
$\{\mathbf{C} \mid \mathbf{C} \in \mathbf{X}\}$	The set of all **C** such that **C** is an element of **X** (sometimes, a colon : is used in place of |).
$\{\mathbf{C} \mid \mathbf{C} \in \mathbf{X}, \mathbf{C} \notin \mathbf{D}\}$	The set of all **C** such that **C** is an element of **X** and **C** is not an element of **D**
$\mathbf{X} \backslash \mathbf{D}$	Shorthand notation to denote that **D** is the complement set of **X**. In other words, $\mathbf{X} \backslash \mathbf{D} = \{\mathbf{C} \mid \mathbf{D} \subset \mathbf{X}, \mathbf{C} \in \mathbf{X}, \mathbf{C} \notin \mathbf{D}\}$.
$\{\mathbf{r}(\mathbf{C}) \in \mathbb{R}^n \mid \mathbf{C} \in \mathbf{D}\}$	The set of all $\mathbf{r}(\mathbf{C})$, where $\mathbf{r}(\mathbf{C})$ is an element of $\mathbb{R}^n$, such that **C** is an element of **D**

B.2.1.4 Vector Spaces and Subspaces The *vector space* $\mathbb{R}^n$ is the space of all vectors with *n* components.[1] All vectors belonging to the space $\mathbb{R}^3$ may be drawn in x–y–z space, for instance. Suppose that two vectors $\mathbf{x}, \mathbf{y} \in \mathbb{R}^n$ are available. Taking linear combinations of **x** and **y** will result in vectors that are also in $\mathbb{R}^n$.

A *subspace* of a vector space is a nonempty subset that is also a vector space. That is, vector subspaces also obey the laws of vector addition and scalar multiplication. If **x** and **y** are two vectors that lie in a vector subspace, then linear combinations of **x** and **y** will produce vectors that also lie in the subspace. For example, linear combinations of vectors $[0, 0, 1]^T$ and $[2, 0, 0]^T$ produce vectors that lie in a two-dimensional subspace (a plane) in $\mathbb{R}^3$; linear combinations of vector $[5, 0.2, -3, 1, 8]^T$ lie in a one-dimensional subspace (a line) in $\mathbb{R}^5$.

B.2.1.5 The Span of a Matrix and Linear Independence A set of *k* vectors $\{\mathbf{x}_1, \mathbf{x}_2, \ldots, \mathbf{x}_k\}$ is said to be *linearly independent* if it is not possible to take linear combinations of all *k* vectors

$$\lambda_1\mathbf{x}_1 + \lambda_2\mathbf{x}_2 + \cdots + \lambda_k\mathbf{x}_k = 0$$

such that the result produces zero, other than when $\lambda_1 = \lambda_2 = \cdots = \lambda_k = 0$. Conversely, when it is possible to have nonzero values for any λ_i and still produce 0, then the vectors $\mathbf{x}_1, \mathbf{x}_2, \ldots, \mathbf{x}_k$ is said to be *linearly dependent*.

Geometrically, two vectors **x** and **y** are linearly independent when **x** and **y** do not lie on the same line (they are not scalar multiples of each other). Linearly independent vectors may be used to generate (span) a vector space. For example, vectors $\mathbf{a}_1, \mathbf{a}_2 \in \mathbb{R}^4$ will span a two-dimensional subspace in $\mathbb{R}^4$ if $\mathbf{a}_1$ and $\mathbf{a}_2$ are linearly independent, and a one-dimensional subspace in $\mathbb{R}^4$ if they are linearly dependent.

[1]For example, $[c_1, c_2, c_3] \in \mathbb{R}^3$, $[y_1, y_2, y_3, y_4, y_5] \in \mathbb{R}^5$, and so on.

B.2.1.6 Rank The *rank* of a matrix **A**, rank(**A**), is a positive integer that counts the number of independent rows or columns in **A**. The *row rank* of **A** counts the number of independent rows in **A**, whereas the *column rank* of **A** counts the number of independent columns in **A**. A fundamental result of linear algebra is that the row and column rank of a matrix are always equal, and thus $\text{rank}(\mathbf{A}) = \text{rank}(\mathbf{A}^T)$.

Since rank is closely linked to the number of independent rows or columns in a matrix, rank(**A**) is a measure of the dimension of the vector space generated (or spanned) by the rows or columns of **A**.

B.2.1.7 Basis A basis for a vector space **V** is a collection of linearly independent vectors that span **V**. For example, the vectors $[1, 0, 0]^T$ and $[0, 0, -1]^T$ are linearly independent and span a two-dimensional vector subspace space in $\mathbb{R}^3$. (Linear combinations of the vectors generate other vectors that lie in a plane in $\mathbb{R}^3$.) Similarly, columns of the 5×5 identity matrix **I** are a basis for the vector space $\mathbb{R}^5$.

B.2.1.8 Row and Column Space The *column space* of a matrix **A** is the vector space generated by all linear combinations of the column vectors of **A**. Hence, the column space of **A** is equal to the span of the columns of **A**. Similarly, the *row space* of matrix **A** is the vector space that is generated by all combinations of the row vectors of **A**. The dimension of the column space is thus equal to the number of linearly independent column vectors in **A**, whereas the dimension of the row space of **A** is equal to the number of linearly independent row vectors in **A**, which are both equal to the rank of **A**, rank(**A**). Hence, the dimension of the column space of

A is equal to the dimension of the row space of **A**. Consider the matrix:

$$\mathbf{A} = \begin{bmatrix} 1 & 0 \\ 2 & -1 \\ 3 & 4 \end{bmatrix}$$

Linear combinations of the columns of **A** ($[1, 2, 3]^T$ and $[0, -1, 4]^T$) produce vectors that span a two-dimensional vector subspace (a plane) in $\mathbb{R}^3$. Similarly, linear combinations of the rows of **A** ($[1, 0]^T$, $[2, -1]^T$ and $[3, 4]^T$) span the entire two-dimensional vector space in $\mathbb{R}^2$. Note that vectors in the column space of **A** reside in $\mathbb{R}^3$, whilst vectors in the row space of **A** reside in $\mathbb{R}^2$, yet the dimensions of both the column and row space of **A** are both two.

B.2.1.9 Null Space All vectors **x** that solve the matrix equation

$$\mathbf{Ax} = \mathbf{0}$$

form a vector space called the *nullspace* of **A**, or null(**A**). If **a** is a row in **A**, and **x** is vector in the nullspace of **A**, then $\mathbf{ax} = 0$, and it follows that **a** and **x** are orthogonal to each other. Hence, the matrix equation $\mathbf{Ax} = \mathbf{0}$ suggests that all rows of **A** are orthogonal to **x**, and thus the vector space spanned by **x** is orthogonal to the row space of **A**. Similarly, the vector space of all **x** that satisfies the relation

$$\mathbf{A}^T\mathbf{x} = 0$$

is orthogonal to the column space of **A** (since **x** is part of the nullspace of $\mathbf{A}^T$).

B.2.2 Properties of Determinants

The determinant of a matrix **A**, Det(**A**), is useful in analyzing the uniqueness of a solution for a system of linear equations. Determinants arise in AR theory when computing conditions for critical CSTRs and DSRs. A number of properties of determinants are provided in the following text. Many of these properties are used in Chapters 6 and 7.

1. The determinant of $\mathbf{A}^T$ is equal to the determinant of **A**: $\text{Det}(\mathbf{A}^T) = \text{Det}(\mathbf{A})$.
2. Switching rows or columns in **A** changes the sign of Det(**A**).
3. A scalar t multiplied by a row in **A** is equal to t multiplied by Det(**A**). Similarly, a scalar can be factored out of a row in **A**, to give t multiplied by the determinant. Hence, $\text{Det}(t\mathbf{A}) = t^n\text{Det}(\mathbf{A})$, if **A** is a $n \times n$ matrix.
4. The determinant of a matrix product is distributive: $\text{Det}(\mathbf{AB}) = \text{Det}(\mathbf{A})\text{Det}(\mathbf{B})$.
5. Row operations on **A** do not affect the value of its determinant. A multiple of a row or column in **A** added to another row or column in **A** does not affect Det(**A**).
6. $\text{Det}(\mathbf{A}) = 0$ if two or more rows or columns in **A** are equal.
7. $\text{Det}(\mathbf{A}) = 0$ if all entries in a row or column in **A** are 0.

B.3 THE COMPLEMENT PRINCIPLE

B.3.1 Introduction

The complement principle serves an important role in the development of two fundamental theorems in AR theory, which are described in Chapter 6. We briefly describe the principle here, which is an adaptation from Feinberg and Hildebrandt (1997).

CONCEPT: Closures

The complement principle uses the idea of closures of a convex set **X**, conv(**X**). The *closure* of con(**X**), cl conv(**X**), is the smallest *closed* subset of conv(**X**). The closure of conv(**X**) is used to represent the set of all points in conv(**X**) including any points that might only be obtained in the limit of a process.[2] cl conv(**X**) is used to include points in conv(**X**) that might not be physically achievable (i.e., equilibrium points).

B.3.2 Description: A Generalized Mole Balance over a Reactor Sub-Network

Suppose that a steady-state reactor network is available. Let $\mathbf{C} = [c_1, c_2, \ldots, c_n]^T \in \mathbb{R}^n$ be a concentration vector representing a stream composition that is produced within the network. Let $\mathbf{r}(\mathbf{C}) = [r_1(\mathbf{C}), r_2(\mathbf{C}), \ldots, r_n(\mathbf{C})]^T \in \mathbb{R}^n$ represent the species rate function associated with the kinetics of the system and **F** be a set of feed concentrations. Let **X** be a set of positive reactor effluent concentrations in $\mathbb{R}^n$ where $\{\mathbf{C} \mid \mathbf{C} \in \mathbf{X}\}$. We can partition **X** into two smaller subsets as follows:

1. A set **D** that is a subset of **X**, $\mathbf{D} \subset \mathbf{X}$, which does *not* contain concentrations in **F**.
2. The remaining concentrations of **X** that form the *complement set* with respect to **X**, **X\D**.

Suppose we can find vectors such that

$$\mathbf{C}_{\text{in}} \in \text{cl conv}(\mathbf{X} \backslash \mathbf{D})$$
$$\mathbf{C}_{\text{out}} \in \text{cl conv}(\mathbf{D})$$
$$\bar{\mathbf{r}} \in \text{cl conv}\{\mathbf{r}(\mathbf{C}) \in \mathbb{R}^n \mid \mathbf{C} \in \mathbf{D}\}$$

[2]This can be thought of as the set **X** including its limit points (e.g., equilibrium points).

Feinberg and Hildebrandt (1997) show that if there is a positive scalar τ that satisfies the relation

$$\mathbf{C}_{\text{out}} - \mathbf{C}_{\text{in}} = \tau\bar{\mathbf{r}}$$

Then **X** is said to be consistent with the complement principle for the prescribed **r**(**C**) and **F**, and **X** is a set of achievable concentrations in a steady-state process involving only reaction and mixing.

Consider an arbitrary network of chemical reactors called the *grand network*. Let us denote the set of achievable concentrations associated with this network by **X**, where **X** must contain positive values. The convex hull of the points in **X** is given by conv(**X**). The complement principle describes a relationship between sections of the grand network to the reaction kinetics of the system.

Let us investigate a sub-network of the grand network. Let the set of concentrations associated with this sub-network be given by **D**. The complement set of **X** excluding **D** is then **X\D**. **D** can be made arbitrarily large, encompassing a number of physical reactors and streams of the network. The only requirement that we impose on **D** is that it contains at least one extreme point of conv(**X**), and that it does not contain any concentrations associated with the feed set **F**. Points contained in **X\D** must therefore contain the feed set **F**.

B.3.2.1 Inflows into D Since the concentrations in **D** initiate from a physical part of the network that is separate from **F**, it must be that the streams entering the reactor network associated with **D** originate from concentrations in **X\D**. Let the total volumetric flow of all streams in the network belonging to **D** be represented by the quantity Q, and the combined molar concentration *into* this network be given by the vector $\mathbf{C}_{\text{in}}$. The total molar flow rate of species into the region is then given by the product $Q\mathbf{C}_{\text{in}}$. Specifically, $\mathbf{C}_{\text{in}}$ must belong to the set in the complement network as follows:

$$\mathbf{C}_{\text{in}} \in \text{cl conv}(\mathbf{X}\backslash\mathbf{D})$$

B.3.2.2 Outflows Out of D Since there is material that flows into the sub-network from set **D**, then by mass balance there is material that flows out of **D**. These might originate from physically separate parts of the network belonging to **D**. The total volumetric flow from all outgoing streams is thus also Q. A combined outgoing concentration vector, given here by $\mathbf{C}_{\text{out}}$, may also be written for the outflows. The total molar flow rate out from **D** is then given by the product of $Q\mathbf{C}_{\text{out}}$. Moreover, since **D** has been partitioned to contain at least one extreme point of **X** and does not contain any points belonging to **F**, then effluent concentrations from **D** must be as a result of reactions participating in **D** specifically. Concentrations in $\mathbf{C}_{\text{out}}$ must therefore be as a result of operations derived from **D** and not **X\D**. $\mathbf{C}_{\text{out}}$ is hence an element of the sub-network and not from the complement network.

$$\mathbf{C}_{\text{out}} \in \text{cl conv}(\mathbf{D})$$

B.3.2.3 Reactions Occurring within D Since $\mathbf{C}_{\text{out}}$ is formed as a result of one or many reactions occurring in **D**, a nonnegative reactor volume must also be present. Let V denote the combined reactor volume obtained from all reactive sections belonging to **D**. In a manner similar to how inflows and outflows of concentration vectors are expressed, a combined reaction rate vector may be written for **D**. We denote by $\bar{\mathbf{r}}$ the combined *average* reaction rate associated with **D**. Since outgoing concentrations must ultimately originate from **D**, then the rate function must also be evaluated with points from **D**:

$$\bar{\mathbf{r}} \in \text{cl conv}\{\mathbf{r}(\mathbf{C}) \in \mathbb{R}^n | \mathbf{C} \in \mathbf{D}\}$$

In other words, $\bar{\mathbf{r}}$ must be an element of the closed set, whose rate functions **r**(**C**) are evaluated using concentrations **C** that belong to the sub-network **D**.

B.3.2.4 Mass Balance between D and X\D The total molar production rate of all species is then given by V**r**. A steady-state mass balance between networks **D** and **X\D** may be written, giving

$$Q(\mathbf{C}_{\text{out}} - \mathbf{C}_{\text{in}}) = V\bar{\mathbf{r}}$$

Combining V and Q to give the average residence time τ of the system,

$$\mathbf{C}_{\text{out}} - \mathbf{C}_{\text{in}} = \tau\bar{\mathbf{r}}$$

Observe that this expression resembles the standard CSTR equation. Note that **D** may be defined as *any* sub-network of **X** provided **D** contains an extreme point of conv(**X**), and it does not contain any feed concentrations.

SIDE NOTE: An additional property of the AR

Properties of the AR have already been discussed in Chapters 3 and 4. An additional property is worth mentioning now the complement principle has been introduced. This might be regarded as a final known property of the AR, stated simply as follows next.

Property 8: The AR must be consistent with the complement principle.

It is clear that if a set of achievable concentrations is obtained by some grand network, then it may be partitioned in a manner that satisfies the complement principle.

This is because the collection of attainable concentrations must contain a feed set and must also be convex.

Note that this property must hold even for candidate regions that are not the true AR. The complement principle serves as further necessary constraint on the AR. The AR must satisfy the principle, but simply being consistent with the principle is insufficient to determine if the true AR has been identified.

REFERENCES

Feinberg, M., Hildebrandt, D., 1997. Optimal reactor design from a geometric viewpoint-I. Universal properties of the attainable region. Chem. Eng. Sci. 52, 1637–1665.

Lay, D.C., 2012. Linear Algebra and Its Applications. Addison-Wesley, Boston.

Strang, G., 2003. Introduction to Linear Algebra. Wellesley-Cambridge Press, Wellesley.

APPENDIX C

COMPANION SOFTWARE AND WEBSITE

C.1 INTRODUCTION

C.1.1 The Python Programming Language

Python is a popular open source programming language, which is commonly used amongst scientists and engineers. We have chosen to use Python for creating companion software for this book, as opposed to conventional scientific software such as MATLAB, for two noteworthy reasons as follows:

1. Accessibility: Python is free to use without restriction. It is freely available and easy to obtain on the Internet. Python is also well supported within the scientific and engineering community, which allows for easy troubleshooting of problematic code.
2. Jupyter notebooks: We have found the use of Jupyter notebooks to be a convenient way of sharing work and code in a compact and reproducible manner. Jupyter notebooks are easy to update and adapt over time compared to a static CD-ROM.

C.1.2 Jupyter Notebooks

Jupyter notebooks (previously known as IPython notebooks) are web applications that are best described as interactive documents containing code (i.e., written in a programming language such as Python), formatted explanatory text, interactive visualizations, and displayed mathematics, as shown in Figure C.1.

We have found Jupyter notebooks to be very useful for the demonstration and dissemination of work, because they are self-contained (meaning that there is no additional software to install) and they provide an interactive element (allowing users to modify and rerun code examples). These qualities of interactive Jupyter notebooks cannot be easily replicated through conventional static text and illustrations. Moreover, interactive controls such as sliders and buttons (called widgets) may be embedded into Jupyter notebooks, which allows for convenient description of core concepts that are much easier to explain via interactive demonstration than through written text.

Jupyter notebooks are executed in a standard web browser, and many Jupyter notebook files may be downloaded off the Internet and run locally on a computer. Jupyter notebooks are supported on many operating systems (i.e., PC, Mac, and Linux), which further allows for easy cross-platform collaboration.

C.1.3 Companion Website

As part of this book, we have created a number of Jupyter notebooks that assist in explaining core concepts in AR theory, as well as additional problems and demonstrations, which could not be included in the book. These files may be downloaded from the ARs companion website: http://attainableregions.com. Additional information and important links related to the book may also be found on the Wiley website: http://www.wiley.com.

The AR companion website is intended to be used in conjunction with the book content. We aim to periodically update the website with new content and amendments to problems.

Attainable Region Theory: An Introduction to Choosing an Optimal Reactor, First Edition. David Ming, David Glasser, Diane Hildebrandt, Benjamin Glasser, and Matthew Metzger.

Companion website: http://attainableregions.com

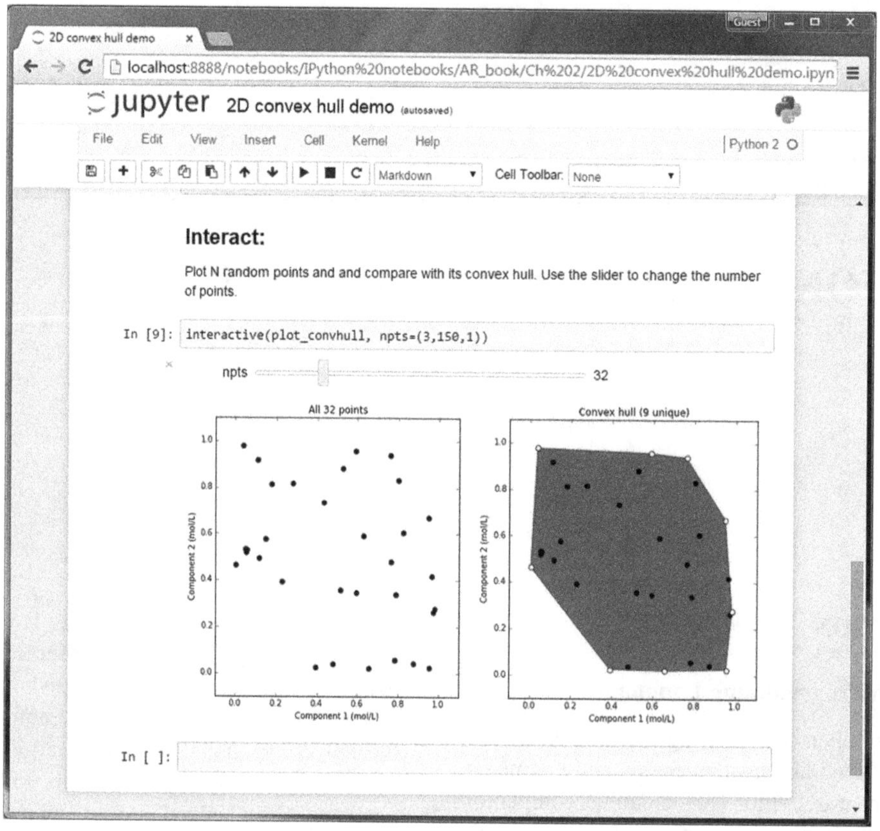

Figure C.1 A Jupyter notebook describing convex hulls.

C.2 OBTAINING PYTHON AND JUPYTER

C.2.1 Anaconda

Anaconda is a Python distribution, developed by Continuum Analytics, containing a number of common Python packages that are catered for scientific computing and data analysis. The Anaconda distribution also contains the Python interpreter, and hence there is no need to have Python installed on your computer prior to installing Anaconda. Jupyter is also bundled with the Anaconda distribution. An integrated development environment (IDE), called Spyder, is included for editing and executing Python code.

All of the required Python packages used in the companion Jupyter notebooks are included as part of the default Anaconda distribution, which makes for easy installation and setup from a single download. We highly recommend Anaconda for this reason. Anaconda supports a number of Python versions (2.7 and 3.5 at the time of writing), and it is updated frequently. Additional information about Anaconda may be found on the Continuum Analytics website.

C.2.2 Downloading and Installing the Anaconda Distribution

- Anaconda may be downloaded, free of charge, from the Continuum Analytics website: http://continuum.io
- Ensure that you select an Anaconda version that is appropriate for your operating system—Linux, Max, and PC versions of Anaconda are supported in both 32-bit and 64-bit varieties.
- Once the appropriate Anaconda distribution has been selected, download the installation file and follow the installer instructions to install Anaconda on your machine.

C.2.3 Downloading Jupyter Notebooks

Jupyter notebooks are hosted on the AR companion website.

- Visit the attainable regions website: http://attainableregions.com
- Instructions for downloading the Jupyter notebook example files can be found on the website.
- Download the notebook files to your computer and store them in a directory that is visible to the Jupyter notebook viewer.
    - The default directory for Jupyter notebooks, installed by Anaconda, is found in your local user directory. The default directory can also be edited to your preference.
    - For example, the default directory for Jupyter notebooks running Windows 7 is C:\Users\User_Name\ if the account name for the machine is "User_Name."

- A shortcut to the Jupyter notebook app should be installed by Anaconda. Launching the Jupyter notebook will open a new tab in your web browser displaying a file explorer that views the Jupyter notebook default path.
- Navigate to the directory of the downloaded AR notebook files and click on the link to open the notebook. This will open a new tab displaying the notebook.
- You are now ready to use the notebook. Experiment with the code and contribute by submitting your own files to the AR theory project website.

Note: Most notebooks are completely isolated and will not require installation of additional Python packages, provided that Anaconda is installed. However, some packages may be updated over time and newer software may be available. Checking for updates to Python packages via Anaconda's package manager, conda, is recommended.

INDEX

Attainable Region Theory: An Introduction to Choosing an Optimal Reactor, First Edition. David Ming, David Glasser, Diane Hildebrandt, Benjamin Glasser, and Matthew Metzger.

Companion website: http://attainableregions.com

www.ingramcontent.com/pod-product-compliance
Lightning Source LLC
LaVergne TN
LVHW082114200926
848484LV00008B/1950